# Principles
# of Foundation
# Engineering

# Principles of Foundation Engineering

## Braja M. Das

Civil Engineering Department
The University of Texas at El Paso

**Brooks/Cole Engineering Division**
Monterey, California 93940

**Brooks/Cole Engineering Division**
A Division of Wadsworth, Inc.

Printed in the United States of America

10  9  8  7  6  5  4  3  2

**Library of Congress Cataloging in Publication Data**

Das, Braja M., [date]
    Principles of foundation engineering.

    Includes bibliographical references and index.
    1. Foundations.    I. Title.
TA775.D227    1984    624.1'5    83-7859
ISBN 0-534-03052-1

Sponsoring Editor: Ray Kingman
Production Services Coordinator: Bill Murdock
Manuscript Editor and Production: Stacey C. Sawyer, San Francisco
Interior Design: Albert Burkhardt
Cover Design: Albert Burkhardt
Illustrations: Ryan Cooper
Typesetting: Interactive Composition Corporation
Printing and Binding: The Maple-Vail Book Manufacturing Group

In the memory of my father;
        and
to Janice and Valerie

# Preface

During the past three decades, soil mechanics and foundation engineering have developed rapidly. Intensive research and observation in the field and the laboratory have refined and improved the science of foundation design. This text on principles of foundation design uses materials recently published in various geotechnical journals. Although foundation design on reinforced earth is virtually in its infancy, the text introduces this concept to students. As the state of the art develops, the materials on this topic, and others, will be updated in future editions.

*Principles of Foundation Engineering* is primarily intended for undergraduate civil engineering students. The first chapter on geotechnical properties of soil and natural soil deposits reviews the topics covered in the introductory soil mechanics course, which is a prerequisite for the foundation engineering course. The text comprises twelve chapters with examples and problems, two appendices, and an answer section for selected problems. The chapters are mostly devoted to the geotechnical aspects of foundation design. Appendix A covers reinforced concrete design of shallow foundations and retaining walls, whereas Appendix B gives the conversion factors from SI to English units and also English to SI units. Systéme International (SI) units and English units are used in the text. Equations and tables in the main text are in SI units. The English equivalents of the equations and tables presented in the main text are given in Appendix B.

Because the text introduces civil engineering students to the application of the fundamental concepts of foundation analysis and design, the mathematical derivations of some equations are not always presented. Instead, just the final form of the equation is given. However, each chapter includes a list of references for further information and study.

Each chapter includes example problems that will help students understand the application of various equations and graphs. A number of practice problems are also given at the end of each chapter. Answers to half of these problems are at the back of the book.

Foundation analysis and design, as my colleagues in the geotechnical engineering area well know, is not just a matter of using theories, equations, and graphs from a textbook. Soil profiles found in nature are seldom homogeneous, elastic, and isotropic. The educated judgment needed to properly apply the theories, equations, and graphs to the evaluation of soils and foundation design cannot be overemphasized or completely taught by any textbook. Field experience must supplement classroom work.

## Acknowledgments

I am grateful to my wife for typing the entire manuscript several times and for preparing the original graphs and figures. Her apparently inexhaustible energy was my primary source of inspiration for completing the manuscript.

During the time I was preparing the manuscript, I lost my father. The moral and spiritual support provided me by Rev. Bill Fox of Belleville, Illinois, and Dr. Haskell Monroe, president of The University of Texas at El Paso, made it possible for me to complete the manuscript on time.

I also wish to acknowledge Professors Paul C. Hassler of The University of Texas at El Paso; Gerald R. Seeley of Tri-State University, Indiana; and Ronald B. McPherson of New Mexico State University for their help, support, and continuous encouragement during the preparation of the manuscript. Two of my former graduate students—Said Larbi-Cherif, presently of Navarro and Associates, El Paso, and Henry Ng, presently of Southwest Tension Company, El Paso—were extremely helpful with many suggestions and ideas. I sincerely appreciate their efforts.

Thanks also to the staff of Brooks/Cole Engineering Division for their interest and patience during the preparation and production of the manuscript.

As a final note, I welcome suggestions from students and instructors for use in further editions.

Braja M. Das

# Contents

Chapter **2** _____

## Subsurface Exploration                                          62

Chapter **3** _____

## Shallow Foundations                                             101

# Chapter 4

## Mat Foundations                                                          172

# Chapter 5

## Lateral Earth Pressure and Retaining Walls                                207

Chapter **6** _____

**Sheet Pile Walls in Waterfront Structures                    267**

Chapter **7** _____

**Braced Cuts                                                307**

# Chapter 8

## Pile Foundations                                                              330

# Chapter 9

## Drilled-Pier and Caisson Foundations                                         416

# Chapter 10

## Foundations on Difficult Soils    452

# Chapter 11

## Reinforced Earth Structures    480

# Chapter **12**

## Soil Improvement                                                                               **506**

## Appendix **A**
## Reinforced Concrete Design                                                                     **545**

## Appendix **B**
## Conversion Factors, Equations, and Tables in English Units                                      **578**

# Appendices

# Geotechnical Properties of Soil and Natural Soil Deposits

## 1.1

### Introduction

The design of foundations of structures such as buildings, bridges, and dams generally requires a knowledge of such factors as (a) the load that will be transmitted by the superstructure to the foundation system, (b) the requirements of the local building code, (c) the behavior and stress-related deformability of soils that will support the foundation system, and (d) the geological conditions of the soil under consideration. To a foundation engineer, the last two factors listed are extremely important, because they concern soil mechanics.

The geotechnical properties of a soil—such as the grain-size distribution, plasticity, compressibility, and shear strength—can be determined by proper laboratory testing. And, recently, emphasis has been placed on *in situ* determination of strength and deformation properties of soil, because this process avoids the sample disturbances that occur during field exploration. However, under certain circumstances, all of the needed parameters cannot be determined, or are not determined because of economic or other reasons. In such cases, the engineer must make certain assumptions regarding the properties of the soil. In order to assess the accuracy of the soil parameters—whether they were made in the laboratory and the field or were assumed—the engineer must have a good grasp of the basic principles of soil mechanics. At the same time, he or she must realize that the natural soil deposits on which foundations are constructed are not homogeneous in most cases. Thus the engineer must have a thorough understanding of the geology of the area—that is, the origin and nature of soil stratification and also the ground water conditions. Foundation engineering is a clever combination of soil mechanics, engineering geology, and proper judgment derived from past experience. To a certain extent, it may be called an "art."

When determining which foundation is the most economical, the engineer must consider the superstructure load, the subsoil conditions, and the desired tolerable settlement. In general, foundations of buildings and bridges may be divided into two major categories: (1) *shallow foundations*, and (2) *deep foundations*. *Spread footings*, *wall footings*, and *mat foundations* are all shallow foundations. In most shallow foundations, the *depth of embedment can be equal to or less than three to four times the width of the foundation*. *Pile* and *caisson* foundations are deep foundations. These are used when topsoil layers have poor load-bearing capacity and when use of shallow foundations will cause considerable damage and/or instability to the structure(s). The problems relating to shallow foundations and mat foundations are considered in Chapters 3 and 4, respectively. Chapter 8 discusses pile foundations, and Chapter 9 examines caissons.

This chapter is divided into two major parts: (1) basic geotechnical properties of soils, and (2) natural soil deposits. The first part reviews topics such as grain-size distribution, plasticity, soil classification, effective stress, consolidation, and shear strength parameters. It is assumed that the reader has already been exposed to these concepts in a basic soil-mechanics course. The second part deals with the formation of soil by water, wind, glaciers, and so on and discusses the characteristics of natural soil deposits. This subject is covered in detail in engineering geology courses.

## Geotechnical Properties of Soils

### 1.2

Grain-Size Distribution

In any soil mass, the size of various soil grains varies greatly. To properly classify a soil, one must know the *grain-size distribution* in it. The grain-size distribution of *coarse-grained* soil is generally determined by means of *sieve analysis*. For a *fine-grained* soil, the grain-size distribution can be obtained by means of *hydrometer analysis*. The fundamental features of these analyses will be presented in this section. For detailed descriptions, readers are referred to any soil mechanics laboratory manual (for example, Das, 1982).

Sieve Analysis

Sieve analysis is conducted by taking a measured amount of dry, well-pulverized soil. The soil is passed through a stack of sieves with a pan at the bottom. The amount of soil retained on each sieve is measured, and the cumulative percentage of soil passing through each sieve is determined. This is generally referred to as *percent finer*. Following is a list of U.S. sieve numbers and the corresponding size of their hole openings. These sieves are commonly used for the analysis of soil for foundation design purposes.

| U.S. sieve no. | Sieve opening, mm |
| --- | --- |
| 4 | 4.75 |
| 10 | 2.00 |
| 20 | 0.85 |
| 40 | 0.425 |
| 60 | 0.25 |
| 100 | 0.150 |
| 200 | 0.075 |

The percent finer for each sieve determined by a sieve analysis is plotted on *semilogarithmic graph paper,* as shown in Figure 1.1. Note that the grain diameter, *D*, is plotted on the *logarithmic scale,* and the percent finer is plotted on the *arithmetic scale.*

Two parameters can be determined from the grain-size distribution curves of coarse-grained soils: (1) the *uniformity coefficient* ($C_u$) and (2) the *coefficient of gradation* ($C_z$). These coefficients can be defined as

$$C_u = \frac{D_{60}}{D_{10}}  \tag{1.1}$$

$$C_z = \frac{D_{30}{}^2}{(D_{60})(D_{10})}  \tag{1.2}$$

where $D_{10}$, $D_{30}$, and $D_{60}$ are the diameters corresponding to percents finer than 10, 30, and 60%, respectively.

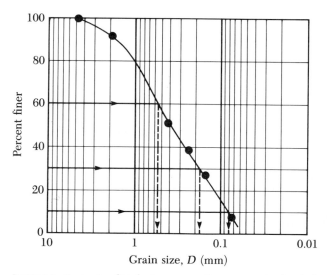

**Figure 1.1**  Grain-size distribution curve of a coarse-grained soil obtained from sieve analysis

For the grain-size distribution curve shown in Figure 1.1, $D_{10} = 0.08$ mm, $D_{30} = 0.17$ mm, and $D_{60} = 0.57$ mm. Thus, the values of $C_u$ and $C_z$ are as follows:

$$C_u = \frac{0.57}{0.08} = 7.13$$

$$C_z = \frac{0.17^2}{(0.57)(0.08)} = 0.63$$

Parameters ($C_u$ and $C_z$) are used in the *Unified Soil Classification System* (Section 1.8).

### Hydrometer Analysis

Hydrometer analysis is conducted on the principle of sedimentation of soil particles in water. In this test, one uses 50 grams of dry, pulverized soil. A *deflocculating* agent is always added to the soil. The most common deflocculating agent used for hydrometer analysis is 125 cc of 4% solution of sodium hexametaphosphate. The soil is allowed to soak for at least 16 hours in the deflocculating agent. After the soaking period, distilled water is added, and the soil-deflocculating agent mixture is thoroughly agitated. The sample is then transferred to a 1000-ml glass cylinder. More distilled water is added to the cylinder to fill it up to the 1000-ml mark, and then the mixture is again thoroughly agitated. A hydrometer is placed in the cylinder to measure—usually

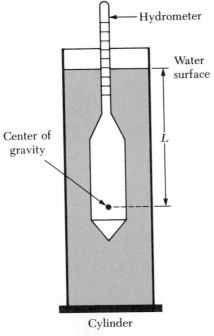

**Figure 1.2**  Hydrometer analysis

over a 24-hour period—the specific gravity of the soil-water suspension in the vicinity of its bulb (Figure 1.2). Hydrometers are calibrated to show the amount of soil that is still in suspension at any given time, $t$. The largest diameter of the soil particles still in suspension at time $t$ can be determined by Stoke's law:

$$D = \sqrt{\frac{18\eta}{(G_s - 1)\gamma_w}} \quad \sqrt{\frac{L}{t}} \tag{1.3}$$

where $D$ = diameter of the soil particle
$\quad G_s$ = specific gravity of soil solids
$\quad \eta$ = viscosity of water
$\quad \gamma_w$ = unit weight of water
$\quad L$ = effective length (that is, length measured from the water surface in the cylinder to the center of gravity of the hydrometer; see Figure 1.2)
$\quad t$ = time

Soil particles having diameters larger than those calculated by Eq. (1.3) would have settled beyond the zone of measurement. In this manner, with hydrometer readings taken at various times, the soil *percent finer* than a given diameter $D$ can be calculated, and a grain-size distribution plot can be prepared.

## 1.3
## Soil-Separate Size Limits

Several organizations have attempted to develop the size limits for *gravel, sand, silt,* and *clay* based on the grain sizes present in soils. Table 1.1 presents the size limits as recommended by the Massachusetts Institute of Technology (MIT), the American Association of State Highway and Transportation Officials (AASHTO), and the Unified (Corps of Engineers, Department of the Army and Bureau of Reclamation) systems. From Table 1.1 it can be seen that soil particles smaller than 0.002 mm have been classified as *clay*. However, it should be kept in mind that clays by nature have a cohesive property and can be rolled into a thread when moist. This is caused by the presence of *clay minerals* such

**Table 1.1** Soil-Separate Size Limits

| Classification system | Grain size (mm) |
|---|---|
| MIT | Gravel: 100 mm to 2 mm<br>Sand: 2 mm to 0.06 mm<br>Silt: 0.06 mm to 0.002 mm<br>Clay: <0.002 mm |
| Unified | Gravel: 75 mm to 4.75 mm<br>Sand: 4.75 mm to 0.075 mm<br>Silt and clay (fines): <0.075 mm |
| AASHTO | Gravel: 75 mm to 2 mm<br>Sand: 2 mm to 0.05 mm<br>Silt: 0.05 mm to 0.002 mm<br>Clay: <0.002 mm |

as *kaolinite, illite,* and *montmorillonite*. In contrast, some minerals such as *quartz* and *feldspar* may be present in a soil in particle sizes as small as clay minerals. But these particles will not have the cohesive property of clay minerals. Hence, they are called *clay-size particles*, not *clay particles*.

## 1.4

### Weight-Volume Relationships

In nature, soils are three-phase systems consisting of solid soil particles, water, and air (or gas). In order to develop the *weight-volume relationships* for a soil, the three phases can be separated as shown in Figure 1.3a. Based on this, the volume relationships can be defined as follows:

*Void Ratio, e,* is the ratio of the volume of voids to the volume of soil solids in a given soil mass and can be written as

$$e = \frac{V_v}{V_s} \tag{1.4}$$

where  $V_v$ = volume of voids
$V_s$ = volume of soil solids

*Porosity, n,* is ratio of the volume of voids to the volume of the soil specimen, or

$$n = \frac{V_v}{V} \tag{1.5}$$

where $V$ = total volume of soil

It can also be seen that

$$n = \frac{V_v}{V} = \frac{V_v}{V_s + V_v} = \frac{\dfrac{V_v}{V_s}}{\dfrac{V_s}{V_v} + \dfrac{V_v}{V_s}} = \frac{e}{1 + e} \tag{1.6}$$

*Degree of Saturation, S,* is the ratio of the volume of water in the void spaces to the volume of voids, and it is generally expressed as a percentage. So

$$S = \frac{V_w}{V_v} \tag{1.7}$$

where $V_w$ = volume of water

Note that, for saturated soils, the degree of saturation is 100%.

The weight relationships are *moisture content, moist unit weight, dry unit weight,* and *saturated unit weight*. They can be defined as follows:

$$\text{Moisture Content} = w(\%) = \frac{W_w}{W_s} \times 100 \tag{1.8}$$

where  $W_s$ = weight of the soil solids
$W_w$ = weight of water

Note: $V_a + V_w + V_s = V$
$W_w + W_s = W$

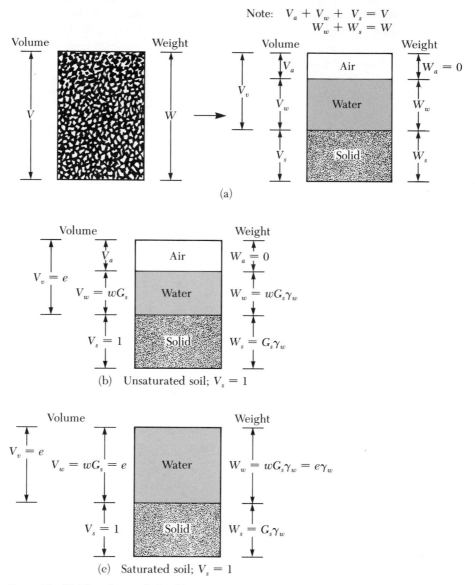

(a)

(b)   Unsaturated soil; $V_s = 1$

(e)   Saturated soil; $V_s = 1$

**Figure 1.3** Weight-volume relationships

$$\text{Moist Unit Weight} = \gamma = \frac{W}{V} \tag{1.9}$$

where $W$ = total weight of the soil specimen = $W_s + W_w$

It is assumed that the weight of air, $W_a$, in the soil mass is negligible.

$$\text{Dry Unit Weight} = \gamma_d = \frac{W_s}{V} \tag{1.10}$$

When a soil mass is completely saturated (that is, all the void volume is occupied by water), the moist unit weight of a soil [Eq. (1.9)] becomes equal to the *saturated unit weight* ($\gamma_{sat}$). So, $\gamma = \gamma_{sat}$ if $V_v = V_w$.

More useful relations can now be developed by considering a representative soil specimen in which the volume of soil solids is equal to *unity* as shown in Figure 1.3b. Note that if $V_s = 1$, from Eq. (1.4), $V_v = e$. The weight of the unit volume of the soil can be given as

$$W_w = G_s\gamma_w$$

where $G_s$ = specific gravity of soil solids
$\gamma_w$ = unit weight of water ($9.81$ kN/m$^3$)

Also, from Eq. (1.8), the weight of water $W_w = wW_s$. Thus, for the soil specimen under consideration, $W_w = wW_s = wG_s\gamma_w$. Now, using the general relation for moist unit weight given in Eq. (1.9)

$$\gamma = \frac{W}{V} = \frac{W_s + W_w}{V_s + V_v} = \frac{G_s\gamma_w(1 + w)}{1 + e} \tag{1.11}$$

Similarly, the dry unit weight [Eq. (1.10)]

$$\gamma_d = \frac{W_s}{V} = \frac{W_s}{V_s + V_v} = \frac{G_s\gamma_w}{1 + e} \tag{1.12}$$

From Eqs. (1.11) and (1.12), note that

$$\gamma_d = \frac{\gamma}{1 + w} \tag{1.13}$$

If a soil specimen is completely saturated as shown in Figure 1.3c

$$V_v = e$$

Also, for this case

$$V_v = \frac{W_w}{\gamma_w} = \frac{wG_s\gamma_w}{\gamma_w} = wG_s$$

Thus

$$e = wG_s \qquad \text{(for saturated soil } only\text{)} \tag{1.14}$$

The saturated unit weight of soil can now be determined as

$$\gamma_{sat} = \frac{W_s + W_w}{V_s + V_v} = \frac{G_s\gamma_w + e\gamma_w}{1 + e} \tag{1.15}$$

Relationships similar to Eqs. (1.11), (1.12), and (1.15) in terms of porosity can also be obtained by considering a representative soil specimen with a unit volume. These relationships are

$$\gamma = G_s\gamma_w(1 - n)(1 + w) \tag{1.16}$$

$$\gamma_d = (1 - n)G_s\gamma_w \tag{1.17}$$

$$\gamma_{sat} = [(1 - n)G_s + n]\gamma_w \tag{1.18}$$

# Example
## 1.1

A representative soil specimen collected from the field weighs 1.8 kN and has a volume of $0.1 \text{ m}^3$. The moisture content as determined in the laboratory is 12.6%. Given $G_s = 2.71$, determine the following:

a. Moist unit weight
b. Dry unit weight
c. Void ratio
d. Porosity
e. Degree of saturation

## Solution

### Part a: Moist Unit Weight

From Eq. (1.9)

$$\gamma = \frac{W}{V} = \frac{1.8 \text{ kN}}{0.1 \text{ m}^3} = \underline{18 \text{ kN/m}^3}$$

### Part b: Dry Unit Weight

From Eq. (1.13)

$$\gamma_d = \frac{\gamma}{1 + w} = \frac{18}{1 + \dfrac{12.6}{100}} = \underline{15.99 \text{ kN/m}^3}$$

### Part c: Void Ratio

From Eq. (1.12)

$$\gamma_d = \frac{G_s \gamma_w}{1 + e}$$

or

$$e = \frac{G_s \gamma_w}{\gamma_d} - 1 = \frac{(2.71)(9.81)}{15.99} - 1 = \underline{0.66}$$

### Part d: Porosity

From Eq. (1.6)

$$n = \frac{e}{1 + e} = \frac{0.66}{1 + 0.66} = \underline{0.398}$$

### Part e: Degree of Saturation

Referring to Figure 1.3b

$$S = \frac{V_w}{V_v} = \frac{w G_s}{e} = \frac{(0.126)(2.71)}{0.66} \times 100 = \underline{51.7\%}$$

## 1.5

### Relative Density

In *granular soils*, the degree of compaction in the field can be measured according to *relative density*, $D_r$, which is defined as

$$D_r = \frac{e_{max} - e}{e_{max} - e_{min}} \tag{1.19}$$

where $e_{max}$ = void ratio of the soil in the loosest state
$e_{min}$ = void ratio in the densest state
$e$ = *in-situ* void ratio

The values of $e_{max}$ and $e_{min}$ are determined in the laboratory in accordance with the test procedures outlined in the American Society for Testing and Materials, *ASTM Standards* (1982) (Test Designation D2040).

The relative density can also be expressed in terms of dry unit weight, or

$$D_r = \left\{ \frac{\gamma_d - \gamma_{d(min)}}{\gamma_{d(max)} - \gamma_{d(min)}} \right\} \frac{\gamma_{d(max)}}{\gamma_d} \tag{1.20}$$

where $\gamma_d$ = *in situ* dry unit weight
$\gamma_{d(max)}$ = dry unit weight in the *densest* state—that is, when the void ratio is $e_{min}$
$\gamma_{d(min)}$ = dry unit weight in the *loosest* state—that is, when the void ratio is $e_{max}$

### Example 1.2

A granular soil (sand) was tested in the laboratory and found to have maximum and minimum void ratios of 0.84 and 0.38, respectively. The value of $G_s$ was determined to be 2.65. A natural soil deposit of the same sand has 9% moisture, and its moist unit weight is 18.64 kN/m³. Determine the relative density of the soil in the field.

**Solution**

From Eq. (1.13)

$$\gamma_d = \frac{\gamma}{1 + w} = \frac{18.64}{1 + \dfrac{9}{100}} = 17.1 \text{ kN/m}^3$$

$$\gamma_d = \frac{G_s \gamma_w}{1 + e}$$

or

$$e = \frac{G_s \gamma_w}{\gamma_d} - 1 = \frac{(2.65)(9.81)}{17.1} - 1 = 0.52$$

From Eq. (1.19)

$$D_r = \frac{e_{max} - e}{e_{max} - e_{min}} = \frac{0.84 - 0.52}{0.84 - 0.38} = 0.696 = 69.6\%$$

## 1.6

### Representative Values of $G_s$, $e$, and $\gamma_d$ for Natural Soils

Except for peat and highly organic soils, the general range of the values of specific gravity of soil solids ($G_s$) found in nature is not great. Table 1.2 gives some representative values.

**Table 1.2**  Specific Gravities of Some Soils

| Soil type | $G_s$ |
|---|---|
| Quartz sand | 2.64–2.66 |
| Silt | 2.67–2.73 |
| Clay | 2.70–2.9 |
| Chalk | 2.60–2.75 |
| Loess | 2.65–2.73 |
| Peat | 1.30–1.9 |

Table 1.3 presents some representative values for the void ratio, dry unit weight, and the moisture content (in a saturated state) of some naturally occurring soils. From this table it can be seen that in most cohesionless soils the void ratio varies from about 0.4 to 0.8. The dry unit weights in these soils generally fall within a range of about 14–19 kN/m³.

**Table 1.3**  Void Ratio, Moisture Content, and Dry Unit Weight for Some Soils

| Type of soil | Void ratio $e$ | Natural moisture content in saturated condition (%) | Dry unit weight $\gamma_d$ (kN/m³) |
|---|---|---|---|
| Loose uniform sand | 0.8 | 30 | 14.5 |
| Dense uniform sand | 0.45 | 16 | 18 |
| Loose angular-grained silty sand | 0.65 | 25 | 16 |
| Dense angular-grained silty sand | 0.4 | 15 | 19 |
| Stiff clay | 0.6 | 21 | 17 |
| Soft clay | 0.9–1.4 | 30–50 | 11.5–14.5 |
| Loess | 0.9 | 25 | 13.5 |
| Soft organic clay | 2.5–3.2 | 90–120 | 6–8 |
| Glacial till | 0.3 | 10 | 21 |

## 1.7

### Atterberg's Limits

When a clayey soil is mixed with an excessive amount of water, it may flow like a *liquid*. If the soil is gradually dried, it will lose moisture. Depending on its moisture content, it will behave like a *plastic, semisolid,* or a *solid* material.

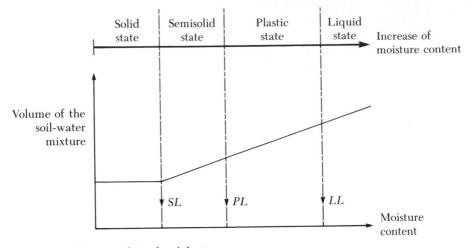

**Figure 1.4**   Definition of Atterberg's limits

The moisture content, in percent, at which the soil changes from a liquid to a plastic stage is defined as the *liquid limit* (*LL*). Similarly, the moisture contents, in percent, at which the soil changes from a plastic to a semisolid state and from a semisolid to a solid state are defined as the *plastic limit* (*PL*) and the *shrinkage limit* (*SL*), respectively. These limits are referred to as *Atterberg's limits* (Figure 1.4).

The liquid limit of a soil is determined by Casagrande's liquid device (ASTM Test Designation D-423) and is defined as the percentage of moisture content at which a groove closure of 12.7 mm (1/2 in.) occurs at 25 blows.

The *plastic limit* is defined as the percentage of moisture content at which the soil crumbles when rolled into a thread of 3.18 mm (1/8 in.) in diameter (ASTM Test Designation D-424).

The *shrinkage limit* is defined as the percentage of moisture content at

**Table 1.4**   Liquid and Plastic Limits for Some Clay Minerals and Soils

| Description | Liquid limit | Plastic limit |
|---|---|---|
| Kaolinite | 35–100 | 25–35 |
| Illite | 50–100 | 30–60 |
| Montmorillonite | 100–800 | 50–100 |
| Boston Blue clay | 40 | 20 |
| Chicago clay | 60 | 20 |
| Louisiana clay | 75 | 25 |
| London clay | 66 | 27 |
| Cambridge clay | 39 | 21 |
| Montana clay | 52 | 18 |
| Mississippi Gumbo | 95 | 32 |
| Loessial soils in north and northwest China | 25–35 | 15–20 |

which the soil does not undergo further volume change with loss of moisture (ASTM Test Designation D-427). This limit is shown in Figure 1.4.

The difference between the liquid limit and the plastic limit of a soil is defined as the *plasticity index (PI)*, or

$$PI = LL - PL \tag{1.21}$$

Table 1.4 gives some representative values of liquid limit and plastic limit for a number of clay minerals and soils. It should, however, be kept in mind that Atterberg's limits for various soils will vary considerably depending on the soil's origin and the nature and amount of clay minerals in it.

## 1.8

## Soil Classification Systems

Soil classification divides soils into a number of groups and subgroups based on common engineering properties such as *grain-size distribution, liquid limit,* and *plastic limit*. The two major classification systems presently in use are (1) the *AASHTO (American Association of State Highway and Transportation Officials) System* and (2) the *Unified System*. The AASHTO classification system is mainly used for *disturbed* soils, as in the case of highway subgrades. It is not used in foundation construction. This section will briefly describe each of these two systems.

### AASHTO System

The AASHTO Soil Classification System was originally proposed by the Highway Research Board's Committee on Classification of Materials for Subgrades and Granular Type Roads (1945). According to the present form of this system, soils can be classified into eight major groups, *A-1* through *A-8*, based on their grain-size distribution, liquid limit, and plasticity indices. Soils listed in the major groups—*A-1*, *A-2*, and *A-3*—are coarse-grained materials, and those in groups *A-4*, *A-5*, *A-6*, and *A-7* are fine-grained materials. Peat, muck, and other highly organic soils are classified under *A-8*. They are identified by visual inspection.

The AASHTO classification system (for soils A-1 through A-7) is presented in Table 1.5. Note that under group A-7 there are two types of soil. For the A-7-5 type, the plasticity index of the soil should be less than or equal to the liquid limit minus 30. For the A-7-6 type, the plasticity index is greater than the liquid limit minus 30.

For qualitative evaluation of the desirability of a given soil as a highway subgrade material, a number referred to as the *group index* has also been developed. The higher the value of the group index for a given soil, the weaker will be the soil's performance as a subgrade. A group index of 20 or more indicates a very poor subgrade material. The formula for group index, *GI*, is

$$GI = (F - 35)[0.2 + 0.005(LL - 40)]$$
$$+ 0.01(F - 15)(PI - 10) \tag{1.22}$$

**Table 1.5**   AASHTO Soil Classification System

| General classification | Granular materials (35% or less of total sample passing No. 200 sieve) | | | | | | | |
|---|---|---|---|---|---|---|---|---|
| Group classification | A-1 | | A-3 | A-2 | | | | |
| | A-1-a | A-1-b | | A-2-4 | A-2-5 | A-2-6 | A-2-7 |
| Sieve analysis (% passing)<br>No. 10 sieve<br>No. 40 sieve<br>No. 200 sieve | 50 max<br>30 max<br>15 max | 50 max<br>25 max | 51 min<br>10 max | 35 max | 35 max | 35 max | 35 max |
| For fraction passing<br>No. 40 sieve<br>Liquid limit ($LL$)<br>Plasticity index ($PI$) | 6 max | | Non-plastic | 40 max<br>10 max | 41 min<br>10 max | 40 max<br>11 min | 41 min<br>11 min |
| Usual types of material | Stone fragments, gravel, and sand | | Fine sand | Silty or clayey gravel and sand | | | |
| Subgrade rating | Excellent to good | | | | | | |

| General classification | Silt-clay materials (More than 35% of total sample passing No. 200 sieve) | | | |
|---|---|---|---|---|
| Group classification | A-4 | A-5 | A-6 | A-7<br>A-7-5[a]<br>A-7-6[b] |
| Sieve analysis (% passing)<br>No. 10 sieve<br>No. 40 sieve<br>No. 200 sieve | 36 min | 36 min | 36 min | 36 min |
| For fraction passing<br>No. 40 sieve<br>Liquid limit ($LL$)<br>Plasticity index ($PI$) | 40 max<br>10 max | 41 min<br>10 max | 40 max<br>11 min | 41 min<br>11 min |
| Usual types of material | Mostly silty soils | | Mostly clayey soils | |
| Subgrade rating | Fair to poor | | | |

[a]If $PI \leq LL - 30$, it is A-7-5
[b]If $PI > LL - 30$, it is A-7-6

where $F$ = percent passing No. 200 sieve, expressed as a whole number
$\quad LL$ = liquid limit
$\quad PI$ = plasticity index

When calculating the group index for a soil belonging to group A-2-6 or A-2-7, use only the partial group index equation relating to the plasticity index:

$$GI = 0.01(F - 15)(PI - 10)$$

The group index is rounded to the nearest whole number and written next to the soil group in parentheses; for example,

A − 4 (5)

Group index

Soil group

## Unified System

The original Unified System of Soil Classification was proposed by A. Casagrande in 1942 and later revised and adopted by the United States Bureau of Reclamation and the Corps of Engineers. This system is presently used in practically all building construction work.

In the Unified System, the following symbols are used for identification.

| Symbol | Description |
|--------|-------------|
| G | Gravel |
| S | Sand |
| M | Silt |
| C | Clay |
| O | Organic silts and clay |
| Pt | Peat and highly organic soils |
| H | High plasticity |
| L | Low plasticity |
| W | Well graded |
| P | Poorly graded |

Table 1.6 shows the present form of this system (pp. 16–17).

## Example
## 1.3

The following laboratory results were obtained for a given soil by conducting sieve analysis and Atterberg's limit tests.

Percent passing No. 200 sieve: 82%

$$LL = 32$$
$$PL = 18$$

Classify the soil by:

    a. AASHTO System
    b. Unified System

**Table 1.6**  Unified System of Classification[a]

| Major divisions | | | Group symbols | Typical names |
|---|---|---|---|---|
| Coarse-Grained Soils<br>More than 50% retained on No. 200 sieve[b] | Gravels<br>50% or more of coarse fraction retained on No. 4 sieve | Clean Gravels | GW | Well-graded gravels and gravel-sand mixtures, little or no fines |
| | | | GP | Poorly graded gravels and gravel-sand mixtures, little or no fines |
| | | Gravels with Fines | GM | Silty gravels, gravel-sand-silt mixtures |
| | | | GC | Clayey gravels, gravel-sand-clay mixtures |
| | Sands<br>More than 50% of coarse fraction passes No. 4 sieve | Clean Sands | SW | Well-graded sands and gravelly sands, little or no fines |
| | | | SP | Poorly graded sands and gravelly sands, little or no fines |
| | | Sands with Fines | SM | Silty sands, sand-silt mixtures |
| | | | SC | Clayey sands, sand-clay mixtures |
| Fine-Grained Soils<br>50% or more passes No. 200 sieve[b] | Silts and Clays<br>Liquid limit 50% or less | | ML | Inorganic silts, very fine sands, rock flour, silty or clayey fine sands |
| | | | CL | Inorganic clays of low to medium plasticity, gravelly clays, sandy clays, silty clays, lean clays |
| | | | OL | Organic silts and organic silty clays of low plasticity |
| | Silts and Clays<br>Liquid limit greater than 50% | | MH | Inorganic silts, micaceous or diatomaceous fine sands or silts, elastic silts |
| | | | CH | Inorganic clays of high plasticity, fat clays |
| | | | OH | Organic clays of medium to high plasticity |
| Highly Organic Soils | | | PT | Peat, muck, and other highly organic soils |

[a]Reprinted with permission from the *Annual Book of ASTM Standards*, Part 19. Copyright 1982, American Society for Testing and Materials, 1916 Race Street, Philadelphia, PA 19103.
[b]Based on the material passing the 3-in. (75-mm) sieve.

**Table 1.6** (Continued)

| | Classification criteria |
|---|---|

The left side has vertical labels:
- Classification on basis of percentage of fines
  - Less than 5% pass No. 200 sieve — GW, GP, SW, SP
  - More than 12% pass No. 200 sieve — GM, GC, SM, SC
  - 5% to 12% pass No. 200 sieve — Borderline classification requiring use of dual symbols

Classification criteria:

$C_u = D_{60}/D_{10}$   Greater than 4

$$C_z = \frac{(D_{30})^2}{D_{10} \times D_{60}}$$   Between 1 and 3

Not meeting both criteria for GW

| | |
|---|---|
| Atterberg limits plot below "A" line or plasticity index less than 4 | Atterberg limits plotting in hatched area are borderline classifications requiring use of dual symbols |
| Atterberg limits plot above "A" line and plasticity index greater than 7 | |

$C_u = D_{60}/D_{10}$   Greater than 6

$$C_z = \frac{(D_{30})^2}{D_{10} \times D_{60}}$$   Between 1 and 3

Not meeting both criteria for SW

| | |
|---|---|
| Atterberg limits plot below "A" line or plasticity index less than 4 | Atterberg limits plotting in hatched area are borderline classifications requiring use of dual symbols |
| Atterberg limits plot above "A" line and plasticity index greater than 7 | |

Plasticity Chart

For classification of fine-grained soils and fine fraction of coarse-grained soils.

Atterberg limits plotting in hatched area are borderline classifications requiring use of dual symbols.

Equation of A line:
$$PI = 0.73(LL - 20)$$

(Plasticity index vs. Liquid limit chart with regions CH, CL, MH & OH, ML & OL, CL-ML, and A line.)

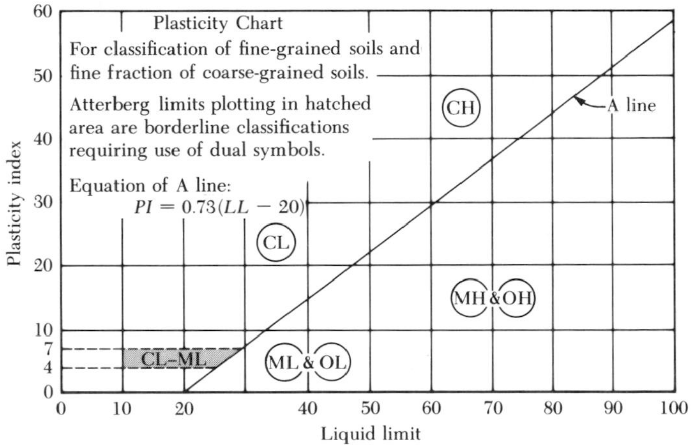

Visual-manual identification, see ASTM Designation D2488.

### Solution

Part a: AASHTO Classification

Because more than 35% is passing the No. 200 U.S. sieve, it is a silty-clay material. Table 1.5 shows that it could be A-4, A-5, A-6, or A-7.

For this soil, $LL = 32$, and $PI = LL - PL = 32 - 18 = 14$. One can see from Table 1.5 that the material is A-6. Again, from Eq. (1.22)

$$GI = (F - 35)[0.02 + 0.005(LL - 40)] + 0.01(F - 15)(PI - 10)$$

For this soil, $F = 82\%$. So

$$GI = (82 - 35)[0.02 + 0.005(32 - 40)] + 0.01(82 - 15)(14 - 10)$$

$$= 10.2 \approx 10$$

So, the soil is A-6(10).

Part b: Unified Classification

Refer to Table 1.6. Because more than 50% of the soil is passing through the No. 200 U.S. sieve, it is a fine-grained soil. Knowing that $LL = 32$ and $PI = 14$, refer to the plasticity chart, which shows that the soil is CL.

## 1.9

### Permeability of Soil

The void spaces that are present between the soil grains allow water to flow through them. In soil mechanics and foundation engineering, one must know how much water is flowing through a given soil in unit time. This knowledge is required to design earth dams, to determine the quantity of seepage under hydraulic structures, and to dewater before and during the construction of foundations. In 1856, Darcy proposed the following equation (Figure 1.5) for calculating the velocity of flow of water through a soil.

$$v = ki \tag{1.23}$$

where $v$ = velocity (unit: cm/sec)
$k$ = coefficient of permeability of soil (unit: cm/sec)
$i$ = hydraulic gradient

The hydraulic gradient ($i$) may be defined as

$$i = \frac{\Delta h}{L} \tag{1.24}$$

where $\Delta h$ = piezometric head difference between the sections at $AA$ and $BB$
$L$ = distance between the sections at $AA$ and $BB$

(*Note:* The sections $AA$ and $BB$ are perpendicular to the direction of flow.)

Darcy's law [Eq. (1.23)] is valid for a wide range of soil types. However, with materials like clean gravel and open-graded rockfills, Darcy's law breaks down because of the turbulent nature of flow through them.

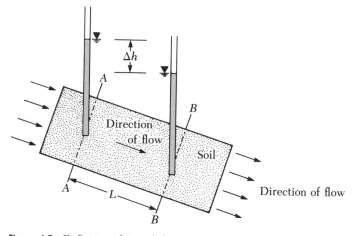

**Figure 1.5** Definition of Darcy's law

The value of the coefficient of permeability of various soils varies greatly. In the laboratory, it can be determined by means of *constant head* or *variable head* permeability tests. The constant head test is more suitable for granular soils. Table 1.7 provides the general range for the values of $k$ for various soils. In granular soils, the value primarily depends on the void ratio. In the past, several equations have been proposed to relate the value of $k$ with the void ratio in the granular soil. They are as follows:

$$\frac{k_1}{k_2} = \frac{e_1^2}{e_2^2} \tag{1.25}$$

$$\frac{k_1}{k_2} = \frac{\left(\dfrac{e_1^2}{1 + e_1}\right)}{\left(\dfrac{e_2^2}{1 + e_2}\right)} \tag{1.26}$$

$$\frac{k_1}{k_2} = \frac{\left(\dfrac{e_1^3}{1 + e_1}\right)}{\left(\dfrac{e_2^3}{1 + e_2}\right)} \tag{1.27}$$

where $k_1$ and $k_2$ are the coefficients of permeability of a given soil at void ratios $e_1$ and $e_2$, respectively.

Hazen (1930) proposed an equation for the coefficient of permeability of fairly uniform sand as

$$k = AD_{10}^2 \tag{1.28}$$

where $k$ is in mm/sec

$\qquad A$ = a constant that varies between 10 and 15

$\qquad D_{10}$ = effective soil size, in mm

**Table 1.7**  Range of the Coefficient of
Permeability for Various Soils

| Type of soil | Coefficient of permeability, $k$ (cm/sec) |
|---|---|
| Medium to coarse gravel | Greater than $10^{-1}$ |
| Coarse to fine sand | $10^{-1}$ to $10^{-3}$ |
| Fine sand, silty sand | $10^{-3}$ to $10^{-5}$ |
| Silt, clayey silt, silty clay | $10^{-4}$ to $10^{-6}$ |
| Clays | $10^{-7}$ or less |

## 1.10

## Steady State Seepage

For most cases of underground seepage of water under hydraulic structures, the flow path changes direction and is not uniform over the entire area
perpendicular to the flow. In such cases, the rate of seepage is determined by
means of a graphical construction that is referred to as *flow net*. The flow net is
based on Laplace's theory of continuity. According to this theory, the flow at any
point $A$ (Figure 1.6) can be represented by the equation

$$k_x \frac{\partial^2 h}{\partial x^2} + k_y \frac{\partial^2 h}{\partial y^2} + k_z \frac{\partial^2 h}{\partial z^2} = 0 \tag{1.29}$$

where   $h$ = hydraulic head at point $A$ (that is, the head of water that a piezometer placed at $A$ would show with the *datum* as the *downstream
water level* as shown in Figure 1.6)

$k_x$, $k_y$, $k_z$ = coefficient of permeability of the soil in $x$, $y$, and $z$ directions,
respectively

For a two-dimensional flow condition as shown in Figure 1.6

$$\frac{\partial^2 h}{\partial^2 y} = 0$$

So, Eq. (1.29) takes the form

$$k_x \frac{\partial^2 h}{\partial x^2} + k_z \frac{\partial^2 h}{\partial z^2} = 0 \tag{1.30}$$

If the soil is isotropic with respect to permeability, $k_x = k_z = k$. So

$$\frac{\partial^2 h}{\partial x^2} + \frac{\partial^2 h}{\partial z^2} = 0 \tag{1.31}$$

The preceding equation represents two orthogonal sets of curves that are known
as *flow lines* and *equipotential lines*. A *flow net* is a combination of a number of
equipotential lines and flow lines. A flow line is the line that a water particle
would follow in traveling from the upstream side to the downstream side. An

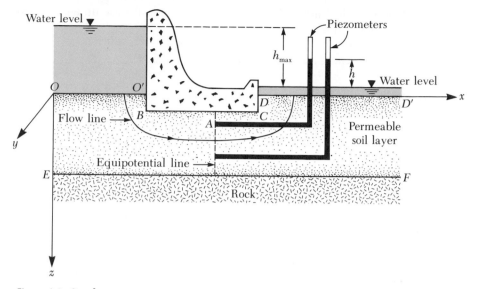

**Figure 1.6** Steady state seepage

equipotential line is the line along which piezometers may be placed to cause the water in them to rise to the same elevation (see Figure 1.6).

In drawing a flow net, one needs to establish the *boundary conditions*. For example, in Figure 1.6 the ground surfaces on the upstream ($OO'$) and downstream ($DD'$) sides are equipotential lines. The bottom of the dam below the ground surface, $O'BCD$, is a flow line. The top of the rock surface, $EF$, is also a flow line. Once the boundary conditions are established, a number of flow lines and equipotential lines are drawn by trial and error such that all of the flow elements in the net have the same length-to-width ratio ($L/B$). In most cases, the $L/B$ ratio is kept as *one*—that is, the flow elements are drawn as squares. This is shown in a flow net drawn in Figure 1.7. Note that all flow lines must intersect all equipotential lines at *right angles*.

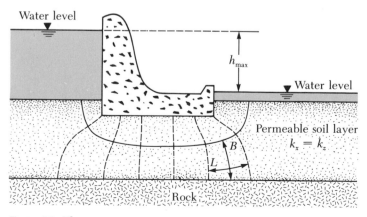

**Figure 1.7** Flow net

Once the flow net is drawn, the seepage in unit time under the structure can be calculated as

$$q = kh_{max} \frac{N_f}{N_d} n \tag{1.32}$$

where $N_f$ = number of flow channels
$N_d$ = number of drops
$n$ = width-to-length ratio of the flow elements in the flow net $(B/L)$
$h_{max}$ = difference in water level between the upstream and downstream sides

Note that the space between two consecutive flow lines is defined as a *flow channel*, and the space between two consecutive equipotential lines is called a *drop*. In Figure 1.7, $N_f = 2$, $N_d = 7$, and $n = 1$. When square elements are drawn in a flow net

$$q = kh_{max} \frac{N_f}{N_d} \tag{1.33}$$

## 1.11
## Filter Design Criteria

In the design of earth structures one often encounters problems caused by the flow of water, such as soil erosion, which may result in structural instability. Erosion is generally prevented by building soil layers that are referred to as *filters* (see Figure 1.8). Two main factors influence the choice of filter material: the grain-size distribution of the filter materials should be such that (a) the soil to be protected is not washed into the filter and (b) excessive hydro-static pressure head is not created in the soil that has a lower coefficient of permeability.

In order to satisfy the preceding conditions, the following requirements must be met.

$$\frac{D_{15(F)}}{D_{85(B)}} < 5 \qquad \text{(to satisfy condition } a\text{)} \tag{1.34}$$

$$\frac{D_{15(F)}}{D_{15(B)}} > 4 \qquad \text{(to satisfy condition } b\text{)} \tag{1.35}$$

In these relations, the subscripts $F$ and $B$ refer to the *filter* and the *base* material (that is, the soil to be protected). $D_{15}$ and $D_{85}$ refer to the diameters through which 15% and 85% of the soil (filter/base, as the case may be) will pass.

The U.S. Department of the Navy (1971) provides some additional require-ments for filter design to satisfy condition (a). They are as follows:

$$\frac{D_{50(F)}}{D_{50(B)}} < 25 \tag{1.36}$$

$$\frac{D_{15(F)}}{D_{15(B)}} < 20 \tag{1.37}$$

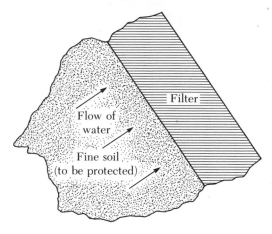

**Figure 1.8** Filter design

An example of filter design using these requirements [Eqs. (1.34)–(1.37)] is given in Example Problem 5.9.

## 1.12

### Effective Stress Concept

Consider the vertical stress at a point $A$ located at a depth $h_1 + h_2$ below the ground surface, as shown in Figure 1.9a. The total vertical stress, $\sigma$, at $A$ can be given as

$$\sigma = h_1\gamma + h_2\gamma_{\text{sat}} \tag{1.38}$$

where $\gamma$ and $\gamma_{\text{sat}}$ are unit weights of soil above and below the ground water table, respectively.

The total stress is carried partially by the *pore water* in the void spaces and partially by the *soil solids* at their points of contact. For example, consider a wavy plane $AB$ drawn through point $A$ (see Figure 1.9a) that passes through the points of contact of soil grains. The plan of this section is shown in Figure 1.9b. The small dots in Figure 1.9b represent the areas in which there is solid-to-solid contact. Let the sum of these areas be equal to $A'$. Hence, the area that is filled by water is equal to $XY - A$. The force carried by the pore water over the area shown in Figure 1.9b is

$$F_w = (XY - A')u \tag{1.39}$$

where $u$ = pore water pressure = $\gamma_w h_2$                  (1.40)

Now let $F_1$, $F_2$, . . . be the forces at the contact points of the soil solids as shown in Figure 1.9a. The sum of the vertical components of these forces over a horizontal area $XY$ can be expressed as

$$F_s = \Sigma F_{1(v)} + F_{2(v)} + \cdots \tag{1.41}$$

where $F_{1(v)}$, $F_{2(v)}$, . . . are vertical components of forces $F_1$, $F_2$, . . . , respectively.

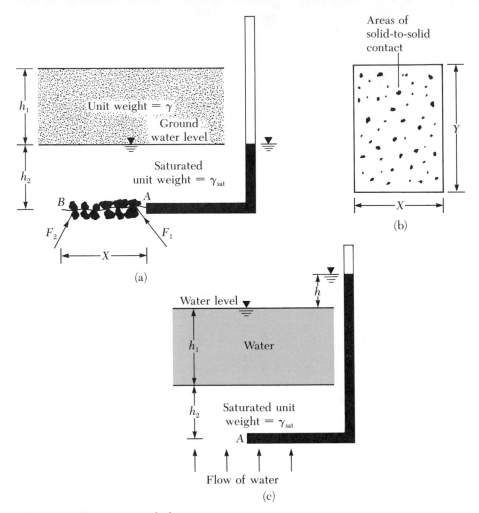

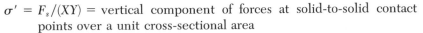

**Figure 1.9**  Effective stress calculation

Based on the principles of statics

$$(\sigma)XY = F_w + F_s$$

or

$$(\sigma)XY = (XY - A')u + F_s$$

So

$$\sigma = (1 - a)u + \sigma' \tag{1.42}$$

where $a = A'/XY$ = fraction of the unit cross-sectional area occupied by solid-to-solid contact

$\sigma' = F_s/(XY)$ = vertical component of forces at solid-to-solid contact points over a unit cross-sectional area

The term $\sigma'$ in Eq. (1.42) is generally referred to as the *effective stress*. Also, the quantity $a$ in Eq. (1.42) is very small. Thus, it can be written that

$$\sigma = u + \sigma' \tag{1.43}$$

Note that the effective stress is a *derived* quantity. Also, because the effective stress $\sigma'$ is related to the contact between the soil solids, it will induce volume changes. It is also responsible for producing *frictional resistance* in soils and rocks. For dry soils, $u = 0$; hence, $\sigma = \sigma'$.

For the problem under consideration in Figure 1.9a, $u = h_2 \gamma_w$ ($\gamma_w$ = unit weight of water). Thus, the effective stress at point $A$ is

$$\sigma' = \sigma - u = (h_1 \gamma + h_2 \gamma_{sat}) - h_2 \gamma_w$$
$$= h_1 \gamma + h_2 (\gamma_{sat} - \gamma_w) = h_1 \gamma + h_2 \gamma' \tag{1.45}$$

where $\gamma'$ = effective or the submerged unit weight of soil
$$= \gamma_{sat} - \gamma_w$$

From Eq. (1.15)

$$\gamma_{sat} = \frac{G_s \gamma_w + e\gamma_w}{1 + e}$$

So

$$\gamma' = \gamma_{sat} - \gamma_w = \frac{G_s \gamma_w + e\gamma_w}{1 + e} - \gamma_w = \frac{\gamma_w (G_s - 1)}{1 + e} \tag{1.46}$$

In the case of the problem in Figure 1.9a and 1.9b, there was *no seepage of water* in soil. Figure 1.9c shows a simple condition in a soil profile where there is an upward seepage. For this case, at point $A$

$$\sigma = h_1 \gamma_w + h_2 \gamma_{sat}$$
$$u = (h_1 + h_2 + h)\gamma_w$$

Thus, from Eq. (1.43)

$$\sigma' = \sigma - u = (h_1 \gamma_w + h_2 \gamma_{sat}) - (h_1 + h_2 + h)\gamma_w$$
$$= h_2 (\gamma_{sat} - \gamma_w) - h\gamma_w = h_2 \gamma' - h\gamma_w$$

or

$$\sigma' = h_2 \left( \gamma' - \frac{h}{h_2} \gamma_w \right) = h_2 (\gamma' - i\gamma_w) \tag{1.47}$$

Note that in Eq. (1.47), $h/h_2$ is the hydraulic gradient, $i$. If the hydraulic gradient is very high, such that $\gamma' - i\gamma_w$ becomes equal to zero, *the effective stress will become equal to zero*. This means that there is no contact stress between the soil particles, and the soil structure will break up. This situation is referred to as the *quicksand condition*, or *failure by heave*. So, for heave

$$i = i_{cr} = \frac{\gamma'}{\gamma_w} = \frac{G_s - 1}{1 + e} \tag{1.48}$$

where $i_{cr}$ = critical hydraulic gradient

For most sandy soils, $i_{cr}$ ranges from 0.9 to 1.1, with an average of about 1.

## Example 1.4

Figure 1.10 shows a soil profile. Determine the total stress, pore water pressure, and effective stress at points $A$, $B$, and $C$. Draw the variation of the effective stress with depth.

### Solution
Determination of Unit Weights of Soil

$$\gamma_{d(\text{sand})} = \frac{G_s \gamma_w}{1 + e} = \frac{(2.65)(9.81)}{1 + 0.6} = 16.25 \text{ kN/m}^3$$

$$\gamma_{\text{sat(clay)}} = \frac{G_s \gamma_w + w G_s \gamma_w}{1 + w G_s}$$

*Note:* For saturated soils, $e = w G_s$ [Eq. (1.14)]; so, for this case, $e = (0.3)(2.7) = 0.81$. So

$$\gamma_{\text{sat(clay)}} = \frac{(2.7)(9.81) + (0.81)(9.81)}{1 + 0.81} = 19.02 \text{ kN/m}^3$$

### Total Stress Calculation

At $A$: $\sigma_A = 0$

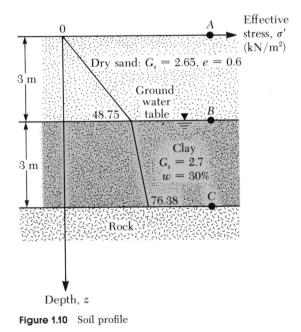

**Figure 1.10**  Soil profile

At B: $\sigma_B = \gamma_{d(sand)} \times 3 = 16.25 \times 3 = \underline{48.75 \text{ kN/m}^2}$

At C: $\sigma_C = \sigma_B + \gamma_{sat(clay)} \times 3 = 48.75 + (19.02)3 = \underline{105.81 \text{ kN/m}^2}$

### Pore Water Pressure Calculation

At A: $u_A = \underline{0}$

At B: $u_B = \underline{0}$

At C: $u_C = 3 \times \gamma_w = 3 \times 9.81 = \underline{29.43 \text{ kN/m}^2}$

### Effective Stress Calculation

At A: $\sigma'_A = \sigma_A - u_A = 0$

At B: $\sigma'_B = 48.75 - 0 = \underline{48.75 \text{ kN/m}^2}$

At C: $\sigma'_C = 105.81 - 29.43 = \underline{76.38 \text{ kN/m}^2}$

The variation of effective stress with depth is shown in Figure 1.10.

## 1.13

## Capillary Rise in Soil

It is a well-known fact that when a capillary tube is placed in water, the water level in the tube rises (Figure 1.11a). This is caused by the *surface tension* effect. According to Figure 1.11a, the pressure at any point $A$ in the capillary tube (with respect to the atmospheric pressure) can be given as

$$u = -\gamma_w z' \qquad (\text{for } z' = 0 \text{ to } h_c)$$

and

$$u = 0 \qquad (\text{for } z' \geq h_c)$$

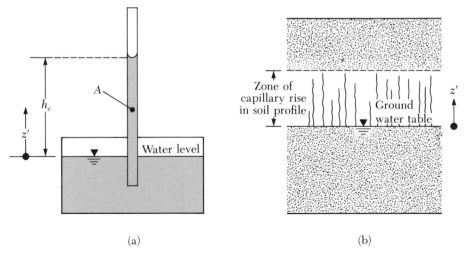

(a)

(b)

**Figure 1.11** Capillary rise

In a given soil mass, the interconnected void spaces can behave like a number of capillary tubes with varying diameters. The surface-tension force may cause water in the soil to rise above the ground water table, as shown in Figure 1.11b. The height of the capillary rise will depend on the diameter of the capillary tubes. The capillary rise will *decrease* with the increase of the tube diameter. Because the capillary tubes in soil have variable diameters, the height of capillary rise will be nonuniform. The pore water pressure at any point in the zone of capillary rise in soil can be approximated as

$$u = -S\gamma_w z'$$  (1.49)

where $S$ = degree of saturation of soil
$z'$ = distance measured above the ground water table

## Example 1.5

Figure 1.12a shows a soil profile. Plot the variation of total stress, pore water pressure, and effective stress with depth.

### Solution
#### Calculation of Unit Weight of Soil

For $z = 0$ m to 2.5 m

$$\gamma_d = \frac{G_s \gamma_w}{1 + e} = \frac{(2.67)(9.81)}{1.5} = 17.46 \text{ kN/m}^3$$

For $z = 2.5$ m to 3.9 m

$$\gamma = \frac{\gamma_w(G_s + Se)}{1 + e} = \frac{(9.81)[2.67 + (0.5)(0.5)]}{1.5} = 19.1 \text{ kN/m}^3$$

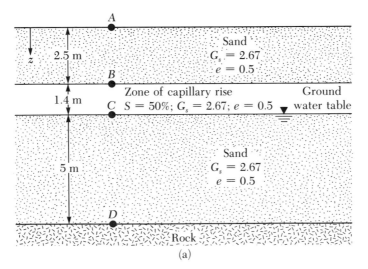

(a)

Figure 1.12a

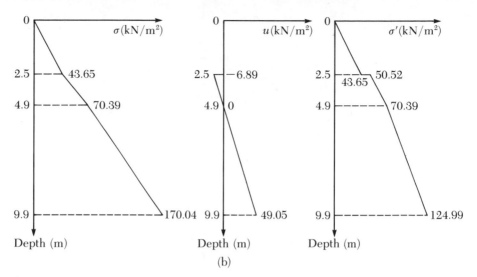

**Figure 1.12b**

For $z = 3.9$ m to $8.9$ m

$$\gamma_{sat} = \frac{\gamma_w(G_s + e)}{1 + e} = \frac{(9.81)(2.67 + 0.5)}{1.5} = \underline{20.73 \text{ kN/m}^3}$$

**Stress Calculation**

At Point $A$:

$$\sigma = \underline{0}$$
$$u = \underline{0}$$
$$\sigma' = \underline{0}$$

At Point $B$:

$$\sigma = (2.5)\gamma_d = (2.5)(17.46) = \underline{43.65 \text{ kN/m}^2}$$
$$u \text{ (immediately above } B) = \underline{0}$$
$$u \text{ (immediately below } B) = -S(1.4)\gamma_w = -(0.5)(1.4)(9.81)$$
$$= \underline{-6.87 \text{ kN/m}^2}$$
$$\sigma' \text{ (immediately above } B) = 43.65 - 0 = \underline{43.65 \text{ kN/m}^2}$$
$$\sigma' \text{ (immediately below } B) = 43.65 - (-6.87) = \underline{50.52 \text{ kN/m}^2}$$

At Point $C$:

$$\sigma = 43.65 + (1.4)(19.1) = \underline{70.39 \text{ kN/m}^2}$$
$$u = \underline{0}$$
$$\sigma' = 70.39 - 0 = \underline{70.39 \text{ kN/m}^2}$$

At Point $D$:

$$\sigma = 70.39 + (5)(20.73) = 174.04 \text{ kN/m}^2$$
$$u = (5)(9.81) = 49.05 \text{ kN/m}^2$$
$$\sigma' = 174.04 - 49.05 = 124.99 \text{ kN/m}^2$$

The variations of $\sigma$, $u$, and $\sigma'$ with depth are plotted in Figure 1.12b.

## 1.14

## Consolidation

When the stress on a saturated clay layer in the field is increased, for example, by the construction of a foundation, the pore water pressure in the clay will increase. Because the coefficients of permeability of clays are very small (see Table 1.7 on p. 20), it will take some time for the excess pore water pressure to gradually dissipate. According to Figure 1.13, if $\Delta p$ is a surcharge at the ground surface over a very large area, the increase of total stress ($\Delta\sigma$) at any depth of the clay layer will be equal to $\Delta p$, or

$$\Delta\sigma = \Delta p$$

However, at time $t = 0$ (that is, immediately after the stress application), the excess pore water pressure at any depth ($\Delta u$) will be equal to $\Delta p$, or

$$\Delta u = \Delta h_i \, \gamma_w = \Delta p \text{ (at time } t = 0)$$

Hence, the increase of effective stress at time $t = 0$ will be

$$\Delta\sigma' = \Delta\sigma - \Delta u = 0$$

Theoretically, at time $t = \infty$, when all the excess pore water pressure in the clay

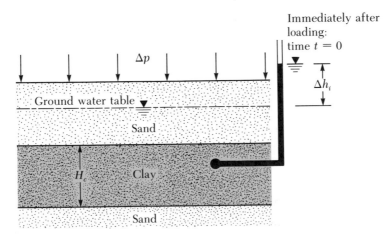

**Figure 1.13**   Principles of consolidation

layer is dissipated as a result of drainage into the sand layers

$$\Delta u = 0 \qquad \text{(at time } t = \infty)$$

This means that the increase of effective stress in the clay layer is

$$\Delta \sigma' = \Delta \sigma - \Delta u = \Delta p - 0 = \Delta p$$

This gradual increase in the effective stress in the clay layer will cause gradual settlement over a period of time, and is referred to as *consolidation*.

Laboratory tests on undisturbed saturated clay specimens can be conducted (ASTM Test Designation D-2435) to determine the consolidation settlement caused by various incremental loadings. The test specimens are usually 63.5 mm (2.5 in.) in diameter and 25.4 mm (1 in.) in height. Specimens are placed inside a brass ring, with one porous stone at the top and one at the bottom of the specimen (Figure 1.14a). Load on the specimen is then applied such that the total stress is equal to $p$. Settlement readings for the specimen are taken for 24 hours. After that, the load on the specimen is doubled and settlement readings taken. At all times during the test the specimen is kept under water. This procedure is continued until the desired limit of stress on the clay specimen is reached.

Based on the laboratory tests, a graph can be plotted showing the variation of the void ratio $e$ at the *end* of consolidation against the corresponding stress $p$ (semilogarithmic graph: $e$ on the arithmetic scale and $p$ on the log scale). The nature of variation of $e$ vs. log $p$ for a clay specimen is shown in Figure 1.14b. During the laboratory tests, after the desired consolidation pressure is reached, the specimen can be gradually unloaded. This will result in the swelling of the specimen. Figure 1.14b also shows the variation of the void ratio during the unloading period.

From the $e$ vs. log $p$ curve shown in Figure 1.14b, three parameters can be determined that will be necessary for calculation of the settlement in the field. They are as follows:

1. *Preconsolidation Pressure, $p_c$*: This is the *maximum past effective overburden pressure* to which the soil specimen has been subjected. It can be determined by using a simple graphical procedure as proposed by Casagrande (1936). This procedure for determination of the preconsolidation pressure, with reference to Figure 1.14b, has five steps:

   a. Determine the point $O$ on the $e$ log $p$ curve that has the highest curvature (that is, the smallest radius of curvature).
   b. Draw a horizontal line $OA$.
   c. Draw a line $OB$ that is tangent to the $e$ log $p$ curve at $O$.
   d. Draw a line $OC$ that bisects the angle $AOB$.
   e. Produce the straight line portion of the $e$ log $p$ curve backwards to intersect $OC$. This is point $D$. The pressure that corresponds to the point $p$ is the preconsolidation pressure, $p_c$.

Natural soil deposits can be *normally consolidated* or *preconsolidated* (or *overconsolidated*). If the present effective overburden pressure $p = p_o$ is equal

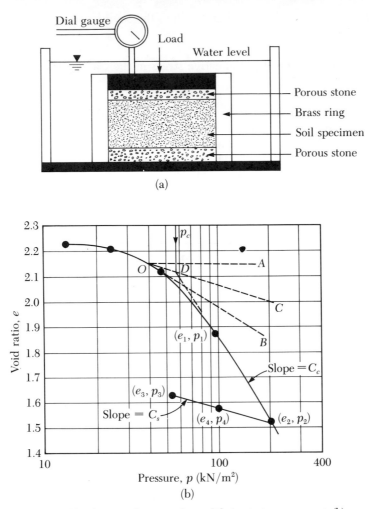

**Figure 1.14** (a) Schematic diagram of consolidation test arrangement; (b) $e$ vs. log $p$ curve for a soft clay from East St. Louis, Illinois

to the preconsolidation pressure $p_c$, the soil is *normally consolidated*. However, if $p_o < p_c$, the soil is *overconsolidated*.

2. *Compression Index, $C_c$*: This is the slope of the straight-line portion (latter part of the loading curve), and it can be determined as

$$C_c = \frac{e_1 - e_2}{\log p_2 - \log p_1} = \frac{e_1 - e_2}{\log\left(\dfrac{p_2}{p_1}\right)} \tag{1.50}$$

where $e_1$ and $e_2$ are the void ratios at the end of consolidation under stresses $p_1$ and $p_2$, respectively.

The *compression index*, as determined from the laboratory $e$-log $p$ curve, will be somewhat different from that encountered in the field. This is primarily due to the remolding of soil to some degree during the field exploration. The nature of variation of the $e$-log $p$ curve in the field for a normally consolidated clay is shown in Figure 1.15. This is generally referred to as the *virgin compression curve*. The virgin curve approximately intersects the laboratory curve at a void ratio of $0.4\,e_o$ (Terzaghi and Peck, 1967). Note that $e_o$ is the void ratio of the clay in the field. Knowing the values of $e_o$ and $p_c$, one can easily construct the virgin curve, as shown in Figure 1.15. The compression index of the virgin curve can now be calculated by using Eq. (1.50).

The value of $C_c$ can vary widely depending on the soil. Skempton (1944) has given an empirical correlation for the compression index in which

$$C_c = 0.009(LL - 10) \tag{1.51}$$

where $LL$ = liquid limit

Several other correlations for the compression index have been summarized by Azzouz, Krizek, and Corotis (1976). Some of these are the following:

$$C_c = 0.01w_n \quad \text{(Chicago clay)} \tag{1.52}$$
$$C_c = 0.208e_o + 0.0083 \quad \text{(Chicago clay)} \tag{1.53}$$
$$C_c = 0.0115w_n \quad \text{(organic soils, peats)} \tag{1.54}$$
$$C_c = 0.0046(LL - 9) \quad \text{(Brazilian clay)} \tag{1.55}$$

where $w_n$ = natural moisture content of soil

In any case, if empirical correlations are used for calculation purposes, a deviation of $\pm 15\%$ from the actual field conditions may be expected.

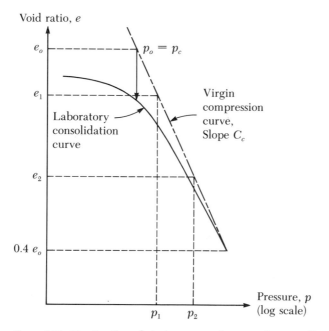

**Figure 1.15** Construction of virgin compression curve for normally consolidated clay

3. *Swelling Index, $C_s$:* This is the slope of the unloading portion of the $e$-log $p$ curve. In Figure 1.14b, it can be defined as

$$C_s = \frac{e_3 - e_4}{\log\left(\dfrac{p_4}{p_3}\right)} \tag{1.56}$$

The value of the swelling index ($C_s$) is, in most cases, ¼ to ⅕ of the compression index. Following are some representative values of $C_s/C_c$ for natural soil deposits.

| Description of soil | $C_s/C_c$ |
|---|---|
| Boston Blue clay | 0.24–0.33 |
| Chicago clay | 0.15–0.3 |
| New Orleans clay | 0.15–0.28 |
| St. Lawrence clay | 0.05–0.1 |

The swelling index determination is important in the estimation of consolidation settlement of *overconsolidated clays* (see Figure 1.16). In the field, depending on the pressure increase, an overconsolidated clay will follow an $e$ vs. log $p$ path *abc*, as shown in Figure 1.16. Note that, in this figure, the point $a$ with coordinates of $p_o$ and $e_o$ corresponds to the field conditions before any pressure increase. The point $b$ corresponds to the preconsolidation pressure ($p_c$) of the clay. The line *ab* is approximately parallel to the laboratory unloading

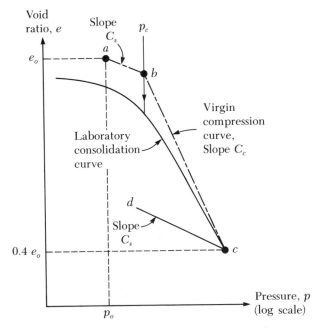

**Figure 1.16**   Construction of field consolidation curve for overconsolidated clay

curve $cd$ (Schmertmann, 1953). Hence, with a knowledge of $e_o$, $p_o$, $p_c$, $C_c$, and $C_s$, one can easily construct the field consolidation curve.

## Calculation of Settlement

The one-dimensional consolidation settlement, caused by an additional load, of a clay layer (Figure 1.17a) having a thickness $H_c$ may be calculated as

$$S = \frac{\Delta e}{1 + e_o} H_c \tag{1.57}$$

where $S$ = settlement
$\Delta e$ = total change of void ratio caused by the additional load application
$e_o$ = the void ratio of the clay before the application of load

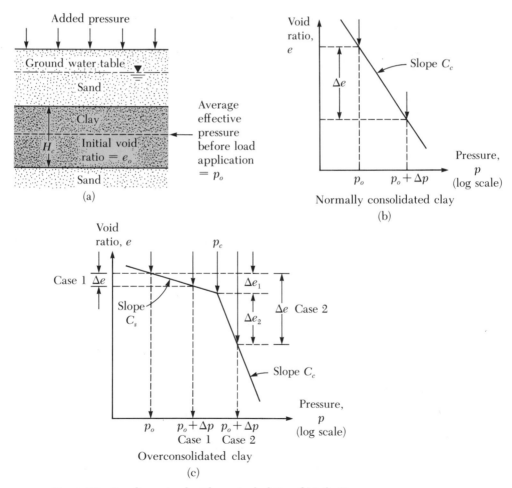

**Figure 1.17**   One-dimensional settlement calculation: (b) is for Eq. (1.58); (c) is for Eqs. (1.61) and (1.63)

For normally consolidated clay, the field $e$ vs. $\log p$ curve will be like the one in Figure 1.17b. If $p_o$ = initial average effective overburden pressure on the clay layer, and $\Delta p$ = average pressure increase on the clay layer caused by the added load, the change of void ratio caused by the load increase is

$$\Delta e = C_c \log \frac{p_o + \Delta p}{p_o} \tag{1.58}$$

Now, combining Eqs. (1.57) and (1.58)

$$S = \frac{C_c H_c}{1 + e_o} \log \frac{p_o + \Delta p}{p_o} \tag{1.59}$$

For overconsolidated clay, the field $e$ vs. $\log p$ curve will be like the one in Figure 1.17c. In this case, depending on the value of $\Delta p$, two conditions may arise. First, if $p_o + \Delta p < p_c$,

$$\Delta e = C_s \log \frac{p_o + \Delta p}{p_o} \tag{1.60}$$

Combining Eqs. (1.57) and (1.60)

$$S = \frac{H_c C_s}{1 + e_o} \log \frac{p_o + \Delta p}{p_o} \tag{1.61}$$

Second, if $p_o < p_c < p_o + \Delta p$,

$$\Delta e = \Delta e_1 + \Delta e_2 = C_s \log \frac{p_c}{p_o} + C_c \log \frac{p_o + \Delta p}{p_o} \tag{1.62}$$

Now, combining Eqs. (1.57) and (1.62)

$$S = \frac{C_s H_c}{1 + e_o} \log \frac{p_c}{p_o} + \frac{C_c H_c}{1 + e_o} \log \frac{p_o + \Delta p}{p_c} \tag{1.63}$$

## Average Degree of Consolidation

Earlier in this section (see Figure 1.13 on p. 30) we saw that consolidation is the result of gradual dissipation of the excess pore water pressure from a clay layer. Pore water pressure dissipation, in turn, increases the effective stress, which induces settlement. Hence, to estimate the degree of consolidation of a clay layer at some time $t$ after the load application, one should know the rate of dissipation of the excess pore water pressure.

Figure 1.18a shows a clay layer of thickness $H_c$ that has highly permeable sand layers at its top and bottom. Let the excess pore water pressure at any point $A$ at any time $t$ after the load application be equal to $\Delta u = (\Delta h)\gamma_w$. For a vertical drainage condition (that is, in the direction of $z$ only) from the clay layer, Terzaghi has derived the following differential equation:

$$\frac{\partial(\Delta u)}{\partial t} = C_v \frac{\partial^2(\Delta u)}{\partial z^2} \tag{1.64}$$

where $C_v$ = coefficient of consolidation

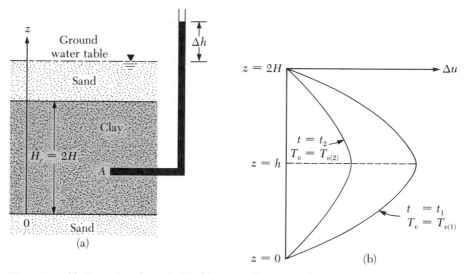

**Figure 1.18** (a) Derivation of Eq. (1.66); (b) nature of variation of $\Delta u$ with time

$$C_v = \frac{k}{m_v \gamma_w} = \frac{k}{\dfrac{\Delta e}{\Delta p(1 + e_{av})} \gamma_w} \qquad (1.65)$$

where $k$ = coefficient of permeability of the clay
$\Delta e$ = total change of void ratio caused by a stress increase of $\Delta p$
$e_{av}$ = average void ratio during consolidation
$m_v$ = volume coefficient of compressibility = $\Delta e / [\Delta p(1 + e_{av})]$

Equation (1.64) can be solved to obtain $\Delta u$ as a function of time $t$ with the following boundary conditions:

1. Because highly permeable sand layers are located at $z = 0$ and $z = H_c$, the excess pore water pressure developed in the clay at those points will be immediately dissipated. So

$\Delta u = 0$ at $z = 0$
$\Delta u = 0$ at $z = H_c = 2H$

where $H$ = length of maximum drainage path (due to two-way drainage condition—that is, at top and bottom of clay)

2. At time $t = 0$

$\Delta u = \Delta u_o$ = initial excess pore water pressure after the load application

With the preceding boundary conditions, Eq. (1.64) yields

$$\Delta u = \sum_{m=0}^{m=\infty} \left[ \frac{2(\Delta u_o)}{M} \sin \left( \frac{Mz}{H} \right) \right] e^{-M^2 T_v} \qquad (1.66)$$

where $M = \dfrac{(2m + 1)\pi}{2}$

$m$ = an integer = 1, 2, . . .

$$T_v = \text{nondimensional time factor} = \frac{C_v t}{H^2} \tag{1.67}$$

The value of $\Delta u$ for various depths (that is, $z = 0$ to $z = 0$ to $2H$) at any given time $t$ (thus, $T_v$) can be calculated from Eq. (1.66). The nature of this variation of $\Delta u$ is shown in Figure 1.18b.

The *average degree of consolidation* of a clay layer can be defined as

$$U = \frac{S_t}{S_{max}} \tag{1.68}$$

where $U$ = average degree of consolidation

$S_t$ = settlement of a clay layer at time $t$ after the load application

$S_{max}$ = maximum consolidation settlement that the clay will undergo under a given loading

If the initial pore water pressure $(\Delta u_o)$ distribution is constant with depth as shown in Figure 1.19a, the average degree of consolidation can also be expressed as

$$U = \frac{S_t}{S_{max}} = \frac{\displaystyle\int_0^{2H} (\Delta u_o)dz - \int_0^{2H} (\Delta u)dz}{\displaystyle\int_0^{2H} (\Delta u_o)dz} \tag{1.69}$$

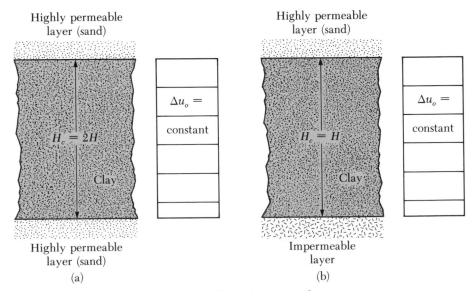

**Figure 1.19**   Drainage condition for consolidation: (a) two-way drainage; (b) one-way drainage

or

$$U = \frac{(\Delta u_o)2H - \int_0^{2H}(\Delta u)dz}{(\Delta u_o)2H} = 1 - \frac{\int_0^{2H}(\Delta u)dz}{2H(\Delta u_o)} \qquad (1.70)$$

Now, combining Eqs. (1.66) and (1.70), one obtains

$$U = \frac{S_t}{S_{\max}} = 1 - \sum_{m=0}^{m=\infty}\left(\frac{2}{M^2}\right)e^{-M^2T_v} \qquad (1.71)$$

The variation of $U$ with $T_v$ can be calculated from Eq. (1.71) and is plotted in Figure 1.20. Note that Eq. (1.71) and, thus, Figure 1.20 are also valid when an impermeable layer is located at the bottom of the clay layer (Figure 1.19b). In such a case, excess pore water pressure dissipation can take place in one direction only. So, the length of the *maximum drainage path* is equal to $H = H_c$.

The variation of $T_v$ with $U$ shown in Figure 1.20 can also be approximated by the following relations:

$$T_v = \frac{\pi}{4}\left(\frac{U\%}{100}\right)^2 \qquad (\text{for } U = 0\text{--}60\%) \qquad (1.72)$$

and

$$T_v = 1.781 - 0.933 \log(100 - U\%) \qquad (\text{for } U > 60\%) \qquad (1.73)$$

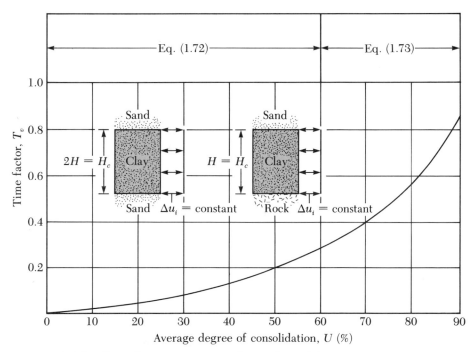

**Figure 1.20** Plot of time factor against average degree of consolidation ($\Delta u_o = $ constant)

## Example
## 1.6

A laboratory consolidation test on a normally consolidated clay showed the following:

| Load, $p$ (kN/m$^2$) | Void ratio at the end of consolidation, $e$ |
|---|---|
| 140 | 0.92 |
| 212 | 0.86 |

The specimen tested was 25.4 mm in thickness drained on both sides. The time required for the specimen to reach 50% consolidation was 4.5 min.

A similar clay layer in the field, 2.8 m thick and drained on both sides, is subjected to similar average pressure increase (that is, $p_o = 140$ kN/m$^2$ and $p_o + \Delta p = 212$ kN/m$^2$). Determine:

a. the expected maximum consolidation settlement in the field;
b. the length of time it will take for the total settlement in the field to reach 40 mm.

### Solution
### Part a

For normally consolidated clay [Eq. (1.50)]

$$C_c = \frac{e_1 - e_2}{\log \left(\dfrac{p_2}{p_1}\right)} = \frac{0.92 - 0.86}{\log \left(\dfrac{212}{140}\right)} = 0.333$$

From Eq. (1.59)

$$S = \frac{C_c H_c}{1 + e_o} \log \frac{p_o + \Delta p}{p_o} = \frac{(0.333)(2.8)}{1 + 0.92} \log \frac{212}{140} = 0.0875 \text{ m}$$

$$= 87.5 \text{ mm}$$

### Part b

From Eq. (1.68), the average degree of consolidation

$$U = \frac{S_t}{S_{\max}} = \frac{40}{87.5}(100) = 45.7\%$$

The coefficient of consolidation, $C_v$, can be calculated from the laboratory test. From Eq. (1.67)

$$T_v = \frac{C_v t}{H^2}$$

For 50% consolidation (Figure 1.20), $T_v = 0.197$, $t = 4.5$ min, $H = H_c/2 = 12.7$ mm. So

$$C_v = T_{50}\frac{H^2}{t} = \frac{(0.197)(12.7)^2}{4.5} = 7.061 \text{ mm}^2/\text{min}$$

Again, for field consolidation, $U = 45.7\%$. From Eq. (1.72)

$$T_v = \frac{\pi}{4}\left(\frac{U\%}{100}\right)^2 = \frac{\pi}{4}\left(\frac{47.5}{100}\right)^2 = 0.177$$

But

$$T_v = \frac{C_v t}{H^2}$$

or

$$t = \frac{T_v H^2}{C_v} = \frac{0.177\left(\dfrac{2.8 \times 1000}{2}\right)^2}{7.061} = 49132 \text{ min} = \underline{34.1 \text{ days}}$$

## 1.15
## Shear Strength

The shear strength of a soil ($s$), in terms of effective stress, can be given by the equation

$$s = c + \sigma' \tan \phi \tag{1.74}$$

where $\sigma' = $ effective normal stress on plane of shearing
$\quad c = $ cohesion, or apparent cohesion
$\quad \phi = $ angle of friction

Equation (1.74) is referred to as the *Mohr-Coulomb failure criteria*. The value of $c$ for sands and normally consolidated clays is equal to zero. For overconsolidated clays, $c > 0$.

For most day-to-day work, the shear strength parameters of a soil (that is, $c$ and $\phi$) are determined by two standard laboratory tests. They are (a) the *direct shear test* and (b) the *triaxial test*. Following is a brief description of each test.

### Direct Shear Test

Dry sand can be conveniently tested by direct shear tests. These tests are conducted by placing the sand in a shear box that is split into two halves (Figure 1.21 a). A normal load is first applied to the specimen. After that, a shear force is applied to the top half of the shear box to cause failure in soil. The normal and shear stresses at failure can be determined as

$$\sigma' = \frac{N}{A}$$

$$s = \frac{R}{A}$$

where $A = $ area of the failure plane in soil—that is, the area of cross section of the shear box.

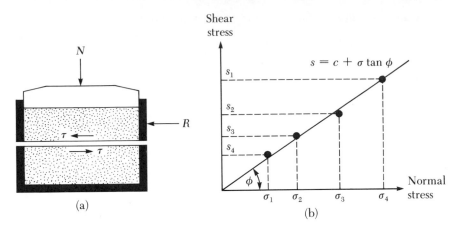

**Figure 1.21**   Direct shear test in sand: (a) schematic diagram of test equipment; (b) plot of test results to obtain the friction angle, $\phi$

Several tests of this type can be conducted by varying the normal load. The angle of friction of the sand can be determined by plotting a graph of $s$ vs. $\sigma'$ as shown in Figure 1.21 b, or

$$\phi = \tan^{-1}\left(\frac{s}{\sigma'}\right) \tag{1.75}$$

For sands, the angle of friction usually ranges from 26° to 45°. It increases with the relative density of compaction. The approximate range of relative density of compaction and the corresponding range of the angle of friction for various sands are given in Table 2.4 on page 74.

### Triaxial Tests

Triaxial compression tests can be conducted on sands and clays. Figure 1.22a shows a schematic diagram of the triaxial test arrangement. It essentially consists of placing a soil specimen confined by a rubber membrane inside a lucite chamber. An all-around confining pressure ($\sigma_3$) is applied to the specimen by means of the chamber fluid (generally water or glycerin). An added stress ($\Delta\sigma$) can also be applied to the specimen in the axial direction to cause failure ($\Delta\sigma = \Delta\sigma_f$ at failure). Drainage from the specimen can be allowed or stopped, depending on the test condition. For clays, three main types of test can be conducted by means of triaxial equipment:

1. Consolidated-drained test (CD test)
2. Consolidated-undrained test (CU test)
3. Unconsolidated-undrained test (UU test)

Table 1.8 on p. 44 gives a summary of these three tests.
For *consolidated-drained tests*, at failure

Major principal effective stress $= \sigma_3 + \Delta\sigma_f = \sigma_1 = \sigma_1'$

Minor principal effective stress $= \sigma_3 = \sigma_3'$

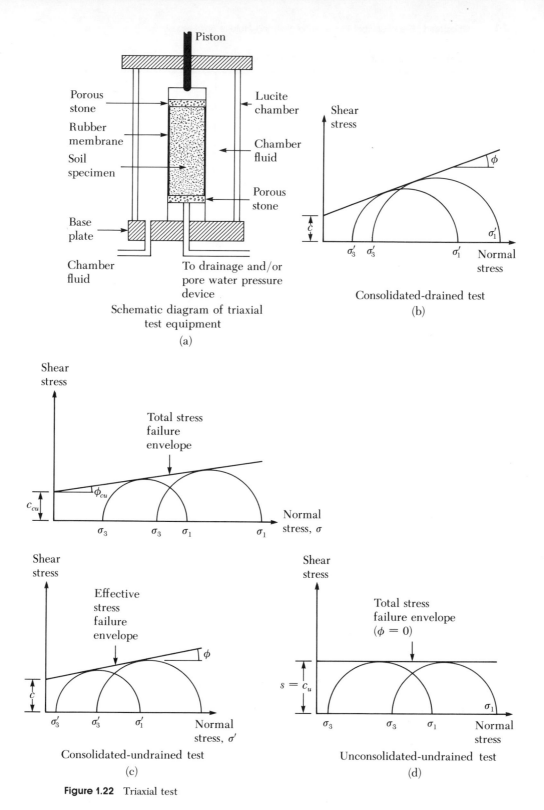

Piston

Porous stone

Rubber membrane

Soil specimen

Base plate

Chamber fluid

Lucite chamber

Chamber fluid

Porous stone

To drainage and/or pore water pressure device

Schematic diagram of triaxial test equipment

(a)

Shear stress

$\sigma_3'$  $\sigma_3'$   $\sigma_1'$  Normal stress

Consolidated-drained test

(b)

Shear stress

Total stress failure envelope

$\phi_{cu}$

$c_{cu}$

$\sigma_3$  $\sigma_3$  $\sigma_1$   $\sigma_1$  Normal stress, $\sigma$

Shear stress

Effective stress failure envelope

$\phi$

$c$

$\sigma_3'$  $\sigma_3'$  $\sigma_1'$   Normal stress, $\sigma'$

Consolidated-undrained test

(c)

Shear stress

Total stress failure envelope ($\phi = 0$)

$s = c_u$

$\sigma_3$  $\sigma_3$  $\sigma_1$   Normal stress

$\sigma_1$

Unconsolidated-undrained test

(d)

**Figure 1.22**  Triaxial test

43

**Table 1.8**  Summary of Triaxial Tests on Saturated Clays

| Test type | Step 1 | Step 2 |
|-----------|--------|--------|
| | $\sigma_3$ (diagram) | $\Delta\sigma$, $\sigma_3$ (diagram) |
| Consolidated-drained | Apply chamber pressure, $\sigma_3$. Allow complete drainage, so pore water pressure $(u = u_a)$ developed is zero. | Apply axial stress, $\Delta\sigma$, slowly. Allow drainage, so pore water pressure $(u = u_d)$ developed through application of $\Delta\sigma$ is zero. At failure, $\Delta\sigma = \Delta\sigma_f$; total pore water pressure $u_f = u_a + u_d = 0$. |
| Consolidated-undrained | Apply chamber pressure, $\sigma_3$. Allow complete drainage, so pore water pressure $(u = u_a)$ developed is zero. | Apply axial stress, $\Delta\sigma$. Do not allow drainage $(u = u_d \neq 0)$. At failure, $\Delta\sigma = \Delta\sigma_f$; pore water pressure $u = u_f = u_a + u_d = 0 + u_{d(f)}$. |
| Unconsolidated-undrained | Apply chamber pressure, $\sigma_3$. Do not allow drainage, so pore water pressure $(u = u_a)$ developed through application of $\sigma_3$ is not zero. | Apply axial stress, $\Delta\sigma$. Do not allow drainage $(u = u_d \neq 0)$. At failure $\Delta\sigma = \Delta\sigma_f$; pore water pressure $u = u_f = u_a + u_{d(f)}$. |

Several tests of this type can be conducted on a number of clay specimens by changing $\sigma_3$. The shear strength parameters $(c$ and $\phi)$ can now be determined by plotting Mohr's circle at failure, as shown in Figure 1.22b, and drawing a common tangent to the Mohr's circles. This is the *Mohr-Coulomb failure envelope*. (*Note:* For normally consolidated clay, $c \approx 0$.) It can be shown that, at failure

$$\sigma_1' = \sigma_3' \tan^2\left(45 + \frac{\phi}{2}\right) + 2c \tan\left(45 + \frac{\phi}{2}\right) \tag{1.76}$$

For *consolidated-undrained* tests, at failure

Major principal total stress $= \sigma_3 + \Delta\sigma_f = \sigma_1$

Minor principal total stress $= \sigma_3$

Major principal effective stress $= (\sigma_3 + \Delta\sigma_f) - u_f = \sigma_1'$

Minor principal effective stress $= \sigma_3 - u_f = \sigma_3'$

A number of tests of this type can be conducted on several soil specimens by changing $\sigma_3$. The total-stress Mohr's circles at failure can now be plotted as shown in Figure 1.22 c, and then a common tangent can be drawn to define the *failure envelope*. This is the *total stress failure envelope* and can be defined by the equation

$$s = c_{cu} + \sigma \tan \phi_{cu} \tag{1.77}$$

where $c_{cu}$ and $\phi_{cu}$ are the *consolidated undrained cohesion* and *angle of friction*, respectively. (Note: $c_{cu} \approx 0$ for normally consolidated clays.)

Similarly, effective-stress Mohr's circles at failure can be drawn to determine the *effective stress failure envelopes* (Figure 1.22c). This will follow the relation as given by Eq. (1.74).

Kenney (1959) has given a correlation between the friction angle, $\phi$, and the plasticity index (*PI*) of normally consolidated clays based on the observations of over 60 soils. This correlation is shown in Figure 1.23. Based on the average plot, the value of $\phi$ generally decreases from about 37–38° with a plasticity index of about 10, to about 25° with a plasticity index of about 100. The consolidated undrained friction angle ($\phi_{cu}$) of normally consolidated saturated clays generally ranges from 5° to 20°.

For *unconsolidated-undrained* triaxial tests,

$$\text{Major principal total stress} = \sigma_3 + \Delta\sigma_f = \sigma_1$$

$$\text{Minor principal total stress} = \sigma_3$$

The total-stress Mohr's circle at failure can now be drawn as shown in Figure 1.22d. It can be shown that, for saturated clays, the value of $\sigma_1 - \sigma_3 = \Delta\sigma_f$ is a constant, irrespective of the chamber confining pressure, $\sigma_3$ (this is also shown in Figure 1.22d). The tangent to these Mohr's circles will be a horizontal line, which is called the $\phi = 0$ condition. The shear stress for this condition can be given as

$$s = c_u = \frac{\Delta\sigma_f}{2} \tag{1.78}$$

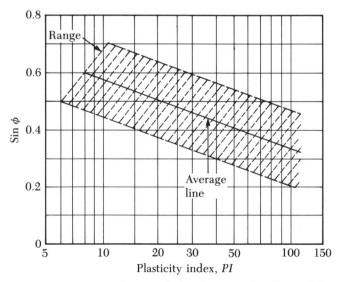

**Figure 1.23** Variation of sin $\phi$ with the plasticity index of several clays (redrawn after Kenney, 1959)

The pore pressure developed in the soil specimen during the unconsolidated-undrained triaxial test is equal to

$$u = u_a + u_d \tag{1.79}$$

The pore pressure $u_a$ is the contribution of the hydrostatic chamber pressure, $\sigma_3$. So, it can be written as

$$u_d = B\sigma_3 \tag{1.80}$$

where $B$ = Skempton's pore pressure parameter

Similarly, the pore pressure $u_d$ is the result of added axial stress, $\Delta\sigma$. The magnitude of $u_d$ can be given by the equation

$$u_d = A\,\Delta\sigma \tag{1.81}$$

where $A$ = Skempton's pore pressure parameter

However

$$\Delta\sigma = \sigma_1 - \sigma_3 \tag{1.82}$$

Combining Eqs. (1.79), (1.80), (1.81), and (1.82)

$$u = u_a + u_d = B\sigma_3 + A(\sigma_1 - \sigma_3) \tag{1.83}$$

The pore water pressure parameter $B$ in saturated soils is equal to one. So,

$$u = \sigma_3 + A(\sigma_1 - \sigma_3) \tag{1.84}$$

The value of the pore water pressure parameter $A$ at failure will vary with the type of soil. Following is a general range of the values of $A$ at failure for various types of clayey soil encountered in nature:

| Type of soil | $A$ at failure |
|---|---|
| Sandy clays | 0.5–0.7 |
| Normally consolidated clays | 0.5–1 |
| Overconsolidated clays | −0.5–0 |

## 1.16

## Unconfined Compression Test

The *unconfined compression test* (Figure 1.24a) is a special type of *unconsolidated-undrained triaxial test* in which the confining pressure $\sigma_3 = 0$, as shown in Figure 1.24b. In this test, an axial stress, $\Delta\sigma$, is applied to the specimen to cause failure (that is, $\Delta\sigma = \Delta\sigma_f$). The corresponding Mohr's circle is shown in Figure 1.24b. Note that, for this case

Major principal total stress = $\Delta\sigma_f = q_u$

Minor principal total stress = 0

The axial stress at failure, $\Delta\sigma_f = q_u$, is generally referred to as the *unconfined*

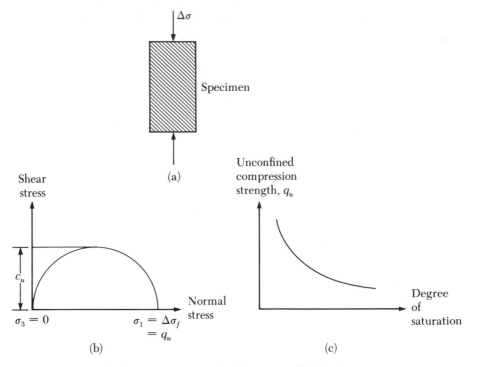

**Figure 1.24** Unconfined compression test: (a) soil specimen; (b) Mohr's circle for the test; (c) variation of $q_u$ with the degree of saturation

*compression strength*. The shear strength of saturated clays under this condition ($\phi = 0$) can be given by Eq. (1.74) as

$$s = c_u = \frac{q_u}{2} \tag{1.85}$$

The unconfined compression strength can be used as an indicator to relate the consistency of clays. This is shown in Table 2.3 on page 73.

Unconfined compression tests are sometimes conducted on unsaturated soils. With the void ratio of a soil specimen remaining constant, the unconfined compresion strength rapidly decreases with the degree of saturation (Figure 1.24c).

# Natural Soil Deposits

## 1.17
## General Soil Deposits

Most of the soils that cover the earth are formed by the weathering of various rocks. There are two general types of weathering: (1) *mechanical weathering*, and (2) *chemical weathering*.

*Mechanical weathering* is the process by which rocks are broken down into smaller and smaller pieces by physical forces. These physical forces may be running water, wind, ocean waves, glacier ice, frost action, and expansion and contraction caused by gain and loss of heat.

*Chemical weathering* is the process of chemical decomposition of the original rock. In the case of mechanical weathering, the rock breaks down into smaller pieces without a change of chemical composition. However, in chemical weathering, the original material may be changed to something entirely different. For example, the chemical weathering of feldspar can produce clay minerals.

The soil that is produced by the weathering process of rocks can be transported by physical agents to other places. These soil deposits are called *transported soils*. In contrast, some soils stay in the place of their formation and cover the rock surface from which they derive. These soils are referred to as *residual soils*.

Based on the *transporting agent*, transported soils can be subdivided into three major categories:

1. *Alluvial*, or *fluvial*: deposited by running water
2. *Glacial*: deposited by glacier action
3. *Aeolian*: deposited by wind action

In addition to transported and residual soils, there are *peats* and *organic soils*, which derive from the decomposition of organic materials.

The following sections discuss the general physical characteristics of residual soils, the three types of transported soil, and organic soils.

## 1.18

### Residual Soil

Residual soil deposits are common in humid tropics, Hawaii, and the southeastern United States. The nature of a residual soil deposit will generally depend on the parent rock. When hard rocks such as granite and gneiss undergo weathering, most of the materials are likely to remain in place. These soil deposits generally have a top layer of clayey or silty-clay material followed by silty and/or sandy soil layers. These are generally followed by a partially weathered rock and then sound bedrock. The depth of the sound bedrock may vary widely from place to place, even within a distance of a few meters. Figure 1.25 shows the log of a boring in a residual soil deposit derived from the weathering of granite.

In contrast to hard rocks, there are some chemical rocks, such as limestone, that are chiefly made up of calcite ($CaCo_3$) mineral. Chalk and dolomite have large concentrations of dolomite minerals [$Ca\ Mg(Co_3)_2$]. These rocks have large amounts of soluble materials, some of which are removed by ground water, leaving behind the insoluble fraction of the rock. Residual soils that derive from chemical rocks do not possess a gradual transition zone to the bedrock as seen in Figure 1.25. The residual soils derived from the weathering of limestone-like rocks are mostly red in color. Although uniform in kind, the depth of weathering

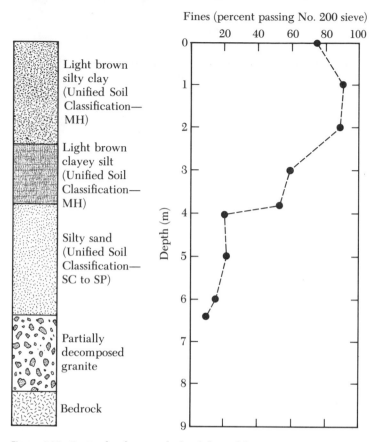

Light brown
silty clay
(Unified Soil
Classification—
MH)

Light brown
clayey silt
(Unified Soil
Classification—
MH)

Silty sand
(Unified Soil
Classification—
SC to SP)

Partially
decomposed
granite

Bedrock

**Figure 1.25** Boring log for a residual soil derived from granite

may vary greatly. The residual soils immediately above the bedrock may be normally consolidated. Large foundations with heavy loads may be susceptible to large consolidation settlements on these soils.

## 1.19

## Alluvial Deposits

Alluvial soil deposits derive from the action of streams and rivers. They can be divided into two major categories: (1) *braided-stream deposits* and (2) deposits caused by the *meandering belt of streams*.

### Deposits from Braided Streams

Braided streams are high-gradient, rapidly flowing streams. They are highly erosive and contain large amounts of sediment load. Because of the high bed load, a minor change in the velocity of flow will cause deposit of sediments. By this process, these streams may build up a complex tangle of converging and diverging channels separated by sandbars and islands.

Depth (m)

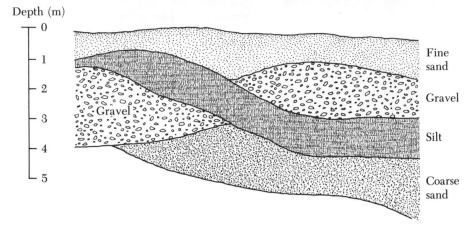

**Figure 1.26**  Cross section of a braided-stream deposit

The deposits formed from braided streams are very irregular in stratification with a wide range of grain sizes. Figure 1.26 shows a cross section of such a deposit. These deposits share several characteristics:

**1.** The grain sizes usually range from gravel to silt. Clay-size particles are generally *not* found in these deposits.

**2.** Although there is a wide variation in the grain size, the soil in a given pocket or lens is rather uniform.

**3.** At any given depth, the void ratio and unit weight may vary over a wide range within a lateral distance of a few meters. This variation can be seen during a soil-exploration program (Chapter 2) for construction of a foundation of a structure. The standard penetration resistance ($N$-value; see Chapter 2) at a given depth obtained from various bore holes will be highly irregular and varies over a wide range.

Alluvial deposits can be found in several parts of the western United States, such as Southern California, Utah, and basin and range provinces of Nevada. Also, a large amount of sediment originally derived from the Rocky Mountain range was carried eastward to form the alluvial deposits of the Great Plains. On a smaller scale, this type of natural soil deposit, left by braided streams, can be encountered locally.

## Meandering Belt Deposits

The term *meander* is derived from the Greek word *maiandros*, which means "bend." Mature streams in a valley curve back and forth in a set of hairpin bends. The valley floor in which a river meanders is referred to as the *meandering belt*. In a meandering river, the soil from the bank is continually eroded from the points where it is concave in shape and deposited at points where the bank is convex in shape, as shown in Figure 1.27. These deposits are called *point bar deposits*, and they usually consist of sand and silt-size particles. Sometimes, during the process of erosion and deposition, the river abandons a

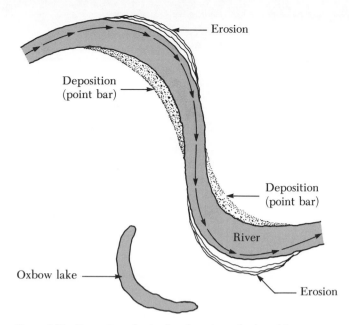

Erosion

Deposition
(point bar)

Deposition
(point bar)

River

Oxbow lake

Erosion

**Figure 1.27**  Formation of point bar deposits and oxbow lake in a meandering stream

meander and cuts a shorter path. The abandoned meander, when filled with
water, is called an *oxbow lake* (see Figure 1.27).

During the flooding seasons, river water floods low-lying areas. The sand
and silt-size particles carried by the river water are deposited along the banks
to form ridges known as *levees* (Figure 1.28). Finer soil particles consisting of

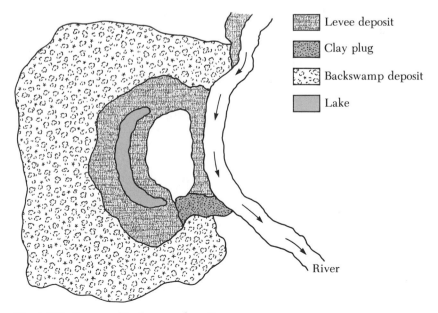

Levee deposit

Clay plug

Backswamp deposit

Lake

River

**Figure 1.28**  Levee and backswamp deposit

**Table 1.9** Properties of Deposits within the Mississippi Alluvial Valley[a]

| Environment | Soil texture | Natural water content (%) | Liquid limit | Plasticity index | Shear strength | |
|---|---|---|---|---|---|---|
| | | | | | Cohesion[b] (kN/m²) | Angle of friction (deg) |
| Natural levee | Clay (CL) | 25–35 | 35–45 | 15–25 | 17–57 | 0 |
| | Silt (ML) | 15–35 | NP[c]–35 | NP–5 | 9–33 | 10–35 |
| Point bar | Silt (ML) and silty sand (SM) | 25–45 | 30–55 | 10–25 | 0–41 | 25–35 |
| Abandoned channel | Clay (CL, CH) | 30–95 | 30–100 | 10–65 | 14–57 | 0 |
| Backswamps | Clay (CH) | 25–70 | 40–115 | 25–100 | 19–120 | 0 |
| Swamp | Organic clay (OH) | 100–265 | 135–300 | 100–165 | — | — |

[a]After Kolb and Shockley, 1959
[b]Rounded off
[c]NP = nonplastic

silts and clays are carried by the water farther into the flooded plains. These particles settle at different rates to form what is referred to as *backswamp deposits* (Figure 1.28). These clays may be highly plastic. Table 1.9 gives the properties of soil deposits found in natural levees, point bars, abandoned channels, backswamps, and swamps within the Mississippi alluvial valley.

## 1.20

### Glacial Deposits

During the great Pleistocene Ice Age, large areas of the earth were covered with glaciers. The glaciers advanced and retreated with time. During their advance, the glaciers carried large amounts of sand, silt, clay, gravel, and boulders. *Drift* is a general term usually applied to the deposits laid down by glaciers. Unstratified deposits laid down by glaciers when they melt are referred to as *till*. The physical characteristics of till may vary from glacier to glacier.

The land forms that developed from the deposits of till are called *moraines*. A *terminal moraine* (Figure 1.29) is a ridge of till that marks the maximum limit of advance of a glacier. *Recessional moraines* are ridges of till developed behind the terminal moraine at varying distances apart. They are the result of temporary stabilization of the glacier during the recessional period. The till deposited by the glacier between the moraines is referred to as *ground moraine* (Figure 1.29). Ground moraines constitute large areas in the central United States and are called *till plains*.

The sand, silt, and gravel that are carried by the melting water from the front of a glacier are called *outwash*. In a pattern similar to the braided-stream

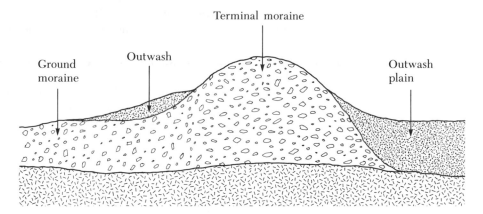

**Figure 1.29** Terminal moraine, ground moraine, and outwash plain

deposits, the melted water deposits the outwash, forming *outwash plains* (Figure 1.29). They are usually called *glaciofluvial deposits*.

The range of grain sizes present in a given till varies greatly. Figure 1.30 compares the grain-size distribution of *glacial-till* and *dune sand* (see Section 1.21). The amount of clay-size fractions present and the plasticity indices of tills also vary widely. During field-exploration programs (Chapter 2), erratic values of standard penetration resistances may also be expected.

Glacial water also carries with it silts and clays. The water finds its way to many basins and forms lakes. The silt particles initially tend to settle to the bottom of the lake when the water is still. During the winter, when the top of the

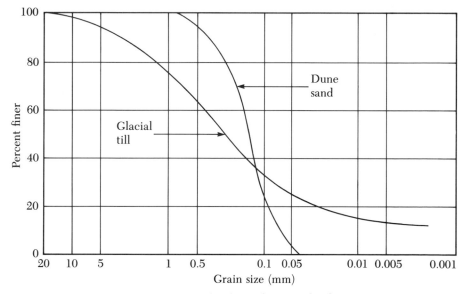

**Figure 1.30** Comparison of the grain-size distribution between glacial till and dune sand

lake freezes, the suspended clay particles gradually settle to the bottom of the lake. Again, during the summer, the snow on the lake melts. The supply of fresh water, loaded with sediments, repeats the process. As a result, the soil that is formed from such a deposit has alternate layers of silt and clay. This soil is called *varved clay*. The varves are usually a few millimeters thick; however, in some instances they can be 50–100 mm thick. Varved clays can be found in the Northeast and the Pacific Northwest of the United States. They are mostly normally consolidated and sensitive. The coefficient of permeability in the vertical direction is usually several times smaller than that in the horizontal direction. The load-bearing capacity of these deposits is quite low, and high settlement of structures with shallow foundations may be anticipated.

## 1.21

### Aeolian Soil Deposits

Wind is also a major transporting agent in the formation of soil deposits. When large areas of sand lie exposed, wind can blow it away and redeposit it at some other place. Deposits of wind-blown sand generally take the shape of *dunes* (Figure 1.31). When dunes are formed, the sand is blown over the crest by the wind. Beyond the crest, the sand particles roll down the slope. This process tends to form a *compact sand deposit* on the *windward side* and a rather *loose deposit* on the *leeward side*. Dunes can be found along the southern and eastern shores of Lake Michigan, the Atlantic Coast, the southern coast of California, and at various places along the coasts of Oregon and Washington. Sand dunes can also be found in the alluvial and rocky plains of the western United States. Following are some of the typical properties of *dune sand:*

**1.** The grain-size distribution of the sand at any given place is surprisingly uniform. This can be attributed to the sorting action of the wind.

**2.** The general grain size decreases with the distance from the source because the wind carries the small particles farther than the large ones.

**3.** The relative density of sand deposited on the windward side of dunes may be as high as 50–65% and may decrease to about 0–15% on the leeward side.

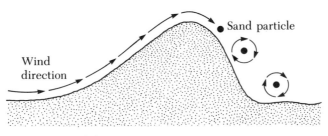

**Figure 1.31**  Sand dune

*Loess* is an aeolian deposit consisting of silt and silt-size particles. The grain-size distribution of loess is rather uniform. The cohesion of loess is generally derived from a clay coating over the silt-size particles, which contributes to a stable soil structure in an unsaturated state. The cohesion may also be the result of the precipitation of chemicals leached by rain water. Loess is a *collapsing* soil, because when the soil becomes saturated, it loses its binding strength between the soil particles. Special precautions need to be taken for construction of foundations over loessial deposits (see Chapter 10). There are extensive deposits of loessial soils in the United States—mostly in the midwestern states, such as Iowa, Missouri, Illinois, and Nebraska.

## 1.22

### Organic Soil

Organic soils are usually found in low-lying areas where the ground water table is near or above the ground surface. The presence of a high ground water table helps in the growth of aquatic plants that, when decomposed, form organic soil. This type of soil deposit is usually encountered in coastal areas and in glaciated regions. Organic soils show the following characteristics:

1. The natural moisture content may range from 200% to 300%.
2. They are highly compressible.
3. Laboratory tests have shown that, under loads, a large amount of settlement is derived from secondary consolidation.

## 1.23

### Some Local Terms for Soils

Soils are sometimes referred to by local terms. Following are a few of these terms with a brief description of each.

1. *Caliche:* a Spanish word derived from the Latin word *calix,* meaning *lime.* It is mostly found in the desert southwest of the United States. It is a mixture of sand, silt, and gravel bonded together by *calcareous deposits.* The calcareous deposits are brought to the surface by a net upward migration of water. The water evaporates in the high local temperature. Because of the sparse rainfall, the carbonates are not washed out of the top layer of soil.
2. *Gumbo:* a highly plastic, clayey soil.
3. *Adobe:* a highly plastic, clayey soil found in the southwestern United States.
4. *Terra Rossa:* residual soil deposits that are red in color and derive from limestone and dolomite.
5. *Muck:* organic soil with a very high moisture content.
6. *Muskeg:* organic soil deposit.
7. *Saprolite:* residual soil deposit derived from mostly insoluble rock.
8. *Loam:* a mixture of soil grains of various sizes, such as sand, silt, and clay.

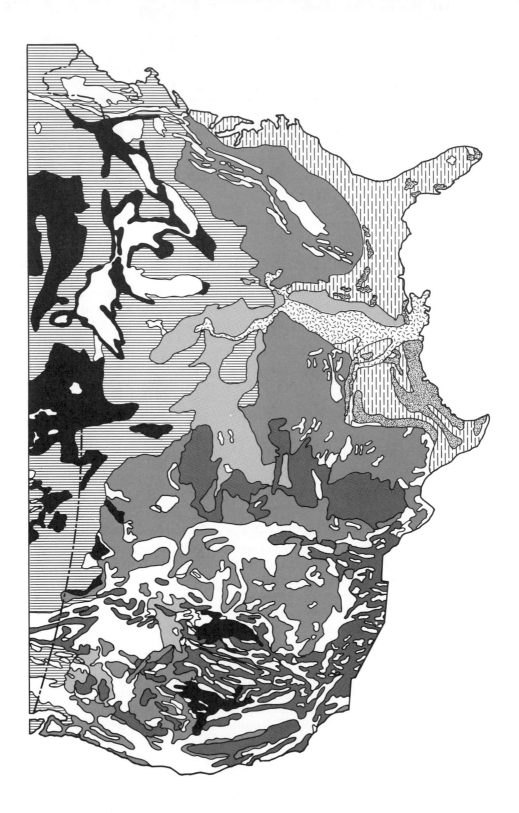

*Glacial Soils*

▤ Young and old drift, including associated sands and gravel

■ Lacustrine deposits, predominantly silts and clays

*Loessial Soils*

▨ Silts and very fine sands

*Soils of the Coastal Plain*

▥ Sand-clay; interbedded and mixed sands, gravels, clays, and silts; gravel and sand; or sand

▦ Clay

*Soils of the Filled Valleys and Great Plains Outwash Mantle*

■ Predominantly sands and gravels with silts; sandy clays and clays

*Residual Soils*

▩ All types

*Recent Alluvium*

▨ Predominantly silts and clays

□ Nonsoil areas

**Figure 1.32** Soil deposits of the United States (adapted from *Foundation Engineering*, 2nd ed., by R. B. Peck, W. E. Hanson, and T. H. Thornburn. Copyright 1974 by John Wiley and Sons. Reprinted by permission)

**9.** *Laterite:* characterized by the accumulation of iron oxide ($Fe_2 O_3$) and aluminum oxide ($Al_2 O_3$) near the surface, and the leaching of silica. Lateritic soils in Central America contain about 80–90% of clay and silt-size particles. In the United States, lateritic soils can be found in the southeastern states, such as Alabama, Georgia, and the Carolinas.

Figure 1.32 on the previous two pages shows the general nature of the various soil deposits encountered in the United States.

## Problems

**1.1** A moist soil has a void ratio of 0.7. The moisture content of the soil is 12%. If $G_s = 2.7$, determine (a) porosity, (b) degree of saturation, and (c) dry unit weight in $lb/ft^3$.

**1.2** For the soil described in Problem 1.1:
a. What would be the saturated unit weight in $lb/ft^3$?
b. How much water, in $lb/ft^3$, needs to be added to the soil for complete saturation?
c. What would be the moist unit weight in $lb/ft^3$ when the degree of saturation is 70%?

**1.3** A soil specimen has a volume of $0.05 m^3$ and a mass of 87.5 kg. Given: $w = 15\%$, $G_s = 2.68$. Determine (a) void ratio, (b) porosity, (c) dry unit weight, (d) moist unit weight, and (e) degree of saturation.

**1.4** A saturated soil specimen has $w = 36\%$ and $\gamma_d = 85.43 \, lb/ft^3$. Determine (a) void ratio, (b) porosity, (c) specific gravity of soil solids, and (d) saturated unit weight (in $lb/ft^3$).

**1.5** The laboratory test results of a sand are as follows: $e_{max} = 0.91$, $e_{min} = 0.48$, and $G_s = 2.67$. What would be the dry and moist unit weights of this sand when compacted at a moisture content of 10% to a relative density of 65%?

**1.6** For a granular soil, given $\gamma = 17 \, kN/m^3$, $D_r = 60\%$, $w = 8\%$, and $G_s = 2.66$. For this soil, if $e_{min}$ is 0.4, what would be the $e_{max}$? What would be the dry unit weight in the loosest state?

**1.7** The laboratory test results of six soils are given in the following table. Classify the soils by the AASHTO Soil Classification System, and give the group indices.

| | Sieve analysis—percent passing | | | | | |
|---|---|---|---|---|---|---|
| | Soil | | | | | |
| Sieve No. | A | B | C | D | E | F |
| 4 | 92 | 100 | 100 | 95 | 100 | 100 |
| 10 | 48 | 60 | 98 | 90 | 91 | 82 |
| 40 | 28 | 41 | 82 | 79 | 80 | 74 |
| 200 | 13 | 33 | 72 | 64 | 30 | 55 |
| Liquid limit | 31 | 38 | 56 | 35 | 43 | 35 |
| Plastic limit | 26 | 25 | 31 | 26 | 29 | 21 |

**1.8** Classify the soils given in Problem 1.7 by the Unified Soil Classification System.

**1.9** The permeability of a sand was tested in the laboratory at a void ratio of 0.56 and was determined to be 0.14 cm/sec. Estimate the coefficient of permeability of this sand at a void ratio of 0.79 by using Eqs. (1.25), (1.26), and (1.27).

**1.10** A soil profile is shown in Figure P1.10. Determine the total stress, pore water pressure, and effective stress at A, B, C, and D.

**1.11** A sandy soil ($G_s$ = 2.66), in its densest and loosest states, has void ratios of 0.42 and 0.97, respectively. Estimate the range of the critical hydraulic gradient in this soil at which quicksand condition might occur.

**1.12** A normally consolidated clay layer 8.53 ft thick has a void ratio of 1.3. Given: $LL$ = 41 and average effective stress on the clay layer = 1720 lb/ft². How much consolidation settlement would the clay layer undergo if the average effective stress on the clay layer is increased to 2510 lb/ft² as the result of the construction of a foundation?

**1.13** Refer to Problem 1.12. Assume the clay layer to be preconsolidated. Given: $p_c$ = 1980 lb/ft². Assume $C_s = \frac{1}{4}C_c$. Estimate the consolidation settlement.

**1.14** Refer to Figure P1.10. The clay is normally consolidated. A laboratory consolidation test on the clay gave the following results:

| Pressure (kN/m²) | Void ratio |
|------------------|------------|
| 100              | 0.905      |
| 200              | 0.815      |

a. Calculate the average effective stress on the clay layer.
b. Determine the compression index, $C_c$.
c. If the average effective stress on the clay layer is increased ($p_o + \Delta p$) to 115 kN/m², what would be the total consolidation settlement?

**1.15** Refer to Problem 1.14c. For the clay soil, given $C_v$ = 5.6 mm²/min, how long will it take to reach half the consolidation settlement? (*Note:* The clay layer in the field is drained on one side only.)

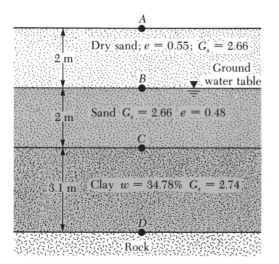

**Figure P1.10**

**1.16** A clay soil specimen, 1 in. in thickness (drained on top and bottom), was tested in the laboratory. For a given load increment, the time for 50% consolidation was 5 min 20 sec. How long will it take for 50% consolidation of a similar clay layer in the field that is 8.2 ft thick and drained on one side only?

**1.17** A direct shear test was conducted on a dry sand. The results are as follows:

Area of the specimen = 2 in. $\times$ 2 in.

| Normal force (lb) | Shear force at failure (lb) |
|---|---|
| 33 | 20.67 |
| 55.17 | 35.8 |
| 66.16 | 40.2 |

Draw a graph of shear stress at failure vs. normal stress and determine the soil friction angle.

**1.18** A consolidation-drained triaxial test on a normally consolidated clay yielded the following results:

All-around confining pressure = $\sigma_3$ = 20 lb/in.$^2$

Added axial stress at failure = $\Delta\sigma$ = 40 lb/in.$^2$

Determine the shear stress parameters.

**1.19** Following are the results of two consolidated-drained triaxial tests on a clay:

Test I: $\sigma_3$ = 82.8 kN/m$^2$; $\sigma_{1(failure)}$ = 329.2 kN/m$^2$

Test II: $\sigma_3$ = 165.6 kN/m$^2$; $\sigma_{1(failure)}$ = 558.6 kN/m$^2$

Determine the shear strength parameters—that is, $c$ and $\phi$.

**1.20** A consolidated-undrained triaxial test was conducted on a saturated normally consolidated clay. Following are the test results:

$\sigma_3$ = 13 lb/in.$^2$

$\sigma_{1(failure)}$ = 32 lb/in.$^2$

Pore water pressure at failure = $u_f$ = 5.5 lb/in.$^2$

Determine $c_u$, $\phi_{cu}$, $c$, and $\phi$.

# References

American Society for Testing and Materials (1982). *ASTM Standards, Part 19*, Philadelphia, Pa.

Azzouz, A. S., Krizek, R. J., and Corotis, R. B. (1976). "Regression Analysis of Soil Compressibility," *Soils and Foundations*, Vol. 16, No. 2, pp. 19–29.

Casagrande, A. (1936). "Determination of the Preconsolidation Load and Its Practical Significance," *Proceedings*, First International Conference on Soil Mechanics and Foundation Engineering, Cambridge, Mass., Vol. 3, pp. 60–64.

Darcy, H. (1856). *Les Fontaines Publiques de la Ville de Dijon*, Paris.

Das, B. M. (1982). *Soil Mechanics Laboratory Manual*, Engineering Press, San Jose, Calif.

Department of the Navy (1971). "Design Manual—Soil Mechanics, Foundations and Earth Structures," *NAVFAC DM-7*, Washington, D.C.

Hazen, A. (1930). "Water Supply," *American Civil Engineers Handbook*, Wiley, New York.

Highway Research Board (1945). *Report of the Committee on Classification of Materials for Subgrades and Granular Type Roads*, Vol. 25, pp. 375–388.

Kenney, T. C. (1959). "Discussion," *Journal of the Soil Mechanics and Foundations Division*, American Society of Civil Engineers, Vol. 85, No. SM3, pp. 67–69.

Kolb, C. R., and Shockley, W. G. (1959). "Mississippi Valley Geology: Its Engineering Significance," *Proceedings*, American Society of Civil Engineers, Vol. 124, pp. 633–656.

Peck, R. B., Hanson, W. E., and Thornburn, T. H. (1974). *Foundation Engineering*, 2nd ed., Wiley, New York.

Schmertmann, J. H. (1953). "Undisturbed Consolidation Behavior of Clay," *Transactions*, American Society of Civil Engineers, Vol. 120, p. 1201.

Skempton, A. W. (1944). "Notes on the Compressibility of Clays," *Quarterly Journal of Geological Society*, London, Vol. C, pp. 119–135.

Terzaghi, K., and Peck, R. B. (1967) *Soil Mechanics in Engineering Practice*, Wiley, New York.

C H A P T E R

2

# Subsurface Exploration

## 2.1

### Introduction

To design a foundation that will support adequate structural load, one must understand the nature of the soil(s) that will support the foundation. The process of determining the layers of natural soil deposits that will underlie a proposed structure and their physical properties is generally referred to as *subsurface exploration*. The purpose of a soil-exploration program is to obtain information that will aid the geotechnical engineer in the following:

1. Selection of the type and the depth of foundation suitable for a given structure
2. Evaluation of the load-bearing capacity of the foundation
3. Estimation of the probable settlement of a structure
4. Determination of potential foundation problems (for example, expansive soil, collapsible soil, sanitary landfill, and so on)
5. Establishment of ground water table
6. Prediction of lateral earth pressure for structures like retaining walls, sheet pile bulkheads, and braced cuts
7. Establishment of construction methods for changing subsoil conditions

Subsurface exploration may also be necessary when additions and alterations to existing structures are contemplated.

Foundation engineers should always remember that the soil at a given site is frequently nonhomogeneous; that is, the formation of the soil profile may be variable. Soil-mechanics theories concern idealized conditions. The application of these theories to foundation-engineering problems involves judicious evaluation of the information on site conditions and soil parameters obtained from field exploration programs. Good professional judgment constitutes a major part of geotechnical engineering, and it comes with practice.

## 2.2

### Subsurface Exploration Program

Subsurface exploration comprises several steps, such as collection of preliminary information, reconnaissance, and site investigation. This section briefly covers each of these topics.

### Collection of Preliminary Information

This step includes obtaining information regarding the type of structure to be built and its general use. For the construction of buildings, the approximate column loads and their spacings, and the local building-code and basement requirements should be known. In the construction of bridges, one should determine the span length as well as the loading on piers and abutments.

A general idea of the topography and the type of soil to be encountered near and around the proposed site can be obtained from the following sources:

1. United States Geological Survey maps.
2. State government geological survey maps.
3. United States Department of Agriculture's agronomy maps.
4. Agronomy maps published by the agriculture departments of various states.
5. The hydrological information published by the United States Corps of Engineers. These include the records of stream flow, high flood levels, tidal records, and so on.
6. Highway department soils manuals published by several states.

The information collected from these sources can be extremely helpful in the proper planning of site investigation. In some cases, substantial savings may be realized by anticipating problems that may be encountered later in the exploration program.

### Reconnaissance

The engineer should always make a trip to the site for a visual inspection. The inspection is intended to obtain the following information:

1. The general topography of the site, possible existence of drainage ditches, abandoned dumps of debris or other materials. Also, evidence of creep of slopes and deep, wide shrinkage cracks at regularly spaced intervals may be indicative of expansive soils.
2. Soil stratification from deep cuts, such as those made for construction of nearby highways and railroads.
3. Type of vegetation at the site. In several instances, the vegetation indicates the nature of the soil. For example, a mesquite cover in central Texas can be an indication of the existence of expansive clays that can cause possible foundation problems.
4. High-water marks on nearby buildings and bridge abutments.
5. Ground water levels—these can be determined from checking nearby wells.

**6.** Types of construction nearby and existence of any cracks in walls or other problems.

The nature of stratification and physical properties of the soil nearby can also be obtained from any available soil-exploration reports for existing structures.

## Site Investigation

This phase of the exploration program consists of planning, making some test boreholes, and collecting soil samples at desired intervals of the boreholes for subsequent observation and laboratory tests. The approximate required minimum depth of the borings should be predetermined. The estimated depths can be changed during the drilling operation, depending on the subsoil encountered. To determine the approximate minimum depth of boring, engineers may use the following rule (American Society of Civil Engineers, 1972):

**1.** Determine the net increase of stress, $\Delta\sigma$, under a foundation with depth as shown in Figure 2.1. (The general equations for estimation of stress increase are given in Chapter 3.)

**2.** Estimate the variation of the vertical effective stress, $\sigma'_v$, with depth.

**3.** Determine the depth, $D = D_1$, at which the stress increase $\Delta\sigma$ is equal to $(1/10)q$ ($q$ = estimated net stress on the foundation).

**4.** Determine the depth, $D = D_2$, at which $\Delta\sigma/\sigma'_v = 0.05$.

**5.** Unless bedrock is encountered, the smaller of the two depths, $D_1$ and $D_2$, just determined is the approximate minimum depth of boring required.

Table 2.1 shows the minimum depths of borings for buildings based on the preceding rule.

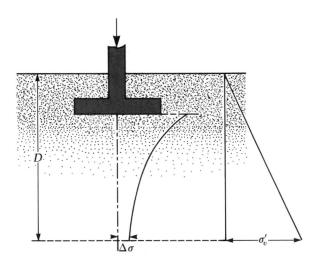

**Figure 2.1**   Determination of the minimum depth of boring

**Table 2.1** Depth of Boring[a]

| Building width (m) | Number of stories | | | | |
|---|---|---|---|---|---|
| | 1 | 2 | 4 | 8 | 16 |
| | Boring depth (m) | | | | |
| 30.5 | 3.4 | 6.1 | 10.1 | 16.2 | 24.1 |
| 61.0 | 3.7 | 6.7 | 12.5 | 20.7 | 32.9 |
| 122.0 | 3.7 | 7.0 | 13.7 | 24.7 | 41.5 |

[a] From *Introductory Soil Mechanics and Foundations*, 3rd ed., by George B. Sowers and George F. Sowers. Copyright © 1970 by Macmillan Publishing Company, Inc. Reprinted by permission.

For hospitals and office buildings, Sowers and Sowers (1970) also use the following rule to determine boring depth:

$$D_b = 3S^{0.7} \qquad \text{(for light steel or narrow concrete buildings)} \qquad (2.1)$$

$$D_b = 6S^{0.7} \qquad \text{(for heavy steel or wide concrete buildings)} \qquad (2.2)$$

where $D_b$ = depth of boring, in meters
$S$ = number of stories

When deep excavations are anticipated, the depth of boring should be at least 1.5 times the depth of excavation.

Sometimes subsoil conditions are such that the foundation load may have to be transmitted to the bedrock. The minimum depth of core boring into the bedrock is about 3 m. If the bedrock is irregular or weathered, the core borings may have to be extended to greater depths.

There are no hard and fast rules for the spacing of the boreholes. Table 2.2 gives some general guidelines for borehole spacing. These spacings can be increased or decreased, depending on the subsoil condition. If various soil strata are more or less uniform and predictable, the number of boreholes can be reduced.

The ultimate cost of the structure should also be taken into account while making decisions regarding the extent of the field exploration. The exploration cost generally should be in the range of 0.1–0.5% of the cost of the structure.

**Table 2.2** Approximate Spacing of Boreholes

| Type of project | Spacing (m) |
|---|---|
| Multistory building | 10–30 |
| One story industrial plants | 20–60 |
| Highways | 250–500 |
| Residential subdivision | 250–500 |
| Dams and dikes | 40–80 |

Soil borings can be done by several methods, such as *auger boring, wash boring, percussion drilling,* and *rotary drilling.* The following section discusses these methods of advancing the test holes in the field and collecting soil samples.

## 2.3

### Exploratory Borings in the Field

*Auger boring* is the simplest method of making exploratory boreholes. Figure 2.2 shows two types of hand auger—the *post hole auger* and the *helical auger.* Hand augers cannot be used for advancing holes to depths exceeding 3–5 m. However, they can be used for soil exploration work for some highways and small structures. *Portable power-driven helical augers* (76.2 mm to 304.8 mm in diameter) are available for making boreholes to greater depths. The soil samples that are obtained from such boring operations are highly disturbed. In several noncohesive soils or soils having low cohesion, the walls of the boreholes will not stand unsupported. In such circumstances, a metal pipe is used as a *casing* to prevent the soil from caving in.

When power is available, *continuous-flight augers* are probably the most common method used for advancing a borehole. The power for drilling is delivered by truck- or tractor-mounted drilling rigs. Boreholes up to about 60–70 m can be easily made by this method. Continuous-flight augers are available in sections of about 1–2 m with the stem either solid or hollow. Some

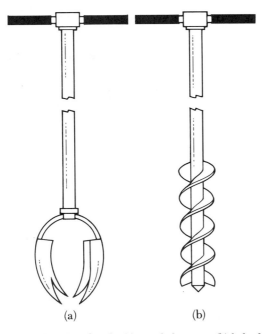

(a)                    (b)

**Figure 2.2**   Hand tools: (a) post hole auger; (b) helical auger

of the commonly used solid stem augers have outside diameters of 66.68 mm (2⅝ in.), 82.55 mm (3¼ in.), 101.6 mm (4 in.), and 114.3 mm (4½ in.). Common hollow stem augers commercially available have dimensions of 63.5 mm ID and 158.75 mm OD (2.5 in. × 6.25 in.), 69.85 mm ID and 177.8 OD (2.75 in. × 7 in.), 76.2 mm ID and 203.2 OD (3 in. × 8 in.), and 82.55 mm ID and 228.6 mm OD (3.25 in. × 9 in.).

In making the boreholes, the tip of the auger is attached to a cutter head (Figure 2.3). During the drilling operation (Figure 2.4), section after section of auger can be added and the hole extended downwards. The flights of the augers bring the loose soil from the bottom of the hole to the surface. During the drilling, the driller can detect changes in soil type by noting changes in the speed and the sound of drilling. When solid stem augers are used, it is necessary

**Figure 2.3**  Carbide-tipped cutting head on auger flight attached with bolt (courtesy of William B. Ellis, El Paso Engineering and Testing, Inc., El Paso, Texas)

**Figure 2.4**  Drilling with continuous-flight augers (courtesy of Danny R. Anderson, Danny R. Anderson Consultants, El Paso, Texas)

to withdraw the auger at regular intervals to obtain soil samples and also to conduct other operations such as standard penetration tests. Hollow stem augers have a distinct advantage over solid stem augers in that they do not have to be removed at frequent intervals for sampling or other tests. As shown schematically in Figure 2.5, the outside of the hollow stem auger acts as a casing. A removable plug is attached to the bottom of the auger by means of a center rod. During the drilling, the plug can be pulled out with the auger in place, and soil sampling and standard penetration tests can be performed. When hollow stem augers are used in sandy soils below the ground water table,

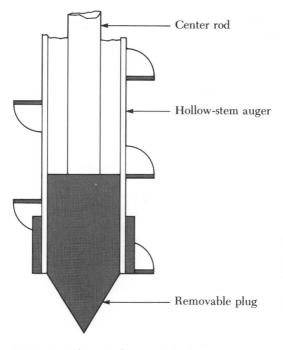

Center rod

Hollow-stem auger

Removable plug

**Figure 2.5** Schematic diagram of the hollow stem auger with removable plug

the sand may be pushed several feet into the stem of the auger by excess hydrostatic pressure immediately after the removal of the plug. In such conditions, the plug should not be used and, instead, water inside the hollow stem should be maintained at a higher level than the ground water table.

*Wash boring* is another method of advancing boreholes. In this method, a casing about 2–3 m long is driven into the ground. The soil inside the casing is then removed by means of a chopping bit that is attached to a drilling rod. Water is forced through the drilling rod, and it goes out at a very high velocity through the holes located at the bottom of the chopping bit (Figure 2.6). The water and the chopped soil particles rise upward in the drill hole and overflow at the top of the casing through a T-connection. The wash water is collected in a container. The casing can be extended with additional pieces as the borehole progresses; however, this is not required if the borehole will stay open and not cave in.

*Rotary drilling* is a procedure by which rapidly rotating drilling bits attached to the bottom of drilling rods cut and grind the soil and advance the borehole down. There are several types of drilling bit. Rotary drilling can be used in sand, clay, and rocks (unless badly fissured). Water, or *drilling mud*, is forced down the drilling rods to the bits, and the return flow forces the cuttings to the surface. Boreholes with diameters ranging from 50.8 mm to 203.2 mm (2 in. to 8 in.) can be easily made by this technique. The drilling mud is a slurry of water and bentonite. This is generally used when the soil encountered is such

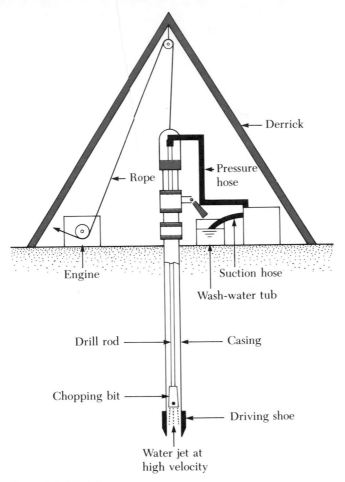

**Figure 2.6**  Wash boring

that boreholes are likely to cave in. When soil samples are needed, the drilling rod is raised and the drilling bit is replaced by a sampler.

*Percussion drilling* is an alternate method to advance a borehole, particularly through hard soil and rock. In this technique, a heavy drilling bit is raised and lowered to chop the hard soil. The chopped soil particles are brought up by circulation of water. Percussion drilling may require casing.

## 2.4

### Procedures for Sampling Soil

Two types of soil sample can be obtained during subsurface exploration: *disturbed* and *undisturbed*. The disturbed but representative samples can be generally used for the following types of laboratory test.

1. Grain-size analysis
2. Determination of liquid and plastic limits
3. Specific gravity of soil solids
4. Organic content determination
5. Classification of soil

The disturbed soil samples cannot, however, be used for consolidation or shear strength tests. Undisturbed soil samples must be obtained for these types of laboratory test. The next sections discuss various types of soil sampling procedure.

## Split-Spoon Sampling

Split-spoon samplers can be used in the field to obtain soil samples that are generally disturbed but still representative. A section of a *standard split-spoon sampler* is shown in Figure 2.7(a). It consists of a tool-steel driving shoe, a steel tube that is split longitudinally in half, and a coupling at the top. The coupling connects the sampler to the drill rod. The standard split tube has an inside diameter of 34.93 mm (1⅜ in.) and an outside diameter of 50.8 mm (2 in.);

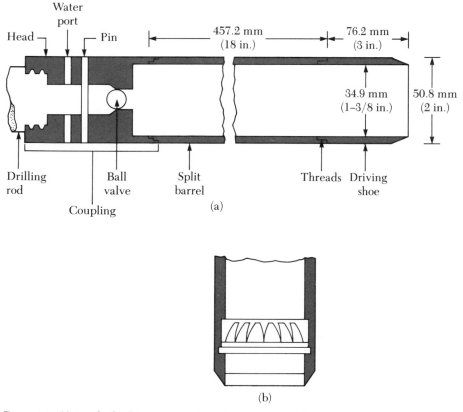

**Figure 2.7** (a) Standard split-spoon sampler; (b) spring core catcher

however, samplers having inside and outside diameters up to 63.5 mm (2½ in.) and 76.2 mm (3 in.), respectively, are also available. When a borehole is extended to a given depth, the drill tools are removed and the sampler is lowered to the bottom of the borehole. The sampler is driven into the soil by hammer blows given at the top of the drill rod. The standard weight of the hammer is 622.72 N (140 lb), and, for each blow, the hammer drops a distance of 0.762 m (30 in.). The number of blows required for spoon penetration of three 152.4-mm (6-in.) intervals are recorded. The number of blows required for the last two intervals are added to give the *standard penetration number* at that depth. This is generally referred to as the *N-value* (American Society for Testing and Materials, 1982, Designation D-1586-67). The sampler is then withdrawn, and the shoe and the coupling are removed. The soil sample that is recovered from the tube is then placed in a glass bottle and transported to the laboratory.

The degree of disturbance for a soil sample is usually expressed by an equation

$$A_R(\%) = \frac{D_o^2 - D_i^2}{D_i^2} (100) \tag{2.3}$$

where $A_R$ = area ratio
  $D_o$ = outside diameter of the sampling tube
  $D_i$ = inside diameter of the sampling tube

When the area ratio is 10% or less, the sample is generally considered to be undisturbed. For a standard split-spoon sampler

$$A_R(\%) = \frac{(50.8)^2 - (34.93)^2}{(34.93)^2} (100) = 111.5\%$$

Hence, these samples are highly disturbed. Split-spoon samples are generally taken at intervals of about 1.53 m (5 ft).

When the material encountered on the field is sand (particularly fine sand below the water table), sample recovery by a split-spoon sampler may be difficult. In that case, a device such as a *spring core catcher* may have to be placed inside the split spoon (Figure 2.7b).

Besides obtaining soil samples, standard penetration tests provide several useful correlations. For example, the consistency of clayey soils can often be estimated from the standard penetration number ($N$). This is shown in Table 2.3.

In granular soils, the $N$-value is affected by the effective overburden pressure, $\sigma_v'$. For that reason, the $N$-value obtained from field exploration under different effective overburden pressures should be changed to correspond to a standard value of $\sigma_v'$. The current generally accepted correlation is that given by Bazaraa (1967) and Peck and Bazaraa (1969), which may be stated as follows:

$$N_{cor} = \frac{4N_F}{1 + 0.04177\sigma_v'} \qquad \text{(for } \sigma_v' < 71.82 \text{ kN/m}^2) \tag{2.4}$$

$$N_{cor} = \frac{4N_F}{3.25 + 0.01044\sigma_v'} \qquad \text{(for } \sigma_v' > 71.82 \text{ kN/m}^2) \tag{2.5}$$

**Table 2.3** Consistency of Clays and Approximate Correlation to the Standard Penetration Number, $N$

| Standard penetration number, $N$ | Consistency | Unconfined compression strength, $q_u$ (kN/m²) |
|---|---|---|
| 0–2 | Very soft | 0–25 |
| 2–5 | Soft | 25–50 |
| 5–10 | Medium stiff | 50–100 |
| 10–20 | Stiff | 100–200 |
| 20–30 | Very stiff | 200–400 |
| >30 | Hard | >400 |

and

$$N_{\text{cor}} = N_F \qquad \text{(for } \sigma'_v = 71.82 \text{ kN/m}^2\text{)} \qquad (2.6)$$

where $N_{\text{cor}}$ = corrected $N$-value to a standard $\sigma'_v = 71.28$ kN/m²
$N_F$ = $N$-value obtained from the field
$\sigma'_v$ = effective overburden pressure in kN/m²

Peck, Hanson, and Thornburn (1974) have recommended the following method for correcting the standard penetration numbers obtained from the field (corrected to a standard $\sigma'_v = 95.6$ kN/m²).

$$N_{\text{cor}} = 0.77 N_F \log\left(\frac{20}{0.0105\sigma'_v}\right) \qquad \text{(for } \sigma'_v \geq 23.9 \text{ kN/m}^2\text{)} \qquad (2.7)$$

Note that when $\sigma'_v = 95.6$ kN/m², Eq. (2.7) yields

$$N_{\text{cor}} = N_F \qquad \text{(for } \sigma'_v = 95.6 \text{ kN/m}^2\text{)} \qquad (2.8)$$

For $\sigma'_v < 23.9$ kN/m²

$$N_{\text{cor}} = C_N N_F \qquad (2.9)$$

where $C_N$ is a correction factor, the variation of which with $\sigma'_v$ is given in the following table.

| $\sigma'_v$ (kN/m²) | $C_N$ |
|---|---|
| 0 | 2 |
| 6 | 1.8 |
| 15 | 1.6 |

The variation of $N_{\text{cor}}/N_F$ with $\sigma'_v$ as obtained from Eqs. (2.4) through (2.6) is given in Figure 2.8. Similar variations of $N_{\text{cor}}/N_F$ as obtained from Eqs. (2.7) through (2.9) are shown in Figure 2.9. This text uses Eqs. (2.4) through (2.6) to solve problems.

An approximate relation between the standard penetration number and the relative density of sand is given in Table 2.4.

When the standard penetration resistance values are used in the preceding

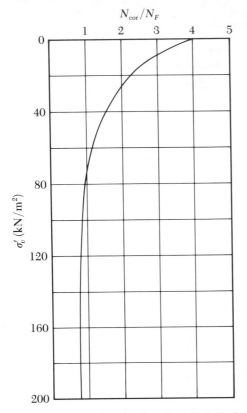

**Figure 2.8**  Plot of $N_{cor}/N_F$ from Eqs. (2.4), (2.5), and (2.6)

correlations (Tables 2.3 and 2.4) for estimation of soil parameters, the following should be noted:

1. The equations are approximate.
2. Because the soil is not homogeneous, there may be a wide variation of the $N$-values obtained from a given borehole.
3. In soil deposits that contain large boulders and gravels, standard penetration numbers may be erratic and unreliable.

**Table 2.4**  Relation between $N$-values, Relative Density, and Angle of Friction in Sands

| Standard penetration number, $N$ | Approximate relative density, $D_r$ (%) | Approximate angle of friction of soil, $\phi$ (deg) |
|---|---|---|
| 0–5 | 0–5 | 26–30 |
| 5–10 | 5–30 | 28–35 |
| 10–30 | 30–60 | 35–42 |
| 30–50 | 60–95 | 38–46 |

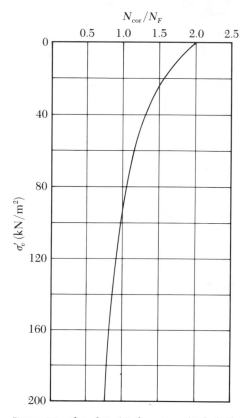

**Figure 2.9** Plot of $N_{cor}/N_F$ from Eqs. (2.7), (2.8), and (2.9)

Although approximate, with correct interpretation the standard penetration test provides a good evaluation of soil properties.

### Scraper Bucket

When soil deposits are sand mixed with pebbles, it may not be possible to obtain samples by a split spoon with a spring core catcher, because the pebbles may prevent the springs from closing. In such cases, a scraper bucket may be used to obtain disturbed representative samples from the field (Figure 2.10a). The scraper bucket has a driving point and can be attached to a drilling rod. The sampler is driven down into the soil and rotated, and the scrapings from the side fall into the bucket.

### Thin Wall Tube

Thin wall tubes are sometimes referred to as *Shelby tubes*. They are made out of seamless steel and are commonly used to obtain undisturbed clayey soils. The commonly used thin wall tube samplers have outside diameters of 50.8 mm (2 in.) and 76.2 mm (3 in.). The bottom end of the tube is sharpened. The tubes can be attached to drilling rods (Figure 2.10b). The drilling rod with the sampler attached to it is lowered to the bottom of the borehole and the sampler is pushed

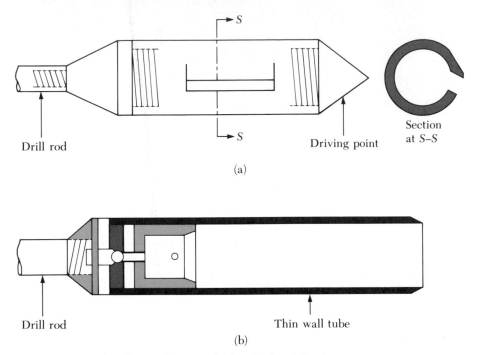

**Figure 2.10**  Sampling devices: (a) scraper bucket; (b) thin wall tube;
(c) and (d) piston sampler

into the soil. The soil sample inside the tube is then pulled out. The two ends
of the sampler are sealed, and it is sent back to the laboratory for testing.

Samples obtained in this manner may be used for consolidation or shear
tests. A thin wall tube with 50.8 mm (2 in.) diameter has an inside diameter of
about 47.63 mm (1–⅞ in.). This gives an area ratio

$$A_R(\%) = \frac{D_o^2 - D_i^2}{D_o^2}(100) = \frac{(50.8)^2 - (47.63)^2}{(47.63)^2}(100) = 13.75\%$$

Note that by increasing the diameters of samples, one increases the cost of
obtaining them.

## Piston Sampler

When undisturbed soil samples become larger than 76.2 mm (3 in.) in
diameter, they tend to fall out of the sampler. Piston samplers are particularly
useful in such conditions. There are several types of piston sampler; however,
the sampler proposed by Osterberg (1952) is the most advantageous (see Figure
2.10c and d). It consists of a thin wall tube with a piston. Initially, the piston
closes the end of the thin wall tube. The sampler is lowered to the bottom of
the borehole (Figure 2.10c), and then the thin wall tube is pushed into the soil
hydraulically, past the piston. After that, the pressure is released through a hole
in the piston rod (Figure 2.10d). To a large extent, the presence of the piston
prevents distortion in the sample by not letting the soil squeeze into the

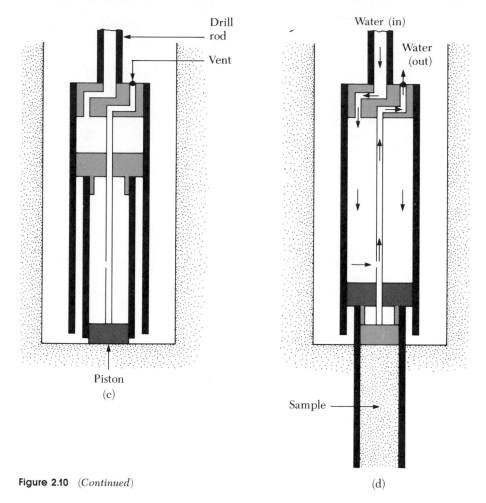

Figure 2.10 (*Continued*)

sampling tube very fast and by not admitting excess soil. Samples obtained in this manner are consequently less disturbed than those obtained by Shelby tubes.

## 2.5

### Observation of Ground Water Tables

The presence of a ground water table near a foundation has a large effect on the foundation's load-bearing capacity and settlement, among other things. The ground water level will also change seasonally. In many cases, it may become necessary to establish the highest and the lowest possible levels of ground water during the life of a project.

If ground water is encountered in a borehole during a field exploration, it should be recorded. In soils with high coefficients of permeability, the level of ground water in a borehole will stabilize after about 24 hours of completion of the boring. The depth of the water table can then be recorded by lowering a chain or tape into the borehole.

In highly impermeable layers, the ground water level in a borehole may not stabilize for several weeks. In such cases, if accurate ground water level measurements are required, a *piezometer* can be used. A piezometer basically consists of a porous stone with a plastic standpipe attached to it. The general nature of the placement of a piezometer in a borehole is shown in Figure 2.11.

For silty soils, Hvorslev (1949) proposed a technique to determine the ground water level (see Figure 2.12). This technique uses the following steps:

**1.** Bail out water in the borehole to a level below the estimated ground water table.

**2.** Observe the water levels in the borehole at times

$$t = 0$$
$$t = t_1$$
$$t = t_2$$

and

$$t = t_3$$

Note that $t_1 - 0 = t_1 - t_2 = t_2 - t_3 = \Delta t$

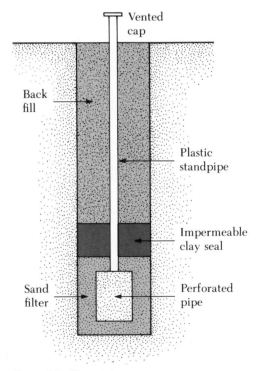

**Figure 2.11**   Piezometer

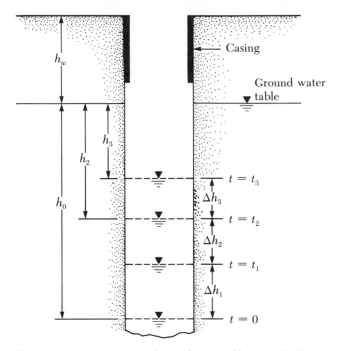

**Figure 2.12** Determination of ground water table—Eq. (2.10)

**3.** Calculate $\Delta h_1$, $\Delta h_2$, and $\Delta h_3$ (see Figure 2.12).
**4.** Calculate

$$h_0 = \frac{\Delta h_1^2}{\Delta h_1 - \Delta h_2} \tag{2.10a}$$

$$h_2 = \frac{\Delta h_2^2}{\Delta h_1 - \Delta h_2} \tag{2.10b}$$

$$h_3 = \frac{\Delta h_3^2}{\Delta h_2 - \Delta h_3} \tag{2.10c}$$

**5.** Plot the values $h_0$, $h_2$, and $h_3$ above the water levels observed at times $t = 0$, $t_2$, and $t_3$, respectively, to determine the final water level in the borehole.

## Example 2.1

Refer to Figure 2.12. For a borehole, given: $h_w + h_o = 9.5$ m

$\Delta t = 24$ hrs

$\Delta h_1 = 0.9$ m

$\Delta h_2 = 0.70$ m

$\Delta h_3 = 0.54$ m

Make the necessary calculations and locate the ground water level.

**Solution**

Using Eq. (2.10),

$$h_o = \frac{\Delta h_1^2}{\Delta h_1 - \Delta h_2} = \frac{0.9^2}{0.9 - 0.70} = 4.05 \text{ m}$$

$$h_2 = \frac{\Delta h_2^2}{\Delta h_1 - \Delta h_2} = \frac{0.7^2}{0.9 - 0.7} = 2.45 \text{ m}$$

$$h_3 = \frac{\Delta h_3^2}{\Delta h_2 - \Delta h_3} = \frac{0.54^2}{0.7 - 0.54} = 1.82 \text{ m}$$

Figure 2.13 shows a plot of the preceding calculations as well as the estimated ground water table. From this figure it can be seen that $h_w = 5.5$ m.

## 2.6

### Vane Shear Test

Vane shear is a type of test (ASTM D-2573) that may be used during the drilling operation to determine the in-situ undrained shear strength of clay soils—particularly soft clays. The vane shear apparatus consists of four blades on the end of a rod, as shown in Figure 2.14a. The vanes of the apparatus are

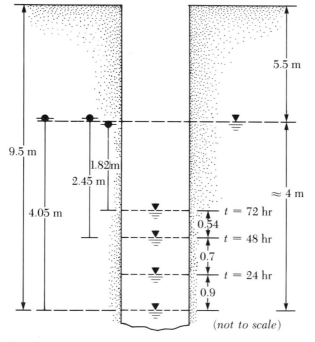

**Figure 2.13**

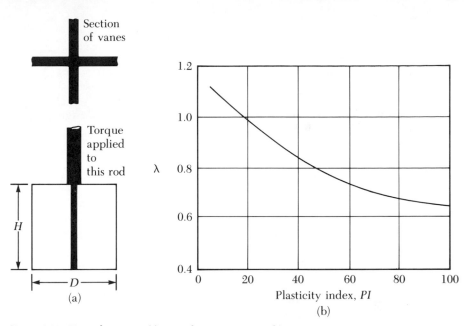

**Figure 2.14** Vane shear test: (a) vane shear equipment; (b) correction factor, $\lambda$—Eq. (2.12) [(b) after Bjerrum, 1972]

pushed into the soil at the bottom of a borehole without disturbing the soil appreciably. Torque is applied at the top of the rod to rotate the vanes. This will induce failure in a soil of cylindrical shape surrounding the vanes. The maximum torque applied can be related to the undrained strength of a clayey soil as

$$T = c_u \pi D^2 \left( \frac{H}{2} + \frac{D}{6} \right)$$

or

$$c_u = \frac{T}{\pi D^2 \left( \frac{H}{2} + \frac{D}{6} \right)} \tag{2.11}$$

where $T$ = maximum torque applied
$c_u$ = undrained shear strength ($\phi = 0$ concept)
$D$ = diameter of the vanes
$H$ = height of the vanes

The height-to-diameter ratio ($H/D$) of the vanes is usually kept at 2. The U.S. Bureau of Reclamation uses vanes with $D \times H$ dimensions of 50.8 mm × 101.6 mm (2 in. × 4 in.), 76.2 mm × 152.4 mm (3 in. × 6 in.), and 101.6 mm × 203.2 mm (4 in. × 8 in.). AASHTO Designation T-223 recommends the following dimensions of vanes for field tests.

| Casing size | $H$ (mm) | $D$ (mm) | Blade thickness (mm) |
|---|---|---|---|
| AX | 38.1 | 76.2 | 1.6 |
| BX | 50.8 | 101.6 | 1.6 |
| NX | 63.5 | 137.0 | 3.2 |

Bjerrum (1972) has recommended that, for actual design purposes, the field vane shear values should be corrected as

$$c_{u(\text{corrected})} = \lambda c_{u(\text{field})} \tag{2.12}$$

where $\lambda$ = correction factor

The value of the correction factor, $\lambda$, varies with plasticity index of soil and is shown in Figure 2.14b.

## 2.7
### Dutch Cone Penetration Test

The *Dutch cone penetration test* (see Figure 2.15) is a technique by which a 60° cone with a base area of 10 cm² is pushed into the ground at a steady rate of about 20 mm/sec and the resistance to the penetration (the point resistance of the cone) and the frictional resistance of the casing to the soil around it are measured. This test is also called the *static penetration test,* and no boreholes are necessary to perform it. Refined Dutch cone penetrometers are also available to measure the point resistance of the cone at a certain depth and the frictional resistance of the soil immediately above the point. The frictional resistance is obtained by means of a sleeve of limited length.

The point resistance, $q_c$, obtained from Dutch cone penetration tests has been related to the soil friction angle of granular soils and also to the consistency of cohesive soils (similar to that obtained for standard penetration tests; see Tables 2.3 and 2.4). The ratio of $q_c/N$ ($q_c$ in kN/m²; $N$ = standard penetration resistance) for sands generally varies in a range of 400 to 600; for a gravelly deposit, it can be in a range of 800 to 1000.

The cone penetration test is more commonly used in Europe than the United States. Schmertmann (1975) has provided an extensive bibliography of the published literature on cone penetration tests.

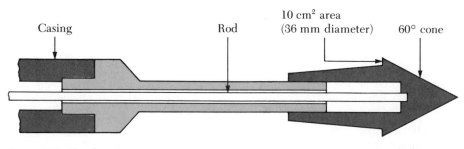

**Figure 2.15**  Dutch cone penetrometer

## 2.8

### Coring of Rocks

When a rock layer is encountered during a drilling operation, rock coring may become necessary. For coring of rocks, a *core barrel* is attached to a drilling rod. A *coring bit* is attached to the bottom of the core barrel (Figure 2.16). The cutting elements may be diamond, tungsten, carbide, and so on. Table 2.5 gives a summary of the various types of core barrel and their sizes as well as the compatible drill rods commonly used for foundation exploration.

The coring is advanced by rotary drilling. Water is circulated through the drilling rod during coring, and the cutting is washed out.

Two types of core barrel are available: the *single-tube core barrel* (Figure 2.16a) and the *double-tube core barrel* (Figure 2.16b). Rock cores obtained by single-tube core barrels can be highly disturbed and fractured because of torsion. (Note also that rock cores that are smaller than the BX size tend to fracture during the coring process.)

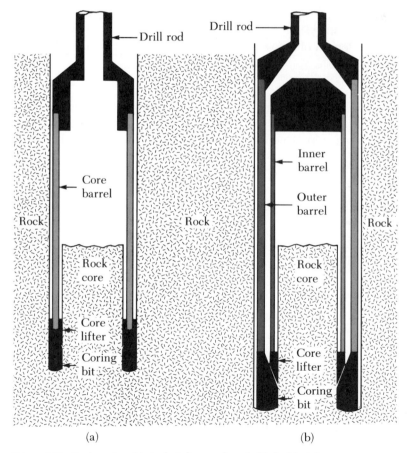

**Figure 2.16** Rock coring: (a) single-tube core barrel; (b) double-tube core barrel

**Table 2.5**   Standard Size and Designation of Casing, Core Barrel, and Compatible Drill Rod

| Casing and core barrel designation | Outside diameter of core barrel bit | | Drill rod designation |
|:---:|:---:|:---:|:---:|
| | (mm) | (in.) | |
| EX | 36.51 | $1\frac{7}{16}$ | E |
| AX | 47.63 | $1\frac{7}{8}$ | A |
| BX | 58.74 | $2\frac{5}{16}$ | B |
| NX | 74.61 | $2\frac{15}{16}$ | N |

When the core samples are recovered, the depth of recovery should be properly recorded for further evaluation in the laboratory. Based on the length of the rock core recovered from each run, the following quantities may be calculated for a general evaluation of the rock quality encountered. They are

a. $\text{recovery ratio} = \dfrac{\text{length of core recovered}}{\text{theoretical length of rock cored}}$     (2.13)

b. rock quality designation $(RQD) =$

$$\frac{\Sigma \text{ length of recovered pieces equal to or larger than } 101.6 \text{ m } (4 \text{ in.})}{\text{theoretical length of rock cored}}$$

(2.14)

A recovery ratio of one will indicate the presence of intact rock; for highly fractured rocks, the recovery ratio may be 0.5 or smaller. Table 2.6 presents the general relationship (Deere, 1963) between the $RQD$ and the *in-situ* rock quality.

**Table 2.6**   Relation between *in-Situ* Rock Quality and $RQD$

| $RQD$ | Rock quality |
|:---|:---|
| 0–0.25 | Very poor |
| 0.25–0.5 | Poor |
| 0.5 –0.75 | Fair |
| 0.75–0.9 | Good |
| 0.9 –1 | Excellent |

## 2.9

### Preparation of Boring Logs

The detailed information gathered from each borehole is presented in a graphical form that is referred to as the *boring log*. As the borehole is advanced downward in the field, the driller should generally record the following information in a standard log.

**1.** Name and address of the drilling company
**2.** Driller's name

**Table 2.5**   (Continued)

| Outside diameter of drill rod | | Diameter of borehole | | Diameter of core sample | |
|---|---|---|---|---|---|
| (mm) | (in.) | (mm) | (in.) | (mm) | (in.) |
| 33.34 | 1$\frac{5}{16}$ | 38.1 | 1$\frac{1}{2}$ | 22.23 | $\frac{7}{8}$ |
| 41.28 | 1$\frac{5}{8}$ | 50.8 | 2 | 28.58 | 1$\frac{1}{8}$ |
| 47.63 | 1$\frac{7}{8}$ | 63.5 | 2$\frac{1}{2}$ | 41.28 | 1$\frac{5}{8}$ |
| 60.33 | 2$\frac{3}{8}$ | 76.2 | 3 | 53.98 | 2$\frac{1}{8}$ |

   **3.** Job description and number
   **4.** Number and type of boring
   **5.** Date of boring
   **6.** Subsurface stratification. This can be obtained by visual observation of the soil brought out by auger, split-spoon sampler, and thin wall Shelby tube sampler.
   **7.** Elevation of ground water table and date observed
   **8.** Standard penetration resistance and the depth of *SPT*
   **9.** Number, type, and depth of soil sample collected
   **10.** In case of rock coring, type of core barrel used should be recorded. For each run, the actual length of coring, length of core recovery, and the *RQD* should also be carefully noted.

None of this information should *ever* be left to memory, because this often results in erroneous boring logs.

   After completion of the necessary laboratory tests, the geotechnical engineer prepares a new finished log that includes notes from the driller's field log and the results of tests conducted in the laboratory. Figure 2.17 shows a typical boring log. These logs have to be attached to the final soil exploration report submitted to the client. Note that Figure 2.17 also lists the classifications of the soils in the left-hand column, along with the description of each soil. This is based on the Unified Soil Classification System.

**2.10**
─────────────────────────────────────────────────────────
Determination of Coefficient of Permeability in the Field

   There are several types of field test now available to determine the coefficient of permeability of soil. Two fairly easy test procedures described by the United States Bureau of Reclamation (1974) are the *open end test* and the *packer test*. Following are brief descriptions of these tests.

Open End Test

   The first step in the open end test (Figure 2.18) is to advance a borehole to the desired depth. A casing is then driven to extend to the bottom of the borehole. Water is supplied at a constant rate from the top of the casing, and it escapes at the bottom of the borehole. The water level in the casing must remain constant. Once the steady state of water supply is established, the coefficient of permeability can be determined as

Boring Log

Name of the Project   Two-story apartment building

Location   Johnson & Olive St.      Date of Boring   March 2, 1982

Boring No.   3      Type of   Hollow stem auger      Ground Elevation   60.8 m
                    Boring

| Soil description | Depth (m) | Soil sample type and number | N | $w_n$ (%) | Comments |
|---|---|---|---|---|---|
| Light brown clay (fill) | | | | | |
| | 1 | | | | |
| Silty sand (SM) | 2 | SS-1 | 9 | 8.2 | |
| | 3 | SS-2 | 12 | 17.6 | $LL = 38$ |
| °G.W.T. ▽ 3.5 m | 4 | | | | $PI = 11$ |
| Light gray silty clay (ML) | 5 | ST-1 | | 20.4 | $LL = 36$ $q_u = 112 \ kN/m^2$ |
| | 6 | SS-3 | 11 | 20.6 | |
| Sand with some gravel (SP) | 7 | | | | |
| End of boring @ 8 m | 8 | SS-4 | 27 | 9 | |

$N$ = standard penetration number (below/304.8 mm)      °Ground water table
$w_n$ = natural moisture content                         observed after one
$LL$ = liquid limit; $PI$ = plasticity index            week of drilling
$q_u$ = unconfined compression strength
SS = split-spoon sample; ST = Shelby tube sample

**Figure 2.17**  A typical boring log

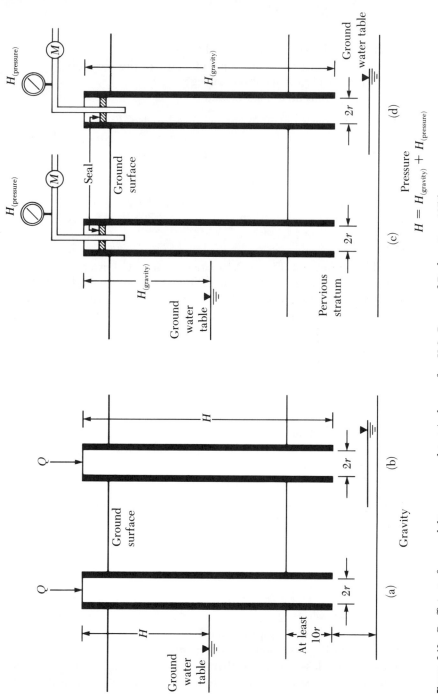

**Figure 2.18** Coefficient of permeability—open end test (redrawn after U.S. Bureau of Reclamation, 1974)

$$k = \frac{Q}{5.5rH} \tag{2.15}$$

where $k$ = coefficient of permeability
$\quad\quad Q$ = constant rate of supply of water to the borehole
$\quad\quad r$ = inside radius of the casing
$\quad\quad H$ = differential head of water

Any type of consistent units may be used in Eq. (2.15).

The head, $H$, has been defined in Figure 2.18. Note that for pressure tests (Figure 2.18c and d) the value of $H$ is given as

$$H = H_{(gravity)} + H_{(pressure)} \tag{2.16}$$

The pressure head, $H_{(pressure)}$, given in Eq. (2.16) is expressed in meters (or feet) of water (1 kN/m$^2$ = 0.102 m; 1 lb/in.$^2$ = 2.308 ft).

## Packer Test

The packer test (Figure 2.19) can be conducted in a portion of the borehole during the drilling or after the drilling has been completed. Water to the portion of the borehole under test is supplied under pressure at a constant rate. The coefficient of permeability can be determined from the following equations:

$$k = \frac{Q}{2\pi LH} \log_e \left(\frac{L}{r}\right) \quad \text{(for } L \geq 10r) \tag{2.17}$$

$$k = \frac{Q}{2\pi LH} \sinh^{-1} \frac{L}{2r} \quad \text{(for } 10r > L \geq r) \tag{2.18}$$

where $k$ = coefficient of permeability
$\quad\quad Q$ = constant rate of flow into the hole
$\quad\quad L$ = length of portion of the hole under test
$\quad\quad r$ = radius of the hole
$\quad\quad H$ = differential pressure head

Note that the differential pressure head is the sum of the gravity head ($H_{(gravity)}$) and the pressure head ($H_{(pressure)}$).

The packer test is primarily used to determine the permeability of rock. However, as mentioned previously, it can also be used in soils.

## 2.11

## Geophysical Exploration

Several types of geophysical exploration technique are now available for a rapid evaluation of subsoil characteristics. They permit rapid coverage of large areas and are less expensive than conventional exploration by drilling. However, in many cases, definitive interpretation of the results is difficult. For that reason, these techniques should be used for preliminary works only. The following three sections cover three types of geophysical exploration technique: *seismic refraction survey*, *cross-hole seismic survey*, and *resistivity survey*.

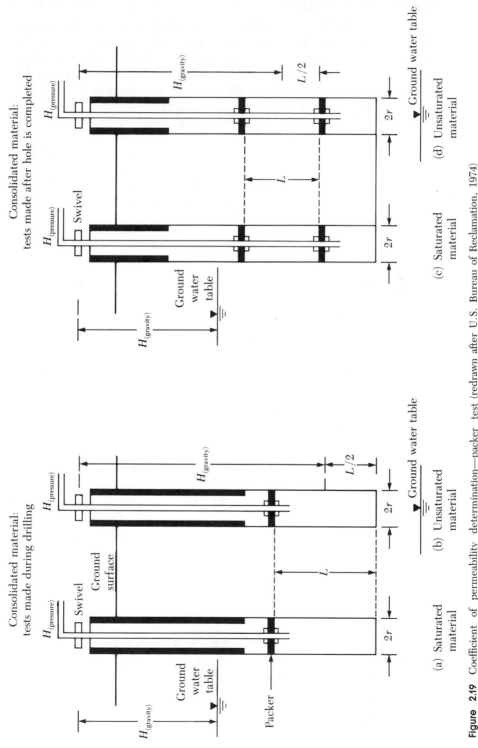

**Figure 2.19** Coefficient of permeability determination—packer test (redrawn after U.S. Bureau of Reclamation, 1974)

## Seismic Refraction Survey

Seismic refraction surveys are useful in obtaining preliminary information about the thickness of the layering of various soils at a given site and also the depth to rock or hard soil. Refraction surveys are conducted by impacting the surface, such as at point $A$ in Figure 2.20a, and observing the first arrival of the disturbance (stress waves) at several points away from point $A$ (such as $B$, $C$, $D$, . . .). The impact can be created by a hammer blow or by a small explosive charge. The observations of the first arrival of disturbances to arrive at various points can be recorded by geophones.

The impact on the ground surface creates two types of *stress wave*: *P-waves* (or *plane waves*) and *S-waves* (or *shear waves*). The $P$-waves travel faster than $S$-waves; hence, the first arrival of the disturbances will be related to the velocities of the $P$-waves in various layers. The velocity of $P$-waves in a medium can be given by the equation

$$ v = \sqrt{\frac{E}{\left(\frac{\gamma}{g}\right)}} \sqrt{\frac{(1 - \mu)}{(1 - 2\mu)(1 + \mu)}} \tag{2.19} $$

where $E$ = Young's modulus of the medium
$\gamma$ = unit weight of the medium
$g$ = acceleration due to gravity
$\mu$ = Poisson's ratio

To determine the velocity ($v$) of $P$-waves in various layers and the thicknesses of those layers, use the following procedure:

**1.** Obtain the times of first arrival, $t_1$, $t_2$, $t_3$, . . . , at various distances, $x_1$, $x_2$, $x_3$, . . . , from the field.

**2.** Plot a graph of time ($t$) vs. distance ($x$). The graph will be like the one in Figure 2.20b.

**3.** Determine the slopes of the lines $ab$, $bc$, $cd$, . . . .

Slope of $ab = \dfrac{1}{v_1}$

Slope of $bc = \dfrac{1}{v_2}$

Slope of $cd = \dfrac{1}{v_3}$

where $v_1$, $v_2$, $v_3$, . . . , are the $P$-wave velocities in layers I, II, III, . . . , respectively (Figure 2.20a).

**4.** Determine the thickness of the top layer as

$$ Z_1 = \frac{1}{2} \sqrt{\frac{v_2 - v_1}{v_2 + v_1}} \cdot x_c \tag{2.20} $$

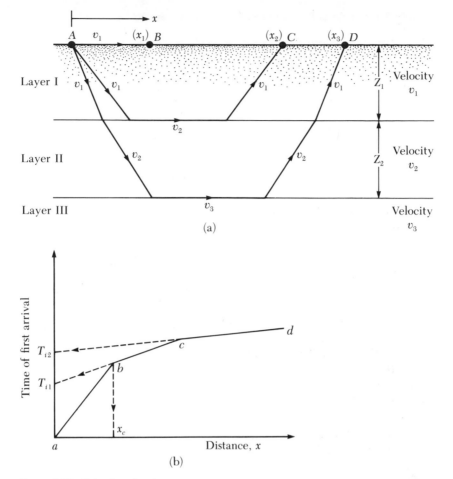

**Figure 2.20** Seismic refraction survey

The value of $x_c$ can be obtained from the plot, as shown in Figure 2.20b.

**5.** Determine the thickness of the second layer $Z_2$, shown in Figure 2.20a, as

$$Z_2 = \frac{1}{2}\left[ T_{i2} - 2Z_1 \frac{\sqrt{v_3^2 - v_1^2}}{v_3 \cdot v_1} \right] \frac{v_3 \cdot v_2}{\sqrt{v_3^2 - v_2^2}} \qquad (2.21)$$

where $T_{i2}$ is the time intercept of the line $cd$ in Figure 2.20b extended backward.

For detailed derivations of these equations and other related information, refer to Dobrin (1960) and Das (1983).

Knowing the velocities of $P$-waves in various layers gives one an idea of the types of soil or rock that are present below the ground surface. The range of the

**Table 2.7** Range of *P*-Wave Velocity in Various Soils and Rocks

| Type of soil or rock | P-wave velocity (m/sec) |
|---|---|
| *Soil* | |
| Sand, dry silt, and fine-grained top soil | 200 to 1,000 |
| Alluvium | 500 to 2,000 |
| Compacted clays, clayey gravel, and dense clayey sand | 1,000 to 2,500 |
| Loess | 250 to 750 |
| *Rock* | |
| Slate and shale | 2,500 to 5,000 |
| Sandstone | 1,500 to 5,000 |
| Granite | 4,000 to 6,000 |
| Sound limestone | 5,000 to 10,000 |

*P*-wave velocity that is generally encountered in various types of soil and rock at shallow depths is given in Table 2.7.

In analyzing the results of a refraction survey, two limitations need to be kept in mind:

1. The basic equations for the refraction survey—that is, Eqs. 2.20 and 2.21—are based on the assumption that the *P*-wave velocity $v_1 < v_2 < v_3$ . . . .

2. When a soil is saturated below the ground water table, the *P*-wave velocity may be deceptive. *P*-waves can travel through water with a velocity of about 1500 m/sec. For dry, loose soils, the velocity may be well below 1500 m/sec. However, in a saturated condition, the waves will travel through water present in the void spaces with a velocity of about 1500 m/sec. If the presence of ground water is not known, the *P*-wave velocity may be erroneously interpreted to be indicative of a stronger material (for example, sandstone) than what is present *in situ*.

## Example 2.2

The results of a refraction survey at a site are given in the following table. Determine the velocity of *P*-waves and the thickness of the material encountered.

| Distance from the source of disturbance (m) | Time of first arrival (sec × 10³) |
|---|---|
| 2.5 | 11.2 |
| 5 | 23.3 |
| 7.5 | 33.5 |
| 10 | 42.4 |
| 15 | 50.9 |
| 20 | 57.2 |
| 25 | 64.4 |
| 30 | 68.6 |
| 35 | 71.1 |
| 40 | 72.1 |
| 50 | 75.5 |

## Solution

### Velocity

In Figure 2.21, the times of first arrival are plotted against the distance from the source of disturbance. The plot has three straight-line segments. The velocity of the top three layers can now be calculated in the following manner:

$$\text{Slope of segment } 0a = \frac{1}{v_1} = \frac{\text{time}}{\text{distance}} = \frac{23 \times 10^{-3}}{5.25}$$

or

$$v_1 = \frac{5.25 \times 10^3}{23} = 228 \text{ m/sec (top layer)}$$

$$\text{Slope of segment } ab = \frac{1}{v_2} = \frac{13.5 \times 10^{-3}}{11}$$

or

$$v_2 = \frac{11 \times 10^3}{13.5} = 814.8 \text{ m/sec (middle layer)}$$

$$\text{Slope of segment } bc = \frac{1}{v_3} = \frac{14.75 \times 10^{-3}}{3.5}$$

or

$$v_3 = 4214 \text{ m/sec (third layer)}$$

Comparing the velocities obtained here with those given in Table 2.7, it appears that the third layer is a *rock layer*.

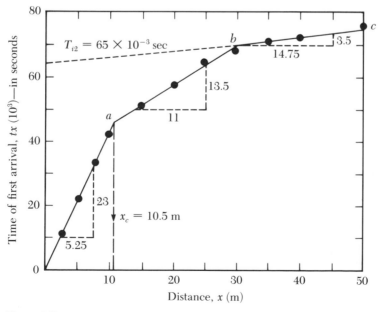

**Figure 2.21**

Thickness of layers

From Figure 2.21, $x_c = 10.5$ m. So

$$Z_1 = \frac{1}{2} \sqrt{\frac{v_2 - v_1}{v_2 + v_1}} \, x_c \qquad [\text{Eq. (2.20)}]$$

So

$$Z_1 = \frac{1}{2} \sqrt{\frac{814.8 - 228}{814.8 + 228}} \times 10.5 = 3.94 \text{ m}$$

Again, from Eq. (2.21)

$$Z_2 = \frac{1}{2} \left[ T_{i2} - \frac{2Z_1 \sqrt{v_3^2 - v_1^2}}{(v_3 \cdot v_1)} \right] \frac{(v_3)(v_2)}{\sqrt{v_3^2 - v_2^2}}$$

The value of $T_{i2}$ can be determined from Figure 2.21 as $65 \times 10^{-3}$ sec. So

$$Z_2 = \frac{1}{2} \left[ 65 \times 10^{-3} - \frac{2(3.94) \sqrt{(4214)^2 - (228)^2}}{(4214)(228)} \right] \frac{(4214)(814.8)}{\sqrt{(4214)^2 - (814.8)^2}}$$

$$= \frac{1}{2} (0.065 - 0.0345)830.47 = 12.66 \text{ m}$$

Hence, the rock layer is located at a depth of $Z_1 + Z_2 = 3.94 + 12.66 = 16.60$ m measured from the ground surface.

## Cross-Hole Seismic Survey

The velocity of shear waves created as the result of an impact to a given soil layer can be effectively determined by *cross-hole seismic survey* (Stokoe and Woods, 1972). The principle of this technique is illustrated in Figure 2.22, which shows two holes drilled into the ground at a certain distance, $L$, apart. A vertical impulse is created at the bottom of one borehole by means of an impulse rod. The shear waves thus generated are recorded by a vertically sensitive transducer. The velocity of shear waves, $v_s$, can be calculated as

$$v_s = \frac{L}{t} \tag{2.22}$$

where $t$ = travel time of shear waves

The shear modulus of the soil at the depth of the test can be determined from $v_s$ as

$$V_s = \sqrt{\frac{G}{\left(\frac{\gamma}{g}\right)}}$$

$$G = \frac{v_s^2 \, \gamma}{g} \tag{2.23}$$

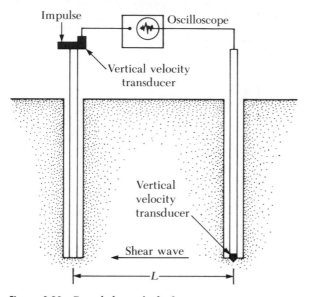

**Figure 2.22**  Cross-hole method of seismic survey

where $G$ = shear modulus of soil
 $\gamma$ = soil unit weight
 $g$ = acceleration due to gravity

The values of shear modulus are useful in the design of foundations to support vibrating machineries and the like.

## Resistivity Survey

Another geophysical method for subsoil exploration is the *electrical resistivity survey*. The electrical resistivity, $\rho$, of any conducting material having a length, $L$, and an area of cross section, $A$, can be defined as

$$\rho = \frac{RA}{L} \tag{2.24}$$

where $R$ = electrical resistance

The unit of resistivity is generally expressed as *ohm-centimeter* or *ohm-meter*. The resistivity of various soils depends primarily on the moisture content and also on the concentration of dissolved ions. Saturated clays have a very low resistivity; in contrast, dry soils and rocks have a high resistivity. The general range of resistivity generally encountered in various soils and rocks is given in Table 2.8.

The most common procedure for measuring electrical resistivity of a soil profile makes use of four electrodes that are driven into the ground, spaced equally along a straight line. This is generally referred to as the *Wenner method*

**Table 2.8**  Representative Values
of Resistivity

| Material | Resistivity (ohm-m) |
|---|---|
| Sand | 500–1500 |
| Clays, saturated silt | 0– 100 |
| Clayey sand | 200– 500 |
| Gravel | 1500–4000 |
| Weathered rock | 1500–2500 |
| Sound rock | >5000 |

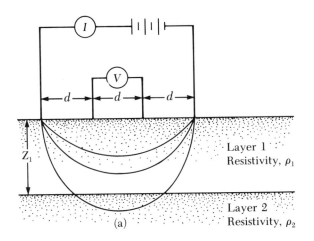

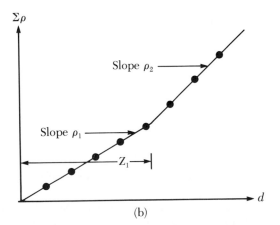

**Figure 2.23**  Electrical resistivity survey: (a) Wenner method; (b) empirical method for determination of resistivity and thickness of each layer

(Figure 2.23a). The two outside electrodes are used to send an electrical current, $I$, (usually a DC current with nonpolarizing potential electrodes) into the ground. The electrical current is typically in the range of 50–100 milliamperes. The voltage drop, $V$, is measured between the two inside electrodes. If the soil profile is homogeneous, its electrical resistivity can be expressed as

$$\rho = \frac{2\pi dV}{I} \tag{2.25}$$

In most cases, the soil profile may consist of various layers with different resistivities. In that case, Eq. (2.25) will yield the *apparent resistivity*. In order to obtain the *actual resistivity* of various layers and their thicknesses, an empirical method may be used. This involves conducting a number of tests at various electrode spacings (that is, $d$ is changed). The sum of the apparent resistivities (that is, $\Sigma \rho$) is plotted against the spacing $d$, as shown in Figure 2.23b. The plot thus obtained will have relatively straight segments. The slope of these straight segments will give the resistivity of individual layers. The thicknesses of various layers can be estimated as shown in Figure 2.23b.

The resistivity survey is particularly useful in locating gravel deposits within a fine-grained soil.

## 2.12

### Soil Exploration Report

At the end of all subsoil exploration programs, the soil and/or rock specimens collected from the field are subjected to visual observation and appropriate laboratory testing (the basic soil tests have been described in Chapter 1). After the compilation of all of the required information, a soil exploration report is prepared for the use of the design office and for reference during future construction work. Although the details and sequence of information in the report may vary to some degree depending on the structure under consideration and the person compiling the report, each report should include the following:

**1.** The scope of the investigation.

**2.** A description of the proposed structure for which the subsoil exploration has been conducted.

**3.** A description of the location of the site. This should include structure(s) nearby, drainage conditions of the site, nature of vegetation on the site and surrounding it, and any other feature(s) unique to the site.

**4.** Geological condition of the site.

**5.** Details of the field exploration—that is, number of borings, depths of borings, type of boring, and so on.

**6.** General description of the subsoil conditions as determined from soil specimens and from related laboratory tests, standard penetration resistance and cone penetration resistance, and so on.

**7.** Ground water table conditions.

**8.** Foundation recommendations. These should include the type of foundation recommended, allowable bearing pressure, and any special construction procedure that may be needed. Alternate foundation design procedures should also be discussed in this portion of the report.

**9.** Conclusions and limitations of the investigations.

Following are the graphic presentations that need to be attached to the report:

**1.** Site-location map

**2.** A plan showing the location of the borings with respect to the proposed structures and those existing nearby

**3.** Boring logs

**4.** Laboratory test results

**5.** Other special graphic presentations

The exploration reports should be well planned and documented. They will help in solving several questions and foundation problems that may arise at a later stage of design and construction.

## Problems

**2.1** A Shelby tube has an outside diameter of 3 in. and an inside diameter of 2.874 in. What is the area ratio of the Shelby tube?

**2.2** Refer to Figure 2.12 on p. 79. The borehole is in a silty soil. Given: $h_w + h_o = 12.24$ m, $t_1 = 24$ hrs, $t_2 = 48$ hrs, $t_3 = 72$ hrs, $\Delta h_1 = 1.2$ m, $\Delta h_2 = 0.86$ m, $\Delta h_3 = 0.6$ m. Determine $h_w$ (the depth of ground water table).

**2.3** A vane shear test was conducted in a saturated clay. The height and the diameter of the vane were 4 in. and 2 in. respectively. During the test, the maximum torque applied was 12.4 lb-ft. Determine the undrained shear strength of the clay.

**2.4** The clay soil described in Problem 2.3 has a liquid limit of 64 and a plastic limit of 29. What should be the corrected undrained shear strength of this clay for design purposes? (Refer to Figure 2.14b on p. 81.)

**2.5** During a field exploration, coring of rock was required. The core barrel was advanced 6 ft during the coring. The length of the core recovered was 3.2 ft. What was the recovery ratio?

**2.6** A boring log in a sandy soil is shown in Figure P2.6. The unit weight of soil above the ground water table is 18.08 kN/m$^3$ and below the ground water table is 19.34 kN/m$^3$. Using Eqs. (2.4) through (2.6), determine the corrected standard penetration numbers.

**2.7** For the sandy soil encountered in Problem 2.6, estimate the relative density and the angle of friction.

**2.8** An open end permeability test was conducted in a borehole (refer to Figure 2.18a on p. 87). The inside diameter of the casing was 1.5 in. The differential head of water was 30.18 ft. In order to maintain a constant head $H = 30.18$ ft, a constant water supply rate of $6.53 \times 10^{-2}$ ft$^3$/min was required. Calculate the coefficient of permeability of soil.

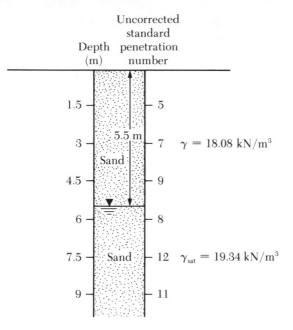

Uncorrected
standard
Depth    penetration
(m)      number

**Figure P2.6**

**2.9** The *P*-wave velocity in a given soil is 5500 ft/sec. Assuming Poisson's ratio to be 0.3, calculate the Young's modulus of the soil. Assume the unit weight of soil to be equal to 112 lb/ft³.

**2.10** The results of a refraction survey (Figure 2.20a on p. 91) at a site are given in the following table. Determine the thickness and the *P*-wave velocity of the materials encountered.

| Distance from the source of disturbance (m) | Time of first arrival of *P*-waves (sec × 10³) |
|:---:|:---:|
| 2.5 | 5.08 |
| 5 | 10.16 |
| 7.5 | 15.24 |
| 10 | 17.01 |
| 15 | 20.02 |
| 20 | 24.2 |
| 25 | 27.1 |
| 30 | 28.0 |
| 40 | 31.1 |
| 50 | 33.9 |

## References

American Association of State Highway and Transportation Officials (1978). *Standard Specifications for Transportation Materials and Methods of Sampling and Testing*, Part II, Washington, D.C.

American Society for Testing and Materials (1982). *Annual Book of ASTM Standards*, Part 19, Philadelphia.

American Society of Civil Engineers (1972). "Subsurface Investigation for Design and Construction of Foundations of Buildings," *Journal of the Soil Mechanics and Foundations Division*, American Society of Civil Engineers, Vol. 98, No. SM5, pp. 481–490.

Bazaraa, A. R. (1967). *Use of Standard Penetration Test for Estimating Settlements of Shallow Foundations on Sand*, Ph.D. Thesis, University of Illinois, Urbana.

Bjerrum, L. (1972). "Embankments on Soft Ground," *Proceedings of the Specialty Conference*, American Society of Civil Engineers, Vol. 2, pp. 1–54.

Das, B. M. (1983). *Fundamentals of Soil Dynamics*, Elsevier Science Publishing Co., New York.

Deere, D. U. (1963). "Technical Description of Rock Cores for Engineering Purposes," *Felsmechanik und Ingenieurgeologie*, Vol. 1, No. 1, pp. 16–22.

Dobrin, M. B. (1960). *Introduction to Geophysical Prospecting*, McGraw-Hill, New York.

Hvorslev, M. J. (1949). *Subsurface Exploration and Sampling of Soils for Civil Engineering Purposes*, Waterways Experiment Station, Vicksburg, Miss.

Osterberg, J. O. (1952). "New Piston-Type Soil Sampler," *Engineering News-Record*, April 24.

Peck, R. B., and Bazaraa, A. S. (1969). "Discussion on Settlement of Spread Footings on Sand," *Journal of the Soil Mechanics and Foundations Division*, American Society of Civil Engineers, Vol. 95, No. SM3, pp. 905–909.

Peck, R. B., Hanson, W. E., and Thornburn, T. H. (1974). *Foundation Engineering*, 2nd ed. Wiley, New York.

Schmertmann, J. H. (1975). "The Measurement of *In-Situ* Shear Strength," *Proceedings*, American Society of Civil Engineers' Specialty Conference on *In-Situ* Measurement of Soil Properties, Raleigh, N.C., Vol. 2, pp. 57–138.

Sowers, G. B., and Sowers, G. F. (1970). *Introductory Soil Mechanics and Foundations*, 3rd ed., Macmillan, New York.

Stokoe, K. H., and Woods, R. D. (1972). "*In Situ* Shear Wave Velocity by Cross-Hole Method," *Journal of Soil Mechanics and Foundations Division*, American Society of Civil Engineers, Vol. 98, No. SM5, pp. 443–460.

United States Bureau of Reclamation (1974). *Design of Small Dams*, 2nd edition, U.S. Government Printing Press, Washington, D.C.

# Shallow Foundations

## 3.1

### Introduction

To perform satisfactorily, shallow foundations must have two main characteristics:

1. The foundation should be safe against shear failure in the soil that supports it.

2. The foundation should not undergo excessive settlement. (The term *excessive* is relative, because the degree of settlement allowable for a given structure is dependent on several considerations.) This chapter discusses in detail the evaluation of the safe load-bearing capacity and settlement of shallow foundations.

## Ultimate Bearing Capacity of Shallow Foundations

## 3.2

### Ultimate Bearing Capacity

Consider a strip foundation resting on the surface of a dense sand or stiff cohesive soil, as shown in Figure 3.1a. Let the width of the foundation be equal to $B$. Now, if load is gradually applied on the foundation, the settlement of the foundation will also increase. The variation of the load per unit area on the foundation ($q$) with the foundation settlement is also shown in Figure 3.1a. At a certain point, when the load per unit area becomes equal to $q_u$, a sudden failure in the soil supporting the foundation will take place, and the failure surface in the soil will extend to the ground surface. This load per unit area, $q_u$,

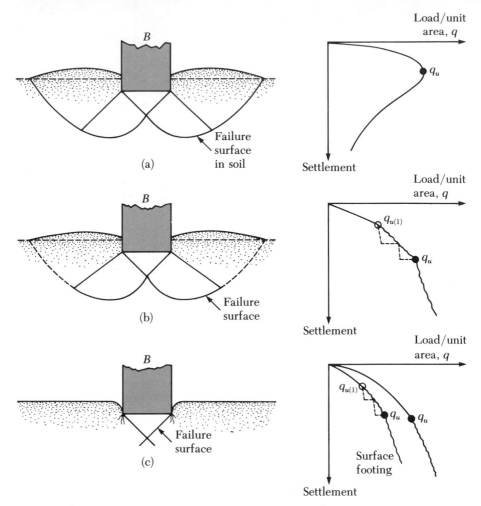

**Figure 3.1**  Nature of bearing capacity failure in soil: (a) general shear
failure; (b) local shear failure; (c) punching shear failure (redrawn after
Vesic, 1973)

is usually referred to as the *ultimate bearing capacity of the foundation*. When
this type of sudden failure in soil takes place, accompanied by the extension of
the failure surface to the ground surface, it is called the *general shear failure*.

In contrast, if the foundation under consideration is resting on sand or
clayey soil of medium compaction (Figure 3.1b), an increase of load on the
foundation will also be accompanied by an increase of settlement. However, in
this case the failure surface in the soil will gradually extend outward from the
foundation, as shown by the solid lines in Figure 3.1b. When the load per unit
area on the foundation becomes equal to $q_{u(1)}$, the foundation movement will be
accompanied by sudden jerks. A considerable movement of the foundation is
then required for the failure surface in soil to extend to the ground surface (this

part is shown by the broken lines in Figure 3.1b). The load per unit area at which this happens is the *ultimate bearing capacity*, $q_u$. Beyond this point, an increase of load will be accompanied by a large increase of foundation settlement. The load per unit area of the foundation, $q_{u(1)}$, is referred to as the *first failure load* (Vesic, 1963). Note that a peak value of $q$ is not realized in the case of this type of failure. This is called the *local shear failure* in soil.

If the foundation is supported by a fairly loose soil, the load-settlement plot will be like the one in Figure 3.1c. In this case, the failure surface in soil will not extend to the ground surface. Beyond the ultimate failure load, $q_u$, the load-settlement plot will be steep and practically linear. This type of failure in soil is called the *punching shear failure*.

Based on experimental results, Vesic (1973) has proposed a relationship for the mode of bearing capacity failure of foundations resting on sands. This is shown in Figure 3.2, which uses the following notations:

$D_r$ = relative density of sand

$D_f$ = depth of foundation measured from the ground surface

$$B* = \frac{2BL}{B + L} \tag{3.1}$$

where $B$ = width of foundation
    $L$ = length of foundation
(*Note*: $L$ is always greater than $B$.)

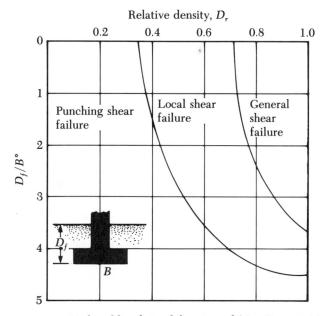

**Figure 3.2** Modes of foundation failure in sand (after Vesic, 1973)

For square foundations, $B = L$; for circular foundations, $B = L =$ diameter. So

$$B* = B \qquad (3.2)$$

For foundations located at a shallow depth (that is, small $D_f/B*$), the ultimate load may occur at a foundation settlement of 4–10% of $B$. This is true when general shear failure in soil occurs; however, in the case of local or punching shear failure, the ultimate load may occur at settlements of 15–25% of the width of foundation ($B$).

## 3.3

### Terzaghi's Bearing Capacity Theory

Terzaghi (1943) was the first to present a comprehensive theory for the evaluation of the ultimate bearing capacity of rough shallow foundations. According to this theory, a foundation is *shallow* if the depth, $D_f$ (Figure 3.3), of the foundation is less than or equal to the width of the foundation. Later investigators, however, have suggested that foundations with $D_f$ equal to 3–4 times the width of the foundation may be defined as *shallow foundations*.

Terzaghi suggested that for a *continuous*, or *strip, foundation* (that is, width-to-length ratio of the foundation is equal to zero), the failure surface in soil at ultimate load may be assumed to be similar to that shown in Figure 3.3. (Note that this is the case of general shear failure as defined in Figure 3.1a.) It may also be assumed that the effect of soil above the bottom of the foundation may be replaced by an equivalent surcharge equal to $q = \gamma D_f$ (where $\gamma =$ unit weight of soil). The failure zone under the foundation can be separated into three parts (see Figure 3.3):

1. The *triangular zone ACD* immediately under the foundation
2. The *radial shear zones ADF* and *CDE*, with the curves *DE* and *DF* being arcs of a logarithmic spiral
3. Two triangular *Rankine passive zones AFH* and *CEG*

It is assumed that the angles *CAD* and *ACD* are equal to the soil friction angle, $\phi$. Note that with the replacement of the soil above the bottom of the foundation by an equivalent surcharge $q$, the shear resistance of the soil along the failure surfaces *GI* and *HJ* was neglected.

Using the equilibrium analysis, Terzaghi expressed the ultimate bearing capacity in the form

$$q_u = cN_c + qN_q + \frac{1}{2}\gamma BN_\gamma \qquad \text{(strip foundation)} \qquad (3.3)$$

where    $c =$ cohesion of soil
$\gamma =$ unit weight of soil
$q = \gamma D_f$
$N_c, N_q, N_\gamma =$ bearing capacity factors that are nondimensional and are only functions of the soil friction angle, $\phi$

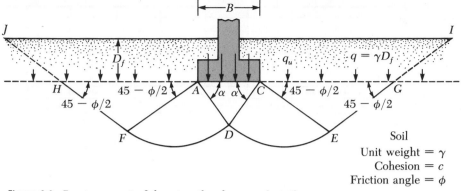

**Figure 3.3** Bearing capacity failure in soil under a rough rigid continuous foundation

The bearing capacity factors, $N_c$, $N_q$, and $N_\gamma$ are defined by the following equations:

$$N_c = \cot \phi \left[ \frac{e^{2(3\pi/4 - \phi/2)\tan \phi}}{2 \cos^2 \left( \frac{\pi}{4} + \frac{\phi}{2} \right)} - 1 \right] \tag{3.4}$$

$$N_q = \frac{e^{2(3\pi/4 - \phi/2)\tan \phi}}{2 \cos^2 \left( 45 + \frac{\phi}{2} \right)} \tag{3.5}$$

$$N_\gamma = \frac{1}{2} \left( \frac{K_{p\gamma}}{\cos^2 \phi} - 1 \right) \tan \phi \tag{3.6}$$

where $K_{p\gamma}$ = passive pressure coefficient

The variations of the bearing capacity factors defined by Eq. (3.3) are given in Figure 3.4.

For estimation of the ultimate bearing capacity of *square* or *circular foundations*, Eq. (3.1) may be modified to the following forms:

$$q_u = 1.3cN_c + qN_q + 0.4\gamma BN_\gamma \quad \text{(square foundation)} \tag{3.7}$$

and

$$q_u = 1.3cN_c + qN_q + 0.3\gamma BN_\gamma \quad \text{(circular foundation)} \tag{3.8}$$

In Eq. (3.7), $B$ is equal to the dimension of each side of the foundation; in Eq. (3.8), $B$ is equal to the diameter of the foundation.

For foundations that exhibit the local shear failure mode in soils, Terzaghi suggested the following modifications to Eqs. (3.3), (3.7), and (3.8).

$$q_u = \frac{2}{3} cN_c' + qN_q' + \frac{1}{2} \gamma BN_\gamma' \quad \text{(strip foundation)} \tag{3.9}$$

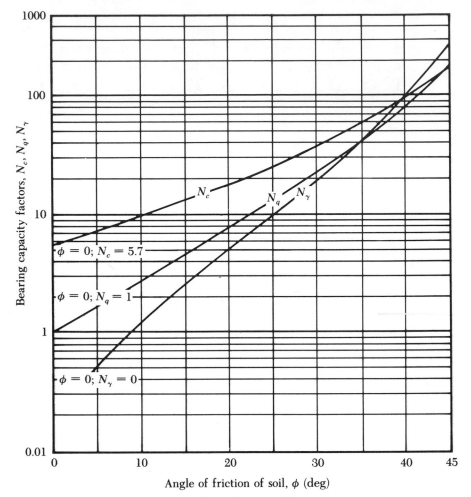

**Figure 3.4**  Terzaghi's bearing capacity factors for general shear failure—Eq. (3.3)

$$q_u = 0.867cN'_c + qN'_q + 0.4\gamma BN'_\gamma \quad \text{(square foundation)} \quad (3.10)$$

$$q_u = 0.867cN'_c + qN'_q + 0.3\gamma BN'_\gamma \quad \text{(circular foundation)} \quad (3.11)$$

$N'_c$, $N'_q$, and $N'_\gamma$ are the *modified bearing capacity factors*. They can be calculated by using the bearing capacity factor equations (for $N_c$, $N_q$, and $N_\gamma$) by replacing $\phi$ by $\phi' = \tan^{-1}(2/3 \tan \phi)$. The variation of $N'_c$, $N'_q$, and $N'_\gamma$ with the soil friction angle, $\phi$, is given in Figure 3.5.

Based on the experience of several model tests, the author has observed (also see Ismael and Vesic, 1981) that the value of $N'_q$ for shallow foundations in granular soil is underestimated by Terzaghi's modified procedure. It is better represented by the equation given by Vesic (1963):

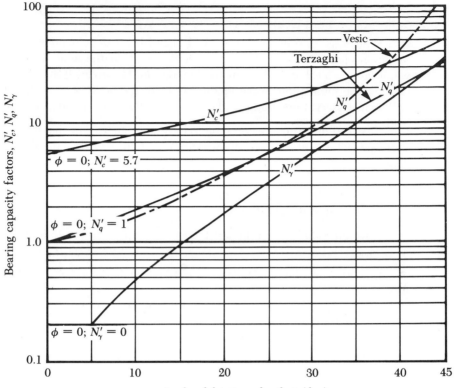

**Figure 3.5** Terzaghi's modified bearing capacity factors: $N'_c$, $N'_q$, $N'_\gamma$; and Vesic's $N'_q$—Eq. (3.12)

$$N'_q = (e^{3.8\phi \tan \phi})\tan^2\left(45 + \frac{\phi}{2}\right) \tag{3.12}$$

The values of $N'_q$ as calculated from Eq. (3.12) are given in Table 3.1. For comparison purposes, these values are also plotted in Figure 3.5.

**Table 3.1** Values of $N'_q$ [Eq. (3.12)]

| $\phi$ (deg) | $N'_q$ |
|:---:|:---:|
| 0 | 1.0 |
| 5 | 1.2 |
| 10 | 1.6 |
| 15 | 2.2 |
| 20 | 3.3 |
| 25 | 5.3 |
| 30 | 9.5 |
| 35 | 18.7 |
| 40 | 42.5 |
| 45 | 115.0 |

## 3.4

## Modification of Bearing Capacity Equations for Water Table

Equations (3.3) and (3.7) to (3.11) have been developed for the determination of the ultimate bearing capacity based on the assumption that the water table is located well below the foundation. However, if the water table is close to the foundation, some modifications of the bearing capacity equations will be necessary, depending on the location of the water table (see Figure 3.6):

*Case I*: If the ground water table is located in such a way that $0 \leq D_1 < D_f$, then the factor $q$ in the bearing capacity equations will take the form

$$q = \text{effective surcharge} = D_1\gamma + D_2(\gamma_{\text{sat}} - \gamma_w) \qquad (3.13)$$

where $\gamma_{\text{sat}}$ = saturated unit weight of soil
$\gamma_w$ = unit weight of water

Also, the value of $\gamma$ in the last term of the equations has to be replaced by $\gamma' = \gamma_{\text{sat}} - \gamma_w$.

*Case II*: For ground water table located in such a way that $0 \leq d \leq B$, the factor

$$q = \gamma D_f \qquad (3.14)$$

The factor $\gamma$ in the last term of the bearing capacity equations must be replaced by the factor

$$\overline{\gamma} = \gamma' + \frac{d}{B}(\gamma - \gamma') \qquad (3.15)$$

The preceding modifications are based on the assumption that there is no seepage force in the soil.

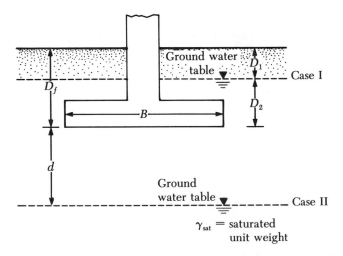

**Figure 3.6**  Modification of bearing capacity equations for water table

*Case III:* When the ground water table is located such that $d \geq B$, the water will have no effect on the ultimate bearing capacity.

## Example 3.1

A square foundation is 1.5 m × 1.5 m in plan. The soil supporting the foundation has a friction angle of $\phi = 20°$ and $c = 15.2$ kN/m². The unit weight of soil, $\gamma$, is 17.8 kN/m³. Determine the allowable gross load on the foundation with a factor of safety (*FS*) of 4. Assume the depth of the foundation $(D_f)$ to be one meter, and general shear failure occurs in soil.

### Solution

From Eq. (3.7)

$$q_u = 1.3cN_c + qN_q + 0.4\gamma BN_\gamma$$

From Figure 3.4, for $\phi = 20°$

$$N_c = 17.7$$
$$N_q = 7.4$$
$$N_\gamma = 5$$

Thus

$$q_u = (1.3)(15.2)(17.7) + (1 \times 17.8)(7.4) + (0.4)(17.8)(1.5)(5)$$
$$= 349.75 + 131.72 + 53.4 = 534.87 \approx 535 \text{ kN/m}^2$$

So, allowable load per unit area of the foundation =

$$q_{all} = \frac{q_u}{FS} = \frac{535}{4} = 133.75 \text{ kN/m}^2$$

Thus, the total allowable gross load

$$Q = (133.75)B^2 = (133.75)(1.5 \times 1.5) = 300.9 \approx 300 \text{ kN}$$

## Example 3.2

Repeat Example Problem 3.1, assuming local shear failure occurs in the soil supporting the foundation.

### Solution

From Eq. (3.10)

$$q_u = 0.867cN'_c + qN'_q + 0.4\gamma BN'_\gamma$$

From Figure 3.5, for $\phi = 20°$

$$N'_c = 12$$
$$N'_q = 4$$
$$N'_\gamma = 1.7$$

So

$$q_u = (0.867)(15.2)(12) + (1 \times 17.8)(4) + (0.4)(17.8)(1.5)(1.7)$$
$$= 158.1 + 71.2 + 18.2 = 247.5 \text{ kN/m}^2$$

$$q_{\text{all}} = \frac{247.5}{4} = 61.9 \text{ kN/m}^2$$

Allowable gross load $= Q = (q_{\text{all}})(B^2) = (61.9)(1.5^2) = \underline{139 \text{ kN}}$

## 3.5

## General Bearing Capacity Equation

The ultimate bearing capacity equations presented in Eqs. (3.3), (3.7), and (3.8) are for continuous, square, and circular foundations only. They do not address the case of rectangular foundations ($0 < B/L < 1$). Also, the equations do not take into account the shearing resistance along the failure surface in soil located above the bottom of the foundation (portion of the failure surface marked as *GI* and *HJ* in Figure 3.3 on p. 105). In addition, the load on the foundation may be inclined. In order to take all these shortcomings into account, the following form of general bearing capacity equation has been suggested by Meyerhof (1963).

$$q_u = cN_cF_{cs}F_{cd}F_{ci} + qN_qF_{qs}F_{qd}F_{qi} + \frac{1}{2}\gamma BN_\gamma F_{\gamma s}F_{\gamma d}F_{\gamma i} \tag{3.16}$$

where $c$ = cohesion
$q$ = effective stress at the level of the bottom of foundation
$\gamma$ = unit weight of soil
$B$ = width of foundation ( = diameter for a circular foundation)
$F_{cs}, F_{qs}, F_{\gamma s}$ = shape factors
$F_{cd}, F_{qd}, F_{\gamma d}$ = depth factors
$F_{ci}, F_{qi}, F_{\gamma i}$ = load inclination factors
$N_c, N_q, N_\gamma$ = bearing capacity factors

The equations for determination of the various factors given in Eq. (3.16) will be briefly summarized in the following sections. Note that the original equation for ultimate bearing capacity is derived only for plane-strain case (that is, for continuous foundations). The shape, depth, and load inclination factors are empirical factors based on experimental data.

### Bearing Capacity Factors

Based on laboratory and field studies done on bearing capacity, it appears now that the basic nature of the failure surface in soil as suggested by Terzaghi is correct (Vesic, 1973). However, the angle $\alpha$ as shown in Figure 3.3 is closer to $45 + \phi/2$ instead of $\phi$. If this change is accepted, the values of $N_c$, $N_q$, and $N_\gamma$ for a given soil friction angle will also change from those given in Figure 3.4. With $\alpha = 45 + \phi/2$, the relations for $N_c$ and $N_q$ can be derived as

$$N_q = \tan^2\left(45 + \frac{\phi}{2}\right)e^{\pi \tan \phi} \tag{3.17}$$

$$N_c = (N_q - 1)\cot\phi \tag{3.18}$$

The equation for $N_c$ given by Eq. (3.18) was originally derived by Prandtl (1921), and the relation for $N_q$ [Eq. (3.17)] was presented by Reissner (1924). Caquot and Kerisel (1953) and Vesic (1973) have given the relation for $N_\gamma$ as

$$N_\gamma = 2(N_q + 1)\tan\phi \tag{3.19}$$

The variation of the preceding bearing capacity factors with soil friction angles is given in Table 3.2.

## Shape Factors

The equation for shape factors ($F_{cs}$, $F_{qs}$, and $F_{\gamma s}$) has been recommended by De Beer (1970). They are

$$F_{cs} = 1 + \left(\frac{B}{L}\right)\left(\frac{N_q}{N_c}\right) \tag{3.20}$$

$$F_{qs} = 1 + \left(\frac{B}{L}\right)\tan\phi \tag{3.21}$$

and

$$F_{\gamma s} = 1 - 0.4\left(\frac{B}{L}\right) \tag{3.22}$$

where $L$ = length of the foundation ($L > B$)

The shape factors are empirical relations based on extensive laboratory tests.

## Depth Factors

Hansen (1970) proposed the equations for depth factors:

$$F_{cd} = 1 + 0.4\left(\frac{D_f}{B}\right) \tag{3.23}$$

$$F_{qd} = 1 + 2\tan\phi(1 - \sin\phi)^2\frac{D_f}{B} \tag{3.24}$$

$$F_{\gamma d} = 1 \tag{3.25}$$

Equations (3.23) and (3.24) are valid for $D_f/B \leq 1$. For depth of embedment-to-foundation width ratio greater than one ($D_f/B > 1$), the preceding equations have to be modified:

$$F_{cd} = 1 + (0.4)\tan^{-1}\left(\frac{D_f}{B}\right) \tag{3.26}$$

$$F_{qd} = 1 + 2\tan\phi(1 - \sin\phi)^2\tan^{-1}\left(\frac{D_f}{B}\right) \tag{3.27}$$

$$F_{\gamma d} = 1 \tag{3.28}$$

The factor, $\tan^{-1}(D_f/B)$ is in radians when used in Eqs. (3.26) and (3.27).

**Table 3.2**  Bearing Capacity Factors[a]

| $\phi$ | $N_c$ | $N_q$ | $N_\gamma$ | $N_q/N_c$ | $\tan \phi$ |
|---|---|---|---|---|---|
| 0 | 5.14 | 1.00 | 0.00 | 0.20 | 0.00 |
| 1 | 5.38 | 1.09 | 0.07 | 0.20 | 0.02 |
| 2 | 5.63 | 1.20 | 0.15 | 0.21 | 0.03 |
| 3 | 5.90 | 1.31 | 0.24 | 0.22 | 0.05 |
| 4 | 6.19 | 1.43 | 0.34 | 0.23 | 0.07 |
| 5 | 6.49 | 1.57 | 0.45 | 0.24 | 0.09 |
| 6 | 6.81 | 1.72 | 0.57 | 0.25 | 0.11 |
| 7 | 7.16 | 1.88 | 0.71 | 0.26 | 0.12 |
| 8 | 7.53 | 2.06 | 0.86 | 0.27 | 0.14 |
| 9 | 7.92 | 2.25 | 1.03 | 0.28 | 0.16 |
| 10 | 8.35 | 2.47 | 1.22 | 0.30 | 0.18 |
| 11 | 8.80 | 2.71 | 1.44 | 0.31 | 0.19 |
| 12 | 9.28 | 2.97 | 1.69 | 0.32 | 0.21 |
| 13 | 9.81 | 3.26 | 1.97 | 0.33 | 0.23 |
| 14 | 10.37 | 3.59 | 2.29 | 0.35 | 0.25 |
| 15 | 10.98 | 3.94 | 2.65 | 0.36 | 0.27 |
| 16 | 11.63 | 4.34 | 3.06 | 0.37 | 0.29 |
| 17 | 12.34 | 4.77 | 3.53 | 0.39 | 0.31 |
| 18 | 13.10 | 5.26 | 4.07 | 0.40 | 0.32 |
| 19 | 13.93 | 5.80 | 4.68 | 0.42 | 0.34 |
| 20 | 14.83 | 6.40 | 5.39 | 0.43 | 0.36 |
| 21 | 15.82 | 7.07 | 6.20 | 0.45 | 0.38 |
| 22 | 16.88 | 7.82 | 7.13 | 0.46 | 0.40 |
| 23 | 18.05 | 8.66 | 8.20 | 0.48 | 0.42 |
| 24 | 19.32 | 9.60 | 9.44 | 0.50 | 0.45 |
| 25 | 20.72 | 10.66 | 10.88 | 0.51 | 0.47 |
| 26 | 22.25 | 11.85 | 12.54 | 0.53 | 0.49 |
| 27 | 23.94 | 13.20 | 14.47 | 0.55 | 0.51 |
| 28 | 25.80 | 14.72 | 16.72 | 0.57 | 0.53 |
| 29 | 27.86 | 16.44 | 19.34 | 0.59 | 0.55 |
| 30 | 30.14 | 18.40 | 22.40 | 0.61 | 0.58 |
| 31 | 32.67 | 20.63 | 25.99 | 0.63 | 0.60 |
| 32 | 35.49 | 23.18 | 30.22 | 0.65 | 0.62 |
| 33 | 38.64 | 26.09 | 35.19 | 0.68 | 0.65 |
| 34 | 42.16 | 29.44 | 41.06 | 0.70 | 0.67 |
| 35 | 46.12 | 33.30 | 48.03 | 0.72 | 0.70 |
| 36 | 50.59 | 37.75 | 56.31 | 0.75 | 0.73 |
| 37 | 55.63 | 42.92 | 66.19 | 0.77 | 0.75 |
| 38 | 61.35 | 48.93 | 78.03 | 0.80 | 0.78 |
| 39 | 67.87 | 55.96 | 92.25 | 0.82 | 0.81 |
| 40 | 75.31 | 64.20 | 109.41 | 0.85 | 0.84 |
| 41 | 83.86 | 73.90 | 130.22 | 0.88 | 0.87 |
| 42 | 93.71 | 85.38 | 155.55 | 0.91 | 0.90 |
| 43 | 105.11 | 99.02 | 186.54 | 0.94 | 0.93 |
| 44 | 118.37 | 115.31 | 224.64 | 0.97 | 0.97 |
| 45 | 133.88 | 134.88 | 271.76 | 1.01 | 1.00 |
| 46 | 152.10 | 158.51 | 330.35 | 1.04 | 1.04 |
| 47 | 173.64 | 187.21 | 403.67 | 1.08 | 1.07 |
| 48 | 199.26 | 222.31 | 496.01 | 1.12 | 1.11 |
| 49 | 229.93 | 265.51 | 613.16 | 1.15 | 1.15 |
| 50 | 266.89 | 319.07 | 762.89 | 1.20 | 1.19 |

[a]After Vesic (1973)

## Inclination Factors

Meyerhof (1963) and Hanna and Meyerhof (1981) suggested the following inclination factors for use in Eq. (3.16).

$$F_{ci} = F_{qi} = \left(1 - \frac{\beta^{\circ}}{90^{\circ}}\right)^2 \tag{3.29}$$

and

$$F_{\gamma i} = \left(1 - \frac{\beta}{\phi}\right)^2 \tag{3.30}$$

where $\beta$ = inclination of the load on the foundation with respect to the vertical

Hansen (1970) has presented a more elaborate set of equations for the determination of the inclination factors:

$$F_{ci} = F_{qi} - \frac{(1 - F_{qi})}{(N_q - 1)} \tag{3.31}$$

$$F_{qi} = \left[1 - \frac{(0.5)(Q_u)\sin \beta}{Q_u \cos \beta + BLc \cot \phi}\right]^5 \tag{3.32}$$

$$F_{\gamma i} = \left[1 - \frac{(0.7)(Q_u)\sin \beta}{Q_u \cos \beta + BLc \cot \phi}\right]^5 \tag{3.33}$$

In this chapter, we will use Eqs. (3.29) and (3.30) for the determination of inclination factors.

## Net Ultimate Bearing Capacity

The net ultimate bearing capacity is defined as the ultimate pressure per unit area of the foundation that can be supported by the soil in excess of the pressure caused by the surrounding soil at the foundation level. Assuming that the difference between the unit weight of concrete used in the foundation and the unit weight of soil surrounding is negligible

$$q_{net(u)} = q_u - q \tag{3.34}$$

where $q_{net(u)}$ = net ultimate bearing capacity

## General Comments

When the water table is present at or near the foundation, the factors $q$ and $\gamma$ given in the general bearing capacity equation will need modifications. The procedure for modifying them is the same as that described in Section 3.4.

For undrained loading conditions ($\phi = 0$ concept) in clayey soils, the general load bearing capacity equation [Eq. (3.16)] takes the form (vertical load)

$$q_u = cN_cF_{cs}F_{cd} + q \tag{3.35}$$

Hence, the net ultimate bearing capacity is equal to (vertical load)

$$q_{net(u)} = q_u - q = cN_cF_{cs}F_{cd} \tag{3.36}$$

Skempton (1951) has proposed the following equation for the net ultimate bearing capacity for clayey soils ($\phi = 0$ condition), which is similar to Eq. (3.36)

$$q_{net(u)} = 5c\left(1 + 0.2\frac{D_f}{B}\right)\left(1 + 0.2\frac{B}{L}\right) \tag{3.37}$$

## Example 3.3

A square column foundation has to carry a gross allowable total load of 150 kN. The depth of the foundation is 0.7 m. The load is inclined at an angle of 20° to the vertical (Figure 3.7). Determine the width of the foundation, $B$. Use Eq. (3.16) and a factor of safety of 3.

### Solution

With $c = 0$, the ultimate bearing capacity becomes

$$q_u = qN_qF_{qs}F_{qd}F_{qi} + \frac{1}{2}\gamma BN_\gamma F_{\gamma s}F_{\gamma d}F_{\gamma i} \tag{3.16}$$

$$q = (0.7)(18) = 12.6 \text{ kN/m}^2$$
$$\gamma = 18 \text{ kN/m}^3$$

From Table 3.2, for $\phi = 30°$

$$N_q = 18.4$$
$$N_\gamma = 22.4$$

$$F_{qs} = 1 + \left(\frac{B}{L}\right)\tan\phi = 1 + 0.577 = 1.577$$

$$F_{\gamma s} = 1 - 0.4\left(\frac{B}{L}\right) = 0.6$$

$$F_{qd} = 1 + 2\tan\phi(1 - \sin\phi)^2\frac{D_f}{B} = 1 + \frac{(0.289)(0.7)}{B} = 1 + \frac{0.202}{B}$$

$$F_{\gamma d} = 1$$

$$F_{qi} = \left(1 - \frac{\beta°}{90°}\right)^2 = \left(1 - \frac{20}{90}\right)^2 = 0.605$$

$$F_{\gamma i} = \left(1 - \frac{\beta°}{\phi}\right)^2 = \left(1 - \frac{20}{30}\right)^2 = 0.11$$

Hence

$$q_u = (12.6)(18.4)(1.577)\left(1 + \frac{0.202}{B}\right)(0.605) + (0.5)(18)(B)(22.4)(0.6)(1)(0.11)$$

$$= 221.2 + \frac{44.68}{B} + 13.3B \tag{a}$$

Thus

$$q_{all} = \frac{q_u}{3} = 73.73 + \frac{14.89}{B} + 4.43B \tag{b}$$

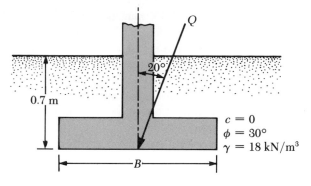

**Figure 3.7**

Given $Q$ = total allowable load = $q_{all} \times B^2$ or

$$q_{all} = \frac{150}{B^2} \qquad \text{(c)}$$

Equating the right-hand sides of Eqs. (b) and (c)

$$\frac{150}{B^2} = 73.73 + \frac{14.89}{B} + 4.43B$$

By trial and error, $B \approx 1.3$ m

## 3.6
## The Factor of Safety

In the preceding problems, in order to calculate the gross allowable load bearing capacity of shallow foundations, a factor of safety (*FS*) has been applied to the gross ultimate bearing capacity, that is

$$q_{all} = \frac{q_u}{FS} \qquad (3.38)$$

However, some practicing engineers prefer to use a factor of safety as

$$\frac{\text{net stress increase}}{\text{on soil}} = \frac{\text{net ultimate bearing capacity}}{FS} \qquad (3.39)$$

Net ultimate bearing capacity has been defined in Eq. (3.34) as

$$q_{net(u)} = q_u - q$$

Substituting this equation in Eq. (3.39), one obtains

net stress increase on soil

$$= \text{load from the superstructure per unit area of the foundation}$$

$$= q_{all(net)} = \frac{q_u - q}{FS} \qquad (3.40)$$

The factor of safety as defined by Eq. (3.40) may be referred to as the *net allowable bearing capacity*. This should be kept at least about 3 in all cases.

Another type of factor of safety for the bearing capacity of shallow foundations is often used. This is the factor of safety with respect to shear failure ($FS_{shear}$). In most cases, a value of $FS_{shear} = 1.4$–$1.6$ is desirable along with a *minimum* factor of safety of 3–4 against gross or net ultimate bearing capacity. In order to calculate the net allowable load on the basis of a given $FS_{shear}$, the following procedure should be adopted:

1. Let $c$ and $\phi$ be the cohesion and the angle of friction of soil, and let $FS_{shear}$ be the required factor of safety with respect to shear failure. So, the developed cohesion and the angle of friction can be given as

$$c_d = \frac{c}{FS_{shear}} \tag{3.41}$$

$$\phi_d = \tan^{-1}\left(\frac{\tan \phi}{FS_{shear}}\right) \tag{3.42}$$

2. The gross allowable bearing capacity can now be calculated according to Eqs. (3.3), (3.7), (3.8) or the general bearing capacity equation [Eq. (3.16)] using $c_d$ and $\phi_d$ as the shear strength parameters of the soil. For example, the gross allowable bearing capacity of a continuous foundation according to Terzaghi's equation can be written as

$$q_{all} = c_d N_c + q N_q + \tfrac{1}{2}\gamma B N_\gamma \tag{3.43}$$

where $N_c$, $N_q$, and $N_\gamma$ = bearing capacity factors for friction angle, $\phi_d$

3. The net allowable bearing capacity is thus

$$q_{net(all)} = q_{all} - q = c_d N_c + q(N_q - 1) + \tfrac{1}{2}\gamma B N_\gamma \tag{3.44}$$

## Example 3.4

Refer to Example Problem 3.1. Determine the net allowable load for the foundation using the definition of factor of safety given by Eq. (3.40). Use $FS = 4$.

### Solution

From Example Problem 3.1.

$$q_u = 535 \text{ kN/m}^2$$

$$q = (1)(17.8) = 17.8 \text{ kN/m}^2$$

$$q_{net(all)} = \frac{q_u - q}{FS} = \frac{535 - 17.8}{4} = 129.3 \text{ kN/m}^2$$

Hence

$$Q_{net(all)} = (129.3)(1.5)(1.5) \approx \underline{291 \text{ kN}}$$

## Example
## 3.5

Refer to Example Problem 3.1. Determine the net allowable load for the foundation using a $FS_{shear} = 1.5$.

### Solution

Given $c = 15.2 \text{ kN/m}^2$ and $\phi = 20°$,

$$c_d = \frac{c}{FS_{shear}} = \frac{15.2}{1.5} = 10.13 \text{ kN/m}^2$$

$$\phi_d = \tan^{-1}\left[\frac{\tan\phi}{FS_{shear}}\right] = \tan^{-1}\left[\frac{\tan 20}{1.5}\right] = 13.64°$$

$$q_{all(net)} = 1.3c_d N_c + q(N_q - 1) + 0.4\gamma B N_\gamma$$

For $\phi = 13.64°$, the values of the bearing capacity factors are (Figure 3.4):

$$N_\gamma = 2$$
$$N_q = 3.8$$
$$N_c = 12$$

So

$$q_{all(net)} = 1.3(10.13)(12) + (17.8)(3.8 - 1) + (0.4)(17.8)(1.5)(2)$$
$$= 229.23 \text{ kN/m}^2$$

So

$$Q_{all(net)} = (229.23)(1.5)(1.5) = \underline{515.8 \text{ kN}}$$

*Note*: There appears to be a large discrepancy between the results of Example Problems 3.4 (or 3.1) and 3.5. By trial and error, it can be shown that when $FS_{shear}$ is about 2, the results will be approximately equal.

## 3.7
## Eccentrically Loaded Foundations

In several instances, foundations are subjected to moments in addition to the vertical load, as shown in Figure 3.8a. In such cases, the distribution of pressure by the foundation on the soil is not uniform. The distribution of pressure can be given by

$$q_{max} = \frac{Q}{BL} + \frac{6M}{B^2L} \tag{3.45}$$

and

$$q_{min} = \frac{Q}{BL} - \frac{6M}{B^2L} \tag{3.46}$$

where $Q$ = total vertical load
$M$ = moment on the foundation

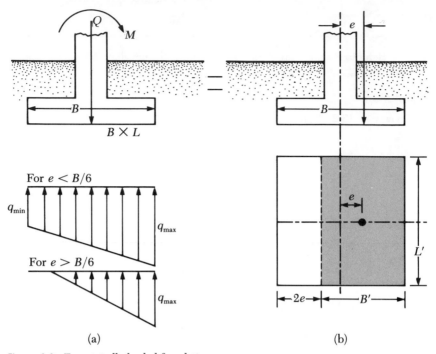

**Figure 3.8** Eccentrically loaded foundations

The factor of safety of such types of loading against bearing capacity failure can be evaluated by using the procedure suggested by Meyerhof (1953), which is generally referred to as the *effective area* method. The following is Meyerhof's step-by-step procedure for determination of the ultimate load that the soil can support and the factor of safety against bearing capacity failure.

1. Figure 3.8b shows a force system equivalent to that shown in Figure 3.8a. The distance $e$ is the eccentricity, or

$$e = \frac{M}{Q} \tag{3.47}$$

Substituting Eq. (3.47) in Eqs. (3.45) and (3.46) gives

$$q_{max} = \frac{Q}{BL}\left(1 + \frac{6e}{B}\right) \tag{3.48a}$$

and

$$q_{min} = \frac{Q}{BL}\left(1 - \frac{6e}{B}\right) \tag{3.48b}$$

Note that, in these equations, when the eccentricity, $e$, becomes equal to $B/6$, $q_{min}$ is equal to zero. For $e > B/6$, $q_{min}$ will be negative, which means that tension will develop. Because soil cannot take any tension, there will be a separation between the foundation and the soil underlying it. The nature of the pressure distribution on the soil will be as shown in Figure 3.8a. The value of $q_{max}$ can be given by the expression

$$q_{max} = \frac{4Q}{3L(B - 2e)} \tag{3.49}$$

2. Determine the effective dimensions of the foundation as

$B' = $ effective width $= B - 2e$

$L' = $ effective length $= L$

Note that, if the eccentricity is in the direction of the length of the foundation, the value of $L'$ would be equal to $L - 2e$. The value of $B'$ would be equal to $B$. The smaller of the two dimensions (that is, $L'$ and $B'$) is the effective width of the foundation.

3. Use Eq. (3.16) for the ultimate bearing capacity as

$$q'_u = cN_cF_{cs}F_{cd}F_{ci} + qN_qF_{qs}F_{qd}F_{qi} + \tfrac{1}{2}\gamma B'N_\gamma F_{\gamma s}F_{\gamma d}F_{\gamma i} \tag{3.50}$$

For the evaluation of $F_{cs}$, $F_{qs}$, $F_{\gamma s}$, $F_{ci}$, $F_{qi}$, and $F_{\gamma i}$, Equations (3.20) to (3.22) and Eqs. (3.29) to (3.33) have to be used with *effective length* and *effective width* dimensions in place of $L$ and $B$, respectively.

For determination of $F_{cd}$, $F_{qd}$, and $F_{\gamma d}$, use Equations (3.23) to (3.28) (*do not replace $B$ with $B'$*).

4. The total ultimate load that the foundation can sustain is

$$Q_{ult} = q'_u (B')(L') \tag{3.51}$$

5. The factor of safety against bearing capacity failure is given as

$$FS = \frac{Q_{ult}}{Q} \tag{3.52}$$

As we can see, eccentricity tends to decrease the load-bearing capacity of a foundation. In such cases, it is probably advantageous to place the foundation columns off center, as shown in Figure 3.9. This, in effect, produces a centrally loaded foundation with uniformly distributed pressure.

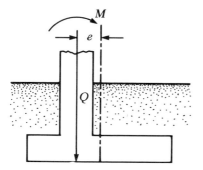

**Figure 3.9** Foundation of columns with off-center loading

## 3.8

### Some Special Cases of Ultimate Bearing Capacity

The bearing capacity equations presented in the preceding sections involve cases in which the soil supporting the foundation is homogeneous and extends to a considerable depth. It was assumed that cohesion, angle of friction, and unit weight of soil remained constant for the bearing capacity analysis. However, in practice, layered soil profiles are often encountered. In such instances, the failure surface at ultimate load may extend through two or more soil layers. Determination of ultimate bearing capacity in layered soils can be made in only a limited number of cases. Two of these cases will be discussed in this section.

### Foundations on Layered Clay

Reddy and Srinivasan (1967) have derived the equation for the bearing capacity of foundations on layered clay soils, as shown in Figure 3.10a. For undrained loading ($\phi = 0$ condition), let $c_1$ and $c_2$ be the shear strength of the upper and lower clay layers, respectively. In such a case, the ultimate bearing capacity of a foundation can be given as [similar to Eq. (3.35)]

$$q_u = c_1 N_c F_{cs} F_{cd} + q \tag{3.53}$$

The relationships for $F_{cs}$ and $F_{cd}$ are the same as given in Eqs. (3.20), (3.23), and (3.26). For layered soils, the value of the bearing capacity factor, $N_c$, is not a constant. It is a function of $c_2/c_1$ and $z/B$ (*note: z =* depth measured from the bottom of the foundation to the interface of the two clay layers). The variation of $N_c$ is given in Figure 3.10b. It can be seen from this figure that, if the lower layer of clay is softer than the top one (that is, $c_2/c_1 < 1$), the value of the bearing capacity factor ($N_c$) is lower than when the soil is not layered (that is, when $c_2/c_1 = 1$). This means that the ultimate bearing capacity is reduced by the presence of a softer clay layer below the top layer.

### Foundations on Sand Overlying Soft Clay

In some cases, foundations are constructed on sand layers that overlie soft clay soils. If the thickness of the sand layer under the foundation is relatively small, the failure surface may extend into the soft clay layer. This is shown in the left half of Figure 3.11. However, if the sand layer under the foundation is large, the failure surface will lie entirely in the sand layer, as shown in the right half of Figure 3.11. According to Meyerhof (1974), in this case the ultimate bearing capacity of a continuous foundation can be given by

$$q_u = cN_c + \gamma H^2 \left( 1 + \frac{2D_f}{H} \right) K_s \frac{\tan \phi}{B} + \gamma D_f \tag{3.54}$$

with a maximum of

$$q_u = \tfrac{1}{2} \gamma B N_\gamma + \gamma D_f N_q \tag{3.55}$$

where $\phi$ = angle of friction of top sand layer
$\gamma$ = unit weight of sand
$K_s$ = punching shear resistance coefficient

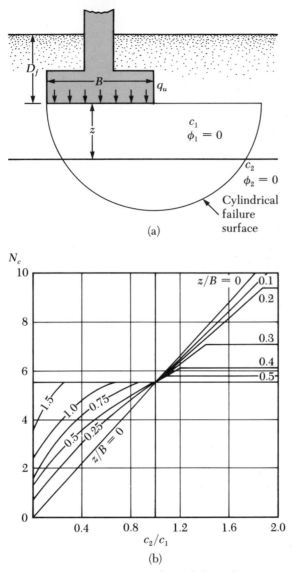

**Figure 3.10** Bearing capacity on layered clay soils—$\phi = 0$ (Figure 3.10b redrawn after Reddy and Srinivasan, 1967)

$N_\gamma$ and $N_q$ correspond to the angle of friction, $\phi$, for sand (Table 3.2). *Note*: for $\phi = 0$ condition, $N_c = 5.14$ as determined from Table 3.3.

For rectangular foundations

$$q_u = \left(1 + 0.2\frac{B}{L}\right) cN_c + \left(1 + \frac{B}{L}\right) \gamma H^2\left(1 + \frac{2D_f}{H}\right) K_s \frac{\tan\phi}{B} + \gamma D_f$$

$$(3.56)$$

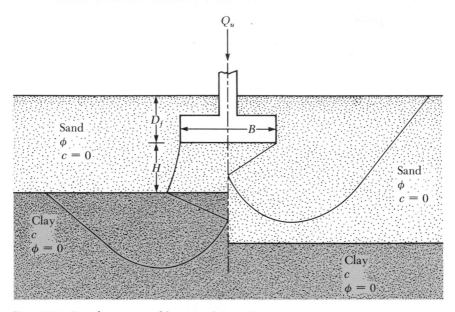

**Figure 3.11** Foundation on sand layer overlying soft clay

with a maximum of

$$q_u = \frac{1}{2}\left(1 - 0.4\frac{B}{L}\right)\gamma BN_\gamma + \gamma D_f N_q \tag{3.57}$$

The variation of the punching shear resistance factor, $K_s$, is given in Table 3.3. At this point, it needs to be pointed out that Eqs. (3.55) and (3.57) are estimates of the values of $q_u$ for continuous and rectangular foundations, respectively, in the upper sand layer. This corresponds to the condition shown in the right half of Figure 3.11.

**Table 3.3** Values of $K_s$ for Use in Eqs. (3.54) and (3.56)

| Friction angle of sand, $\phi$ (deg) | $K_s$ |
|---|---|
| 20 | 1.89 |
| 25 | 2.22 |
| 30 | 3.06 |
| 35 | 4.45 |
| 40 | 6.95 |
| 45 | 11.12 |
| 50 | 19.15 |

## Example
## 3.6

Refer to Figure 3.10a. A foundation 1.5 m × 1 m is located at a depth $(D_f)$ of 1 m in a clay. A softer clay layer is located at a depth $(z)$ of 1 m measured from the bottom of the foundation. Given:

For top clay layer: undrained shear strength = 120 kN/m²
unit weight = 16.8 kN/m³
For bottom clay layer: undrained shear strength = 48 kN/m²
unit weight = 16.2 kN/m³

Determine the gross allowable load for the foundation with a factor of safety of 4.

### Solution

From Eq. (3.53)

$$q_u = c_1 N_c F_{cs} F_{cd} + q$$

$$c_1 = 120 \text{ kN/m}^2$$

$$q = \gamma D_f = (16.8)(1) = 16.8 \text{ kN/m}^2$$

$$\frac{c_2}{c_1} = \frac{48}{120} = 0.4; \frac{z}{B} = \frac{1}{1} = 1$$

From Figure 3.10b, for $z/B = 1$ and $c_2/c_1 = 0.4$, the value of $N_c$ is equal to 4.6.

$$F_{cs} = 1 + \left(\frac{B}{L}\right)\left(\frac{N_q}{N_c}\right) = 1 + \left(\frac{1}{1.5}\right)\left(\frac{1}{4.6}\right) = 1.145$$

$$F_{cd} = 1 + 0.4\frac{D_f}{B} = 1 + 0.4\left(\frac{1}{1}\right) = 1.4$$

Thus

$$q_u = (120)(4.6)(1.145)(1.4) + 16.8 = 884.8 + 16.8 = 901.6 \text{ kN/m}^2$$

So

$$q_{all} = \frac{q_u}{FS} = \frac{901.6}{4} = 225.4 \text{ kN/m}^2$$

Total allowable load $= (q_{all})(B \times L) = (225.4)(1 \times 1.5) = \underline{338.1 \text{ kN}}$

# Settlement of Shallow Foundations

## 3.9

Types of Foundation Settlement

Foundation settlement under load can be classified according to two major types: *immediate*, or *elastic*, *settlement*, $S_e$, and *consolidation settlement*, $S_c$. Elastic settlement of a foundation takes place during or immediately after the construction of the structure. Consolidation settlement is time dependent and takes place as the result of extrusion of the pore water from the void spaces of saturated clayey soils. The total settlement of a foundation is the sum of the elastic settlement and the consolidation settlement.

Consolidation settlement comprises two phases: *primary* consolidation settlement and *secondary* consolidation settlement. The fundamentals of primary consolidation settlement have been explained in detail in Section 1.14. Secondary consolidation settlement occurs after completion of the primary consolidation that is caused by slippage and reorientation of soil particles under sustained load. Primary consolidation settlement is more significant than secondary settlement in inorganic clays and silty clay soils. However, in organic soils, secondary consolidation settlement is more significant.

The settlement of foundations discussed in Section 3.2 for bearing capacity tests was primarily the elastic type. The procedure for calculating each type of foundation settlement is discussed in more detail in the following sections.

## 3.10

### Elastic Settlement

Figure 3.12 shows a shallow foundation subjected to a net force per unit area equal to $q_o$. Let the Poisson's ratio and the Young's modulus of the soil supporting it be $\mu_s$ and $E_s$, respectively. Theoretically, if $D_f = 0$, $H = \infty$ and the foundation is perfectly flexible, the elastic settlement can be expressed as (Harr, 1966)

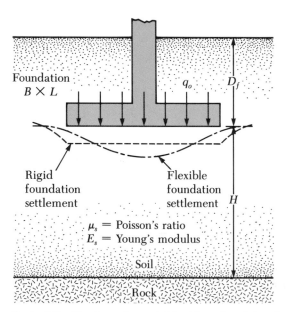

**Figure 3.12**  Elastic settlement of flexible and rigid foundations

$$S_e = \frac{Bq_o}{E_s} (1 - \mu_s^2) \frac{\alpha}{2} \quad \text{(corner of the flexible foundation)} \tag{3.58}$$

$$S_e = \frac{Bq_o}{E_s} (1 - \mu_s^2)\alpha \quad \text{(center of the flexible foundation)} \tag{3.59}$$

where $\alpha = \dfrac{1}{\pi}\left[ ln\left(\dfrac{\sqrt{1 + m^2} + m}{\sqrt{1 + m^2} - m}\right) + mln\left(\dfrac{\sqrt{1 + m^2} + 1}{\sqrt{1 + m^2} - 1}\right)\right]$ (3.60)

$m = B/L$ (3.61)
$B$ = width of foundation
$L$ = length of foundation

The values of $\alpha$ for various length-to-width $(L/B)$ ratios are shown in Figure 3.13. The average elastic settlement for a flexible foundation can also be expressed as

$$S_e = \frac{Bq_o}{E_s} (1 - \mu_s^2)\alpha_{av} \quad \text{(average for flexible foundation)} \tag{3.62}$$

Figure 3.13 also shows the values of $\alpha_{av}$ for various types of foundation.
However, if the foundation shown in Figure 3.12 is rigid, the elastic settlement will be modified and can be expressed as

$$S_e = \frac{Bq_o}{E_s}(1 - \mu_s^2)\alpha_r \quad \text{(rigid foundation)} \tag{3.63}$$

The values of $\alpha_r$ for various types of foundation are given in Figure 3.13.
The preceding equations for elastic settlement have been obtained by integrating the strain at any given depth below the foundations for limits of

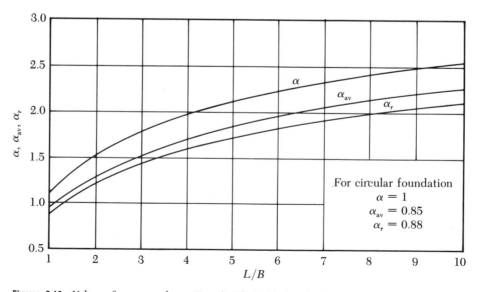

**Figure 3.13** Values of $\alpha$, $\alpha_{av}$, and $\alpha_r$—Eqs. (3.58), (3.59), (3.62), (3.63)

$z = 0$ to $z = \infty$. If an incompressible layer of rock is located at a limited depth, the actual settlement may be less than that calculated by the preceding equations. However, if the depth $H$ in Figure 3.12 is greater than about $2B$ to $3B$, the actual settlement would not change considerably. Also note that the deeper the embedment, $D_f$, the less is the total elastic settlement.

## 3.11
Elastic Settlement of Foundations on Saturated Clay

Janbu, Bjerrum, and Kjaernsli (1956) proposed an equation for evaluation of the average elastic settlement of flexible foundations on saturated clay soils (Poisson's ratio, $\mu_s = 0.5$). Referring to Figure 3.14 for notations, this equation can be written as

$$S_e = A_1 A_2 \frac{q_o B}{E_s} \tag{3.64}$$

where $A_1$ is a function of $H/B$ and $L/B$, and $A_2$ is a function of $D_f/B$.

Christian and Carrier (1978) have modified the values of $A_1$ and $A_2$ to some extent, and these are presented in Figure 3.14.

## 3.12
Elastic Settlement of Sandy Soil:
Use of Strain Influence Factor

Elastic settlement of granular soils can also be evaluated by use of a semi-empirical *strain influence factor* (Figure 3.15) proposed by Schmertmann and

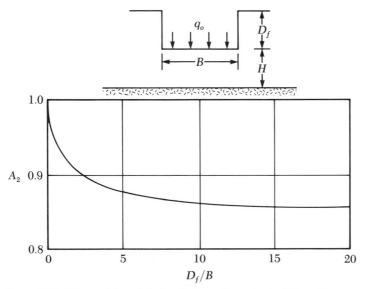

**Figure 3.14**  Values of $A_1$ and $A_2$ for elastic settlement calculation—Eq. (3.64) (after Christian and Carrier, 1978)

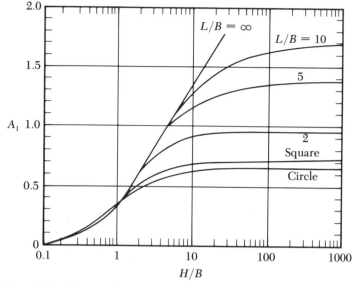

**Figure 3.14** (Continued)

Hartman (1978). According to this method, the elastic settlement can be given by the equation

$$S_e = C_1 C_2 (\bar{q} - q) \sum_0^{2B} \frac{I_z}{E_s} \Delta z \qquad (3.65)$$

where $I_z$ = strain influence factor
$\quad C_1$ = a correction factor for the depth of foundation
$\qquad$ embedment = $1 - 0.5[q/(\bar{q} - q)]$
$\quad C_2$ = a correction factor to account for creep in soil
$\qquad$ = $1 + 0.2 \log$ (time in years/0.1)
$\quad \bar{q}$ = stress at the level of the foundation

The variation of the strain influence factor with depth below the foundation is shown in Figure 3.15a. Note that, for square or circular foundations,

$\quad I_z = 0.1$ at $z = 0$
$\quad I_z = 0.5$ at $z = 0.5B$
$\quad I_z = 0$ at $z = 2B$

Similarly, for foundations with $L/B \geq 10$,

$\quad I_z = 0.2$ at $z = 0$
$\quad I_z = 0.5$ at $z = B$
$\quad I_z = 0$ at $z = 4B$

where $B$ = width of the foundation and $L$ = length of the foundation. For values of $L/B$ between 1 and 10, necessary interpolations can be made.

In order to use Eq. (3.65), one first needs to evaluate the approximate variation of Young's modulus with depth (Figure 3.15b). This can be done by

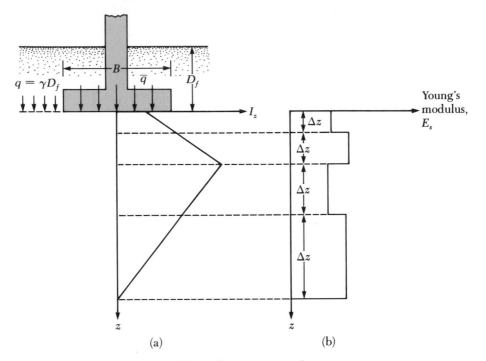

**Figure 3.15** Elastic settlement calculation by using strain influence factor

using the standard penetration numbers or cone penetration resistances (Chapter 2). Then the soil layer can be divided into several layers up to a depth of $z = 2B$ (or $4B$ as the case may be), and the immediate settlement of each layer can be estimated. The sum of the settlement of all layers is equal to $S_e$. The principle of settlement calculation using this procedure is demonstrated in Example Problem 3.7.

## 3.13

### Range of Elastic Parameters for Soils

Sections 3.10– 3.12 presented the equations for the calculation of elastic settlement of foundations. These equations contain the elastic parameters, such as $E_s$ and $\mu_s$. If the laboratory test results for these parameters are not available, certain realistic assumptions have to be made. Table 3.4 shows the approximate range of the elastic parameters for various soils.

Several investigators have correlated the values of Young's modulus with the standard penetration number $(N)$ and the cone penetration resistance $(q_c)$. A list of these correlations has been compiled by Mitchell and Gardner (1975). Schmertmann (1970) has indicated that Young's modulus of sand can be given by the following correlations.

**Table 3.4** Elastic Parameters of Various Soils

| Type of soil | Young's modulus, $E_s$ | | Poisson's ratio, $\mu_s$ |
|---|---|---|---|
| | $MN/m^2$ | $lb/in.^2$ | |
| Loose sand | 10.35– 24.15 | 1,500– 3,500 | 0.20–0.40 |
| Medium dense sand | 17.25– 27.60 | 2,500– 4,000 | 0.25–0.40 |
| Dense sand | 34.50– 55.20 | 5,000– 8,000 | 0.30–0.45 |
| Silty sand | 10.35– 17.25 | 1,500– 2,500 | 0.20–0.40 |
| Sand and gravel | 69.00–172.50 | 10,000–25,000 | 0.15–0.35 |
| Soft clay | 2.07– 5.18 | 300– 750 | |
| Medium clay | 5.18– 10.35 | 750– 1,500 | 0.20–0.50 |
| Stiff clay | 10.35– 24.15 | 1,500– 3,500 | |

$$E_s(kN/m^2) = 766N \tag{3.66}$$

$$E_s = 2q_c \tag{3.67}$$

where $N$ = standard penetration number

$\quad q_c$ = static cone penetration resistance

*Note*: Any consistent set of units can be used in Eq. (3.67).

The Young's modulus of normally consolidated clays can be estimated as

$$E_s = 250c \text{ to } 500c \tag{3.68a}$$

For overconsolidated clays

$$E_s = 750c \text{ to } 1000c \tag{3.68b}$$

where $c$ = undrained cohesion of clayey soil

## Example 3.7

Figure 3.16a shows a shallow foundation on a deposit of sandy soil that is 3 m × 3 m in plan. The actual variation of the values of Young's modulus with depth determined by using the standard penetration numbers and Eq. (3.66) are also shown in Figure 3.16a. Using the strain influence factor method, estimate the elastic settlement of the foundation after five years of construction.

### Solution

By observing the actual variation of Young's modulus with depth, one can plot an estimated idealized form of the variation of $E_s$, as shown in Figure 3.16a. Figure 3.16b shows the plot of the strain influence factor. The following table can now be prepared.

| Depth (m) | $\Delta z$ (m) | $E_s$ (kN/m²) | Average $I_z$ | $\dfrac{I_z}{E_s} \cdot \Delta z$ (m³/kN) |
|---|---|---|---|---|
| 0–1 | 1 | 8,000 | 0.233 | $0.291 \times 10^{-4}$ |
| 1.0–1.5 | 0.5 | 10,000 | 0.433 | $0.217 \times 10^{-4}$ |
| 1.5–4 | 2.5 | 10,000 | 0.361 | $0.903 \times 10^{-4}$ |
| 4.0–6 | 2 | 16,000 | 0.111 | $0.139 \times 10^{-4}$ |
| | | | | $\Sigma = 1.55 \times 10^{-4}$ |

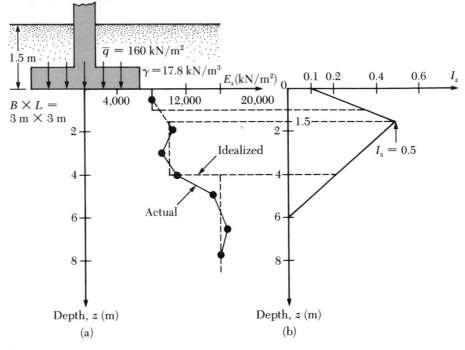

**Figure 3.16**

$$C_1 = 1 - 0.5 \left( \frac{q}{\bar{q} - q} \right) = 1 - 0.5 \left[ \frac{17.8 \times 1.5}{160 - (17.8 \times 1.5)} \right] = 0.9$$

$$C_2 = 1 + 0.2 \log \left( \frac{5}{0.1} \right) = 1.34$$

Hence

$$S_e = C_1 \cdot C_2 (\bar{q} - q) \sum_0^{2B} \frac{I_z}{E_s} \cdot \Delta z$$

$$= (0.9)(1.34)[160 - (17.8 \times 1.5)](1.55 \times 10^{-4})$$

$$= 249.2 \times 10^{-4} \text{ m} \approx 24.9 \text{ mm}$$

## 3.14
## Elastic Settlement of Eccentrically Loaded Foundations

The elastic settlement calculation procedure described in Sections 3.10, 3.11, and 3.12 relate to the settlement of centrally loaded foundations. An eccentrically loaded foundation will undergo vertical settlement and rotation as shown in Figure 3.17. Prakash (1981) has suggested a procedure for determining the settlement and rotation of a foundation under such loading conditions. This procedure is as follows:

**1.** Let the applied total load on the foundation ($Q$) and the load eccentricity ($e$) be known, and let it be required to determine the settlements $S_{e(1)}$ and $S_{e(2)}$ and the rotation angle, $t$ (see Figure 3.17 for notations).

**2.** The ultimate load, $Q_{ult(e)}$, that the foundation can sustain can be evaluated by using Eq. (3.51) [Section 3.7; note the change of notation from $Q_{ult}$ to $Q_{ult(e)}$].

**3.** Determine the factor of safety for the eccentrically loaded foundation as

$$FS = \frac{Q_{ult(e)}}{Q} = F_1 \tag{3.69}$$

**4.** Determine the ultimate load $Q_{ult(e=0)}$ for the same foundation with eccentricity $e = 0$ [centrally loaded foundation; Eq. (3.16)].

**5.** Determine:

$$\frac{Q_{ult(e=0)}}{F_1} = Q_{(e=0)} \tag{3.70}$$

Note that $Q_{(e=0)}$ is the allowable load for the foundation with a factor of safety $FS = F_1$ for central loading condition.

**6.** For the load $Q_{(e=0)}$ on the foundation, estimate the settlement by using the techniques presented in Sections 3.10–3.12. Let the settlement determined by any one of the methods be equal to $S_{e(e=0)}$.

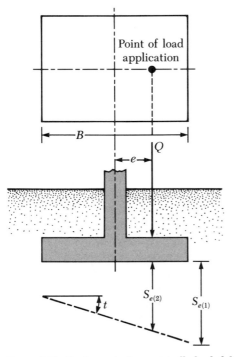

**Figure 3.17**   Settlement of eccentrically loaded foundation

**7.** Now use the following equations to calculate $S_{e(1)}$, $S_{e(2)}$, and $t$:

$$S_{e(1)} = S_{e(e=0)}\left[1 + 2.31\left(\frac{e}{B}\right) - 22.61\left(\frac{e}{B}\right)^2 + 31.54\left(\frac{e}{B}\right)^3\right] \qquad (3.71)$$

↑
Step 6

$$S_{e(2)} = S_{e(e=0)}\left[1 - 1.63\left(\frac{e}{B}\right) - 2.63\left(\frac{e}{B}\right)^2 + 5.83\left(\frac{e}{B}\right)^3\right] \qquad (3.72)$$

$$t = \sin^{-1}\left[\frac{S_{e(1)} - S_{e(2)}}{\dfrac{B}{2} - e}\right] \qquad (3.73)$$

Equations (3.71) and (3.72) are valid up to a limit of $e/B \le 0.4$. The use of this procedure is illustrated in the following Example Problem.

## Example 3.8

A square foundation is shown in Figure 3.18. It is subjected to a load of 180 kN and a moment of 27 kN-m. Determine the settlement of the foundation [$S_{e(1)}$, $S_{e(2)}$, and $t$] according to the method presented in Section 3.14.

### Solution
Step 1

Given $Q = 180$ kN and moment $= M = 27$ kN-m, load eccentricity $= e = M/Q = 27/180 = 0.15$ m.

Step 2

Determination of $Q_{ult(e)}$: From Eq. (3.50), with $c = 0$

$$q_u' = qN_qF_{qs}F_{qd} + \tfrac{1}{2}\gamma B'N_\gamma F_{\gamma s}F_{\gamma d}$$
$$q = (0.7)(18) = 12.6 \text{ kN/m}^2$$

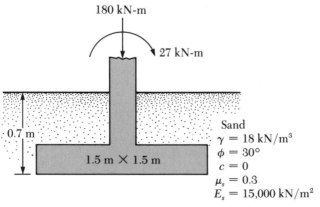

180 kN-m

27 kN-m

0.7 m

1.5 m × 1.5 m

Sand
$\gamma = 18$ kN/m³
$\phi = 30°$
$c = 0$
$\mu_s = 0.3$
$E_s = 15{,}000$ kN/m²

**Figure 3.18**

For $\phi = 30°$, from Table 3.2, $N_q = 18.4$ and $N_\gamma = 22.4$.

$$B' = 1.5 - 2(0.15) = 1.2 \text{ m}$$

$$L' = 1.5 \text{ m}$$

$$F_{qs} = 1 + \frac{B'}{L'} \tan \phi = 1 + \left(\frac{1.2}{1.5}\right) \tan 30° = 1.462$$

$$F_{dq} = 1 + 2 \tan \phi (1 - \sin \phi)^2 \frac{D_f}{B} = 1 - \frac{(0.289)(0.7)}{1.5} = 1.135$$

$$F_{\gamma s} = 1 - 0.4 \left(\frac{B'}{L'}\right) = 1 - 0.4 \left(\frac{1.2}{1.5}\right) = 0.68$$

$$F_{\gamma d} = 1$$

So

$$q'_u = (12.6)(18.4)(1.462)(1.135) + \tfrac{1}{2}(18)(1.2)(22.4)(0.68)(1)$$
$$= 384.3 + 164.50 = 548.8 \text{ kN/m}^2$$

Hence

$$Q_{ult(e)} = B'L'(q'_u) = (1.2)(1.5)(548.8) = 988 \text{ kN}$$

Step 3

Determination of factor of safety, $F_1$: From Eq. (3.69)

$$\text{Factor of safety} = F_1 = \frac{Q_{ult(e)}}{Q} = \frac{988 \text{ kN}}{180} = 5.49$$

Step 4

Determination of $Q_{ult(e=0)}$: From Eq. (3.16), because $c = 0$,

$$q_u = q N_q F_{qs} F_{qd} + \tfrac{1}{2} \gamma B N_\gamma F_{\gamma s} F_{\gamma d}$$
$$q = 12.6 \text{ kN/m}^2$$

For $\phi = 30°$ (from Table 3.2), $N_q = 18.4$ and $N_\gamma = 22.4$

$$F_{qs} = 1 + \frac{B}{L} \tan \phi = 1 + \left(\frac{1.5}{1.5}\right) \tan 30 = 1.577$$

$$F_{dq} = 1 + 2 \tan \phi (1 - \sin \phi)^2 \frac{D_f}{B} = 1 + \frac{(0.289)(0.7)}{1.5} = 1.135$$

$$F_{\gamma s} = 1 - 0.4 \left(\frac{B}{L}\right) = 1 - 0.4 \left(\frac{1.5}{1.5}\right) = 0.6$$

$$F_{\gamma d} = 1$$

$$q_u = (12.6)(18.4)(1.577)(1.135) + (\tfrac{1}{2})(18)(1.5)(22.4)(0.6)(1)$$
$$= 414.97 + 181.44 = 596.41 \text{ kN/m}^2$$

So

$$Q_{ult(0)} = (596.41)(1.5 \times 1.5) = 1342 \text{ kN}$$

Step 5

Determination of $Q_{(e=0)}$: From Eq. (3.70)

$$Q_{(e=0)} = \frac{Q_{\text{ult}(e=0)}}{F_1} = \frac{1342}{5.49} = 244.4 \text{ kN/m}^2$$

Step 6

Determination of $S_{e(e=0)}$: From Eq. (3.63)

$$S_{e(e=0)} = \frac{B(Q_{e=0})}{E_s(B \times L)}(1 - \mu_s^2)\alpha_r$$

For $L/B = 1$, $\alpha_r \approx 0.82$ (Figure 3.13). Given $\mu_s = 0.3$ and $E_s = 15,000 \text{ kN/m}^2$,

$$S_{e(e=0)} = \frac{(1.5)(244.4)}{(15,000)(1.5 \times 1.5)}(1 - 0.3^2)0.82 = 0.0081 \text{ m} = 8.1 \text{ mm}$$

Step 7

Determination of $S_{e(1)}$, $S_{e(2)}$, and $t$: From Eqs. (3.71), (3.72), and (3.73)

$$S_{e(1)} = 8.1[1 + (2.32)(90.1) - 22.61(0.1)^2 + (31.54)(0.1)^3] = 8.39 \text{ mm}$$

$$S_{e(2)} = 8.1[1 - (1.63)(0.1) - 2.63(0.1)^2 + 5.83(0.1)^3] = 6.62 \text{ mm}$$

$$t = \sin^{-1}\left[\frac{S_{e(1)} - S_{e(2)}}{\dfrac{B}{2} - e}\right] = \sin^{-1}\left[\frac{8.39 - 6.62}{750 - 150}\right] = 0.17°$$

## 3.15

## Consolidation Settlement

As mentioned before, consolidation settlement is time dependent, and it occurs in saturated clayey soils when they are subjected to increased load caused by foundation construction (Figure 3.19). The one-dimensional consolidation settlement equations are given in Chapter 1 as

$$S_c = \frac{C_c H_c}{1 + e_o}\log\frac{p_o + \Delta p_{av}}{p_o} \quad \text{(for normally consolidated clays)} \quad (1.60)$$

$$S_c = \frac{C_s H_c}{1 + e_o}\log\frac{p_o + \Delta p_{av}}{p_o} \quad \begin{array}{l}\text{(for overconsolidated clays} \\ \text{with } p_o + \Delta p_{av} < p_c)\end{array} \quad (1.61)$$

$$S_c = \frac{C_s H_c}{1 + e_o}\log\frac{p_c}{p_o} + \frac{C_c H_c}{1 + e_o}\log\frac{p_o + \Delta p_{av}}{p_c}$$

$$\begin{array}{l}\text{(for overconsolidated clays} \\ \text{with } p_o < p_c < p_o + \Delta p_{av})\end{array} \quad (1.63)$$

where $p_o$ = average effective pressure on the clay layer before the construction of the foundation

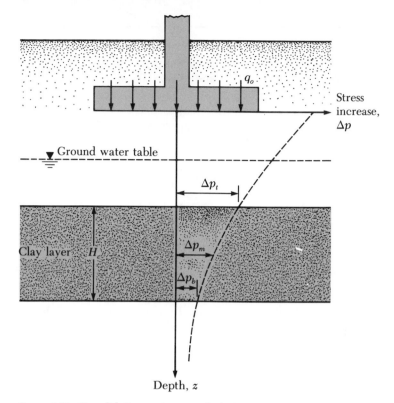

**Figure 3.19** Consolidation settlement calculation

$\Delta p_{av}$ = average increase of pressure on the clay layer caused by the foundation construction

$p_c$ = preconsolidation pressure

$e_o$ = initial void ratio of the clay layer

$C_c$ = compression index

$C_s$ = swelling index

$H_c$ = thickness of the clay layer

The procedures for the determination of the compression and swelling indexes have been discussed in Chapter 1.

Note that the increase of pressure, $\Delta p$, on the clay layer is not constant with depth. The magnitude of $\Delta p$ will decrease with the increase of depth measured from the bottom of the foundation. However, the average increase of pressure can be approximated by the equation

$$\Delta p_{av} = \tfrac{1}{6}(\Delta p_t + 4\,\Delta p_m + \Delta p_b)  \tag{3.74}$$

where $\Delta p_t$, $\Delta p_m$, and $\Delta p_b$ are the pressure increases at the *top*, *middle*, and *bottom* of the clay layer that are caused by the foundation construction.

The method of determination of the pressure increase caused by various types of foundation load is discussed in Section 3.17.

**3.16**

Skempton-Bjerrum Modification for Consolidation Settlement

The consolidation settlement calculation presented in the preceding section is based on Eqs. (1.60), (1.61), and (1.63). These equations, as shown in Chapter 1, are based on one-dimensional laboratory consolidation tests. The underlying assumption for these equations is that the increase of pore water pressure $(\Delta u)$ immediately after the load application is equal to the increase of stress $(\Delta p)$ at any depth. For this case

$$S_{c(oed)} = \int \frac{\Delta e}{1 + e_o} \, dz = \int m_v \, \Delta p_{(1)} \, dz \qquad (3.75)$$

where $S_{c(oed)}$ = consolidation settlement calculated by using Eqs. (1.60), (1.61), and (1.63)

$\Delta p_{(1)}$ = vertical stress increase (note the change of notation from $\Delta p$ as given in Section 3.15)

$m_v$ = volume coefficient of compressibility (see Chapter 1)

In the field, however, when load is applied over a limited area on the ground surface, this assumption will not be correct. Consider the case of a circular foundation on a clay layer as shown in Figure 3.20. The vertical and the horizontal stress increase at a point in the clay layer immediately below the center of the foundation are $\Delta p_{(1)}$ and $\Delta p_{(3)}$, respectively. For a saturated clay, the pore water pressure increase at that depth can be given as (Chapter 1)

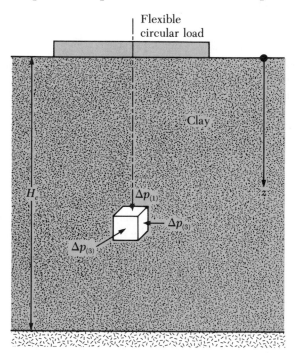

**Figure 3.20**  Circular foundation on a clay layer

$$\Delta u = \Delta p_{(3)} + A[\Delta p_{(1)} - \Delta p_{(3)}] \tag{3.76}$$

where $A$ = pore water pressure parameter (see Chapter 1). For this case, one can write that

$$S_c = \int m_v \, \Delta u \, dz = \int (m_v)\{\Delta p_{(3)} + A[\Delta p_{(1)} - \Delta p_{(3)}]\} \, dz \tag{3.77}$$

Combining Eqs. (3.75) and (3.77)

$$K_{cir} = \frac{S_c}{S_{c(oed)}} = \frac{\displaystyle\int_0^{H_c} m_v \, \Delta u \, dz}{\displaystyle\int_0^{H_c} m_v \, \Delta p_{(1)} \, dz} = A + (1 - A) \left[ \frac{\displaystyle\int_0^{H_c} \Delta p_{(3)} \, dz}{\displaystyle\int_0^{H_c} \Delta p_{(1)} \, dz} \right] \tag{3.78}$$

where $K_{cir}$ = settlement ratio for circular foundations

The settlement ratio for a continuous foundation ($K_{str}$) can be determined in a manner similar to that for a circular foundation.) The variation of $K_{cir}$ and $K_{str}$ with $A$ and $H_c/B$ is given in Figure 3.21. (*Note:* $B$ = diameter of a circular foundation, and $B$ = width of a continuous foundation.)

Following is the procedure for determination of consolidation settlement according to Skempton and Bjerrum (1957).

**1.** Determine the consolidation settlement using the procedure outlined in Section 3.15. This is $S_{c(oed)}$. (Note the change of notation from $S_c$.)

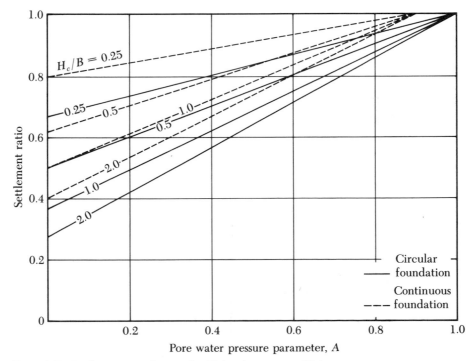

**Figure 3.21** Settlement ratio for circular ($K_{cir}$) and continuous ($K_{str}$) foundations

**2.** Determine the pore water pressure parameter, $A$.
**3.** Determine $H_c/B$.
**4.** Obtain the settlement ratio—in this case, from Figure 3.20.
**5.** Calculate the actual consolidation settlement as

$$S_c = S_{c(oed)} \times \text{settlement ratio} \tag{3.79}$$

$$\uparrow$$
$$\text{Step 1}$$

This technique is generally referred to as the *Skempton-Bjerrum modification* for consolidation settlement calculation.

## 3.17

### Vertical Stress Increase in a Soil Mass Caused by Foundation Load (For Consolidation Settlement Calculation)

#### Stress Due to a Concentrated Load

In 1885, Boussinesq developed the mathematical relationships for the determination of the normal and shear stresses at any point inside *homogeneous, elastic,* and *isotropic* mediums due to a *concentrated point load* located at the surface, as shown in Figure 3.22. According to his analysis, the *vertical stress increase* ($\Delta p$) at point $A$ (Figure 3.22) caused by the point load of magnitude $P$ can be given as

$$\Delta p = \frac{3P}{2\pi z^2 \left[ 1 + \left( \dfrac{r}{z} \right)^2 \right]^{5/2}} \tag{3.80}$$

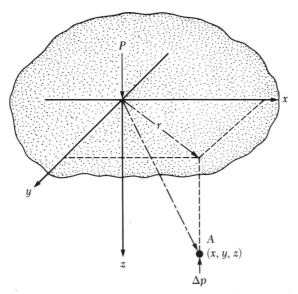

**Figure 3.22** Vertical stress at a point, $A$, caused by a point load on the surface

where $r = \sqrt{x^2 + y^2}$

    $x,y,z$ = coordinates of the point $A$

    Note that Eq. (3.80) is not a function of the Poisson's ratio of the soil.

## Stress Due to a Circularly Loaded Area

    The Boussinesq equation [Eq. (3.80)] can also be used to determine the vertical stress below the center of a flexible circularly loaded area, as shown in Figure 3.23a. Let the radius of the loaded area be $B/2$, and let $q_o$ be the uniformly distributed load per unit area. In order to determine the stress increase at a point $A$ that is located at a depth $z$ below the center of the circular area, consider an elementary area on the circle, as shown in Figure 3.23a. The load on this elementary area can be given to be equal to $q_o r \, d\theta \, dr$. This load can be considered as a point load. The stress increase at point $A$ caused by this load can be determined by using Eq. (3.80):

$$dp = \frac{3(q_o r \, d\theta \, dr)}{2\pi z^2 \left[1 + \left(\dfrac{r}{z}\right)^2\right]^{5/2}} \tag{3.81}$$

Thus, the total increase of stress caused by the entire loaded area can be obtained by integration of the above equation, or

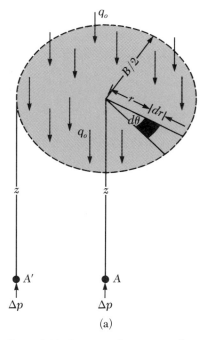

**Figure 3.23**  Increase of pressure under a uniformly loaded flexible circular area

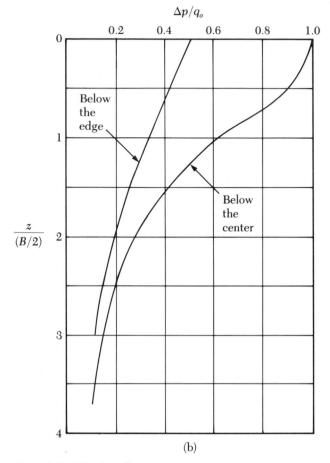

$$\Delta p/q_o$$

(b)

**Figure 3.23**  (Continued)

$$\Delta p = \int dp = \int_{\theta=0}^{\theta=2\pi} \int_{r=0}^{r=B/2} \frac{3(q_o r\, d\theta\, dr)}{2\pi z^2 \left[1 + \left(\dfrac{r}{z}\right)^2\right]^{5/2}}$$

$$= q_o \left\{ 1 - \frac{1}{\left[1 + \left(\dfrac{B}{2z}\right)^2\right]^{3/2}} \right\} \qquad (3.82)$$

Based on the preceding equation, Figure 3.23b shows a plot of $\Delta p/q_o$ against $z/(B/2)$. A similar integration could be performed to obtain the stress increase at a point $A'$ (Figure 3.23a) located at a depth $z$ below the loaded area. The variation of $\Delta p/q_o$ for the point $A'$ is also shown in Figure 3.23b.

## Stress Below a Rectangular Area

Using the integration technique of Boussinesq's equation, the vertical stress at a given point $A$ below a flexible rectangular loaded area (Figure 3.24a)

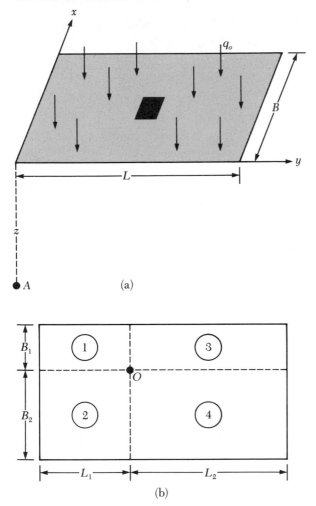

**Figure 3.24** Determination of stress below a flexible rectangular loaded area

can also be evaluated. In order to do that, consider an elementary area $dA = dx \cdot dy$ on the flexible loaded area. If the load per unit area is $q_o$, then the total load on the elementary area can be given as

$$dP = q_o \, dx \, dy \tag{3.83}$$

This elementary load, $dP$, can be treated as a point load. The increase of vertical stress at point $A$ caused by $dP$ can be evaluated by using Eq. (3.80). Note that one needs to substitute $dP = q_o \, dx \, dy$ for $P$, and $x^2 + y^2$ for $r^2$, in Eq. (3.80). Thus

$$\text{the stress increase at } A \text{ caused by } dP = \frac{3q_o(dx \cdot dy)z^3}{2\pi(x^2 + y^2 + z^2)^{5/2}}$$

The total stress increase caused by the entire loaded area at point $A$ can now be obtained by integrating the preceding equation:

$$\Delta p = \int_{y=0}^{L} \int_{x=0}^{B} \frac{3q_o(dx\ dz)z^3}{2\pi(x^2 + y^2 + z^2)^{5/2}} = q_oI \tag{3.84}$$

where $\Delta p$ = stress increase at $A$

$$I = \text{influence factor} = \frac{1}{4\pi}\left(\frac{2mn\sqrt{m^2 + n^2 + 1}}{m^2 + n^2 + m^2n^2 + 1} \cdot \frac{m^2 + n^2 + 2}{m^2 + n^2 + 1}\right.$$

$$\left. + \tan^{-1}\frac{2mn\sqrt{m^2 + n^2 + 1}}{m^2 + n^2 + 1 - m^2n^2}\right) \tag{3.85}$$

where

$$m = \frac{B}{z} \tag{3.86}$$

and

$$n = \frac{L}{z} \tag{3.87}$$

The variations of the influence values with $m$ and $n$ are given in Table 3.5 on pages 144–145. For convenience, they are also plotted in Figure 3.25.

The stress increase at any point below a rectangular loaded area can also be found by using Eq. (3.84). This can be demonstrated using Figure 3.24b. To determine the stress at a depth $z$ below point $O$, divide the loaded area into four rectangles. Point $O$ is the common corner to each rectangle. The increase of stress at a depth z below point $O$ caused by each rectangular area can now be calculated by using Eq. (3.84). The total stress increase caused by the entire loaded area can now be given by the expression

$$\Delta p = q_o(I_1 + I_2 + I_3 + I_4) \tag{3.88}$$

where $I_1$, $I_2$, $I_3$, and $I_4$ = the influence values of rectangles 1, 2, 3, and 4, respectively. The use of this technique is illustrated in Example Problem 3.9. Also, Figure 3.26 shows the variation of $\Delta p/q_o$ below the center of rectangular areas with $L/B$ = 1, 1.5, 2, and ∞. This has been calculated from Table 3.5.

### Stress Increase Under a Rectangular Foundation—2:1 Method

In several instances, foundation engineers use an approximate method to determine the increase of stress with depth caused by the construction of a foundation. This is referred to as the *2:1 method* (Figure 3.27). According to this method, the increase of stress at a depth $z$ can be given as

$$\Delta p = \frac{q_o \times B \times L}{(B + z)(L + z)} \tag{3.89}$$

Note that Eq. (3.89) assumes that the stress from the foundation spreads out along lines with a *2 vertical to 1 horizontal slope*.

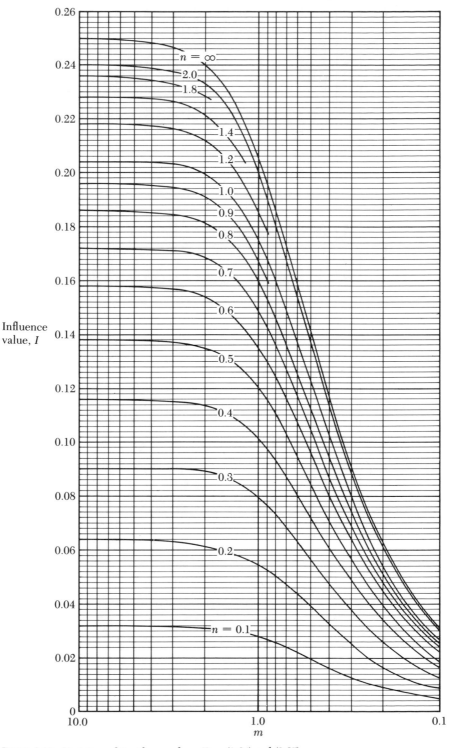

**Figure 3.25** Variation of $I$ with $m$ and $n$—Eqs. (3.84) and (3.85)

**Table 3.5** Variation of Influence Value, $I$ / [Eq. (3.85)][a]

| $m$ | $n$ | | | | | | | | | | | |
|---|---|---|---|---|---|---|---|---|---|---|---|---|
| | 0.1 | 0.2 | 0.3 | 0.4 | 0.5 | 0.6 | 0.7 | 0.8 | 0.9 | 1.0 | 1.2 | 1.4 |
| 0.1 | 0.00470 | 0.00917 | 0.01323 | 0.01678 | 0.01978 | 0.02223 | 0.02420 | 0.02576 | 0.02698 | 0.02794 | 0.02926 | 0.03007 |
| 0.2 | 0.00917 | 0.01790 | 0.02585 | 0.03280 | 0.03866 | 0.04348 | 0.04735 | 0.05042 | 0.05283 | 0.05471 | 0.05733 | 0.05894 |
| 0.3 | 0.01323 | 0.02585 | 0.03735 | 0.04742 | 0.05593 | 0.06294 | 0.06858 | 0.07308 | 0.07661 | 0.07938 | 0.08323 | 0.08561 |
| 0.4 | 0.01678 | 0.03280 | 0.04742 | 0.06024 | 0.07111 | 0.08009 | 0.08734 | 0.09314 | 0.09770 | 0.10129 | 0.10631 | 0.10941 |
| 0.5 | 0.01978 | 0.03866 | 0.05593 | 0.07111 | 0.08403 | 0.09473 | 0.10340 | 0.11035 | 0.11584 | 0.12018 | 0.12626 | 0.13003 |
| 0.6 | 0.02223 | 0.04348 | 0.06294 | 0.08009 | 0.09473 | 0.10688 | 0.11679 | 0.12474 | 0.13105 | 0.13605 | 0.14309 | 0.14749 |
| 0.7 | 0.02420 | 0.04735 | 0.06858 | 0.08734 | 0.10340 | 0.11679 | 0.12772 | 0.13653 | 0.14356 | 0.14914 | 0.15703 | 0.16199 |
| 0.8 | 0.02576 | 0.05042 | 0.07308 | 0.09314 | 0.11035 | 0.12474 | 0.13653 | 0.14607 | 0.15371 | 0.15978 | 0.16843 | 0.17389 |
| 0.9 | 0.02698 | 0.05283 | 0.07661 | 0.09770 | 0.11584 | 0.13105 | 0.14356 | 0.15371 | 0.16185 | 0.16835 | 0.17766 | 0.18357 |
| 1.0 | 0.02794 | 0.05471 | 0.07938 | 0.10129 | 0.12018 | 0.13605 | 0.14914 | 0.15978 | 0.16835 | 0.17522 | 0.18508 | 0.19139 |
| 1.2 | 0.02926 | 0.05733 | 0.08323 | 0.10631 | 0.12626 | 0.14309 | 0.15703 | 0.16843 | 0.17766 | 0.18508 | 0.19584 | 0.20278 |
| 1.4 | 0.03007 | 0.05894 | 0.08561 | 0.10941 | 0.13003 | 0.14749 | 0.16199 | 0.17389 | 0.18357 | 0.19139 | 0.20278 | 0.21020 |
| 1.6 | 0.03058 | 0.05994 | 0.08709 | 0.11135 | 0.13241 | 0.15028 | 0.16515 | 0.17739 | 0.18737 | 0.19546 | 0.20731 | 0.21510 |
| 1.8 | 0.03090 | 0.06058 | 0.08804 | 0.11260 | 0.13395 | 0.15207 | 0.16720 | 0.17967 | 0.18986 | 0.19814 | 0.21032 | 0.21836 |
| 2.0 | 0.03111 | 0.06100 | 0.08867 | 0.11342 | 0.13496 | 0.15326 | 0.16856 | 0.18119 | 0.19152 | 0.19994 | 0.21235 | 0.22058 |
| 2.5 | 0.03138 | 0.06155 | 0.08948 | 0.11450 | 0.13628 | 0.15483 | 0.17036 | 0.18321 | 0.19375 | 0.20236 | 0.21512 | 0.22364 |
| 3.0 | 0.03150 | 0.06178 | 0.08982 | 0.11495 | 0.13684 | 0.15550 | 0.17113 | 0.18407 | 0.19470 | 0.20341 | 0.21633 | 0.22499 |
| 4.0 | 0.03158 | 0.06194 | 0.09007 | 0.11527 | 0.13724 | 0.15598 | 0.17168 | 0.18469 | 0.19540 | 0.20417 | 0.21722 | 0.22600 |
| 5.0 | 0.03160 | 0.06199 | 0.09014 | 0.11537 | 0.13737 | 0.15612 | 0.17185 | 0.18488 | 0.19561 | 0.20440 | 0.21749 | 0.22632 |
| 6.0 | 0.03161 | 0.06201 | 0.09017 | 0.11541 | 0.13741 | 0.15617 | 0.17191 | 0.18496 | 0.19569 | 0.20449 | 0.21760 | 0.22644 |
| 8.0 | 0.03162 | 0.06202 | 0.09018 | 0.11543 | 0.13744 | 0.15621 | 0.17195 | 0.18500 | 0.19574 | 0.20455 | 0.21767 | 0.22652 |
| 10.0 | 0.03162 | 0.06202 | 0.09019 | 0.11544 | 0.13745 | 0.15622 | 0.17196 | 0.18502 | 0.19576 | 0.20457 | 0.21769 | 0.22654 |
| $\infty$ | 0.03162 | 0.06202 | 0.09019 | 0.11544 | 0.13745 | 0.15623 | 0.17197 | 0.18502 | 0.19577 | 0.20458 | 0.21770 | 0.22656 |

[a]After Newmark, 1935.

**Table 3.5** (Continued)

| $m$ | $n$ | | | | | | | | | | |
|---|---|---|---|---|---|---|---|---|---|---|---|
| | 1.6 | 1.8 | 2.0 | 2.5 | 3.0 | 4.0 | 5.0 | 6.0 | 8.0 | 10.0 | $\infty$ |
| 0.1 | 0.03058 | 0.03090 | 0.03111 | 0.03138 | 0.03150 | 0.03158 | 0.03160 | 0.03161 | 0.03162 | 0.03162 | 0.03162 |
| 0.2 | 0.05994 | 0.06058 | 0.06100 | 0.06155 | 0.06178 | 0.06194 | 0.06199 | 0.06201 | 0.06202 | 0.06202 | 0.06202 |
| 0.3 | 0.08709 | 0.08804 | 0.08867 | 0.08948 | 0.08982 | 0.09007 | 0.09014 | 0.09017 | 0.09018 | 0.09019 | 0.09019 |
| 0.4 | 0.11135 | 0.11260 | 0.11342 | 0.11450 | 0.11495 | 0.11527 | 0.11537 | 0.11541 | 0.11543 | 0.11544 | 0.11544 |
| 0.5 | 0.13241 | 0.13395 | 0.13496 | 0.13628 | 0.13684 | 0.13724 | 0.13737 | 0.13741 | 0.13744 | 0.13745 | 0.13745 |
| 0.6 | 0.15028 | 0.15207 | 0.15326 | 0.15483 | 0.15550 | 0.15598 | 0.15612 | 0.15617 | 0.15621 | 0.15622 | 0.15623 |
| 0.7 | 0.16515 | 0.16720 | 0.16856 | 0.17036 | 0.17113 | 0.17168 | 0.17185 | 0.17191 | 0.17195 | 0.17196 | 0.17197 |
| 0.8 | 0.17739 | 0.17967 | 0.18119 | 0.18321 | 0.18407 | 0.18469 | 0.18488 | 0.18496 | 0.18500 | 0.18502 | 0.18502 |
| 0.9 | 0.18737 | 0.18986 | 0.19152 | 0.19375 | 0.19470 | 0.19540 | 0.19561 | 0.19569 | 0.19574 | 0.19576 | 0.19577 |
| 1.0 | 0.19546 | 0.19814 | 0.19994 | 0.20236 | 0.20341 | 0.20417 | 0.20440 | 0.20449 | 0.20455 | 0.20457 | 0.20458 |
| 1.2 | 0.20731 | 0.21032 | 0.21235 | 0.21512 | 0.21633 | 0.21722 | 0.21749 | 0.21760 | 0.21767 | 0.21769 | 0.21770 |
| 1.4 | 0.21510 | 0.21836 | 0.22058 | 0.22364 | 0.22499 | 0.22600 | 0.22632 | 0.22644 | 0.22652 | 0.22654 | 0.22656 |
| 1.6 | 0.22025 | 0.22372 | 0.22610 | 0.22940 | 0.23088 | 0.23200 | 0.23236 | 0.23249 | 0.23258 | 0.23261 | 0.23263 |
| 1.8 | 0.22372 | 0.22736 | 0.22986 | 0.23334 | 0.23495 | 0.23617 | 0.23656 | 0.23671 | 0.23681 | 0.23684 | 0.23686 |
| 2.0 | 0.22610 | 0.22986 | 0.23247 | 0.23614 | 0.23782 | 0.23912 | 0.23954 | 0.23970 | 0.23981 | 0.23985 | 0.23987 |
| 2.5 | 0.22940 | 0.23334 | 0.23614 | 0.24010 | 0.24196 | 0.24344 | 0.24392 | 0.24412 | 0.24425 | 0.24429 | 0.24432 |
| 3.0 | 0.23088 | 0.23495 | 0.23782 | 0.24196 | 0.24394 | 0.24554 | 0.24608 | 0.24630 | 0.24646 | 0.24650 | 0.24654 |
| 4.0 | 0.23200 | 0.23617 | 0.23912 | 0.24344 | 0.24554 | 0.24729 | 0.24791 | 0.24817 | 0.24836 | 0.24842 | 0.24846 |
| 5.0 | 0.23236 | 0.23656 | 0.23954 | 0.24392 | 0.24608 | 0.24791 | 0.24857 | 0.24885 | 0.24907 | 0.24914 | 0.24919 |
| 6.0 | 0.23249 | 0.23671 | 0.23970 | 0.24412 | 0.24630 | 0.24817 | 0.24885 | 0.24916 | 0.24939 | 0.24946 | 0.24952 |
| 8.0 | 0.23258 | 0.23681 | 0.23981 | 0.24425 | 0.24646 | 0.24836 | 0.24907 | 0.24939 | 0.24964 | 0.24973 | 0.24980 |
| 10.0 | 0.23261 | 0.23684 | 0.23985 | 0.24429 | 0.24650 | 0.24842 | 0.24914 | 0.24946 | 0.24973 | 0.24981 | 0.24989 |
| $\infty$ | 0.23263 | 0.23686 | 0.23987 | 0.24432 | 0.24654 | 0.24846 | 0.24919 | 0.24952 | 0.24980 | 0.24989 | 0.25000 |

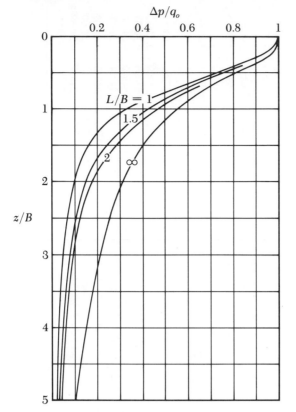

**Figure 3.26**  Increase of stress under the center of a flexible loaded rectangular area

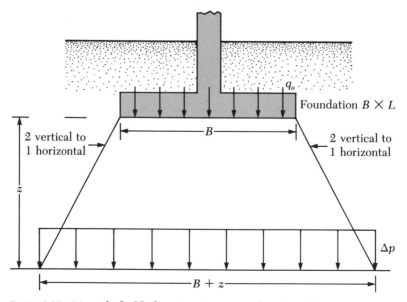

**Figure 3.27**  2:1 method of finding stress increase under a foundation

## Stress Increase Under an Embankment

Figure 3.28 shows the cross section of an embankment of height $H$. This is a two-dimensional loading condition. The vertical stress increase caused by the embankment loading condition can be expressed as

$$\Delta p = \frac{q_o}{\pi}\left[\left(\frac{B_1 + B_2}{B_2}\right)(\alpha_1 + \alpha_2) - \frac{B_1}{B_2}(\alpha_2)\right] \tag{3.90}$$

where $q = \gamma H$

$\gamma$ = unit weight of the embankment soil
$H$ = height of the embankment

$$\alpha_1 \text{ (radians)} = \tan^{-1}\left(\frac{B_1 + B_2}{z}\right) - \tan^{-1}\left(\frac{B_1}{z}\right) \tag{3.91}$$

$$\alpha_2 = \tan^{-1}\left(\frac{B_1}{z}\right) \tag{3.92}$$

For detailed derivation of the equation, see Das (1983). In a simplified form, Eq. (3.90) can be expressed as

$$\Delta p = q_o I' \tag{3.93}$$

where $I'$ = a function of $B_1/z$ and $B_2/z$

The variation of $I'$ with $B_1/z$ and $B_2/z$ is shown in Figure 3.29. Application of this diagram is shown in Example Problem 3.10.

## Stress Increase Due to Any Type of Loading

The increase of vertical stress under any type of flexible loaded area can be easily determined by the use of Newmark's *influence chart* (1942). The chart, in principle, is based on Eq. (3.82), which is for the estimation of vertical stress increase under the center of a circularly loaded area. According to Eq. (3.82)

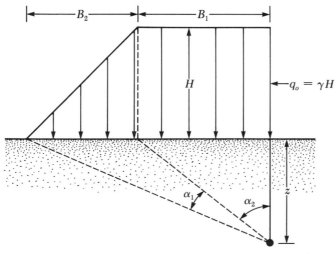

**Figure 3.28**  Embankment loading

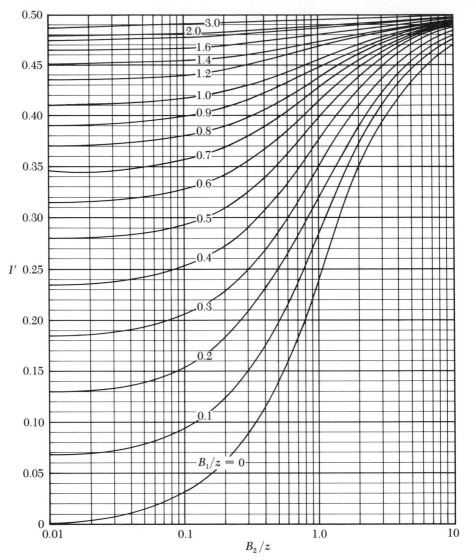

**Figure 3.29**  Influence value of $I'$ for embankment loading (after Osterberg, 1957)

$$\Delta p = q_o \left\{ 1 - \frac{1}{\left[ 1 + \left( \dfrac{B}{2z} \right)^2 \right]^{3/2}} \right\}$$

where $B/2$ = radius of the loaded area = $R$

The preceding equation can be rewritten as

$$\frac{R}{z} = \left[ \left( 1 - \frac{\Delta p}{q_o} \right)^{-2/3} - 1 \right]^{1/2} \tag{3.94}$$

**Table 3.6** Values
of $R/z$ for Various
Values of $\Delta p/q_o$
[Eq. (3.94)]

| $\Delta p/q_o$ | $R/z$ |
|---|---|
| 0 | 0 |
| 0.1 | 0.2698 |
| 0.2 | 0.4005 |
| 0.3 | 0.5181 |
| 0.4 | 0.6370 |
| 0.5 | 0.7664 |
| 0.6 | 0.9174 |
| 0.7 | 1.1097 |
| 0.8 | 1.3871 |
| 0.9 | 1.9084 |
| 1.0 | $\infty$ |

One can now substitute various values of $\Delta p/q_o$ into Eq. (3.94) to obtain corresponding values of $R/z$. Table 3.6 shows the calculated values of $R/z$ for $\Delta p/q_o = 0, 0.1, 0.2, \ldots , 1$.

Using the nondimensional values of $R/z$ shown in Table 3.6, one can draw the concentric circles having radii equal to $R/z$. This is shown in Figure 3.30. Note that the distance $AB$ in Figure 3.30 is equal to unity. The first circle is a point having a radius equal to zero. Similarly, the second circle has a radius of 0.2698 times $AB$. The last circle has a radius of infinity. These circles have been divided by a number of equally spaced radial lines. This is what is referred to as *Newmark's chart*. The influence value ($IV$) of this chart is equal to

$$IV = \frac{1}{\text{number of elements on the chart}} \qquad (3.95)$$

For the chart shown in Figure 3.30, $IV = 1/200 = 0.005$.

Following is a step-by-step procedure using Newmark's chart for determination of vertical stress under a loaded area of any shape.

**1.** Determine the depth $z$ below the loaded area at which the stress is to be determined.

**2.** Adopt a scale $z = \overline{AB}$ (that is, unit length according to Newmark's chart).

**3.** Draw the plan of the loaded area based on the scale adopted in Step 2.

**4.** Place the plan drawn in Step 3 on the Newmark's chart such that the point under which the stress is to be determined is directly above the center of the chart.

**5.** Count the number of elements of the chart that fall inside the plan. Let it be equal to $N$.

**6.** Calculate the stress increase as

$$\Delta p = (IV)(N)(q_o) \qquad (3.96)$$

where $q_o$ = load per unit area on the loaded area

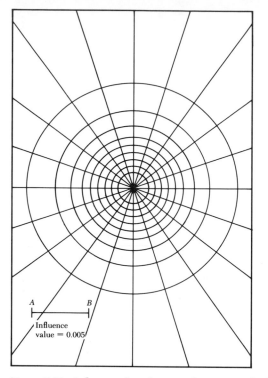

A
B
Influence
value = 0.005

**Figure 3.30** Influence chart for vertical pressure calculation (after Newmark, 1942)

## Example
## 3.9

A flexible rectangular area, 2.5 m × 5 m, is located on the ground surface and loaded with $q_o = 145$ kN/m². Determine the stress increase caused by this loading at a depth of 6.25 m below the center of the rectangular area. Use Eq. (3.84).

### Solution

Refer to Figure 3.24b

$$B_1 = \frac{2.5 \text{ m}}{2} = 1.25 \text{ m}$$

$$L_1 = \frac{5}{2} = 2.5 \text{ m}$$

From Eqs. (3.86) and (3.87)

$$m_1 = \frac{B_1}{z} = \frac{1.25}{6.25} = 0.2$$

$$n_1 = \frac{L_1}{z} = \frac{2.5}{6.25} = 0.4$$

From Table 3.5, for $m_1 = 0.20$ and $n_1 = 0.4$, the value of $I_1 = 0.0328$. Also note that $I_1 = I_2 = I_3 = I_4$. Thus

$$\Delta p = q_o(4I_1) = (145)(4)(0.0328) = 19.024 \text{ kN/m}^2$$

## Example 3.10

An embankment is shown in Figure 3.31a. Determine the stress increase under the embankment at points $A_1$ and $A_2$.

### Solution

$$\gamma H = (17.5)(7) = 122.5 \text{ kN/m}^2$$

**Stress Increase at $A_1$**

Consider the left side of Figure 3.31b. For this, $B_1 = 2.5$ m and $B_2 = 14$ m. So

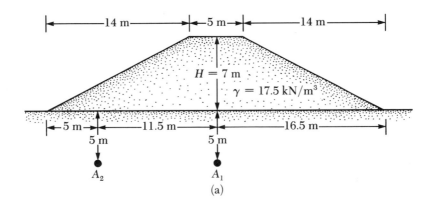

(a)

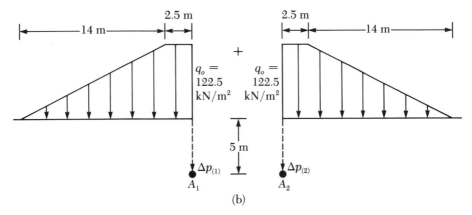

(b)

**Figure 3.31**

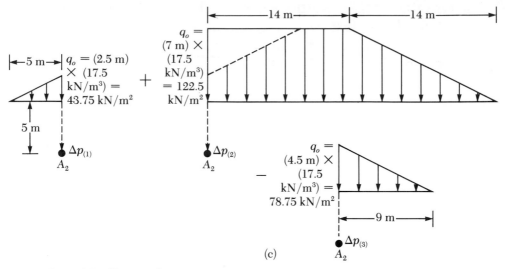

**Figure 3.31** (Continued)

$$\frac{B_1}{z} = \frac{2.5}{5} = 0.5$$

$$\frac{B_2}{z} = \frac{14}{5} = 2.8$$

According to Figure 3.29 on p. 148, in this case, $I' = 0.445$. Because the two sides in Figure 3.31b are symmetrical, the value of $I'$ for the right side will also be 0.445. So

$$\Delta p = \Delta p_1 + \Delta p_2 = q_o[I'_{(\text{left side})} + I'_{(\text{right side})}]$$
$$= 122.5[0.445 + 0.445] = \underline{109.03 \text{ kN/m}^2}$$

**Stress Increase at $A_2$**

Refer to Figure 3.31c. For the left side, $B_2 = 5$ and $B_1 = 0$. So

$$\frac{B_2}{z} = \frac{5}{5} = 1$$

$$\frac{B_1}{z} = \frac{0}{5} = 0$$

According to Figure 3.29, for these values of $B_2/z$ and $B_1/z$, $I' = 0.25$. So

$$\Delta p_1 = 43.75(0.25) = 10.94 \text{ kN/m}^2$$

For the middle section

$$\frac{B_2}{z} = \frac{14}{5} = 2.8$$

$$\frac{B_1}{z} = \frac{14}{5} = 2.8$$

Thus, $I' = 0.495$. So

$$\Delta p_2 = 0.495(122.5) = 60.64 \text{ kN/m}^2$$

For the right side

$$\frac{B_2}{z} = \frac{9}{5} = 1.8$$

$$\frac{B_1}{z} = \frac{0}{5} = 0$$

$I' = 0.335$. So

$$\Delta p_3 = (78.75)(0.335) = 26.38 \text{ kN/m}^2$$

Total stress increase at point $A_2$ is

$$\Delta p = \Delta p_1 + \Delta p_2 - \Delta p_3 = 10.94 + 60.64 - 26.38 = 45.2 \text{ kN/m}^2$$

## Example 3.11

Redo Example Problem 3.9 using Newmark's chart (Figure 3.30).

### Solution

For this problem, $z = 6.25$ m. So, the length $\overline{AB}$ in Figure 3.30 is equal to 6.25 m. With this scale, the plan of the loaded rectangular area can be drawn. Figure 3.32 shows this plan placed over the Newmark's chart such that the center of the loaded area falls above the center of the chart. This is because the stress increase is required at a point that is located immediately below the center of the rectangular area. The number of elements from the influence chart that are inside the plan is about 26. So

$$\Delta p = (IV)(N)(q_o) = (0.005)(26)(145) = 18.85 \text{ kN/m}^2$$

This value of $\Delta p$ is practically the same as that determined in Example Problem 3.9.

## Example 3.12

A foundation 1 m $\times$ 2 m in plan is shown in Figure 3.33. Estimate the total settlement of the foundation.

### Solution
#### Elastic Settlement

The clay layer is located at a depth of 2 m—that is, $2B$ below the foundation. From Figure 3.15 on p. 128, it can be seen that the soil located at a depth $z > 2B$ has very little influence on the elastic settlement. Hence, if Eq. (3.63) is used for the elastic settlement calculation, it is reasonable to use the Young's modulus and Poisson's ratio values of the sand layer. Thus

$$S_e = \frac{Bq_o}{E_s}(1 - \mu_s^2)\alpha_r$$

Given: $q_o = 150 \text{ kN/m}^2$, $E_s = 10,000 \text{ kN/m}^2$, $\mu_s = 0.3$, and $\alpha_r \approx 1.2$ (Figure 3.13b). So

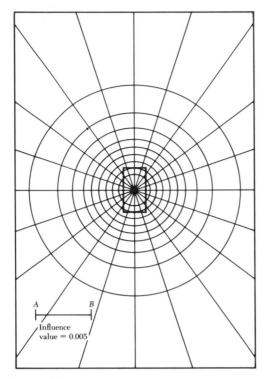

A        B

Influence
value = 0.005

**Figure 3.32**

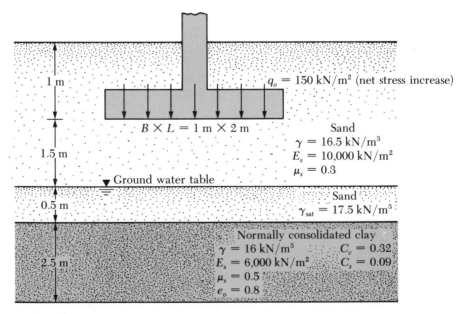

1 m

$q_o = 150\ \text{kN/m}^2$ (net stress increase)

$B \times L = 1\ \text{m} \times 2\ \text{m}$

Sand
$\gamma = 16.5\ \text{kN/m}^3$
$E_s = 10{,}000\ \text{kN/m}^2$
$\mu_s = 0.3$

1.5 m

▽ Ground water table

0.5 m

Sand
$\gamma_{\text{sat}} = 17.5\ \text{kN/m}^3$

Normally consolidated clay
$\gamma = 16\ \text{kN/m}^3$      $C_c = 0.32$
$E_s = 6{,}000\ \text{kN/m}^2$    $C_s = 0.09$
$\mu_s = 0.5$
$e_o = 0.8$

2.5 m

**Figure 3.33**

$$S_e = \frac{(1)(150)}{10,000}(1 - 0.3^2)(1.2) = 0.0163 \text{ m} = \underline{16.38 \text{ mm}}$$

### Consolidation Settlement

The clay is normally consolidated.

$$S_c = \frac{C_c H}{1 + e_o} \log \frac{p_o + \Delta p_{av}}{p_o}$$

$$p_o = (2.5)(16.5) + (0.5)(17.5 - 9.81) + 1.25(16 - 9.81)$$

$$= 41.25 + 3.85 + 7.74 = 52.84 \text{ kN/m}^2$$

From Eq. (3.74)

$$\Delta p_{av} = \tfrac{1}{6}(\Delta p_t + 4\,\Delta p_m + \Delta p_b)$$

Using the 2:1 method

$$\Delta p = \frac{q_o \times B \times L}{(B + z)(L + z)}$$

For the top of the clay layer, $z = 2$ m, so

$$\Delta p_t = \frac{(150)(1)(2)}{(1 + 2)(2 + 2)} = 25 \text{ kN/m}^2$$

Similarly

$$\Delta p_m = \frac{(150)(1)(2)}{(1 + 4.25)(2 + 3.25)} = 13.45 \text{ kN/m}^2$$

and

$$\Delta p_b = \frac{(150)(1)(2)}{(1 + 4.5)(2 + 4.5)} = 8.39 \text{ kN/m}^2$$

Thus

$$\Delta p_{av} = \tfrac{1}{6}[25 + 4(13.45) + 8.39] = 14.53 \text{ kN/m}^2$$

So

$$S_c = \frac{(0.32)(2.5)}{1 + 0.8} \log\left(\frac{52.84 + 14.53}{52.84}\right) = 0.0469 \text{ m} = \underline{46.90 \text{ mm}}$$

Hence, total settlement $= S = S_e + S_c = 16.38 + 46.90 = \underline{63.28 \text{ mm}}$

*Note:* The total settlement just calculated may exceed the tolerable settlement of the foundation. In order to reduce the settlement, the foundation size may be changed so that it will carry the same total load of 300 kN but cause less settlement.

## 3.18
### Allowable Bearing Pressure in Sand Based on Settlement Consideration

Meyerhof (1956) proposed a correlation for the *net allowable bearing pressure* for foundations with the standard penetration resistance ($N$). The net pressure has been defined in Eq. (3.34) as

$$q_{net(all)} = q_{all} - \gamma D_f \qquad (3.97)$$

According to Meyerhof's theory, for one inch of estimated maximum settlement

$$q_{net(all)}(kN/m^2) = 11.98\ N$$

(for $B \leq 1.22$ m and 25.4 mm settlement) $\qquad (3.98a)$

$$q_{net(all)}(kN/m^2) = 7.99\ N \left( \frac{3.28B + 1}{3.28B} \right)^2$$

(for $B > 1.22$ m and 25.4 mm settlement) $\qquad (3.98b)$

where $N$ = corrected standard penetration number
Note that in the preceding equations $B$ is in meters.

Since Meyerhof proposed his correlation, researchers have observed that its results are rather conservative. Bowles (1977) suggested that the net allowable bearing pressure should be increased by about 50%. The modified form of the bearing pressure equations can then be expressed as follows:

$$q_{net(all)}(kN/m^2) = 19.16NF_d \left( \frac{S}{25.4} \right)$$

(for $B \leq 1.22$ m) $\qquad (3.99a)$

$$q_{net(all)}(kN/m^2) = 11.98N \left( \frac{3.28B + 1}{3.28B} \right)^2 F_d \left( \frac{S}{25.4} \right)$$

(for $B > 1.22$ m) $\qquad (3.99b)$

where $F_d$ = depth factor = $1 + 0.33\ (D_f/B) \leq 1.33$ $\qquad (3.100)$
      $S$ = tolerable settlement, in mm

The unit of $B$ is in meters.

The empirical relations just presented may raise some questions: for example, which value of the standard penetration number should be adopted, and what is the effect of the water table on the net allowable bearing capacity? The design value of $N$ should be determined by taking into account the $N$ values for a depth of $2B$ to $3B$, measured from the bottom of the foundation. Many engineers are also of the opinion that the $N$ value should be reduced somewhat if the water table is close to the foundation. However, the author feels that this is not required, because the penetration resistance reflects the location of the water table.

Meyerhof (1956) also prepared empirical relations for the net allowable bearing capacity of foundations based on the cone penetration resistance ($q_c$):

$$q_{net(all)} = \frac{q_c}{15} \text{ (for } B \leq 1.22 \text{ m and settlement of 25.4 mm)} \qquad (3.101a)$$

and

$$q_{net(all)} = \frac{q_c}{25} \left( \frac{3.28B + 1}{3.28B} \right)^2 \text{ (for } B > 1.22 \text{ m and settlement of 25.4 mm)}$$

$$(3.101b)$$

In the preceding two equations, the unit of $B$ is in meters, and the units of $q_{net(all)}$ and $q_c$ are in $kN/m^2$.

The basic philosophy behind the development of these correlations is that, if the maximum settlement is no more than 25.4 mm (1 in.) for any foundation, the differential settlement would be no more than 19.05 mm (3/4 in.). These are probably the allowable limits for most building foundation designs.

## Example 3.13

A shallow square foundation for a column is to be constructed. It must carry a net vertical load of 1000 kN. The foundation soil is sand. The standard penetration numbers obtained from field exploration are given in Figure 3.34. Assume that the depth of the foundation will be 1.5 m and the tolerable settlement is 25.4 mm. Determine the size of the foundation.

### Solution

The field standard penetration numbers need to be corrected by using Eqs. (2.4) to (2.6). This is done in the following table.

| Depth (m) | Field value of N | $\sigma_v'$ (kN/m²) | Corrected N* |
|---|---|---|---|
| 2 | 4 | 31.4 | 7 |
| 4 | 7 | 62.8 | 8 |
| 6 | 12 | 94.2 | 11 |
| 8 | 12 | 125.6 | 11 |
| 10 | 16 | 157.0 | 13 |
| 12 | 13 | 188.4 | 10 |
| 14 | 12 | 206.4 | 9 |
| 16 | 14 | 224.36 | 10 |
| 18 | 18 | 242.34 | 12 |

*rounded off

From the table, it appears that a corrected average $N$ value of about 10 would be appropriate. Using Eq. (3.99b)

$$q_{net(all)} = 11.98N \left( \frac{3.28B + 1}{3.28B} \right)^2 F_d \left( \frac{S}{25.4} \right)$$

Allowable $S = 25.4$ mm and $N = 10$. So

$$q_{net(all)} = 119.8 \left( \frac{3.28B + 1}{3.28B} \right)^2 F_d$$

The following table can now be prepared for trial calculations.

| B (m) | $F_d$* | $q_{net(all)}$ (kN/m²) | $Q = q_{net(all)} \times B^2$ (kN) |
|---|---|---|---|
| 2 | 1.248 | 197.24 | 788.96 |
| 2.25 | 1.22 | 187.19 | 947.65 |
| 2.3 | 1.215 | 185.46 | 981.1 |
| 2.4 | 1.206 | 182.29 | 1050.0 |
| 2.5 | 1.198 | 179.45 | 1121.56 |

*$D_f = 1.5$ m

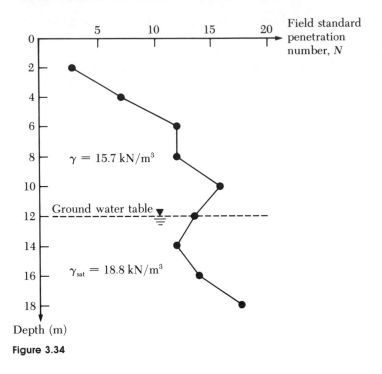

Depth (m)

**Figure 3.34**

Because $Q_{net}$ required is 1000 kN, $B$ will be approximately equal to 2.4 m.

Let us go on to see what the net allowable load is by using the ultimate bearing capacity equation [Eq. (3.16)] with $B = 2.4$ m.

$$q_{net(ult)} = q_{ult} - q = qN_qF_{qs}F_{qd} + \tfrac{1}{2}\gamma BN_\gamma F_{\gamma s}F_{\gamma d} - q$$

For an $N$ value of about 10, the friction angle of soil is about 34° (Table 2.4). For $\phi = 34°$, from Table 3.2

$$N_q = 29.44$$

$$N_\gamma = 41.06$$

$$F_{qs} = 1 + \left(\frac{2.4}{2.4}\right) \tan 34° = 1.675$$

$$F_{\gamma s} = 1 - 0.4\left(\frac{2.4}{2.4}\right) = 0.6$$

$$F_{qd} = 1 + 0.262\left(\frac{1.5}{2.4}\right) = 1.164$$

$$F_{\gamma d} = 1$$

So, $q_{net(ult)} = (15.7 \times 1.5)(29.44)(1.675)(1.164)$
$$+ \tfrac{1}{2}(15.7)(2.4)(41.06)(0.6)(1) - (15.7 \times 1.5)$$
$$= 1351.8 + 464.14 - 23.55 \approx 1792 \text{ kN/m}^2$$
$$Q_{net(ult)} = 1792 \times B^2 = 1792 \times 2.4^2 = 10{,}322 \text{ kN}$$

So, the factor safety required for a tolerable settlement of 25.4 mm is equal to $10,322/1000 = 10.3$. This demonstrates that, in most cases, the design is controlled by the tolerable settlement criterion.

## 3.19

### Field Load Test

The ultimate load-bearing capacity of a foundation, as well as the allowable bearing capacity based on tolerable settlement considerations, can be effectively determined from the field load test. This is generally referred to as the *plate load test* (ASTM Test Designation D-1194-72). The plates that are used for tests in the field are usually made out of steel and are 25 mm thick and 150 mm to 722 mm in diameter. Occasionally, square plates that are 305 mm × 305 mm are also used.

To conduct a plate load test, a hole is excavated with a minimum diameter of $4B$ ($B$ = diameter of the test plate) up to a depth of $D_f$ ($D_f$ = depth of the proposed foundation). The plate is placed at the center of the hole. Load to the plate is applied in steps—about one-fourth to one-fifth of the estimated ultimate load—by means of a jack. A schematic diagram of the test arrangement is shown in Figure 3.35a. During each step load application, the settlement of the plate is observed using dial gauges. At least one hour elapses after the application of each step load before the next load is applied. The test should be conducted until failure, or at least until the plate has gone through 25 mm settlement. Figure 3.35b shows the nature of the load-settlement curve obtained from such tests, from which the ultimate load per unit area can be determined.

For tests in clay

$$q_{u(F)} = q_{u(P)} \qquad (3.102)$$

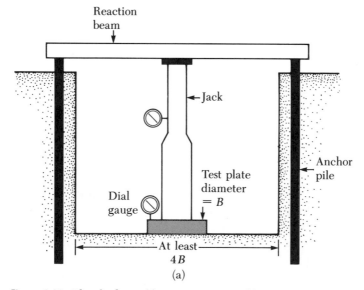

**Figure 3.35** Plate load test: (a) test arrangement; (b) nature of load-settlement curve

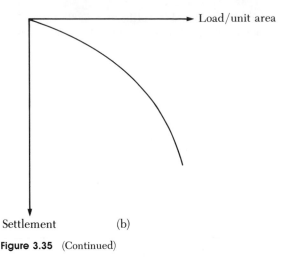

Settlement                    (b)

**Figure 3.35**  (Continued)

where $q_{u(F)}$ = ultimate bearing capacity of the proposed foundation
      $q_{u(P)}$ = ultimate bearing capacity of the test plate

Equation (3.102) implies that the ultimate bearing capacity in clay is practically independent of the size of the plate.

For tests in sandy soils

$$q_{u(F)} = q_{u(P)} \frac{B_F}{B_P}$$  (3.103)

where $B_F$ = width of the foundation
      $B_P$ = width of the test plate

In order to obtain the allowable bearing capacity of a foundation based on settlement considerations, the following equations apply.

For a given intensity of load, $q_o$

$$S_F = S_P \frac{B_F}{B_P}$$  (for clayey soil)  (3.104)

and

$$S_F = S_P \left(\frac{B_F}{B_P}\right)^2 \left(\frac{3.28B_P + 1}{3.28B_F + 1}\right)^2$$  (for sandy soil)  (3.105)

In Eq. (3.105), the units of $B_P$ and $B_F$ are in meters.

Equation (3.105) is based on the works of Terzaghi and Peck (1967). Example Problem 3.14 illustrates its application.

Housel (1929) proposed a different technique for determining the load-bearing capacity of shallow foundations based on settlement consideration. Following are the steps of this procedure:

1. Let it be required to find the dimensions of a foundation that will carry a load of $Q_o$ with a tolerable settlement of $S_{tol}$.

2. Conduct two plate load tests with plates of diameters $B_1$ and $B_2$.

3. From the load-settlement curves obtained in Step 2, determine the total loads on the plates ($Q_1$ and $Q_2$) that correspond to the settlement of $S_{tol}$. For plate No. 1, the total load can be expressed as

$$Q_1 = A_1 m + P_1 n \qquad (3.106)$$

Similarly, for plate No. 2

$$Q_2 = A_2 m + P_2 n \qquad (3.107)$$

where $A_1$, $A_2$ = areas of the plates No. 1 and No. 2, respectively
$P_1$, $P_2$ = perimeters of the plates No. 1 and No. 2, respectively
$m$, $n$ = two constants that correspond to the bearing pressure and perimeter shear, respectively

The values of $m$ and $n$ can be determined by solving Eqs. (3.106) and (3.107).

4. For the foundation to be designed

$$Q_o = Am + Pn \qquad (3.108)$$

where $A$ = area of the foundation
$P$ = perimeter of the foundation

Because $Q_o$, $m$, and $n$ are known, Eq. (3.108) can be solved for determination of foundation width. The application of this procedure is given in Example Problem 3.15.

## Example 3.14

The results of a plate load test in a sandy soil are shown in Figure 3.36. The size of the plate is 0.305 m × 0.305 m. Determine the size of a square column foundation that should carry a load of 2500 kN with a maximum settlement of 25 mm.

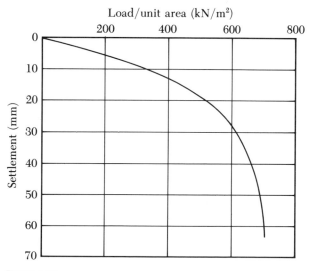

**Figure 3.36**

## Solution

The problem has to be solved by a trial-and-error procedure. Use the following table and Equation (3.105).

| $Q_o$ (kN) (1) | Assume width $B_F$ (m) (2) | $q_o = \dfrac{Q_o}{B_F^2}$ (kN/m²) (3) | $S_P$ corresponding to $q_o$ in Col. 3 (mm) (4) | $S_F$ from Eq. (3.105) (mm) (5) |
|---|---|---|---|---|
| 2500 | 4.0 | 156.25 | 4.0 | 13.80 |
| 2500 | 3.0 | 277.80 | 8.0 | 26.35 |
| 2500 | 3.2 | 244.10 | 6.8 | 22.70 |
| 2500 | 3.1* | 260.10 | 7.2 | 23.86 |

*So, a column footing with dimensions of 3.1 m × 3.1 m will be appropriate.

## Example 3.15

The results of two plate load tests are given in the following table.

| Plate diameter, $B$ (m) | Total load, $Q$ (kN) | Settlement (mm) |
|---|---|---|
| 0.305 | 32.2 | 20 |
| 0.610 | 71.8 | 20 |

A square column foundation has to be constructed to carry a total load of 715 kN. The tolerable settlement is 20 mm. Determine the size of the foundation.

## Solution

Referring to Eqs. (3.106) and (3.107)

$$32.2 = \frac{\pi}{4}(0.305)^2 m + \pi(0.305)n \tag{a}$$

$$71.8 = \frac{\pi}{4}(0.610)^2 m + \pi(0.610)n \tag{b}$$

From the preceding two equations,

$$m = 50.68 \text{ kN/m}^2$$
$$n = 29.75 \text{ kN/m}$$

For the foundation to be designed [Eq. (3.108)]

$$Q_o = Am + Pn$$

or

$$Q_o = B_F{}^2 m + 4B_F n$$

Given $Q_o = 715$ kN. So

$$715 = B_F{}^2(50.68) + 4B(29.75)$$

or

$$50.68B_F{}^2 + 119B_F - 715 = 0$$

From the above equation, $B_F \approx 2.8$ m.

## 3.20
## Presumptive Bearing Capacity

Several building codes (for example, Uniform Building Code, Chicago Building Code, New York City Building Code) specify the allowable bearing capacity of foundations on various types of soil. For minor construction works, they often provide fairly acceptable guidelines. However, these bearing capacity values are primarily based on the *visual* classification of near-surface soils. They generally do not take into consideration such factors as the stress history of the soil, ground water table location, the depth of the foundation, and the tolerable settlement. So, for large construction projects, the codes' presumptive values should be taken as a guide only.

## 3.21
## Tolerable Settlement of Buildings

As has been emphasized in this chapter, settlement analysis plays an important part in the design and construction of foundations. Large settlements of various components of a structure may lead to considerable damage and/or interfere with the proper functioning of the structure. Limited studies have been made to evaluate the conditions for tolerable settlement of various types of structure (for example, Bjerrum, 1963; Burland and Worth, 1974; Grant, Christian, and Vanmarcke, 1974; Polshin and Tokar, 1957; Wahls, 1981). Wahls (1981) has provided an excellent review of these studies.

Figure 3.37 provides the parameters for definition of tolerable settlement. Figure 3.37a is for a structure that has undergone settlement without tilt; Figure 3.37b is for a structure that has undergone settlement with tilt. The parameters are as follows:

$\rho_i$ = total vertical displacement at point $i$

$\delta_{ij}$ = differential settlement between points $i$ and $j$

$\Delta$ = relative deflection

$\omega$ = tilt

$\eta_{ij} = \dfrac{\delta_{ij}}{l_{ij}} - \omega$ = angular distortion

$\dfrac{\Delta}{L}$ = deflection ratio

$L$ = lateral dimension of the structure

Bjerrum (1963) has provided the conditions of *limiting* angular distortion, $\eta$, of various structures (see Table 3.7).

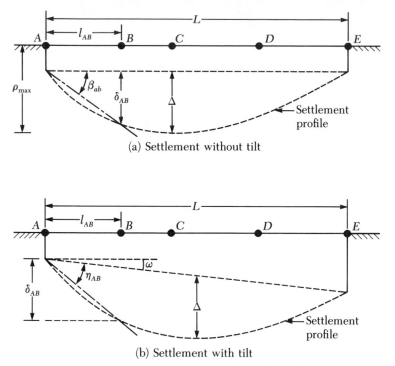

(a) Settlement without tilt

(b) Settlement with tilt

**Figure 3.37**  Parameters for definition of tolerable settlement (redrawn after Wahls, 1981)

Polshin and Tokar (1957) have presented the settlement criteria of the 1955 U.S.S.R. Building Code. These were based on the experience gained from the observation of foundation settlement over 25 years. Tables 3.8 and 3.9 present their findings.

**Table 3.7**  Limiting Angular Distortion As Recommended by Bjerrum[a]

| Category of potential damage | $\eta$ |
|---|---|
| Danger to machinery sensitive to settlement | 1/750 |
| Danger to frames with diagonals | 1/600 |
| Safe limit for no cracking of buildings[b] | 1/500 |
| First cracking of panel walls | 1/300 |
| Difficulties with overhead cranes | 1/300 |
| Tilting of high rigid buildings becomes visible | 1/250 |
| Considerable cracking of panel and brick walls | 1/150 |
| Danger of structural damage to general buildings | 1/150 |
| Safe limit for flexible brick walls, $L/H > 4$[b] | 1/150 |

[a]After Wahls, 1981
[b]Safe limits include a factor of safety.

**Table 3.8** Allowable Settlement Criteria: 1955 U.S.S.R. Building Code[a]

| Type of structure | Sand and hard clay | Plastic clay |
|---|---|---|
| (a) $\eta$ | | |
| Civil- and industrial-building column foundations: | | |
| For steel and reinforced concrete structures | 0.002 | 0.002 |
| For end rows of columns with brick cladding | 0.007 | 0.001 |
| For structures where auxiliary strain does not arise during nonuniform settlement of foundations | 0.005 | 0.005 |
| Tilt of smokestacks, towers, silos, and so on | 0.004 | 0.004 |
| Craneways | 0.003 | 0.003 |
| (b) $\Delta/L$ | | |
| Plain brick walls: | | |
| For multistory dwellings and civil buildings | | |
| at $L/H \leq 3$ | 0.0003 | 0.0004 |
| at $L/H \geq 5$ | 0.0005 | 0.0007 |
| For one-story mills | 0.0010 | 0.0010 |

[a]After Wahls, 1981

**Table 3.9** Allowable Average Settlement for Different Building Types[a]

| Kind of building | Allowable average settlement, in inches (millimeters) |
|---|---|
| Building with plain brick walls | |
| $L/H \geq 2.5$ | 3 (80) |
| $L/H \leq 1.5$ | 4 (100) |
| Building with brick walls, reinforced with reinforced concrete or reinforced brick | 6 (150) |
| Framed building | 4 (100) |
| Solid reinforced concrete foundations of smokestacks, silos, towers, and so on | 12 (300) |

[a]After Wahls, 1981

# Problems

**3.1** A continuous foundation is 4.65 ft wide. Given: $D_f = 3.59$ ft. $\gamma = 109.5$ lb/ft$^3$, $\phi = 26°$, and $c = 585$ lb/ft$^2$. Using Terzaghi's equation, determine the allowable gross vertical load-bearing capacity (factor of safety = 4). Assume that general shear failure occurs in soil.

**3.2** A square column foundation is 2.1 m × 2.1 m in plan. Given: $D_f = 1.4$ m, $\gamma = 15.9$ kN/m$^3$, $\phi = 34°$, and $c = 0$. Using Terzaghi's equation, determine the gross allowable vertical load that the column could carry (factor of safety = 3). Assume that general shear failure occurs in soil.

**3.3** Solve Problem 3.1 using Eq. (3.16).

**3.4** Solve Problem 3.2 using Eq. (3.16).

**3.5** For the foundation given in Problem 3.2, what will be the gross allowable load-bearing capacity if the load is inclined at an angle 10° to the vertical? Use Eqs. (3.16), (3.29), and (3.30).

**3.6** A column foundation is 13.0 ft × 6.5 ft in plan. Given: $D_f = 4.5$ ft, $c = 3200$ lb/ft², $\phi = 0$, and $\gamma = 117$ lb/ft³. What is the net ultimate load that the column could carry? [Use Eq. (3.36).]

**3.7** Repeat Problem 3.6 using Eq. (3.37).

**3.8** An eccentrically loaded foundation is shown in Figure P3.8. Use a factor of safety of 4 and determine the maximum allowable load that the foundation could carry.

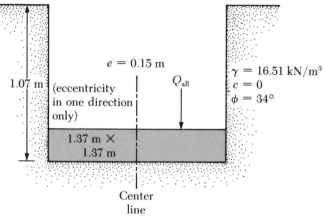

**Figure P3.8**

**3.9** An eccentrically loaded foundation is shown in Figure P3.9. Determine the ultimate load, $Q_u$, that the foundation could carry.

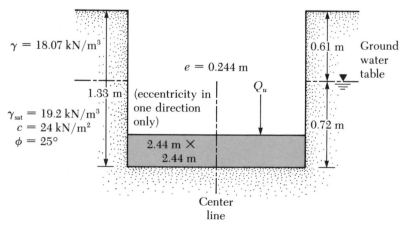

**Figure P3.9**

**3.10** For a square foundation that is $B \times B$ in plan, the following are given: $D_f = 3$ ft; vertical gross allowable load, $Q_{all} = 150{,}000$ lb, $\gamma = 115$ lb/ft³, $\phi = 40°$, $c = 0$, and factor of safety $= 3$. Determine the size of the foundation.

**3.11** A square footing is shown in Figure P3.11. Using a factor of safety of 6, determine the size of the footing.

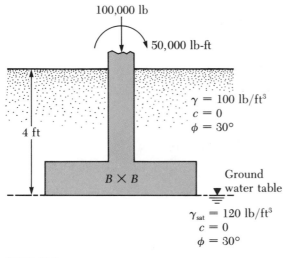

100,000 lb

50,000 lb-ft

$\gamma = 100$ lb/ft³
$c = 0$
$\phi = 30°$

4 ft

$B \times B$

Ground
water table

$\gamma_{sat} = 120$ lb/ft³
$c = 0$
$\phi = 30°$

**Figure P3.11**

**3.12** A continuous footing in a two-layered clay is shown in Figure P3.12. Find the gross allowable bearing capacity. Factor of safety $= 3$.

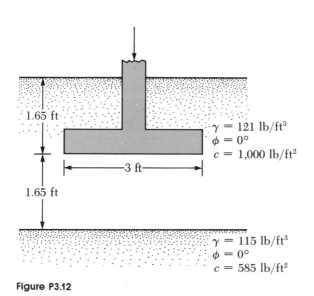

1.65 ft

$\gamma = 121$ lb/ft³
$\phi = 0°$
$c = 1{,}000$ lb/ft²

3 ft

1.65 ft

$\gamma = 115$ lb/ft³
$\phi = 0°$
$c = 585$ lb/ft²

**Figure P3.12**

**3.13** Find the gross ultimate load that the footing shown in Figure P3.13 can carry.

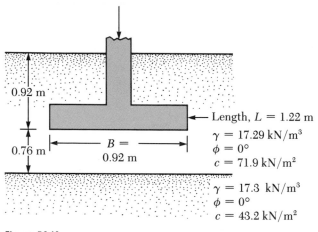

Length, $L = 1.22$ m

$\gamma = 17.29$ kN/m$^3$

$\phi = 0°$

$c = 71.9$ kN/m$^2$

$\gamma = 17.3$ kN/m$^3$

$\phi = 0°$

$c = 43.2$ kN/m$^2$

0.92 m

0.76 m

$B =$

0.92 m

**Figure P3.13**

**3.14** Refer to Figure 3.11. The foundation is 1 m × 2 m in plan. $D_f = 1$ m and $H = 1.5$ m. For the sand layer, $\phi = 35°$, $c = 0$, $\gamma = 17.8$ kN/m$^3$; and, for the clay layer, $\phi = 0$, $c = 60$ kN/m$^2$, $\gamma = 18.2$ kN/m$^3$. Determine the gross allowable load that the foundation could carry. Use a factor of safety of 4 against bearing capacity failure. For bearing capacity factors shown in Eqs. (3.56) and (3.57), use Table 3.2.

**3.15** Redo Example Problem 3.3 for a $FS_{shear} = 1.5$.

**3.16** Refer to Figure 3.12. A foundation that is 10 ft × 6.5 ft in plan is resting on a sand deposit. The net load per unit area at the level of the foundation ($q_o$) is 3000 lb/ft$^2$. For the sand, given $\mu_s = 0.3$, $E_s = 3200$ lb/in.$^2$, $D_f = 2.95$ ft, and $H = 32$ ft. Assuming the foundation to be rigid, determine the elastic settlement that the foundation would undergo. Use Eq. (3.63).

**3.17** Solve Problem 3.16 using Eq. (3.65). For the correction factor, $C_2$, use a time of 5 years for creep.

**3.18** A flexible circular area is subjected to a uniformly distributed load of 2000 lb/ft$^2$. The diameter of the loaded area is 11.5 ft. Determine the stress increase in a soil mass at a point located 32.8 ft below the center of the loaded area.

**3.19** Refer to Figure 3.24b on p. 141, which shows a flexible rectangular area. Given: $B_1 = 3.3$ ft, $B_2 = 6.6$ ft, $L_1 = 6.6$ ft, and $L_2 = 9.9$ ft. If the area is subjected to a uniform load of 2250 lb/ft$^2$ determine the stress increase at a depth of 33 ft located immediately below point $O$.

**3.20** Solve Problem 3.19 using Newmark's chart.

**3.21** A square column foundation is shown in Figure P3.21. Determine the average increase of pressure in the clay layer below the center of the foundation:

a. by using Figure 3.26 on p. 146;
b. by using the 2:1 method (Figure 3.27, p. 146).

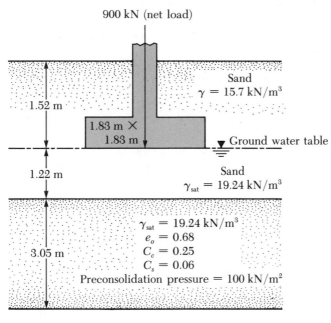

**Figure P3.21**

**3.22** Solve Problem 3.21 using Newmark's chart.

**3.23** Refer to Figure P3.21. Determine the average increase of stress in the clay layer below the corner of the foundation. Use Newmark's chart.

**3.24** Redo Problem 3.23 using Table 3.5 on p. 143–144.

**3.25** Estimate the consolidation settlement of the clay layer shown in Figure P3.21 using the results of Part (a) of Problem 3.21.

**3.26** Estimate the consolidation settlement of the clay layer shown in Figure P3.21 using the results of Part (b) of Problem 3.21.

**3.27** Refer to Problem 2.6 on p. 98. What will be the net allowable bearing capacity of a foundation 2.2 m × 2.2 m in plan? Given: $D_f = 1.3$ m; allowable settlement = 25.4 mm.

**3.28** Two plate load tests with square plates were conducted in the field. At 1-in. settlement, the results are as follows:

| Width of plate (in.) | Load (lb) |
|---|---|
| 12 | 8070 |
| 24 | 25,800 |

What size square footing is required to carry a net load of 236,000 lb at a settlement of 1 in.?

# References

American Society for Testing and Materials (1982). *Annual Book of ASTM Standards*, Part 19, Philadelphia.

Bjerrum, L. (1963). "Allowable Settlement of Structures," *Proceedings*, European Conference on Soil Mechanics and Foundation Engineering, Wiesbaden, Germany, Vol. III, pp. 135–137.

Bowles, J. E. (1977). *Foundation Analysis and Design*, 2nd ed., McGraw-Hill, New York.

Burland, J. B., and Worth, C. P. (1974). "Allowable and Differential Settlement of Structures Including Damage and Soil-Structure Interaction," *Proceedings*, Conference on Settlement of Structures, Cambridge University, England, pp. 611–654.

Caquot, A., and Kerisel, J. (1953). "Sur le terme de surface dans le calcul des fondations en milieu pulverulent," *Proceedings*, Third International Conference on Soil Mechanics and Foundation Engineering, Vol. I, pp. 336–337.

Christian, J. T., and Carrier, W. D. (1978). "Janbu, Bjerrum, and Kjaernsli's Chart Reinterpreted," *Canadian Geotechnical Journal*, Vol. 15, pp. 124–128.

Das, B. M. (1983). *Advanced Soil Mechanics*, McGraw-Hill, New York.

De Beer, E. E. (1970). "Experimental Determination of the Shape Factors and Bearing Capacity Factors of Sand," *Geotechnique*, Vol. 20, No. 4, pp. 387–411.

Grant, R. J., Christian, J. T., and Vanmarcke, E. H. (1974). "Differential Settlement of Buildings," *Journal of the Geotechnical Engineering Division*, American Society of Civil Engineers, Vol. 100, No. GT9, pp. 973–991.

Hanna, A. M., and Meyerhof, G. G. (1981). "Experimental Evaluation of Bearing Capacity of Footings Subjected to Inclined Loads," *Canadian Geotechnical Journal*, Vol. 18, No. 4, pp. 599–603.

Hansen, J. B. (1970). "A Revised and Extended Formula for Bearing Capacity," Danish Geotechnical Institute, *Bulletin 28*, Copenhagen.

Harr, M. E. (1966). *Fundamentals of Theoretical Soil Mechanics*, McGraw-Hill, New York.

Housel, W. S. (1929). "A Practical Method for the Selection of Foundations Based on Fundamental Research in Soil Mechanics," *Research Bulletin No. 13*, University of Michigan, Ann Arbor.

Ismael, N. F., and Vesic, A. S. (1981). "Compressibility and Bearing Capacity," *Journal of the Geotechnical Engineering Division*, American Society of Civil Engineers, Vol. 107, No. GT12, pp. 1657–1676.

Janbu, N., Bjerrum, L., and Kjaernsli, B. (1956). "Veiledning ved losning av fundamentering—soppgaver," *Publication No. 16*, Norwegian Geotechnical Institute, pp. 30–32.

Meyerhof, G. G. (1953). "The Bearing Capacity of Foundations Under Eccentric and Inclined Loads," *Proceedings*, Third International Conference on Soil Mechanics and Foundation Engineering, Zürich, Vol. 1, pp. 440–445.

Meyerhof, G. G. (1956). "Penetration Tests and Bearing Capacity of Cohesionless Soils," *Journal of the Soil Mechanics and Foundations Division*, American Society of Civil Engineers, Vol. 82, No. SM1, pp. 1–19.

Meyerhof, G. G. (1963). "Some Recent Research on the Bearing Capacity of Foundations," *Canadian Geotechnical Journal*, Vol. 1, No. 1, pp. 16–26.

Meyerhof, G. G. (1974). "Ultimate Bearing Capacity of Footings on Sand Layer Overlying Clay," *Canadian Geotechnical Journal*, Vol. 11, No. 2, pp. 224–229.

Mitchell, J. K., and Gardner, W. S. (1975). "*In Situ* Measurement of Volume Change Characteristics," *Proceedings*, Specialty Conference, American Society of Civil Engineers, Vol. 2, pp. 279–345.

Newmark, N. M. (1935). "Simplified Computation of Vertical Pressure in Elastic Foundation," *Circular 24*, University of Illinois Engineering Experiment Station, Urbana.

Newmark, N. M. (1942). "Influence Charts for Computation of Stresses in Elastic Foundations," *Bulletin No. 338*, University of Illinois Engineering Experiment Station, Urbana.

Osterberg, J. O. (1957). "Influence Values for Vertical Stresses in Semi-Infinite Mass Due to Embankment Loading," *Proceedings*, 4th International Conference on Soil Mechanics and Foundation Engineering, Vol. 1, pp. 393–396.

Polshin, D. E. and Tokar, R. A. (1957). "Maximum Allowable Nonuniform Settlement of Structures," *Proceedings*, Fourth International Conference on Soil Mechanics and Foundation Engineering, London, Vol. 1, pp. 402–405.

Prakash, S. (1981). *Soil Dynamics*, McGraw-Hill, New York.

Prandtl, L. (1921). "Über die Eindringungsfestigkeit (Härte) plastischer Baustoffe und die Festigkeit von Schneiden," *Zeitschrift für angewandte Mathematik und Mechanik*, Vol. 1, No. 1, pp. 15–20.

Reddy, A. S., and Srinivasan, R. J. (1967). "Bearing Capacity of Footings on Layered Clay," *Journal of the Soil Mechanics and Foundations Division*, American Society of Civil Engineers, Vol. 93, No. SM2, pp. 83–99.

Reissner, H. (1924). "Zum Erddruckproblem," *Proceedings*, First International Congress of Applied Mechanics, Delft, pp. 295–311.

Schmertmann, J. H. (1970). "Static Cone to Compute Settlement Over Sand," *Journal of the Soil Mechanics and Foundations Division*, American Society of Civil Engineers, Vol. 96, No. SM3, pp. 1011–1043.

Schmertmann, J. H., and Hartman, J. P. (1978). "Improved Strain Influence Factor Diagrams," *Journal of the Geotechnical Engineering Division*, American Society of Civil Engineers, Vol. 104, No. GT8, pp. 1131–1135.

Skempton, A. W. (1951). "The Bearing Capacity of Clays," *Proceedings*, Building Research Congress, London, pp. 180–189.

Skempton, A. W., and Bjerrum, L. (1957). "A Contribution to Settlement Analysis of Foundations in Clay," *Geotechnique*, London, Vol. 7, p. 178.

Terzaghi, K. (1943). *Theoretical Soil Mechanics*, Wiley, New York.

Terzaghi, K., and Peck, R. B. (1967). *Soil Mechanics in Engineering Practice*, 2nd ed., Wiley, New York.

Vesic, A. S. (1963). "Bearing Capacity of Deep Foundations in Sand," *Highway Research Record No. 39*, National Academy of Sciences, pp. 112–153.

Vesic, A. S. (1973). "Analysis of Ultimate Loads of Shallow Foundations," *Journal of the Soil Mechanics and Foundations Division*, American Society of Civil Engineers, Vol. 99, No. SM1, pp. 45–73.

Wahls, H. E. (1981). "Tolerable Settlement of Buildings," *Journal of the Geotechnical Engineering Division*, American Society of Civil Engineers, Vol. 107, No. GT11, pp. 1489–1504.

4

# Mat Foundations

## 4.1

### Introduction

Mat foundations are primarily shallow foundations. They are a type of *combined footing*. There are four major types of combined footing (see Figure 4.1a):

1. *Rectangular Combined Footing:* In several instances, the load to be carried by a column and the soil bearing capacity are such that the standard spread footing design will require the extension of the column foundation beyond the property line. In such a case, two or more columns can be supported on a single rectangular foundation, as shown in Figure 4.1b. If the net allowable soil pressure is known, the size of the foundation ($B \times L$) can be determined in the following manner:

a. Determine the area of the foundation, $A$, as

$$A = \frac{Q_1 + Q_2}{q_{all(net)}} \tag{4.1}$$

where $Q_1$, $Q_2$ = column loads
$q_{all(net)}$ = net allowable soil bearing capacity

b. Determine the location of the resultant of the column loads. Referring to Figure 4.1b,

$$X = \frac{Q_2 L_3}{Q_1 + Q_2} \tag{4.2}$$

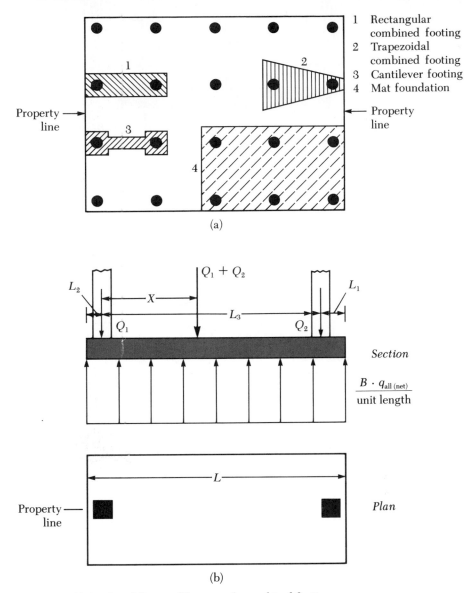

**Figure 4.1** (a) Combined footing; (b) rectangular combined footing; (c) trapezoidal combined footing; (d) cantilever footing

c. For uniform distribution of soil pressure under the foundation, the resultant of the column loads should pass through the centroid of the foundation. Thus

$$L = 2(L_2 + X) \tag{4.3}$$

where $L$ = length of the foundation

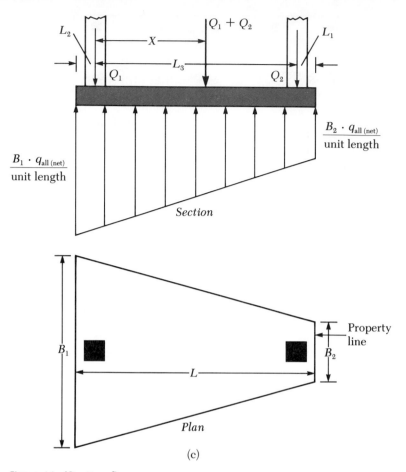

**Figure 4.1** (Continued)

d. Once the length $L$ is determined, the value of $L_1$ can be obtained as

$$L_1 = L - L_2 - L_3 \qquad (4.4)$$

Note that the magnitude of $L_2$ will be known, depending on the extent of the property line.

   e. The width of the foundation can now be obtained as

$$B = \frac{A}{L} \qquad (4.5)$$

   2. *Trapezoidal Combined Footing:* This type of combined footing (Figure 4.1c) is sometimes used where there is not enough space for isolated spread foundation for a column carrying a large load. The size of the foundation that will uniformly distribute pressure on the soil can be obtained in the following manner:

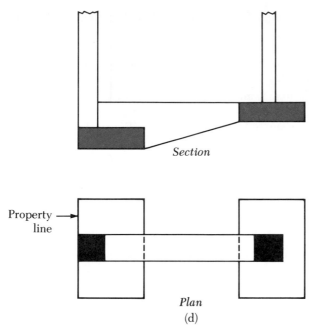

*Section*

Property
line

*Plan*
(d)

**Figure 4.1** (Continued)

a. If the net allowable soil pressure is known, determine the area of the foundation as

$$A = \frac{Q_1 + Q_2}{q_{all(net)}}$$

Referring to Figure 4.1c,

$$A = \frac{B_1 + B_2}{2} L \qquad (4.6)$$

b. Determine the location of the resultant for the column loads as

$$X = \frac{Q_2 L_3}{Q_1 + Q_2}$$

c. From the property of a trapezoid

$$X + L_2 = \left(\frac{B_1 + 2B_2}{B_1 + B_2}\right)\frac{L}{3} \qquad (4.7)$$

With known values of $A$, $L$, $X$, and $L_2$, solve Eqs. (4.6) and (4.7) to obtain $B_1$ and $B_2$. Note that, for a trapezoid

$$\frac{L}{3} < X + L_2 < \frac{L}{2}$$

3. *Cantilever Footing:* This type of combined footing construction uses a *strap beam* to connect an eccentrically loaded column foundation to the foundation of an interior column (Figure 4.1d). Cantilever footings may be used in place of trapezoidal or rectangular combined footings when the allowable soil bearing capacity is high and the distances between the columns are large.

4. *Mat Foundation:* A mat foundation, which is sometimes referred to as a *raft foundation,* is a combined footing that may cover the whole area under a structure supporting several columns and walls (Figure 4.1a). Mat foundations are sometimes preferred for soils that have low load-bearing capacity but that will have to support high column and/or wall loads. In some conditions, spread footings may cover more than half the building area; mat foundations may prove to be more economical. This chapter will discuss mat foundations in detail.

## 4.2

### Common Types of Mat Foundation

There are several types of mat foundation in use nowadays. Some of the common types of construction are schematically shown in Figure 4.2. They are

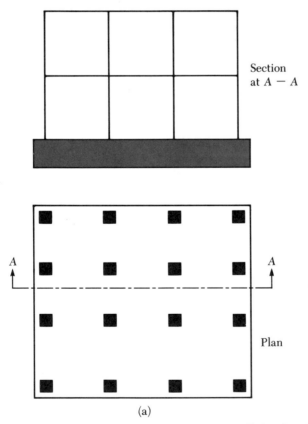

Section at $A - A$

Plan

(a)

**Figure 4.2**  Types of mat foundation: (a) flat plate; (b) flat plate thickened under column; (c) beams and slab; (d) slab with basement wall

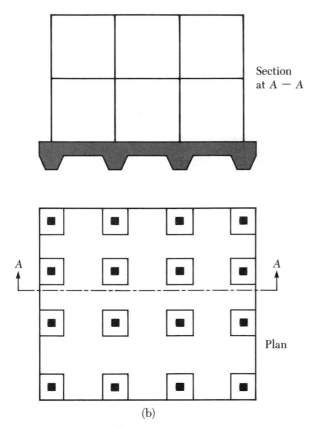

Section
at $A - A$

Plan

(b)

**Figure 4.2** (Continued)

1. Flat plate (Figure 4.2a): The mat is of uniform thickness.
2. Flat plate thickened under columns (Figure 4.2b).
3. Beams and slab (Figure 4.2c, p. 178): The beams run both ways, and the columns are located at the intersection of the beams.
4. Slab with basement walls as a part of the mat (Figure 4.2d, p. 179): The walls act as stiffeners for the mat.

   Mats are sometimes supported by piles. The piles help in reducing the settlement of a structure located over highly compressible soil. Where the ground water table is high, mats are often placed over piles to control buoyancy.

## 4.3
## Bearing Capacity of Mat Foundations

   The *gross ultimate bearing capacity* of a mat foundation can be determined by the same equation used for shallow foundations (see Section 3.5), or

$$q_u = cN_cF_{cs}F_{cd}F_{ci} + qN_qF_{qs}F_{qd}F_{qi} + \tfrac{1}{2}\gamma BN_\gamma F_{\gamma s}F_{\gamma d}F_{\gamma i} \tag{3.16}$$

(Chapter 3 gives the proper values of the bearing capacity factors, as well as the

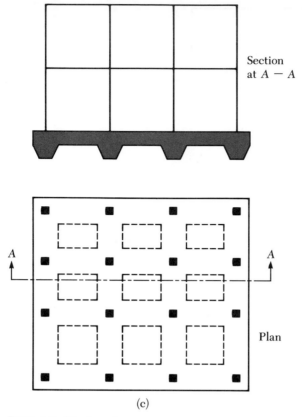

(c)

**Figure 4.2**  (Continued)

shape, depth, and load inclination factors.) The term $B$ in the preceding equation is the smallest dimension of the mat.

The *net ultimate capacity* can be given as

$$q_{\text{all(net)}} = q_u - q \tag{3.34}$$

A suitable factor of safety should be used to calculate the net *allowable* bearing capacity. For rafts on clay, the factor of safety should not be less than 3 under dead load and maximum live load. However, under the most extreme conditions, the factor of safety should be at least 1.75 to 2. For rafts constructed over sand, a factor of safety of 3 should normally be used. Under most working conditions, the factor of safety against bearing capacity failure of rafts on sand is very large.

For saturated clays with $\phi = 0$ and vertical loading condition, Eq. (3.16) gives (note: $N_c = 5.14$, $N_q = 1$, and $N_\gamma = 0$)

$$q_u = c_u N_c F_{cs} F_{cd} + q \tag{4.8}$$

where $c_u$ = undrained cohesion

From Eqs. (3.20) and (3.23), for $\phi = 0$ condition

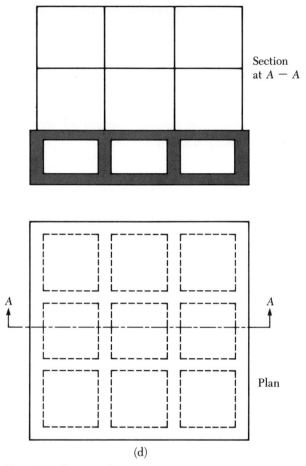

Section
at $A - A$

Plan

(d)

**Figure 4.2** (Continued)

$$F_{cs} = 1 + \frac{B}{L}\left(\frac{N_q}{N_c}\right) = 1 + \left(\frac{B}{L}\right)\left(\frac{1}{5.14}\right) = 1 + \frac{0.195B}{L}$$

and

$$F_{cd} = 1 + 0.4\left(\frac{D_f}{B}\right)$$

Substitution of the preceding shape and depth factors into Eq. (4.8) yields

$$q_u = 5.14c_u\left(\frac{1 + 0.195B}{L}\right)\left(1 + 0.4\frac{D_f}{B}\right) + q \qquad (4.9)$$

Hence, the net ultimate bearing capacity

$$q_{u(net)} = q_u - q = 5.14c_u\left(1 + \frac{0.195B}{L}\right)\left(1 + 0.4\frac{D_f}{B}\right) \qquad (4.10)$$

If a factor of safety of 3 is used, the net allowable soil bearing capacity can be given as

$$q_{\text{all(net)}} = \frac{q_{u(\text{net})}}{FS} = 1.713c_u\left(1 + \frac{0.195B}{L}\right)\left(1 + 0.4\frac{D_f}{B}\right) \tag{4.11}$$

Figure 4.3 shows a plot of $q_{\text{all(net)}}/c_u$ for various values of $L/B$ and $D_f/B$, based on Eq. (4.11).

The net allowable bearing capacity for mats constructed over granular soil deposits can be adequately determined from the standard penetration resistance numbers. Referring to Eq. (3.99b), for shallow foundations

$$q_{\text{all(net)}}(\text{kN/m}^2) = 11.98N\left(\frac{3.28B + 1}{3.28B}\right)^2 F_d\left(\frac{s}{25.4}\right)$$

where $N$ = corrected standard penetration resistance

$B$ = width (m)

$F_d = 1 + 0.33(D_f/B) \le 1.33$

$s$ = settlement, in mm

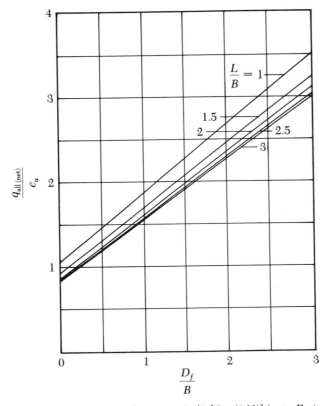

**Figure 4.3**  Plot of $q_{\text{all(net)}}/c_u$ against $D_f/B$ [Eq. (4.11)] (*note:* Factor of safety = 3)

When the width $B$ is large, the preceding equation can be approximated (assuming $3.28B + 1 \approx 3.28B$) as

$$q_{\text{all(net)}}(\text{kN/m}^2) \approx 11.98 N F_d \left(\frac{s}{2.54}\right)$$

$$= 11.98 N \left[1 + 0.33\left(\frac{D_f}{B}\right)\right]\left[\frac{s\,(\text{mm})}{25.4}\right]$$

$$\leq 15.93 N \left[\frac{s\,(\text{mm})}{25.4}\right] \tag{4.12}$$

Note that the original Eqs. (3.99a) and (3.99b) were for a settlement of 25.4 mm with a differential settlement of about 19 mm. However, the widths of the raft foundations are larger than the isolated spread footings. As can be seen from Figure 3.26 on p. 146, the depth of significant stress increase in the soil below a foundation is dependent on the foundation width. Hence, for a raft foundation, the depth of the zone of influence is likely to be much larger than that of a spread footing. This means that the loose soil pockets under a raft may be more evenly distributed, resulting in a smaller differential settlement. Thus it is customary to assume that for a maximum raft settlement of 50.8 mm, the differential settlement would be 19 mm. Using this logic and conservatively assuming $F_d$ to be equal to one, Eq. (4.12) can be approximated as

$$q_{\text{all(net)}}(\text{kN/m}^2) \approx 23.96 N \tag{4.13}$$

The net pressure applied on a foundation can be expressed as (Figure 4.4)

$$q = \frac{Q}{A} - \gamma D_f \tag{4.14}$$

where $Q$ = dead weight of the structure and the live load
        $A$ = area of the raft

Hence, in all cases, $q$ should be less than or equal to $q_{\text{all(net)}}$.

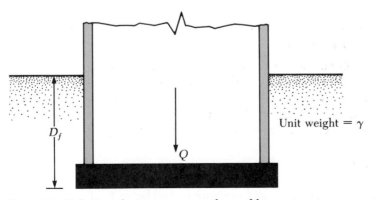

**Figure 4.4**  Definition of net pressure on soil caused by a mat foundation

## 4.4

Differential Settlement of Mats

The American Concrete Institute Committee 436 (1966) has suggested the following method for calculation of the differential settlement of mat foundations. According to this method, the rigidity factor $(K_r)$ can be calculated as

$$K_r = \frac{E' I_b}{E_s B^3} \tag{4.15}$$

where $E'$ = modulus of elasticity of the material used in the structure
$E_s$ = modulus of elasticity of the soil
$B$ = width of foundation
$I_b$ = moment of inertia of the structure per unit length at right angles to $B$

The term $E' I_b$ can be expressed as

$$E' I_b = E' \left( I_F + \sum I_{b'} + \sum \frac{ah^3}{12} \right) \tag{4.16}$$

where   $E' I_F$ = flexural rigidity of the foundation per unit length at right angles to $B$
$\sum E' I_b'$ = flexural rigidity of the framed members
$\sum(E' ah^3/12)$ = flexural rigidity of the shear walls
$a$ = shear wall thickness
$h$ = shear wall height

Based on the value of $K_r$, the ratio $(\delta)$ of the differential settlement to the total settlement can be estimated in the following manner:

1. If $K_r > 0.5$, it can be treated as a rigid mat, and $\delta = 0$.
2. If $K_r = 0.5$, then $\delta \approx 0.1$.
3. If $K_r = 0$, $\delta = 0.35$ for square mats $(B/L = 1)$, and $\delta = 0.5$ for long foundations $(B/L = 0)$.

## 4.5

Compensated Foundations

The settlement of a mat foundation can be reduced by decreasing the net pressure increase on soil, which can be done by increasing the depth of embedment, $D_f$. This increase is particularly important for rafts on soft clays, where large consolidation settlements are expected. From Eq. (4.14), the net average applied pressure on soil is (Figure 4.4)

$$q = \frac{Q}{A} - \gamma D_f$$

For no increase of the net soil pressure on soil below a raft foundation, $q$ should be equal to zero. Thus

$$D_f = \frac{Q}{A\gamma} \tag{4.17}$$

The relation for $D_f$ given in the preceding equation is usually referred to as the depth of a *fully compensated foundation*.

The factor of safety against bearing capacity failure for partially compensated foundations (that is, $D_f < Q/A\gamma$) may be given as

$$FS = \frac{q_{u(net)}}{q} = \frac{q_{u(net)}}{\dfrac{Q}{A} - \gamma D_f} \tag{4.18}$$

For saturated clays, the factor of safety against bearing capacity failure can thus be obtained by substituting Eq. (4.10) into Eq. (4.18):

$$FS = \frac{5.14c_u \left(1 + \dfrac{0.195B}{L}\right)\left(1 + 0.4\dfrac{D_f}{B}\right)}{\dfrac{Q}{A} - \gamma D_f} \tag{4.19}$$

## Example 4.1

Refer to Figure 4.4. The mat has dimensions of 30 m × 40 m. The live load and dead load on the mat are 200 MN. The mat is placed over a layer of soft clay. The unit weight of the clay is $18.75 \ kN/m^3$. Find $D_f$ for a fully compensated foundation.

**Solution**

From Eq. (4.17)

$$D_f = \frac{Q}{A\gamma} = \frac{200 \times 10^3 \ kN}{(30 \times 40)(18.75)} = \underline{8.89 \ m}$$

## Example 4.2

Refer to Example Problem 4.1. Given: $c_u$ for the clay is $12.5 \ kN/m^2$. If the required factor of safety against bearing capacity failure is 3, determine the depth of the foundation.

**Solution**

From Eq. (4.19)

$$FS = \frac{5.14c_u \left(1 + \dfrac{0.195B}{L}\right)\left(1 + 0.4\dfrac{D_f}{B}\right)}{\dfrac{Q}{A} - \gamma D_f}$$

$FS = 3$; $c_u = 12.5 \ kN/m^2$; $B/L = 30/40 = 0.75$; $Q/A = (200 \times 10^3)/(30 \times 40) = 166.67 \ kN/m^2$. Substitution of the preceding values into Eq. (4.19) yields

$$3 = \frac{(5.14)(12.5)[1 + (0.195)(0.75)]\left[1 + 0.4\left(\dfrac{D_f}{30}\right)\right]}{166.67 - (18.75)D_f}$$

$$500.01 - 56.25D_f = 73.65 + 0.982D_f$$
$$426.36 = 57.23D_f$$

or

$$D_f = 7.5 \text{ m}$$

## Example
## 4.3

Consider a mat foundation 30 m × 40 m in plan, as shown in Figure 4.5. The total dead load plus the live load on the raft is 200 MN. Estimate the consolidation settlement at the center of the foundation.

**Solution**

Given: $Q = 200 \times 10^3$ kN. So, net load per unit area is

$$q = \frac{Q}{A} - \gamma D_f = \frac{200 \times 10^3}{30 \times 40} - (15.7)(2) = 166.67 - 31.4 = 135.27 \text{ kN/m}^2$$

From Eq. (3.74), the average pressure increase on the clay layer below the center of the foundation is

$$\Delta p_{av} = \tfrac{1}{6}(\Delta p_t + \Delta p_m + \Delta p_b)$$

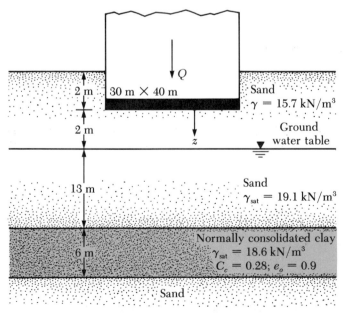

**Figure 4.5**

The values of $\Delta p_t$, $\Delta p_m$, and $\Delta p_b$ can be determined by referring to Figure 3.26 on p. 146. At the *top of the clay layer*

$$\frac{z}{B} = \frac{15}{30} = 0.5$$

$$\frac{L}{B} = \frac{40}{30} = 1.33$$

So, for $z/B = 0.5$ and $L/B = 1.33$

$$\frac{\Delta p_t}{q} = 0.75, \ \Delta p_t = (0.75)(135.27) = 101.45 \ kN/m^2$$

Similarly, for the *middle of the clay layer*

$$\frac{z}{B} = \frac{18}{30} = 0.6$$

$$\frac{L}{B} = 1.33$$

So, $\Delta p_m/q = 0.66$, $\Delta p_m = 89.3 \ kN/m^2$
At the *bottom of the clay layer*

$$\frac{z}{B} = \frac{21}{30} = 0.7$$

$$\frac{L}{B} = 1.33$$

So, $\Delta p_b/q = 0.58$, $\Delta p_b = 75.46 \ kN/m^2$
Hence

$$\Delta p_{av} = \tfrac{1}{6}[101.45 + 4(89.3) + 75.46] = 89 \ kN/m^2$$

From Eq. (1.59), the consolidation settlement

$$S = \frac{C_c H_c}{1 + e_o} \log \frac{p_o + \Delta p_{av}}{p_o}$$

$$p_o = 4(15.7) + 13(19.1 - 9.81) + \tfrac{6}{2}(18.6 - 9.81) = 209.94 \ kN/m^2$$

So

$$S = \frac{(0.28)(6 \times 1000)}{1 + 0.9} \log \left(\frac{209.94 + 89}{209.94}\right) = \underline{135.7 \ mm}$$

*Note:* Similar calculations can be made to obtain the settlement at the corner of the raft. The differential settlement between the center and the corner of the raft calculated in the preceding manner will, in most cases, be higher than the actual differential settlement. This is because of the stiffness of the superstructure.

The reliability of the settlement calculation will depend on several factors. If the clay is normally consolidated and uniform in thickness, the settlement calculation will yield fairly accurate results. If the compressible soil is erratic in formation and contains several types of soil, the calculation will indicate only the maximum magnitude.

## 4.6

### Structural Design of Mat Foundations

The structural design of mat foundations can be carried out by two conventional methods: The *conventional rigid method* and the *approximate flexible method*. Finite difference and finite element methods can also be used; however, this section will cover the basic concepts of the first two design methods.

### Conventional Rigid Method

The conventional rigid method of mat foundation design can be explained in a step-by-step manner with reference to Figure 4.6. The steps are as follows:

**1.** Figure 4.6a shows that the mat has a dimension of $L \times B$. $Q_1$, $Q_2$, $Q_3$, . . . are the column loads. Calculate the total column load as

$$Q = Q_1 + Q_2 + Q_3 \ldots \qquad (4.20)$$

**2.** Determine the pressure on the soil ($q$) below the mat at points $A$, $B$, $C$, $D$, . . . by using the equation

$$q = \frac{Q}{A} \pm \frac{M_y x}{I_y} \pm \frac{M_x y}{I_x} \qquad (4.21)$$

where $A = BL$
$\quad I_x = (1/12)BL^3 =$ moment of inertia about the $x$ axis
$\quad I_y = (1/12)LB^3 =$ moment of inertia about the $y$ axis
$\quad M_x =$ moment of the column loads about the $x$ axis $= Qe_y$
$\quad M_y =$ moment of the column loads about the $y$ axis $= Qe_x$

$e_x$ and $e_y$ are the load eccentricities in the direction of $x$ and $y$. They can be determined by using $x'$, $y'$ coordinates as

$$X' = \frac{Q_1 x_1' + Q_2 x_2' + Q_3 x_3' \ldots}{Q} \qquad (4.22)$$

and

$$e_x = X' - \frac{B}{2} \qquad (4.23)$$

Similarly

$$Y' = \frac{Q_1 y_1' + Q_2 y_2' + Q_3 y_3' \ldots}{Q} \qquad (4.24)$$

and

$$e_y = Y' - \frac{L}{2} \qquad (4.25)$$

**3.** Compare the values of the soil pressures determined in Step 2 with the net allowable soil pressure to check if $q \leq q_{\text{all(net)}}$.

**4.** Divide the mat into several strips in $x$ and $y$ directions (see Figure 4.6a). Let the width of any strip be $B_1$.

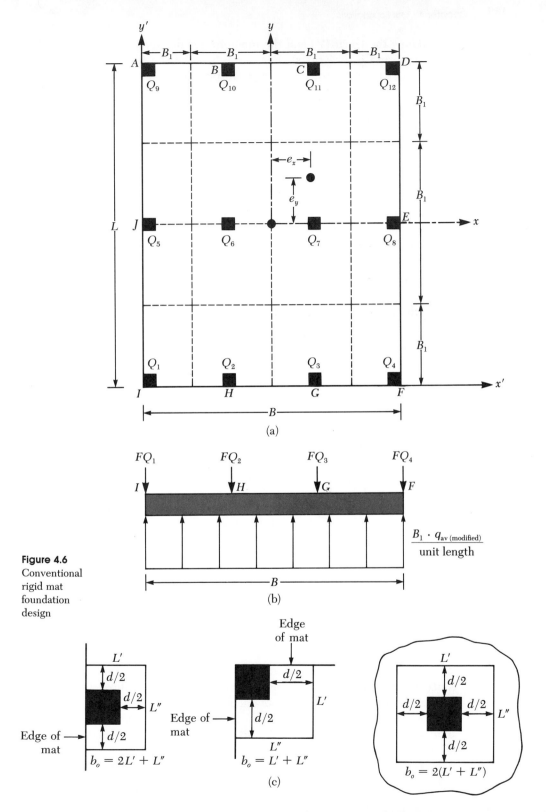

**Figure 4.6**
Conventional rigid mat foundation design

**5.** Draw the shear ($V$) and the moment ($M$) diagrams for each individual strip (in $x$ and $y$ directions). For example, take the bottom strip in the $x$ direction of Figure 4.6a; its average soil pressure can be given as

$$q_{av} \approx \frac{q_I + q_F}{2} \qquad (4.26)$$

where $q_I$ and $q_F$ = soil pressures at points $I$ and $F$ as determined from Step 2

The total soil reaction is equal to $q_{av}B_1B$. Now obtain the total column load on the strip as $Q_1 + Q_2 + Q_3 + Q_4$. The sum of the column loads on the strip will not be equal to $q_{av}B_1B$ because the shear between the adjacent strips has not been taken into account. For this reason, the soil reaction and the column loads need to be adjusted, or

$$\text{average load} = \frac{q_{av}B_1B + (Q_1 + Q_2 + Q_3 + Q_4)}{2} \qquad (4.27)$$

Now, the modified average soil reaction

$$q_{av(\text{modified})} = q_{av}\left(\frac{\text{average load}}{q_{av}B_1B}\right) \qquad (4.28)$$

Also, the column load modification factor is

$$F = \frac{\text{average load}}{Q_1 + Q_2 + Q_3 + Q_4} \qquad (4.29)$$

So, the modified column loads are $FQ_1$, $FQ_2$, $FQ_3$, and $FQ_4$. This modified loading on the strip under consideration is shown in Figure 4.6b. Now the shear and the moment diagram for this strip can be drawn. This procedure can be repeated for all strips in the $x$ and $y$ directions.

**6.** Determine the depth of the mat $d$ (refer to Figure A.1 in Appendix A for the definition of $d$). This can be done by checking for diagonal tension shear near various columns. According to ACI Code 318-77 (Section 11.11.1.2), for the critical section

$$U = b_o d[\phi(0.34)\sqrt{f_c'}] \qquad (4.30)$$

where $U$ = factored column load (MN) = (column load) × (load factor)
$\phi$ = reduction factor (see Appendix A)
$f_c'$ = compressive strength of concrete at 28 days (MN/m²)

The units of $b_o$ and $d$ in the preceding equation are in meters. The expression for $b_o$ in terms of $d$, which depends on the location of the column with respect to the plan of the mat, can be obtained from Figure 4.6c.

**7.** From the moment diagrams of all strips *in a given direction* (that is, $x$ or $y$), obtain the *maximum* positive and negative moments per unit width (that is, $M' = M/B_1$).

**8.** Determine the areas of steel per unit width for positive and negative reinforcement in $x$ and $y$ directions from the following equations [Appendix; Eqs. (A.4) and (A.6)]:

$$M_u = (M')(\text{load factor}) = \phi A_s f_y \left( d - \frac{a}{2} \right) \tag{A.6}$$

and

$$a = \frac{A_s f_y}{0.85 f_c' b} \tag{A.4}$$

where $A_s$ = area of steel per unit width
$f_y$ = yield stress of reinforcement in tension
$M_u$ = factored moment

Example Problems 4.4 and 4.5 illustrate the use of the conventional rigid method of mat foundation design.

## Example
## 4.4

The plan of a mat foundation with column loads is shown in Figure 4.7. Using Eq. (4.21), calculate the soil pressures at points $A$, $B$, $C$, $D$, $E$, and $F$. The size of the mat is 16.5 m × 21.5 m. All columns are 0.5 m × 0.5 m in section. Given: $q_{\text{all(net)}}$ = 60 kN/m². Determine that the soil pressures are less than the net allowable soil bearing capacity.

### Solution

From Eq. (4.21)

$$q = \frac{Q}{A} \pm \frac{M_y x}{I_y} \pm \frac{M_x y}{I_x}$$

$A$ = area of the mat = $(16.5)(21.5) = 354.75 \text{ m}^2$

$$I_x = \frac{1}{12} BL^3 = \frac{1}{12} (16.5)(21.5)^3 = 13,665 \text{ m}^4$$

$$I_y = \frac{1}{12} LB^3 = \frac{1}{12} (21.5)(16.5)^3 \approx 8050 \text{ m}^4$$

$Q$ = sum of the column loads = $350 + 2(400) + 450 + 2(500) + 2(1200)$
$\qquad + 4(1500) = 11,000 \text{ kN}$

$M_y = Q e_x$

$e_x = X' - \dfrac{B}{2}$

$$X' = \frac{Q_1 x_1' + Q_2 x_2' + Q_3 x_3' \cdots}{Q} = \frac{1}{11,000} [(8.25)(500 + 1500 + 1500 + 500)$$

$\qquad + (16.25)(350 + 1200 + 1200 + 450) + (0.25)(400 + 1500 + 1500$

$\qquad + 400)]$

$\qquad = 7.814 \text{ m}$

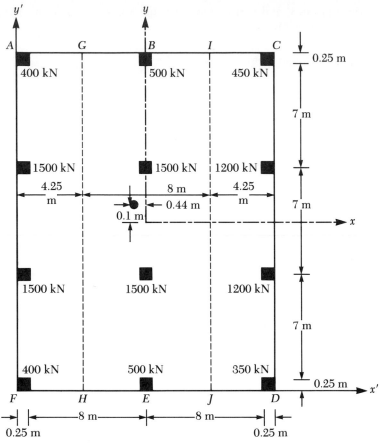

**Figure 4.7**

So

$$e_x = X' - \frac{B}{2} = 7.814 - 8.25 = -0.436 \approx -0.44 \text{ m}$$

Hence, the resultant line of action is located to the left of the center of the mat. So

$$M_y = (11,000)(0.44) = 4840 \text{ kN-m}$$

Similarly

$$M_x = Qe_y$$

$$e_y = \left(Y' - \frac{L}{2}\right)$$

$$Y' = \frac{Q_1y_1' + Q_2y_2' + Q_3y_3' + \dots}{Q} = \frac{1}{11,000}[(0.25)(400 + 500 + 350)$$

$$+ (7.25)(1500 + 1500 + 1200) + (14.25)(1500 + 1500 + 1200)$$

$$+ (21.25)(400 + 500 + 450)]$$

$$= 10.85 \text{ m}$$

So

$$e_y = Y' - \frac{L}{2} = 10.85 - \frac{21.5}{2} = 0.1 \text{ m}$$

The location of the line of action of the resultant column loads is shown in Figure 4.7.

$$M_x = (11,000)(0.1) = 1100 \text{ kN-m}$$

So

$$q = \frac{11,000}{354.75} \pm \frac{(4840)x}{8050} \pm \frac{(1100)y}{13,665} = 31.0 \pm 0.6x \pm 0.086 \text{ (kN/m}^2)$$

### Calculation of Soil Pressure

At A: $q = 31.0 + (0.6)(8.25) + (0.08)(10.75) = 36.81 \text{ kN/m}^2$

At B: $q = 31.0 + (0.6)(0) + (0.08)(10.75) = 31.86 \text{ kN/m}^2$

At C: $q = 31.0 - (0.6)(8.25) + (0.08)(10.75) = 26.91 \text{ kN/m}^2$

At D: $q = 31.0 - (0.6)(8.25) - (0.08)(10.75) = 25.19 \text{ kN/m}^2$

At E: $q = 31.0 + (0.6)(0) - (0.08)(10.75) = 30.14 \text{ kN/m}^2$

At F: $q = 31.0 + (0.6)(8.25) - (0.08)(10.75) = 35.09 \text{ kN/m}^2$

The soil pressures at all points are less than the given value of

$$q_{\text{all(net)}} = 60 \text{ kN/m}^2.$$

## Example 4.5

Divide the mat shown in Figure 4.7 into three strips, such as $AGHF$ ($B_1 = 4.25$ m), $GIJH$ ($B_1 = 8$ m), and $ICDJ$ ($B_1 = 4.25$ m). Using the results of Example Problem 4.4, determine the reinforcement requirements in the $y$ direction. Given: $f_c' = 20.7 \text{ MN/m}^2$ and $f_y = 413.7 \text{ MN/m}^2$. Use a load factor of 1.7.

### Solution
#### Determination of Shear and Moment Diagrams for Strips

*Strip AGHF:* Average soil pressure $= q_{\text{av}} = q_{\text{(at A)}} + q_{\text{(at F)}} = (36.81 + 35.09)/2 = 35.95 \text{ kN/m}^2$

Total soil reaction $= q_{\text{av}}B_1L = (35.95)(4.25)(21.50) = 3285 \text{ kN}$

Total column load on this strip $= 400 + 1500 + 1500 + 400 = 3800 \text{ kN}$

Average load $=$ (total soil reaction $+$ column loads)$/2 = (3285 + 3800)/2 = 3542.5 \text{ kN}$

So, modified average soil pressure $= q_{\text{av(modified)}} = q_{\text{av}}(3542.5/3285) = (35.95)(3542.5/3285) = 38.768 \text{ kN/m}^2$. The column loads can be modified in a similar manner by multiplying factor $F = 3542.5/3800 = 0.9322$.

Figure 4.8a shows the loading on the strip and the corresponding shear and moment diagrams. Note that the column loads shown in this figure have been multiplied by $F = 0.9322$. Also, the load per unit length of the beam is equal to $B_1 q_{\text{av(modified)}} = (4.25)(38.768) = 164.76 \text{ kN/m}$.

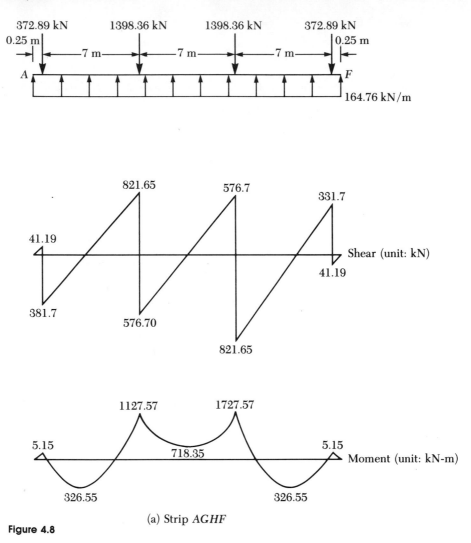

(a) Strip $AGHF$

**Figure 4.8**

*Strip GIJH:* In a similar manner

$$q_{av} = \frac{q_{(at\ B)} + q_{(at\ E)}}{2} = \frac{31.86 + 30.14}{2} = 31.0 \text{ kN/m}^2$$

Total soil reaction = (31)(8)(21.5) = 5332 kN
Total column load = 4000 kN
Average load = (5332 + 4000)/2 = 4666 kN
$q_{av(modified)}$ = 31.0 (4666/5332) = 27.12 kN/m$^2$
$F$ = 4666/4000 = 1.1665
The load shear and moment diagrams are shown in Figure 4.8b.
    *Strip ICDJ:* Figure 4.8c shows the load shear and moment diagrams for this strip.

## Determination of the Thickness of the Mat

For this problem, the critical section for diagonal tension shear will be at the column carrying 1500 kN load at the edge of the mat (Figure 4.8d). So

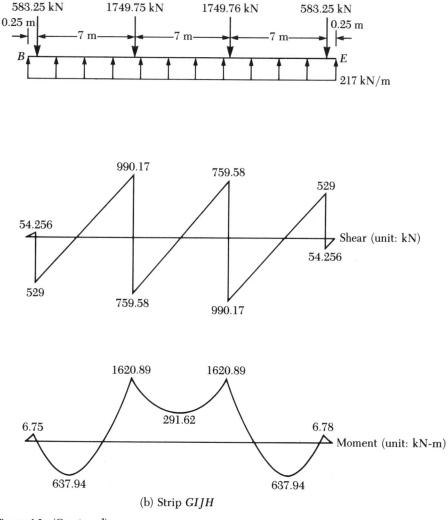

(b) Strip *GIJH*

**Figure 4.8** (Continued)

$$b_o = \left(0.5 + \frac{d}{2}\right) + \left(0.5 + \frac{d}{2}\right) + (0.5 + d) = 1.5 + 2d$$

$$U = (b_o d)[(\phi)(0.34)\sqrt{f_c'}]$$

$$U = (1.7)(1500) = 2550 \text{ kN} = 2.55 \text{ MN}$$

$$2.55 = (1.5 + 2d)(d)[(0.85)(0.34)\sqrt{20.7}]$$

or

$$(1.5 + 2d)d = 1.94$$

$$d \approx 0.68 \text{ m}$$

Assuming a minimum cover of 76 mm over the steel reinforcement and also assuming that the steel bars to be used are 25 mm in diameter, the total thickness of the slab is

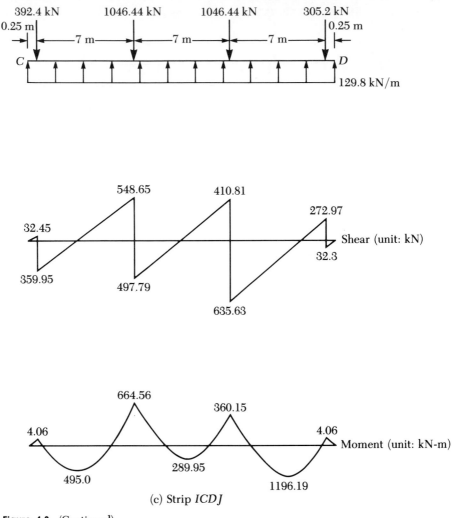

(c) Strip *ICDJ*

**Figure 4.8**  (Continued)

$$h = 0.68 + 0.076 + 0.025 = 0.781 \text{ m} \approx \underline{0.8 \text{ m}}$$

The thickness of this mat will satisfy the wide beam shear condition across the three strips under consideration.

### Determination of Reinforcement

From the moment diagrams shown in Figure 4.8a, b, and c, it can be seen that the maximum positive moment is located in strip *AGHF*, and its magnitude is

$$M' = \frac{1727.57}{B_1} = \frac{1727.57}{4.25} = 406.5 \text{ kN-m/m}$$

Similarly, the maximum negative moment is located in strip *ICDJ*, and its magnitude is $1196.19/B_1 = 1196.19/4.25 = 281.5 \text{ kN-m/m}$.

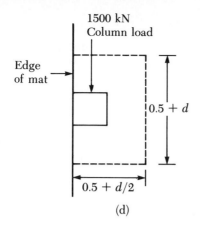

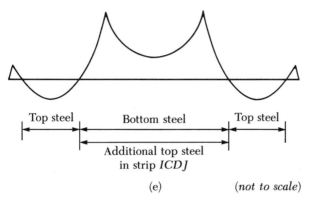

**Figure 4.8** (Continued)

Now, referring to Eq. (A.6) in Appendix A

$$M_u = (M')(\text{load factor}) = \phi A_s f_y \left(d - \frac{a}{2}\right)$$

For the positive moment

$$M_u = (406.5)(1.7) = (\phi)(A_s)(413.7 \times 1000)\left(0.68 - \frac{a}{2}\right)$$

$$\phi = 0.9$$

Also, from Eq. (A.4) in Appendix A

$$a = \frac{A_s f_y}{0.85 f'_c b} = \frac{(A_s)(413.7)}{(0.85)(20.7)(1)} = 23.51 A_s$$

or

$$A_s = 0.0425a$$

So

$$M_u = 691.22 = (0.9)(0.0425a)(413700)\left(0.68 - \frac{0.0425a}{2}\right)$$

or

$$a \approx 0.0645$$

So

$$A_s = (0.0425)(0.0645) = 0.00274 \text{ m}^2/\text{m} = 2740 \text{ mm}^2/\text{m}$$

Hence, use 25 mm-diameter bars at 175 mm center-to-center [$A_s$ provided = (491)(1000/175) = 2805.7 mm²/m].

Similarly, for negative reinforcement

$$M_u = (281.5)(1.7) = \phi A_s (413.7 \times 1000)\left(0.68 - \frac{a}{2}\right)$$

$\phi = 0.9$; $A_s = 0.0425a$. So

$$478.38 = (0.9)(0.0425a)(413.7 \times 1000)\left(0.68 - \frac{0.0425a}{2}\right)$$

Solution of the preceding equation gives $a \approx 0.045$. So

$$A_s = (0.045)(0.0425) = 0.001913 \text{ m}^2/\text{m} = 1913 \text{ mm}^2/\text{m}$$

Hence, use 25-mm diameter bars at 250 mm center-to-center ($A_s$ provided = 1924 mm²).

Because negative moment occurs at midbay of strip *ICDJ*, reinforcement should be provided. This moment is $289.95/4.25 = 68.22$ kN-m/m. Hence

$$M_u = (68.22)(1.7) = (0.9)(0.0425a)(413.7 \times 1000)\left(0.68 - \frac{0.0425a}{2}\right)$$

Solution of the preceding equation gives $a \approx 0.0108$. So

$$A_s = (0.0108)(0.0425) = 0.000459 \text{ m}^2/\text{m} = 459 \text{ mm}^2/\text{m}$$

Provide 16-mm diameter bars at 400 mm center-to-center ($A_s$ provided = 502 mm²).

For general arrangement of the reinforcement see Figure 4.8e. Note that a similar type of analysis needs to be done to determine the reinforcement in the *x* direction of the mat.

## Approximate Flexible Method

In the conventional rigid method of design, the mat is assumed to be infinitely rigid. Also, the soil pressure is distributed in a straight line, and the centroid of the soil pressure is coincidental with the line of action of the resultant column loads (see Figure 4.9a). In the approximate flexible method of design, the soil is assumed to be equivalent to an infinite number of elastic

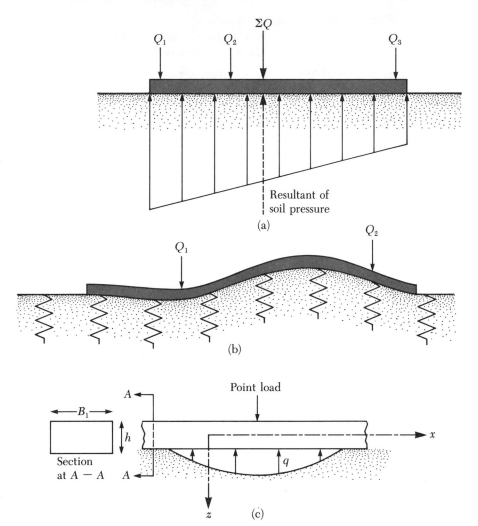

**Figure 4.9** (a) Principles of design by conventional rigid method; (b) principles of approximate flexible method; (c) derivation of Eq. (4.35) for beams on elastic foundation

springs, as shown in Figure 4.9b. This is sometimes referred to as the *Winkler foundation*. The elastic constant of these assumed springs is referred to as the *coefficient of subgrade reaction, k.*

In order to understand the fundamental concepts behind flexible foundation design, consider a beam of width $B_1$ having an infinite length, as shown in Figure 4.9c. The beam is subjected to a single concentrated load $Q$. From the fundamentals of mechanics of materials

$$M = E_F I_F \frac{d^2z}{dx^2} \tag{4.31}$$

where $M$ = moment at any section

$\quad\quad E_F$ = Young's modulus of the material of the beam

$\quad\quad I_F$ = moment of inertia of the cross section of the beam = $(1/12)\, B_1 h^3$ (see Figure 4.9c)

However

$$\frac{dM}{dx} = \text{shear force} = V$$

and

$$\frac{dV}{dx} = q = \text{soil reaction}$$

Hence

$$\frac{d^2 M}{dx^2} = q \tag{4.32}$$

Now, combining Eqs. (4.31) and (4.32)

$$E_F I_F \frac{d^4 z}{dx^4} = q \tag{4.33}$$

However, the soil reaction

$$q = -zk'$$

where $z$ = deflection

$\quad\quad k' = kB_1$

$\quad\quad k$ = coefficient of subgrade reaction (unit—kN/m³)

So

$$E_F I_F \frac{d^4 z}{dx^2} = -zkB_1 \tag{4.34}$$

Solution of the preceding equation yields

$$z = e^{-\alpha x}(A' \cos \beta x + A'' \sin \beta x) \tag{4.35}$$

where $A'$ and $A''$ are constants and

$$\beta = \sqrt[4]{\frac{B_1 k}{4 E_F I_F}} \tag{4.36}$$

The unit of the term $\beta$ as defined by the preceding equation is (length)⁻¹. This is a very important parameter in the determination of whether a mat foundation should be designed by conventional rigid method or approximate flexible method. According to the American Concrete Institute Committee 436 (1966), the design of mats should be done by the conventional rigid method if the spacings of columns in a strip are less than $1.75/\beta$. If the spacings of columns are larger than $1.75/\beta$, the approximate flexible method may be adopted. For example, determine whether or not the mat foundation in Example Problems 4.4 and 4.5 is rigid. Let the value of $k$ for that foundation be 8000 kN/m³. The strips $AGHF$ and $ICDJ$ have the lower value of $B_1$—that is, 4.25 m. So, according to Eq. (4.36)

$$\beta = \sqrt[4]{\frac{B_1 k}{4 E_F I_F}}$$

For this section, $I_F = (1/12)\, B_1 h^3 = (1/12)(4.25)(0.8)^3 = 0.1813 \text{ m}^4$. Also, let $E_F$ be equal to $21 \times 10^6$ kN/m². So

$$\beta = \sqrt[4]{\frac{(4.25)(8000)}{(4)(21 \times 10^6)(0.1813)}} = 0.217 \text{ m}^{-1}$$

So

$$\frac{1.75}{\beta} = \frac{1.75}{0.217} = 8.07 \text{ m}$$

Because the actual spacing of columns is 7 m, which is less than $1.75/\beta$, this mat should be designed by the conventional rigid method.

To perform the analysis for the structural design of a flexible mat, one must know the principles of evaluating the *coefficient of subgrade reaction, k*. Before proceeding with the discussion of the approximate flexible design method, let us discuss this coefficient in more detail.

If a foundation of width $B$ (Figure 4.10) is subjected to a load per unit area of $q$, it will undergo a settlement $\Delta$. The coefficient of subgrade modulus $k$ can be defined as

$$k = \frac{q}{\Delta} \tag{4.37}$$

The unit of $k$ is in kN/m³ (or lb/in.³). The value of the coefficient of subgrade reaction is not a constant for a given soil. It depends on several factors, such as the length ($L$) and width ($B$) of the foundation and also the depth of embedment of the foundation. A comprehensive study of the parameters affecting the coefficient of subgrade reaction has been given by Terzaghi (1955). According to this study, the value of the coefficient of subgrade reaction decreases with the width of the foundation. In the field, load tests can be carried out by means of

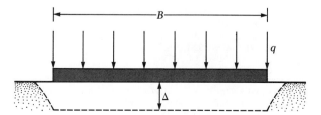

**Figure 4.10** Definition of coefficient of subgrade reaction, $k$.

square plates measuring 0.3 m × 0.3 m (1 ft × 1 ft), and values of $k$ can be calculated. The value of $k$ can be related to large foundations measuring $B \times B$ as follows:

*Foundations on Sandy Soils:*

$$k = k_{0.3} \left( \frac{B + 0.3}{2B} \right)^2 \tag{4.38}$$

where $k_{0.3}$ and $k$ = coefficients of subgrade reaction of footings measuring 0.3 (m) × 0.3 (m) and $B$ (m) × $B$ (m), respectively (unit $kN/m^3$).

*Foundations on Clays:*

$$k \ (kN/m^3) = k_{0.3} \ (kN/m^3) \left[ \frac{0.3 \ (m)}{B \ (m)} \right] \tag{4.39}$$

The definition of $k$ in Eqs. (4.39) is the same as that given in Eq. (4.38).

For rectangular foundations having dimensions of $B \times L$ (for similar soil and $q$)

$$k = \frac{k_{(B \times B)} \left( 1 + \dfrac{B}{L} \right)}{1.5} \tag{4.40}$$

where $k$ = coefficient of subgrade modulus of the rectangular foundation $(L \times B)$

$k_{(B \times B)}$ = coefficient of subgrade modulus of a square foundation having dimension of $B \times B$

The preceding equation indicates that the value of $k$ of a very long foundation with a width $B$ is approximately equal to $0.67 k_{(B \times B)}$.

The Young's modulus of granular soils increases with depth. Because of the fact that the settlement of a foundation is dependent on the Young's modulus, the value of $k$ increases as the depth of the foundation increases.

Following are some typical ranges of value for the coefficient of subgrade reaction $k_{0.3}$ for sandy and clayey soils.

---

| Sand (dry or moist) |
| --- |
| Loose: 8–25 $MN/m^3$ (29–92 $lb/in.^3$) |
| Medium: 25–125 $MN/m^3$ (91–460 $lb/in.^3$) |
| Dense: 125–375 $MN/m^3$ (460–1380 $lb/in.^3$) |

| Sand (saturated) |
| --- |
| Loose: 10–15 $MN/m^3$ (38–55 $lb/in.^3$) |
| Medium: 35–40 $MN/m^3$ (128–147 $lb/in.^3$) |
| Dense: 130–150 $MN/m^3$ (478–552 $lb/in.^3$) |

| Clay |
| --- |
| Stiff ($q_u$ = 100–200 $kN/m^2$): 12–25 $MN/m^3$ (44–92 $lb/in.^3$) |
| Very stiff ($q_u$ = 200–400 $kN/m^2$): 25–50 $MN/m^3$ (92–184 $lb/in.^3$) |
| Hard ($q_u$ > 400 $kN/m^2$): > 50 $MN/m^3$ (> 184 $lb/in.^3$) |

(*Note:* $q_u$ = unconfined compression strength)

Scott (1981) has proposed that for sandy soils, the value of $k_{0.3}$ can be obtained from standard penetration resistance at any given depth as

$$k_{0.3} \ (\text{MN/m}^3) = 1.8N \tag{4.41}$$

where $N = corrected$ standard penetration resistance

For long beams, Vesic (1961) proposed an equation for estimation of subgrade reaction that can be expressed as

$$k' = Bk = 0.65 \ \sqrt[12]{\frac{E_s B^4}{E_F I_F}} \ \frac{E_s}{1 - \mu^2}$$

or

$$k = 0.65 \ \sqrt[12]{\frac{E_s B^4}{E_F I_F}} \ \frac{E_s}{B(1 - \mu^2)} \tag{4.42}$$

where $E_s =$ Young's modulus of soil
$\quad \ \ B =$ foundation width
$\quad \ E_F =$ Young's modulus of foundation material
$\quad \ \ I_F =$ moment of inertia of the cross section of the foundation
$\quad \ \ \mu =$ Poisson's ratio of soil

For most practical purposes, Eq. (4.42) can be approximated as

$$k = \frac{E_s}{B(1 - \mu^2)} \tag{4.43}$$

The coefficient of subgrade reaction is also a very useful parameter in the design of rigid highway and airfield pavements. The pavement with a concrete wearing surface is generally referred to as a *rigid pavement*, and the pavement with an asphaltic wearing surface is called a *flexible pavement*. For a surface load acting on a rigid pavement, the maximum tensile stress occurs at the base of the slab. To estimate the magnitude of the maximum horizontal tensile stress developed at the base of the rigid pavement, elastic solutions involving slabs on Winkler foundations are extremely useful. Some of the early work in this area was done by Westergaard (1926, 1939, 1947).

Now that we have discussed the coefficient of subgrade reaction, we will proceed with the discussion of the approximate flexible method of designing mat foundations. This method, as proposed by the American Concrete Institute Committee 436 (1966) is described below in a step-by-step manner. The design procedure is primarily based on the theory of plates. Its use allows the effects (that is, moment, shear, and deflection) of a concentrated column load in the area surrounding it to be evaluated. If the zones of influence of two or more columns overlap, the method of superposition can be used to obtain the net moment, shear, and deflection at any point.

**1.** Assume a thickness ($h$) for the mat. This can be done according to Step 6 as outlined in the preceding section on the conventional rigid method. (*Note: h* is the *total* thickness of the mat.)

**2.** Determine the flexural rigidity $R$ of the mat as

$$R = \frac{E_F h^3}{12(1 - \mu_F^2)} \tag{4.44}$$

where $E_F$ = Young's modulus of foundation material
$\mu_F$ = Poisson's ratio of foundation material

**3.** Determine the radius of effective stiffness as

$$L' = \sqrt[4]{\frac{R}{k}} \tag{4.45}$$

where $k$ = coefficient of subgrade reaction

The zone of influence of any column load will be in the order of 3 to 4 $L'$.

**4.** Determine the moment in polar coordinate system at a point caused by a column load from the following equations (Figure 4.11a).

$$M_t = \text{tangential moment} = -\frac{Q}{4}\left[ A_1 - \frac{(1 - \mu_F)A_2}{\frac{r}{L'}} \right] \tag{4.46}$$

$$M_r = \text{radial moment} = -\frac{Q}{4}\left[ \mu_F A_1 + \frac{(1 - \mu_F)A_2}{\frac{r}{L'}} \right] \tag{4.47}$$

where $r$ = radial distance from the column load
$Q$ = column load
$A_1$, $A_2$ = functions of $r/L'$

The variations of $A_1$ and $A_2$ with $r/L'$ are shown in Figure 4.11b (for details, see Hetenyi, 1946).

In the cartesian coordinate system (Figure 4.11a)

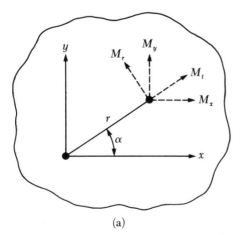

(a)

**Figure 4.11** Approximate flexible method of mat design

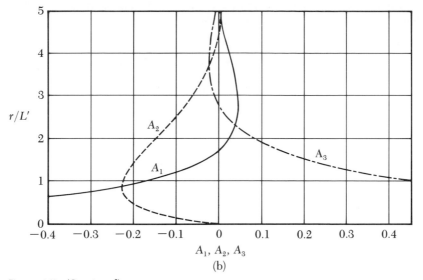

Figure 4.11   (Continued)

$$M_x = M_t \sin^2 \alpha + M_r \cos^2 \alpha \qquad\qquad (4.48)$$
$$M_y = M_t \cos^2 \alpha + M_r \sin^2 \alpha \qquad\qquad (4.49)$$

**5.** Determine the shear force $(V)$ for the unit width of the mat caused by a column load as

$$V = -\frac{Q}{4L'} A_3 \qquad\qquad (4.50)$$

The variation of $A_3$ with $r/L'$ is shown in Figure 4.11b.

**6.** If the edge of the mat is located in the zone of influence of a column, determine the moment and shear along the wedge, assuming that the mat is continuous. Moment and shear opposite in sign to those determined are applied at the edges to satisfy the known conditions.

---

## Problems

**4.1** Determine the net ultimate bearing capacity of a mat foundation measuring 45 ft × 30 ft on a saturated clay with $c_u = 1950$ lb/ft², $\phi = 0$, and $D_f = 6.5$ ft. [Use Eq. (4.10).]

**4.2** What will be the net allowable bearing capacity of a mat foundation with dimensions of 15 m × 10 m constructed over a sand deposit? Given: $D_f = 2$ m, allowable settlement = 30 mm, corrected average penetration number, $N = 10$. [Use Eq. (4.12).]

**4.3** Repeat Problem 4.2 for an allowable settlement of 50.8 mm.

**4.4** Consider a mat foundation with dimensions of 60 ft × 40 ft. The dead and live load on the mat is $10^7$ lb. The mat is to be placed on a clay with $c_u = 850$ lb/ft². The unit weight of the clay is 112 lb/ft³. Find the depth, $D_f$, of the mat for a fully compensated foundation.

**4.5**  For the mat considered in Problem 4.4, what will be the depth of the mat $(D_f)$ for a FS = 3 against bearing capacity failure?

**4.6**  Repeat Problem 4.5 assuming that the undrained cohesion of the clay is 1250 lb/ft².

**4.7**  Consider the mat foundation shown in Figure P4.7. Given: $L$ = 12 m, $B$ = 10 m, $Q$ = 25 MN, $D_f$ = 1.5 m, $x_1$ = 2 m, $x_2$ = 3 m, $x_3$ = 4 m. The clay is normally consolidated. Estimate the consolidation settlement under the center of the mat.

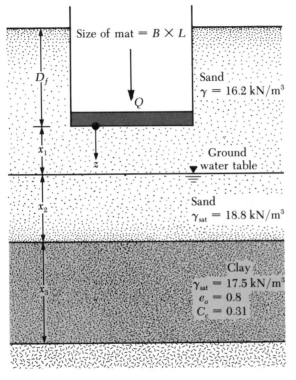

**Figure P4.7**

**4.8**  For the mat foundation described in Problem 4.7, estimate the consolidation settlement under the corner of the mat.

**4.9**  Redo Problem 4.7 assuming that the preconsolidation pressure of the clay is 120 kN/m² and the swelling index is about $\frac{1}{4} C_c$.

**4.10**  Refer to Figure P4.10. For the mat, given: $Q_1$, $Q_3$ = 40 ton, $Q_4$, $Q_5$, $Q_6$ = 60 ton, $Q_2$, $Q_9$ = 45 ton, $Q_7$, $Q_8$ = 50 ton. All columns are 20 in. × 20 in. in cross section. Using the procedure outlined on pp. 186, determine the pressure on the soil at $A$, $B$, $C$, $D$, $E$, $F$, $G$, and $H$.

**4.11**  Refer to the mat foundation described in Problem 4.10. Divide the mat into three strips in the $y$ direction ($B_1$ = 8 ft, 16 ft, and 8 ft). Determine the reinforcement requirement

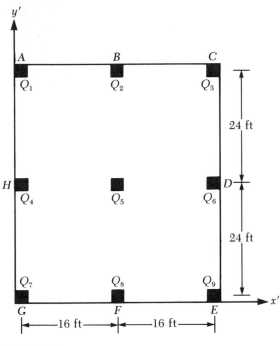

**Figure P4.10**

according to the conventional rigid method. Given: $f'_c = 3000$ lb/in.$^2$ and $f_y = 60,000$ lb/in.$^2$ Use a load factor of 1.6.

**4.12** Refer to Example Problem 4.4. Divide the mat shown in Figure 4.7 into four strips and determine the reinforcement requirement according to the conventional rigid method.

**4.13** From the plate-load test (plate dimension 1 ft × 1 ft) in the field, the coefficient of subgrade reaction of a sandy soil was determined to be 80 lb/in.$^3$ What will be the value of the coefficient of subgrade reaction on the same soil for a foundation with dimensions of 30 ft × 30 ft?

**4.14** Refer to Problem 4.13. If the full-sized foundation has dimensions of 45 ft × 30 ft, what will be the value of the coefficient of subgrade reaction?

## References

American Concrete Institute (1977). *ACI Standard—Building Code Requirements for Reinforced Concrete,* ACI 318-77, Detroit.

American Concrete Institute Committee 436 (1966). "Suggested Design Procedures for Combined Footings and Mats," *Journal of the American Concrete Institute,* Vol. 63, No. 10, pp. 1041–1057.

Hetenyi, M. (1946). *Beams on Elastic Foundations,* University of Michigan Press, Ann Arbor.

Scott, R. F. (1981). *Foundations Analysis,* Prentice-Hall, Englewood Cliffs, N.J.

Terzaghi, K. (1955). "Evaluation of the Coefficient of Subgrade Reactions," *Geotechnique,* Institute of Engineers, London, Vol. 5, No. 4, pp. 197–326.

Vesic, A. S. (1961). "Bending of Beams Resting on Isotropic Elastic Solid," *Journal of the Engineering Mechanics Division,* American Society of Civil Engineers, Vol. 87, No. EM2, pp. 35–53.

Westergaard, H. M. (1926). "Stresses in Concrete Pavements Computed by Theoretical Analysis," *Public Roads,* Vol. 7, No. 12, pp. 23–35.

Westergaard, H. M. (1939). "Stresses in Concrete Runways of Airports," *Proceedings,* Highway Research Board, Vol. 19, pp. 197–205.

Westergaard, H. M. (1947). "New Formulas for Stresses in Concrete Pavements of Airfields," *Proceedings,* American Society of Civil Engineers, Vol. 73, pp. 687–701.

# Lateral Earth Pressure and Retaining Walls

## 5.1

### Introduction

A *retaining wall* is a wall that provides lateral support for a *vertical* or *near-vertical* slope of soil. It is a common structure used in many construction projects. The most common types of retaining wall may be classified as follows:

1. Gravity retaining walls
2. Semigravity retaining walls
3. Cantilever retaining walls
4. Counterfort retaining walls

*Gravity retaining walls* (Figure 5.1a) are constructed with plain concrete or stone masonry. These walls depend on their own weight and any soil resting on the masonry for their stability. This type of construction is not very economical for high walls.

In many cases, a small amount of steel may be used for the construction of gravity walls, thereby minimizing the size of the wall sections. Such walls are generally referred to as *semigravity walls* (Figure 5.1b).

*Cantilever retaining walls* (Figure 5.1c) are made of reinforced concrete that consists of a thin stem and a base slab. This type of wall is economical up to a height of about 8 m.

*Counterfort retaining walls* (Figure 5.1d) are similar to cantilever walls except for the fact that, at regular intervals, they have thin vertical concrete slabs known as *counterforts* that tie the wall and the base slab together. The purpose of the counterforts is to reduce the shear and the bending moments.

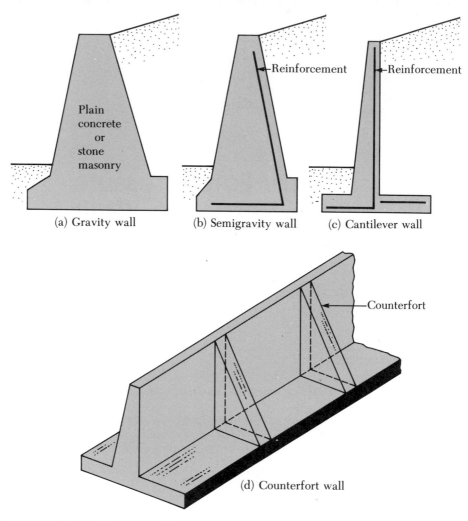

(a) Gravity wall                (b) Semigravity wall          (c) Cantilever wall

(d) Counterfort wall

**Figure 5.1**   Types of retaining wall

To properly design retaining walls, one must know the basic soil parameters—that is, the *unit weight, angle of friction,* and *cohesion*—for the soil retained behind the wall and the soil below the base slab. Knowing the properties of the soil behind the wall enables an engineer to determine the *lateral force* for which the design has to be made.

There are two phases in the design of a retaining wall. First, with the lateral earth pressure known, the structure as a whole is checked for *stability*. This includes checking for possible *overturning, sliding,* and *bearing capacity* failures. Second, each component of the structure is checked for *adequate strength,* and the *steel reinforcement* of each component is determined.

This chapter presents the procedures for determination of lateral earth pressure and retaining-wall stability. Checks for adequate strength of each component of the structures are covered in Appendix A.

## Lateral Earth Pressure

### 5.2

Lateral Earth Pressure At-Rest

Consider a vertical wall of height $H$, as shown in Figure 5.2, retaining a soil having a unit weight of $\gamma$. A uniformly distributed load, $q$/unit area, is also applied at the ground surface. The shear strength, $s$, of the soil may be given by the equation

$$s = c + \sigma' \tan \phi$$

where $c$ = cohesion
$\phi$ = angle of friction
$\sigma'$ = effective normal stress

At any depth $z$ below the ground surface, the vertical stress below the ground surface can be given as

$$\sigma_v = q + \gamma z \tag{5.1}$$

If the *wall is not allowed to move at all* either way from the soil mass or into the soil mass, the lateral pressure at a depth $z$ can be given as

$$\sigma_h = K_o \sigma_v' + u \tag{5.2}$$

where $u$ = pore water pressure
$K_o$ = coefficient of at-rest earth pressure

For granular soil, the relation for $K_o$ is

$$K_o = 1 - \sin \phi \tag{5.3}$$

For normally consolidated clays, the coefficient of earth pressure at-rest can be approximated as (Brooker and Ireland, 1965)

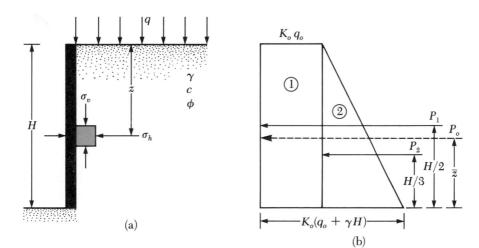

(a)

(b)

**Figure 5.2**  At-rest earth pressure

$$K_o = 0.95 - \sin \phi \tag{5.4}$$

where $\phi$ = drained friction angle

Based on the experimental results of Brooker and Ireland (1965), the value of $K_o$ for normally consolidated clays can be approximately correlated with plasticity index $(PI)$ as

$$K_o = 0.4 + 0.007(PI) \qquad \text{(for } PI \text{ between 0 and 40)} \tag{5.5}$$

and

$$K_o = 0.64 + 0.001(PI) \qquad \text{(for } PI \text{ between 40 and 80)} \tag{5.6}$$

For overconsolidated clays

$$K_{o(\text{overconsolidated})} \approx K_{o(\text{normally consolidated})} \sqrt{OCR} \tag{5.7}$$

where $OCR$ = overconsolidation ratio

With a properly selected value of the at-rest earth pressure coefficient, Eq. (5.2) can be used to determine the variation of lateral earth pressure with depth, $z$. Figure 5.2b shows the variation of $\sigma_h$ with depth for the wall shown in Figure 5.2a. Note that if the surcharge $q = 0$ and the pore water pressure $u = 0$, the pressure diagram will be a triangle. The total force, $P_o$, *per unit length* of the wall given in Figure 5.2a can now be obtained from the area of the pressure diagram given in Figure 5.2b as

$$P_o = P_1 + P_2 = q_o K_o H + \tfrac{1}{2} \gamma H^2 K_o \tag{5.8}$$

where $P_1$ = area of rectangle 1
$P_2$ = area of triangle 2

The location of the line of action of the resultant force, $P_o$, can be obtained by taking the moment about the bottom of the wall. Thus

$$\bar{z} = \frac{P_1 \left( \dfrac{H}{2} \right) + P_2 \left( \dfrac{H}{3} \right)}{P_o} \tag{5.9}$$

If the water table is located at a certain depth, $z < H$, the at-rest pressure diagram shown in Figure 5.2b will have to be somewhat modified. This is shown in Figure 5.3. Let the effective unit weight of soil below the water be equal to $\gamma'$ (that is, $\gamma_{\text{sat}} - \gamma_w$). For this condition

At $z = 0$, $\sigma_h' = K_o \sigma_v' = K_o q$
At $z = H_1$, $\sigma_h' = K_o \sigma_v' = K_o (q + \gamma H_1)$
At $z = H_2$, $\sigma_h' = K_o \sigma_v' = K_o (q + \gamma H_1 + \gamma' H_2)$

Note that in the preceding equations, $\sigma_v'$ and $\sigma_h'$ are effective vertical and horizontal pressures. In order to determine the total pressure distribution on the wall, the hydrostatic pressure has to be added. The hydrostatic pressure, $u$, is equal to zero from $z = 0$ to $z = H_1$; at $z = H_2$, $u = H_2 \gamma_w$. The variation of $\sigma_h'$

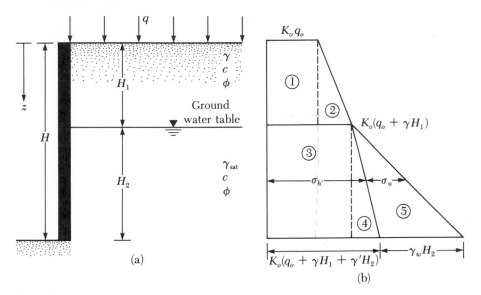

Figure 5.3

and $u$ with depth is shown in Figure 5.3b. Hence the total force per unit length of the wall can be determined from the area of the pressure diagram. Thus

$$P_o = A_1 + A_2 + A_3 + A_4 + A_5$$

where $A$ = area of the pressure diagram

So

$$P_o = K_o q_o H_1 + \tfrac{1}{2} K_o \gamma H_1^2 + K_o (q + \gamma H_1) H_2 + \tfrac{1}{2} K_o \gamma' H_2^2 + \tfrac{1}{2} \gamma_w H_2^2 \quad (5.10)$$

**Example 5.1**

For a retaining wall shown in Figure 5.4a, determine the lateral earth force at rest per unit length of the wall. Also determine the location of the resultant earth pressure.

**Solution**

$$K_o = 1 - \sin \phi = 1 - \sin 30° = 0.5$$

At $z = 0$, $\sigma'_v = 0$; $\sigma'_h = 0$

At $z = 2.5$ m, $\sigma'_v = (16.5)(2.5) = 41.25$ kN/m²;

$$\sigma'_h = K_o \sigma'_v = (0.5)(41.25) = 20.63 \text{ kN/m}^2$$

At $z = 5$ m, $\sigma'_v = (16.5)(2.5) + (19.3 - 9.81)2.5 = 64.98$ kN/m²;

$$\sigma'_h = K_o \sigma'_v = (0.5)(64.98) = 32.49 \text{ kN/m}^2$$

The hydrostatic pressure distribution is as follows:
From $z = 0$ to $z = 2.5$ m, $u = 0$. At $z = 5$ m, $u = \gamma_w(2.5) = (9.81)(2.5) = 24.53$ kN/m². The pressure distribution for the wall is shown in Figure 5.4b.
The total force per unit length of the wall can be determined from the area of the pressure diagram, or

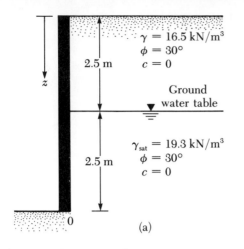

(a)

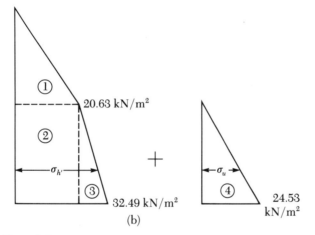

(b)

**Figure 5.4**

$$P_o = \text{Area 1} + \text{Area 2} + \text{Area 3} + \text{Area 4}$$
$$= \tfrac{1}{2}(2.5)(20.63) + (2.5)(20.63) + \tfrac{1}{2}(2.5)(32.49 - 20.63)$$
$$+ \tfrac{1}{2}(2.5)(24.53) = 122.85 \text{ kN/m}$$

The location of the center of pressure measured from the bottom of the wall (point $O$) =

$$\bar{z} = \frac{(\text{Area 1})\left(2.5 + \dfrac{2.5}{3}\right) + (\text{Area 2})\left(\dfrac{2.5}{2}\right) + (\text{Area 3} + \text{Area 4})\left(\dfrac{2.5}{3}\right)}{P_o}$$

$$= \frac{(25.788)(3.33) + (51.575)(1.25) + (14.825 + 30.663)(0.833)}{122.85}$$

$$= \frac{85.87 + 64.47 + 37.89}{122.85} = \underline{1.53 \text{ m}}$$

## 5.3

## Rankine Active Earth Pressure

The lateral earth pressure condition described in Section 5.2 involves walls that do not yield at all. However, if a wall tends to move away from the soil a distance $\Delta x$, as shown in Figure 5.5a, the soil pressure on that wall at any given depth will gradually decrease. For a wall that is *frictionless*, the horizontal stress, $\sigma_h$, at a depth $z$ will be equal to $K_o\sigma_v$ $(= K_o\gamma z)$ when $\Delta x$ is equal to zero. However, with $\Delta x > 0$, $\sigma_h$ will be less than $K_o\sigma_v$.

The Mohr's circles corresponding to wall displacements of $\Delta x = 0$ and $\Delta x > 0$ are shown as circles $a$ and $b$, respectively, in Figure 5.5b. If the displacement of the wall, $\Delta x$, continues to increase, there will be a time when

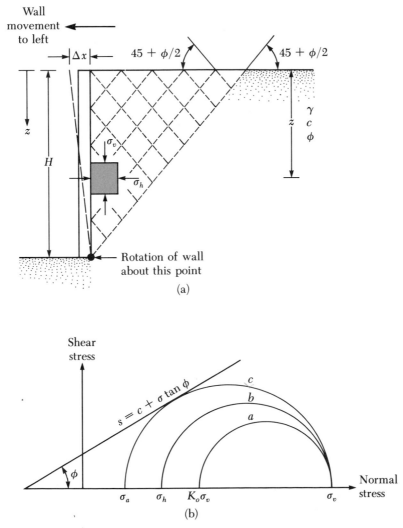

(a)

(b)

**Figure 5.5** Rankine active pressure

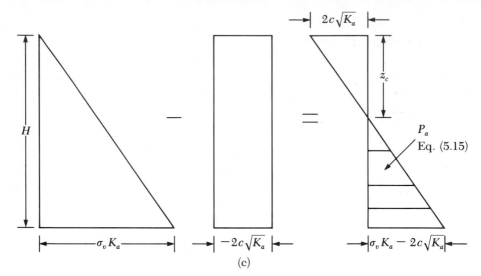

**Figure 5.5** (Continued)

the corresponding Mohr's circle will just touch the Mohr-Coulomb failure envelope defined by the equation

$$s = c + \sigma \tan \phi$$

This circle is marked $c$ in Figure 5.5b. It represents the failure condition in the soil mass; the horizontal stress at this time is equal to $\sigma_a$. This horizontal stress, $\sigma_a$, is referred to as the *Rankine active pressure*. The *slip lines* (failure planes) in the soil mass at this time will make angles of $\pm(45 + \phi/2)$ with the horizontal. This is shown in Figure 5.5a.

Referring to Eq. (1.76), the equation relating the principal stresses for a Mohr's circle that touches the Mohr-Coulomb failure envelope can be given by

$$\sigma_1 = \sigma_3 \tan^2\left(45 + \frac{\phi}{2}\right) + 2c \tan\left(45 + \frac{\phi}{2}\right)$$

For the Mohr's circle $c$ in Figure 5.5b,

$$\text{major principal stress, } \sigma_1 = \sigma_v$$

and

$$\text{minor principal stress, } \sigma_3 = \sigma_a$$

Thus

$$\sigma_v = \sigma_a \tan^2\left(45 + \frac{\phi}{2}\right) + 2c \tan\left(45 + \frac{\phi}{2}\right)$$

$$\sigma_a = \frac{\sigma_v}{\tan^2\left(45 + \frac{\phi}{2}\right)} - \frac{2c}{\tan\left(45 + \frac{\phi}{2}\right)}$$

or

$$\sigma_a = \sigma_v \tan^2 \left(45 - \frac{\phi}{2}\right) - 2c \tan \left(45 - \frac{\phi}{2}\right)$$

$$= \sigma_v K_a - 2c\sqrt{K_a} \tag{5.11}$$

where $K_a = \tan^2 (45 - \phi/2)$ = Rankine active earth pressure coefficient $5.11(a)$

The variation of the active pressure with depth for the wall shown in Figure 5.5a is given in Figure 5.5c. Note that $\sigma_v$ is equal to zero at $z = 0$, and $\sigma_v = \gamma H$ at $z = H$. The pressure distribution shows that at $z = 0$, the active pressure is equal to $-2c\sqrt{K_a}$, indicating tensile stress. This tensile stress decreases with depth and becomes zero at a depth $z = z_c$, or

$$\gamma z_c K_a - 2c\sqrt{K_a} = 0$$

$$z_c = \frac{2c}{\gamma \sqrt{K_a}} \tag{5.12}$$

The depth $z_c$ is usually referred to as the *depth of tensile crack*, because the tensile stress in the soil will eventually cause a crack along the soil-wall interface. Thus the total Rankine active force per unit length of the wall before the tensile crack occurs is equal to

$$P_a = \int_0^H \sigma_a \, dz = \int_0^H \gamma z \cdot K_a \, dz - \int_0^H 2c\sqrt{K_a} \, dz$$

$$= \tfrac{1}{2}\gamma H^2 K_a - 2cH\sqrt{K_a} \quad \text{area as in fig. } 5.5(c) \tag{5.13}$$

After the occurrence of the tensile crack, the force on the wall will be caused only by the pressure distribution between depths $z = z_c$ to $z = H$, as shown by the hatched area in Figure 5.5c, and it can be expressed as

$$P_a = \frac{1}{2} (H - z_c)(\gamma H K_a - 2c\sqrt{K_a}) \tag{5.14}$$

or

$$P_a = \frac{1}{2}\left(H - \frac{2c}{\gamma\sqrt{K_a}}\right)(\gamma H K_a - 2c\sqrt{K_a}) \tag{5.15}$$

For calculation purposes in some retaining wall design problems, a cohesive soil backfill is replaced by an assumed granular soil with a triangular Rankine active pressure diagram with $\sigma_a = 0$ at $z = 0$, and $\sigma_a = \sigma_v K_a - 2c\sqrt{K_a}$ at $z = H$ (see Figure 5.6). In such a case, the assumed active force per unit length of the wall is

$$P_a = \tfrac{1}{2}H(\gamma H K_a - 2c\sqrt{K_a}) = \tfrac{1}{2}\gamma H^2 K_a - cH\sqrt{K_a} \tag{5.16}$$

However, the active earth pressure condition will only be reached if the wall is allowed to yield sufficiently. The amount of wall yield necessary is about $0.001H$ to $0.004H$ for granular soil backfills and about $0.01H$ to $0.04H$ for cohesive soil backfills.

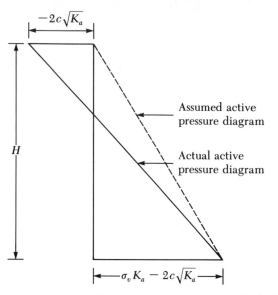

**Figure 5.6** Assumed active pressure diagram for clay backfill behind a retaining wall

## Example
## 5.2

A 6-m high retaining wall is to support a soil with unit weight $\gamma = 17.4$ kN/m³, soil friction angle $\phi = 26°$, and cohesion $c = 14.36$ kN/m². Determine the Rankine active force per unit length of the wall both before and after the tensile crack occurs, and determine the line of action of the resultant in both cases.

### Solution

Given: $\phi = 26°$. So

$$K_a = \tan^2\left(45 - \frac{\phi}{2}\right) = \tan^2(45 - 13) = 0.39$$

$$\sqrt{K_a} = 0.625$$

$$\sigma_a = \gamma H K_a - 2c\sqrt{K_a}$$

Referring to Figure 5.5c,
At $z = 0$, $\sigma_a = -2c\sqrt{K_a} = -2(14.36)(0.625) = -17.95$ kN/m²
At $z = 6$ m, $\sigma_a = (17.4)(6)(0.39) - 2(14.36)(0.625)$
    $= 40.72 - 17.95 = 22.77$ kN/m²

### Active Force Before the Occurrence of Tensile Crack: Eq. (5.13)

$$P_o = \tfrac{1}{2}\gamma H^2 K_a - 2cH\sqrt{K_a}$$

$$= \tfrac{1}{2}(6)(40.72) - (6)(17.95) = 122.16 - 107.7 = 14.46 \text{ kN/m}$$

The line of action of the resultant can be determined by taking the moment of the area of the pressure diagrams about the bottom of the wall, or

$$P_o\bar{z} = (122.16)\left(\frac{6}{3}\right) - (107.7)\left(\frac{6}{2}\right)$$

or

$$\bar{z} = \frac{244.32 - 323.1}{14.46} = \underline{-5.448 \text{ m}}$$

**Active Force After the Occurrence of Tensile Crack: Eq. (5.12)**

$$z_c = \frac{2c}{\gamma\sqrt{K_a}} = \frac{2(14.36)}{(17.4)(0.625)} = 2.64 \text{ m}$$

Using Eq. (5.14)

$$P_a = \tfrac{1}{2}(H - z_c)(\gamma H K_a - 2c\sqrt{K_a}) = \tfrac{1}{2}(6 - 2.64)(22.77) = 38.25 \text{ kN/m}$$

Figure 5.5c shows that the force $P_a = 38.25$ kN/m is the area of the hatched triangle. Hence, the line of action of the resultant will be located at a height of $\bar{z} = (H - z_c)/3$ above the bottom of the wall, or

$$\bar{z} = \frac{6 - 2.64}{3} = \underline{1.12 \text{ m}}$$

*Note*: For most retaining wall construction, a granular backfill is used and so $c = 0$. Thus Example 5.2 is an academic problem; however, it illustrates the basic principles of the Rankine active earth pressure equation.

## Example 5.3

Refer to Figure 5.7a. Assume that the wall can yield sufficiently and determine the Rankine active force per unit length of the wall. Also determine the location of the resultant line of action.

**Solution**

If the cohesion, $c$, is equal to zero

$$\sigma_a = \sigma_v K_a$$

For the top soil layer, $\phi_1 = 30°$, so

$$K_{a(1)} = \tan^2\left(45 - \frac{\phi_1}{2}\right) = \tan^2(45 - 15) = \frac{1}{3}$$

Similarly, for the bottom soil layer, $\phi_2 = 36°$, and

$$K_{a(2)} = \tan^2\left(45 - \frac{36}{2}\right) = 0.26$$

Because of the presence of the water table, the effective lateral pressure and the hydrostatic pressure have to be calculated separately.
At $z = 0$, $\sigma'_v = 0$, $\sigma'_a = 0$
At $z = 3$ m, $\sigma'_v = \gamma z = (16)(3) = 48$ kN/m$^2$

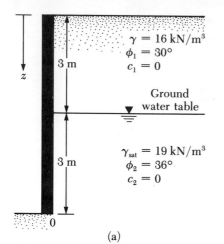

(a)

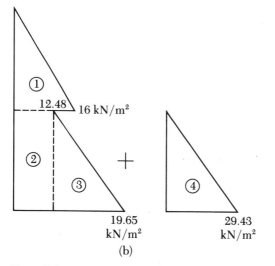

(b)

**Figure 5.7**

At this depth, for the top soil layer

$$\sigma'_a = K_{a(1)}\sigma'_v = (\tfrac{1}{3})(48) = 16 \text{ kN/m}^2$$

Similarly, for the bottom soil layer

$$\sigma'_a = K_{a(2)}\sigma'_v = (0.26)(48) = 12.48 \text{ kN/m}^2$$

At $z = 6$ m, $\sigma'_v = (\gamma)(3) + (\gamma_{sat} - \gamma_w)(3) = (16)(3) + (19 - 9.81)(3)$

$$= 48 + 27.57 = 75.57 \text{ kN/m}^2$$

$$\sigma'_a = K_{a(2)}\sigma'_v = (0.26)(75.57) = 19.65 \text{ kN/m}^2$$

The hydrostatic pressure $u$ is zero from $z = 0$ to $z = 3$ m. At $z = 6$ m, $u =$

$3\gamma_w = 3(9.81) = 29.43$ kN/m$^2$. The pressure distribution diagram is plotted in Figure 5.7b. The force per unit length

$$P_o = \text{Area } 1 + \text{Area } 2 + \text{Area } 3 + \text{Area } 4$$

$$= \tfrac{1}{2}(3)(16) + (3)(12.48) + \tfrac{1}{2}(3)(19.65 - 12.48) + \tfrac{1}{2}(3)(29.43)$$

$$= 24 + 37.44 + 10.76 + 44.15 = 116.35 \text{ kN/m}$$

The distance of the line of action of the resultant from the bottom of the wall ($\bar{z}$) can be determined by taking the moments about the bottom of the wall (point $O$ in Figure 5.7a), or

$$\bar{z} = \frac{(24)\left(3 + \dfrac{3}{3}\right) + (37.44)\left(\dfrac{3}{2}\right) + (10.76)\left(\dfrac{3}{3}\right) + (44.15)\left(\dfrac{3}{3}\right)}{116.35}$$

$$= \frac{96 + 56.16 + 10.76 + 44.15}{116.35} = \underline{1.779 \text{ m}}$$

## 5.4
## Coulomb's Active Earth Pressure

The Rankine active earth pressure calculations discussed in the preceding section were based on the assumption that the wall is frictionless. In 1776, Coulomb proposed a theory to calculate the lateral earth pressure on a retaining wall with granular soil backfill. This theory takes wall friction into consideration.

To apply Coulomb's active earth pressure theory, let us consider a retaining wall with its back face inclined at an angle $\beta$ with the horizontal as shown in Figure 5.8a. The backfill is a granular soil that slopes at an angle $\alpha$ with the horizontal. Also, let $\delta$ be the angle of friction between the soil and the wall (that is, angle of wall friction).

In an active pressure condition, the wall will move away from the soil mass (that is, to the left in Figure 5.8a). Coulomb assumed that, in such a case, the failure surface in the soil mass would be a plane (such as $BC_1$, $BC_2$, . . .). So, to find the active force in our example, consider a possible soil failure wedge $ABC_1$. The forces acting on this wedge, $ABC_1$ (per unit length at right angles to the cross section shown), are as follows:

1. Weight of the wedge, $W$.
2. The resultant, $R$, of the normal and resisting shear forces along the surface, $BC_1$. The force $R$ will be inclined at an angle $\phi$ to the normal drawn to the surface $BC_1$.
3. The active force per unit length of the wall, $P_a$. The force $P_a$ will be inclined at an angle $\delta$ to the normal drawn to the back face of the wall.

For equilibrium purposes, a force triangle can be drawn. This is shown in Figure 5.8b. Note that $\theta_1$ is the angle that $BC_1$ makes with the horizontal. Because the magnitude of $W$ as well as the directions of all three forces are known, the value of $P_a$ can now be determined. In a similar manner, one can determine the active forces of other trial wedges, such as $ABC_2$, $ABC_3$, . . . .

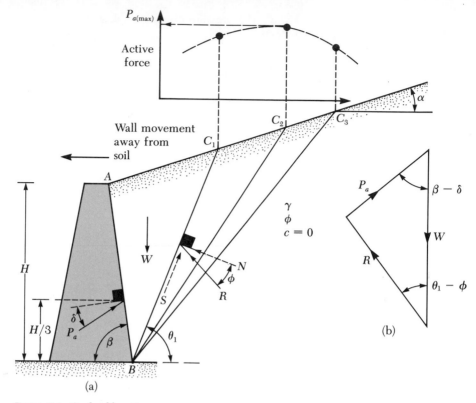

**Figure 5.8**  Coulomb's active pressure

The maximum value of $P_a$ thus determined is Coulomb's active force (see top part of Figure 5.8a). This can be expressed as

$$P_a = \tfrac{1}{2}K_a\gamma H^2 \tag{5.17}$$

where $K_a$ = Coulomb's active earth pressure coefficient

$$= \frac{\sin^2(\beta + \phi)}{\sin^2\beta \cdot \sin(\beta - \delta)\left[1 + \sqrt{\dfrac{\sin(\phi + \delta) \cdot \sin(\phi - \alpha)}{\sin(\beta - \delta) \cdot \sin(\alpha + \beta)}}\right]^2} \tag{5.18}$$

$H$ = height of the wall

The values of the active earth pressure coefficient, $K_a$, for a vertical retaining wall ($\beta = 90°$) with horizontal backfill ($\alpha = 0°$) are given in Table 5.1. Note that the line of action of the resultant ($P_a$) will act at a distance of $H/3$ above the base of the wall and will be inclined at an angle $\delta$ to the normal drawn to the back of the wall.

In the actual design of retaining walls, the value of the wall friction angle, $\delta$, is assumed to be between $\phi/2$ and $2/3\ \phi$. The active earth pressure coefficients for various values of $\phi$, $\alpha$, and $\beta$ with $\delta = 2/3\ \phi$ are given in Table 5.2. This is a very useful table for design considerations.

**Table 5.1** Values of $K_a$ [Eq. (5.18)] for $\beta = 90°$, $\alpha = 0°$

| $\phi$(deg) | $\delta$ (deg) | | | | | |
|---|---|---|---|---|---|---|
| | 0 | 5 | 10 | 15 | 20 | 25 |
| 28 | 0.3610 | 0.3448 | 0.3330 | 0.3251 | 0.3203 | 0.3186 |
| 30 | 0.3333 | 0.3189 | 0.3085 | 0.3014 | 0.2973 | 0.2956 |
| 32 | 0.3073 | 0.2945 | 0.2853 | 0.2791 | 0.2755 | 0.2745 |
| 34 | 0.2827 | 0.2714 | 0.2633 | 0.2579 | 0.2549 | 0.2542 |
| 36 | 0.2596 | 0.2497 | 0.2426 | 0.2379 | 0.2354 | 0.2350 |
| 38 | 0.2379 | 0.2292 | 0.2230 | 0.2190 | 0.2169 | 0.2167 |
| 40 | 0.2174 | 0.2098 | 0.2045 | 0.2011 | 0.1994 | 0.1995 |
| 42 | 0.1982 | 0.1916 | 0.1870 | 0.1841 | 0.1828 | 0.1831 |

## Example 5.4

Consider the retaining wall shown in Figure 5.8a. Given: $H = 4.6$ m; unit weight of soil $= 16.5$ kN/m³; angle of friction of soil $= 30°$; wall friction angle, $\delta = 2/3\ \phi$; soil cohesion, $c = 0$; $\alpha = 0$, and $\beta = 90°$. Calculate the Coulomb's active force per unit length of the wall.

**Solution**

*use eq. 5.18 to find $K_A$*    *analytic method for P.2 root #2*

From Eq. (5.17)

$$P_a = \tfrac{1}{2}\gamma H^2 K_a$$

From Table 5.1, for $\alpha = 0°$, $\beta = 90°$, $\phi = 30°$, and $\delta = 2/3\ \phi = 20°$, $K_a = 0.297$. Hence

$$P_a = \tfrac{1}{2}(16.5)(4.6)^2(0.297) = 51.85 \text{ kN/m}$$

## 5.5

### Rankine Passive Pressure

Figure 5.9a shows a vertical frictionless retaining wall with a horizontal backfill. At a depth $z$, the vertical pressure on a soil element is equal to $\sigma_v = \gamma z$. Initially, if the wall does not yield at all, then the lateral stress at that depth will be equal to $\sigma_h = K_o\sigma_v$. This state of stress is illustrated by the Mohr's circle $a$ in Figure 5.9b. Now, if the wall is pushed into the soil mass by an amount $\Delta x$ as shown in Figure 5.9a, the vertical stress at depth $z$ will stay the same; however, the horizontal stress will increase. Thus $\sigma_h$ will be greater than $K_o\sigma_v$. The state of stress can now be represented by the Mohr's circle $b$ in Figure 5.9b. If the wall moves farther inward (that is, $\Delta x$ is increased still more), the stresses at a depth $z$ will ultimately reach the state represented by Mohr's circle $c$ (Figure 5.9b). Note that this Mohr's circle touches the Mohr-Coulomb failure envelope. This implies that the soil behind the wall will fail by being pushed upward. The horizontal stress, $\sigma_h$, at this point is referred to as the *Rankine passive pressure*, or $\sigma_h = \sigma_p$.

**Table 5.2**  Values of $K_a$ [Eq. (5.18)] (Note: $\delta = \frac{2}{3}\phi$ in all cases)

| $\alpha$ (deg) | $\phi$ (deg) | $\beta$ (deg) | | | | | |
|---|---|---|---|---|---|---|---|
| | | 90 | 85 | 80 | 75 | 70 | 65 |
| 0 | 28 | 0.3213 | 0.3588 | 0.4007 | 0.4481 | 0.5026 | 0.5662 |
| | 30 | 0.2973 | 0.3349 | 0.3769 | 0.4245 | 0.4794 | 0.5435 |
| | 32 | 0.2750 | 0.3125 | 0.3545 | 0.4023 | 0.4574 | 0.5220 |
| | 34 | 0.2543 | 0.2916 | 0.3335 | 0.3813 | 0.4367 | 0.5017 |
| | 36 | 0.2349 | 0.2719 | 0.3137 | 0.3615 | 0.4170 | 0.4825 |
| | 38 | 0.2168 | 0.2535 | 0.2950 | 0.3428 | 0.3984 | 0.4642 |
| | 40 | 0.1999 | 0.2361 | 0.2774 | 0.3250 | 0.3806 | 0.4468 |
| | 42 | 0.1840 | 0.2197 | 0.2607 | 0.3081 | 0.3638 | 0.4303 |
| 5 | 28 | 0.3431 | 0.3845 | 0.4311 | 0.4843 | 0.5461 | 0.6191 |
| | 30 | 0.3165 | 0.3578 | 0.4043 | 0.4575 | 0.5194 | 0.5926 |
| | 32 | 0.2919 | 0.3329 | 0.3793 | 0.4324 | 0.4943 | 0.5678 |
| | 34 | 0.2691 | 0.3097 | 0.3558 | 0.4088 | 0.4707 | 0.5443 |
| | 36 | 0.2479 | 0.2881 | 0.3338 | 0.3866 | 0.4484 | 0.5222 |
| | 38 | 0.2282 | 0.2679 | 0.3132 | 0.3656 | 0.4273 | 0.5012 |
| | 40 | 0.2098 | 0.2489 | 0.2937 | 0.3458 | 0.4074 | 0.4814 |
| | 42 | 0.1927 | 0.2311 | 0.2753 | 0.3271 | 0.3885 | 0.4626 |
| 10 | 28 | 0.3702 | 0.4164 | 0.4686 | 0.5287 | 0.5992 | 0.6834 |
| | 30 | 0.3400 | 0.3857 | 0.4376 | 0.4974 | 0.5676 | 0.6516 |
| | 32 | 0.3123 | 0.3575 | 0.4089 | 0.4683 | 0.5382 | 0.6220 |
| | 34 | 0.2868 | 0.3314 | 0.3822 | 0.4412 | 0.5107 | 0.5942 |
| | 36 | 0.2633 | 0.3072 | 0.3574 | 0.4158 | 0.4849 | 0.5682 |
| | 38 | 0.2415 | 0.2846 | 0.3342 | 0.3921 | 0.4607 | 0.5438 |
| | 40 | 0.2214 | 0.2637 | 0.3125 | 0.3697 | 0.4379 | 0.5208 |
| | 42 | 0.2027 | 0.2441 | 0.2921 | 0.3487 | 0.4164 | 0.4990 |
| 15 | 28 | 0.4065 | 0.4585 | 0.5179 | 0.5869 | 0.6685 | 0.7671 |
| | 30 | 0.3707 | 0.4219 | 0.4804 | 0.5484 | 0.6291 | 0.7266 |
| | 32 | 0.3384 | 0.3887 | 0.4462 | 0.5134 | 0.5930 | 0.6895 |
| | 34 | 0.3091 | 0.3584 | 0.4150 | 0.4811 | 0.5599 | 0.6554 |
| | 36 | 0.2823 | 0.3306 | 0.3862 | 0.4514 | 0.5295 | 0.6239 |
| | 38 | 0.2578 | 0.3050 | 0.3596 | 0.4238 | 0.5006 | 0.5949 |
| | 40 | 0.2353 | 0.2813 | 0.3349 | 0.3981 | 0.4740 | 0.5672 |
| | 42 | 0.2146 | 0.2595 | 0.3119 | 0.3740 | 0.4491 | 0.5416 |
| 20 | 28 | 0.4602 | 0.5205 | 0.5900 | 0.6715 | 0.7690 | 0.8810 |
| | 30 | 0.4142 | 0.4728 | 0.5403 | 0.6196 | 0.7144 | 0.8303 |
| | 32 | 0.3742 | 0.4311 | 0.4968 | 0.5741 | 0.6667 | 0.7800 |
| | 34 | 0.3388 | 0.3941 | 0.4581 | 0.5336 | 0.6241 | 0.7352 |
| | 36 | 0.3071 | 0.3609 | 0.4233 | 0.4970 | 0.5857 | 0.6948 |
| | 38 | 0.2787 | 0.3308 | 0.3916 | 0.4637 | 0.5587 | 0.6580 |
| | 40 | 0.2529 | 0.3035 | 0.3627 | 0.4331 | 0.5185 | 0.6243 |
| | 42 | 0.2294 | 0.2784 | 0.3360 | 0.4050 | 0.4889 | 0.5931 |

For Mohr's circle $c$ in Figure 5.9b, the major principal stress is $\sigma_p$, and the minor principal stress is $\sigma_v$. Substituting them into Eq. (1.76) yields

$$\sigma_p = \sigma_v \tan^2\left(45 + \frac{\phi}{2}\right) + 2c \tan\left(45 + \frac{\phi}{2}\right) \tag{5.19}$$

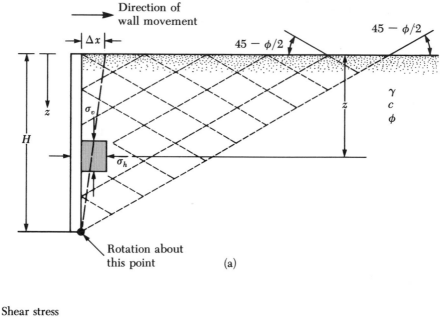

(a)

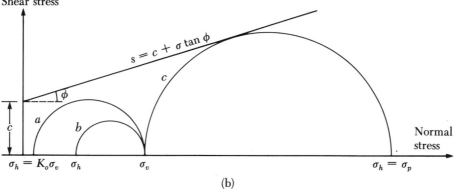

(b)

**Figure 5.9** Rankine passive pressure (continued on p. 224)

Now, let

$$K_p = \text{Rankine passive earth pressure coefficient}$$

$$= \tan^2 \left( 45 + \frac{\phi}{2} \right) \tag{5.20}$$

Hence, from Eq. (5.19)

$$\sigma_p = \sigma_v K_p + 2c \sqrt{K_p} \tag{5.21}$$

Using the preceding equation, Figure 5.9c shows the passive pressure diagram for the wall shown in Figure 5.9a. Note that,

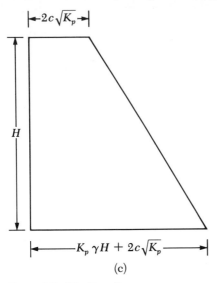

(c)

**Figure 5.9**  (Continued)

$$\text{at } z = 0, \ \sigma_v = 0, \ \sigma_p = 2c\sqrt{K_p}$$

and

$$\text{at } z = H, \ \sigma_v = \gamma H, \ \sigma_p = \gamma H K_p + 2c\sqrt{K_p}$$

The passive force per unit length of the wall can be determined from the area of the pressure diagram, or

$$P_p = \tfrac{1}{2}\gamma H^2 K_p + 2cH\sqrt{K_p} \tag{5.22}$$

The magnitudes of the wall movements, $\Delta x$, required to develop failure under passive conditions are as follows:

| Soil type | Wall movement for passive condition, $\Delta x$ |
|---|---|
| Dense sand | $0.005H$ |
| Loose sand | $0.01H$ |
| Stiff clay | $0.01H$ |
| Soft clay | $0.05H$ |

## Example 5.5

A 3-m high wall is shown in Figure 5.10a. Determine the Rankine passive force per unit length of the wall.

**Solution**

For the top soil layer

$$K_{p(1)} = \tan^2\left(45 + \frac{\phi_1}{2}\right) = \tan^2(45 + 15) = 3$$

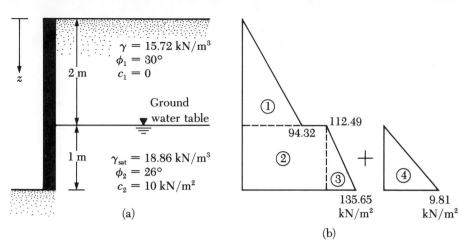

(a)

(b)

**Figure 5.10**

For the bottom soil layer

$$K_{p(2)} = \tan^2\left(45 + \frac{\phi_2}{2}\right) = \tan^2(45 + 13) = 2.56$$

$$\sigma_p = \sigma_v' K_p + 2c\sqrt{K_p}$$

where  $\sigma_v'$ = effective vertical stress

at $z = 0$, $\sigma_v' = 0$, $c_1 = 0$, $\sigma_p = 0$
at $z = 2$ m, $\sigma_v' = (15.72)(2) = 31.44$ kN/m², $c_1 = 0$

So, for the top soil layer

$$\sigma_p = 31.44K_{p(1)} + 2(0)\sqrt{K_p} = 31.44(3) = 94.32 \text{ kN/m}^2$$

At this depth, that is, $z = 2$ m, for the bottom soil layer

$$\sigma_p = \sigma_v' K_{p(2)} + 2c\sqrt{K_{p(2)}} = 31.44(2.56) + 2(10)\sqrt{2.56}$$

$$= 80.49 + 32 = 112.49 \text{ kN/m}^2$$

Again, at $z = 3$ m, $\sigma_v' = (15.72)(2) + (\gamma_{\text{sat}} - \gamma_w)(1)$
$$= 31.44 + (18.86 - 9.81)(1) = 40.49 \text{kN/m}^2$$

Hence

$$\sigma_p = \sigma_v' K_{p(2)} + 2c\sqrt{K_{p(2)}} = 40.49(2.56) + (2)(10)(1.6)$$

$$= 135.65 \text{ kN/m}^2$$

Note that, because a water table is present, the hydrostatic stress, $\sigma_u$, also has to be taken into consideration. For $z = 0$ to 2 m, $u = 0$; at $z = 3$ m, $u = (1)(\gamma_w) = 9.81$ kN/m².

The passive pressure diagram is plotted in Figure 5.10b. The passive force per unit length of the wall can be determined from the area of the pressure diagram as follows:

| Area No. | Area | |
|---|---|---|
| 1 | $(\frac{1}{2})(2)(94.32)$ | = 94.32 |
| 2 | $(112.49)(1)$ | = 112.49 |
| 3 | $(\frac{1}{2})(1)(135.65 - 112.49)$ | = 11.58 |
| 4 | $(\frac{1}{2})(9.81)(1)$ | = 4.905 |
| | $P_p \approx$ | 223.3 kN/m |

## 5.6

## Coulomb's Passive Earth Pressure

Coulomb (1776) also presented an analysis for determination of the passive earth pressure (that is, when the wall moves *into* the soil mass) for walls possessing friction ($\delta$ = angle of wall friction) and retaining a granular backfill material similar to that discussed in Section 5.4.

To understand the determination of Coulomb's passive force, $P_p$, consider the wall shown in Figure 5.11a. As in the case of active pressure, Coulomb assumed that the potential failure surface in soil is a plane. If one pictures a trial failure wedge of soil, such as $ABC_1$, the forces per unit length of the wall acting on the wedge are as follows:

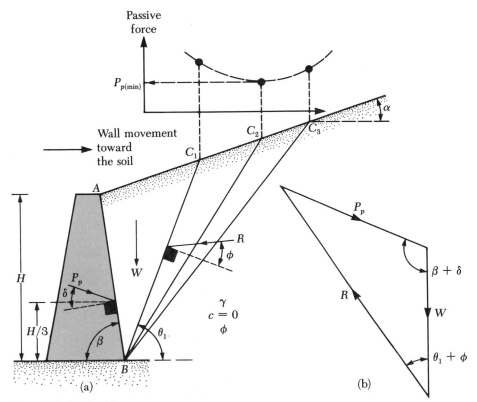

**Figure 5.11**   Coulomb's passive pressure

**1.** The weight of the wedge, $W$
**2.** The resultant, $R$, of the normal and shear forces on the plane $BC_1$
**3.** The passive force, $P_p$

Figure 5.11b shows the force triangle, for equilibrium, for the trial wedge $ABC_1$. From this force triangle, the value of $P_p$ can be determined, because the direction of all three forces as well as the magnitude of one force are known.

Similar force triangles for several trial wedges, such as $ABC_1$, $ABC_2$, $ABC_3$, . . . , can be constructed, and the corresponding values of $P_p$ can be determined. The top part of Figure 5.11a shows the nature of variation of the $P_p$ values for different wedges. The *minimum value of $P_p$* in this diagram is *Coulomb's passive force*. Mathematically, this can be expressed as

$$P_p = \tfrac{1}{2}\gamma H^2 K_p \tag{5.23}$$

where $K_p$ = Coulomb's passive pressure coefficient

$$= \frac{\sin^2(\beta - \phi)}{\sin^2\beta \cdot \sin(\beta + \delta)\left[1 - \sqrt{\dfrac{\sin(\phi + \delta)\,\sin(\phi + \alpha)}{\sin(\beta + \delta)\,\sin(\beta + \alpha)}}\,\right]^2} \tag{5.24}$$

The values of the passive pressure coefficient, $K_p$, for various values of $\phi$ and $\delta$ are given in Table 5.3 ($\beta = 90°$, $\alpha = 0°$).

**Table 5.3** Values of $K_p$ [Eq. (5.24)] for $\beta = 90°$ and $\alpha = 0°$

| | δ (deg) | | | | |
|---|---|---|---|---|---|
| φ (deg) | 0 | 5 | 10 | 15 | 20 |
| 15 | 1.698 | 1.900 | 2.130 | 2.405 | 2.735 |
| 20 | 2.040 | 2.313 | 2.636 | 3.030 | 3.525 |
| 25 | 2.464 | 2.830 | 3.286 | 3.855 | 4.597 |
| 30 | 3.000 | 3.506 | 4.143 | 4.977 | 6.105 |
| 35 | 3.690 | 4.390 | 5.310 | 6.854 | 8.324 |
| 40 | 4.600 | 5.590 | 6.946 | 8.870 | 11.772 |

Note that the resultant passive force $P_p$ will act at a distance of $H/3$ measured from the bottom of the wall and will be inclined at an angle $\delta$ to the normal drawn to the back face of the wall.

## 5.7
## Range of Wall Friction Angle, δ

Retaining walls are generally constructed with masonry or mass concrete. It is always useful for a designer to have a general idea of the range of the wall friction angle, $\delta$, that may be encountered. Table 5.4 shows the general range of the values of $\delta$ for various backfill materials.

**Table 5.4** General Range of Wall Friction Angles for Masonry or Mass Concrete Walls

| Backfill material | Range of δ (deg) |
|---|---|
| Gravel | 27–30 |
| Coarse sand | 20–28 |
| Fine sand | 15–25 |
| Stiff clay | 15–20 |
| Silty clay | 12–16 |

## 5.8

## Comments on the Failure Surface Assumption for Coulomb Pressure Calculation

Coulomb's pressure calculation methods for active and passive pressure have been discussed in Sections 5.4 and 5.6. The fundamental assumption for these analyses is the acceptance of *plane failure surfaces*. However, for walls with friction, this assumption does not apply in practice. The nature of *actual* failure surfaces in the soil mass for active and passive pressure is shown in Figure 5.12a and b, respectively (for a vertical wall with a horizontal backfill). Note that the failure surfaces marked *BC* are curved. The failure surfaces marked *CD* are planes.

Although the actual failure surface in soil for the case of active pressure is somewhat different than that assumed in the calculation of the Coulomb pressure, the results are not greatly different. However, this is not so in the case of passive pressure, when, as the value of δ increases, Coulomb's method of calculation gives increasingly erroneous values of $P_p$. This factor of error could lead to an unsafe condition, because the values of $P_p$ would become higher than the soil could resist. To remedy this situation, several investigations have been done to calculate the value of $P_p$ using the type of failure surface shown in Figure 5.12b. Some of those investigations are listed in the following table.

| Investigator | Assumption for the curved part, *BC* (Figure 5.12b) |
|---|---|
| 1. Packshaw (1969) | circle |
| 2. Caquot and Kerisel (1948) | ellipse |
| 3. James and Bransby (1971) | log spiral |
| 4. Shields and Tolunay (1972, 1973) | log spiral |
| 5. Terzaghi and Peck (1967) | log spiral |

The details of the calculation procedures involved in the studies listed in the table are beyond the scope of this text. However, the results of Shields and Tolunay's work (1973) appear to be reasonable. They show that

$$P_p = \tfrac{1}{2}\gamma H^2 K_p \qquad (5.25)$$

where $K_p$ = passive pressure coefficient

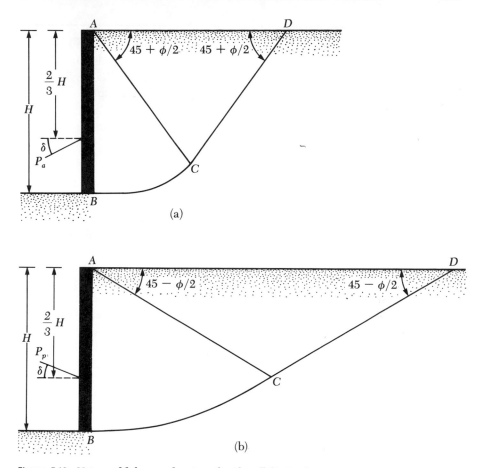

**Figure 5.12** Nature of failure surface in soil with wall friction for:
(a) active pressure case; (b) passive pressure case

The values of $K_p$ for a vertical wall with a horizontal granular backfill (that is, $c = 0$) are given in Table 5.5.

**Table 5.5** Values of $K_p$ [Eq. (5.25)] for a Vertical Wall with Horizontal Cohesionless Soil as Backfill[a]

| $\phi$ (deg) | $\delta$ (deg) | | | | | | | | | |
|---|---|---|---|---|---|---|---|---|---|---|
| | 0 | 5 | 10 | 15 | 20 | 25 | 30 | 35 | 40 | 45 |
| 20 | 2.04 | 2.26 | 2.43 | 2.55 | 2.70 | | | | | |
| 25 | 2.46 | 2.77 | 3.03 | 3.23 | 3.39 | 3.63 | | | | |
| 30 | 3.00 | 3.43 | 3.80 | 4.13 | 4.40 | 4.64 | 5.03 | | | |
| 35 | 3.69 | 4.29 | 4.84 | 5.34 | 5.80 | 6.21 | 6.59 | 7.25 | | |
| 40 | 4.60 | 5.44 | 6.26 | 7.05 | 7.80 | 8.51 | 9.18 | 9.83 | 11.03 | |
| 45 | 5.83 | 7.06 | 8.30 | 9.55 | 10.80 | 12.04 | 13.26 | 14.46 | 15.69 | 18.01 |

[a]After Shields and Tolunay (1973)

## 5.9

### Rankine Active and Passive Earth Pressure for Inclined Granular Backfill

If the backfill of a frictionless retaining wall is a granular soil (that is, $c = 0$) and rises at an angle $\alpha$ with respect to the horizontal (Figure 5.13), the *active earth pressure coefficient, $K_a$,* can be expressed in the form

$$K_a = \cos \alpha \; \frac{\cos \alpha - \sqrt{\cos^2 \alpha - \cos^2 \phi}}{\cos \alpha + \sqrt{\cos^2 \alpha - \cos^2 \phi}} \tag{5.26}$$

where $\phi$ = angle of friction of soil

At any given depth, $z$, the *Rankine active pressure* can be expressed as

$$\sigma_a = \gamma z K_a \tag{5.27}$$

Also, the total force per unit length of the wall ($P_a$) is equal to

$$P_a = \tfrac{1}{2}\gamma H^2 K_a \tag{5.28}$$

Note that, in this case, the direction of the resultant force, $P_a$, is *inclined at an angle $\alpha$ with the horizontal* and intersects the wall at a distance of $H/3$ measured from the base of the wall. Table 5.6 presents the values of $K_a$ (active earth pressure) for various values of $\alpha$ and $\phi$.

In a similar manner, the *Rankine passive earth pressure* for a wall of height $H$ with a granular sloping backfill can be represented by the equation

$$P_p = \tfrac{1}{2}\gamma H^2 K_p \tag{5.29}$$

where $K_p$ = passive earth pressure coefficient

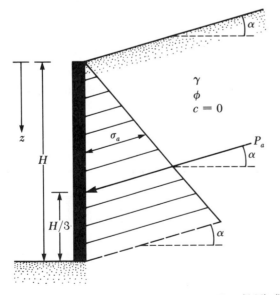

**Figure 5.13**  Notations for active pressure—Eqs. (5.26), (5.27), (5.28)

**Table 5.6** Active Earth Pressure Coefficient, $K_a$ [Eq. (5.26)]

| | $\phi$ (deg) → | | | | | | |
|---|---|---|---|---|---|---|---|
| ↓ $\alpha$ (deg) | 28 | 30 | 32 | 34 | 36 | 38 | 40 |
| 0 | 0.361 | 0.333 | 0.307 | 0.283 | 0.260 | 0.238 | 0.217 |
| 5 | 0.366 | 0.337 | 0.311 | 0.286 | 0.262 | 0.240 | 0.219 |
| 10 | 0.380 | 0.350 | 0.321 | 0.294 | 0.270 | 0.246 | 0.225 |
| 15 | 0.409 | 0.373 | 0.341 | 0.311 | 0.283 | 0.258 | 0.235 |
| 20 | 0.461 | 0.414 | 0.374 | 0.338 | 0.306 | 0.277 | 0.250 |
| 25 | 0.573 | 0.494 | 0.434 | 0.385 | 0.343 | 0.307 | 0.275 |

*Note:* With $\alpha = \phi$, $K_a = \cos \alpha$. So, $\alpha = \phi = 28°$, $K_a = 0.883$
$\alpha = \phi = 30°$, $K_a = 0.866$
$\alpha = \phi = 32°$, $K_a = 0.848$
$\alpha = \phi = 34°$, $K_a = 0.829$
$\alpha = \phi = 36°$, $K_a = 0.809$
$\alpha = \phi = 38°$, $K_a = 0.788$
$\alpha = \phi = 40°$, $K_a = 0.866$

$$= \cos \alpha \cdot \frac{\cos \alpha + \sqrt{\cos^2 \alpha - \cos^2 \phi}}{\cos \alpha - \sqrt{\cos^2 \alpha - \cos^2 \phi}} \qquad (5.30)$$

As in the case of the active force, the resultant force, $P_p$, is inclined at an angle $\alpha$ with the horizontal and intersects the wall at a distance of $H/3$ measured from the bottom of the wall. The values of $K_p$ (passive earth pressure coefficient) for various values of $\alpha$ and $\phi$ are given in Table 5.7.

## 5.10
## Lateral Earth Pressure Due to Surcharge

In several instances, theory of elasticity is used to determine the lateral earth pressure on retaining structures caused by various types of surcharge loading, such as *line loading* (Figure 5.14a) and *strip loading* (Figure 5.14b).

According to the theory of elasticity, the stress at any depth, $z$, on a retaining structure caused by a line load of intensity $q$/unit length (Figure 5.14a) may be given as

$$\sigma = \frac{2q}{\pi H} \cdot \frac{a^2 b}{(a^2 + b^2)^2} \qquad (5.31)$$

where $\sigma =$ horizontal stress at a depth $z = bH$

(See Figure 5.14a for explanations of the terms $a$ and $b$.)

However, because soil is not a perfectly elastic medium, some deviations from Eq. (5.31) may be expected. The modified forms of this equation generally accepted for use with soils are as follows

$$\sigma = \frac{4q}{\pi H} \frac{a^2 b}{(a^2 + b^2)^2} \qquad \text{for } a > 0.4 \qquad (5.32)$$

**Table 5.7**  Passive Earth Pressure Coefficient, $K_p$ [Eq. (5.30)]

| ↓ α (deg) | $\phi$ (deg) → | | | | | | |
|---|---|---|---|---|---|---|---|
| | 28 | 30 | 32 | 34 | 36 | 38 | 40 |
| 0 | 2.770 | 3.000 | 3.255 | 3.537 | 3.852 | 4.204 | 4.599 |
| 5 | 2.715 | 2.943 | 3.196 | 3.476 | 3.788 | 4.136 | 4.527 |
| 10 | 2.551 | 2.775 | 3.022 | 3.295 | 3.598 | 3.937 | 4.316 |
| 15 | 2.284 | 2.502 | 2.740 | 3.003 | 3.293 | 3.615 | 3.977 |
| 20 | 1.918 | 2.132 | 2.362 | 2.612 | 2.886 | 3.189 | 3.526 |
| 25 | 1.434 | 1.664 | 1.894 | 2.135 | 2.394 | 2.676 | 2.987 |

*Note:* With $\alpha = \phi$, $K_p = \cos \alpha$. So, $\alpha = \phi = 28°$, $K_p = 0.883$
$\alpha = \phi = 30°$, $K_p = 0.866$
$\alpha = \phi = 32°$, $K_p = 0.848$
$\alpha = \phi = 34°$, $K_p = 0.829$
$\alpha = \phi = 36°$, $K_p = 0.809$
$\alpha = \phi = 38°$, $K_p = 0.788$
$\alpha = \phi = 40°$, $K_p = 0.766$

and

$$\sigma = \frac{q}{H} \frac{0.203b}{(0.16 + b^2)^2} \qquad \text{for } a \le 0.4 \tag{5.33}$$

Figure 5.14b shows a strip load with an intensity of $q$/unit area located at a distance $b'$ from a wall of height $H$. Based on theory of elasticity, the horizontal stress, $\sigma$, at any depth $z$ on a retaining structure can be given by the equation

$$\sigma = \frac{q}{H} (\beta - \sin \beta \cdot \cos 2\alpha) \tag{5.34}$$

(The angles $\alpha$ and $\beta$ are defined in Figure 5.14b.)

However, in the case of soils, the right side of Eq. (5.34) is doubled to account for the yielding soil continuum, or

$$\sigma = \frac{2q}{\pi} (\beta - \sin \beta \cdot \cos 2\alpha) \tag{5.35}$$

The total force per unit length $(P)$ and the location of the resultant force, $\bar{z}$, due to the *strip loading only* can be expressed as follows (Jarquio, 1981)

$$P = \frac{q}{90} [H(\theta_2 - \theta_1)] \tag{5.36}$$

where $\theta_1 = \tan^{-1}\left(\frac{b'}{H}\right)$ (5.37)

$$\theta_2 = \tan^{-1}\left(\frac{a' + b'}{H}\right) \tag{5.38}$$

$$\bar{z} = H - \frac{H^2(\theta_2 - \theta_1) + (R - Q) - 57.30 \, a'H}{2H(\theta_2 - \theta_1)}$$

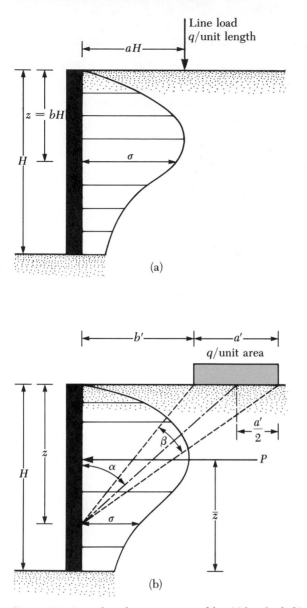

**Figure 5.14** Lateral earth pressure caused by: (a) line load; (b) strip load

$$= \frac{H^2(\theta_2 - \theta_1) - (R - Q) + 57.30\ a'H}{2H(\theta_2 - \theta_1)} \tag{5.39}$$

where $R = (a' + b')^2(90 - \theta_2)$ (5.40)
$\qquad Q = b'^2(90 - \theta_1)$ (5.41)

## Example
## 5.6

Refer to Figure 5.14b. Given: $a' = 2$ m, $b' = 1$ m, $q = 40$ kN/m², and $H = 6$ m.

a. Determine the total pressure on the wall caused by the strip loading only.
b. Determine the location of the center of pressure, $\bar{z}$, measured from the bottom of the wall.

### Solution
#### Part a

Using Eqs. (5.37) and (5.38)

$$\theta_1 = \tan^{-1}\left(\frac{1}{6}\right) = 9.46°$$

$$\theta_2 = \tan^{-1}\left(\frac{2+1}{6}\right) = 26.57°$$

From Eq. (5.36)

$$P = \frac{q}{90}[H(\theta_2 - \theta_1)] = \frac{40}{90}[6(26.57 - 9.46)] = 45.63 \text{ kN/m}$$

#### Part b

Again, from Eq. (5.39)

$$H - \bar{z} = \frac{H^2(\theta_2 - \theta_1) - (R - Q) + 57.30\, a'H}{2H(\theta_2 - \theta_1)}$$

$$R = (a' + b')^2(90 - \theta_2) = (2 + 1)^2(90 - 26.57) = 570.87$$

$$Q = b'^2(90 - \theta_1) = 1^2(90 - 9.46) = 80.54$$

So

$$H - \bar{z} = \frac{6^2(26.57 - 9.46) - (570.87 - 80.54) + 57.30(2)(6)}{(2)(6)(26.57 - 9.46)}$$

$$= \frac{615.96 - 490.33 + 687.6}{205.32} = 3.96 \text{ m}$$

Hence

$$\bar{z} = H - 3.96 = 6.0 - 3.96 = 2.04 \text{ m}$$

## 5.11

## Active Earth Pressure with Earthquake Forces

Coulomb's active earth pressure theory (see Section 5.4) can be extended to take into account the forces caused by an earthquake. Figure 5.15 shows a condition of active pressure with a granular backfill ($c = 0$). Note that the forces acting on the soil failure wedge in Figure 5.15 are essentially the same as those shown in Figure 5.8a, with the addition of $k_h W$ and $k_v W$ in the horizontal and

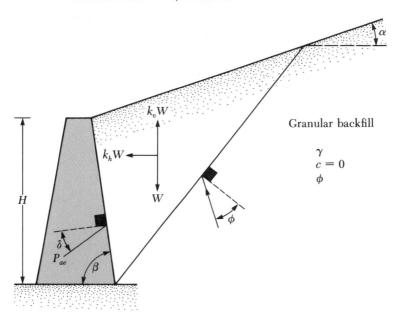

**Figure 5.15** Derivation of Eq. (5.44)

vertical directions, respectively; $k_h$ and $k_v$ may be defined as follows:

$$k_h = \frac{\text{horizontal earthquake acceleration component}}{\text{acceleration due to gravity, } g} \tag{5.42}$$

$$k_v = \frac{\text{vertical earthquake acceleration component}}{\text{acceleration due to gravity, } g} \tag{5.43}$$

Proceeding in a manner similar to that described in Section 5.4, the relation for the active force per unit length of the wall ($P_{ae}$) can be determined as

$$P_{ae} = \tfrac{1}{2}\gamma H^2 (1 - k_v) K_{ae} \tag{5.44}$$

where

$K_{ae}$ = active earth pressure coefficient

$$= \frac{\sin^2(\phi + \beta - \theta')}{\cos\theta' \sin^2\beta \sin(\beta - \theta' - \delta)\left[1 + \sqrt{\dfrac{\sin(\phi + \delta)\sin(\phi - \theta' - \alpha)}{\sin(\beta - \delta - \theta')\sin(\alpha + \beta)}}\right]^2} \tag{5.45}$$

$$\theta' = \tan^{-1}\left[\frac{k_h}{1 - k_v}\right] \tag{5.46}$$

Note that, for no earthquake condition

$k_h = 0$

$k_v = 0$

$\theta' = 0$

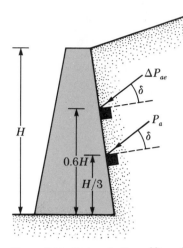

**Figure 5.16**  Determination of line of action of $P_{ae}$

Hence, $K_{ae} = K_a$ [as given by Eq. (5.18)].

Equation (5.44) is usually referred to as the *Mononobe-Okabe* solution. Unlike the case shown in Figure 5.8a, the resultant earth pressure in this situation, as calculated by Eq. (5.44), *does not act* at a distance of $H/3$ measured from the bottom of the wall. In order to obtain the location of the resultant earth pressure, the following procedure may be used.

**1.** Calculate $P_{ae}$ by using Eq. (5.44).
**2.** Calculate $P_a$ by using Eq. (5.17).
**3.** Calculate

$$\Delta P_{ae} = P_{ae} - P_a \tag{5.47}$$

**4.** Assume that $P_a$ acts at a distance of $H/3$ from the bottom of the wall (Figure 5.16).
**5.** Assume that $\Delta P_{ae}$ acts at a distance of $0.6H$ from the bottom of the wall (Figure 5.16).
**6.** Calculate the location of the resultant as

$$\bar{z} = \frac{(0.6H)(\Delta P_{ae}) + \left(\dfrac{H}{3}\right)(P_a)}{P_{ae}} \tag{5.48}$$

## Stability of Retaining Walls

### 5.12

Proportioning Retaining Walls

When designing retaining walls, one must assume some of the dimensions. This is referred to as *proportioning*, which allows the engineer to check trial sections for stability. If the stability checks yield undesirable results, the sections can be changed and rechecked. Figure 5.17 shows the general proportions of various retaining-wall components that can be used for initial checks.

Note that the top of the stem of any retaining wall should not be less than about 0.3 m for proper placement of concrete. The depth, $D$, to the bottom of the base slab should be kept at a minimum of 0.6 m. However, the bottom of the base slab should be kept below the seasonal frost line.

For counterfort retaining walls, the general proportion of the stem and the base slab is the same as for cantilever walls. However, the counterfort slabs may be about 0.3 m thick and spaced at center-to-center distances of $0.3\,H$ to $0.7\,H$.

## 5.13
## Application of Lateral Earth Pressure Theories to Design

The preceding sections have presented the fundamental theories of calculation of lateral earth pressure. To use these theories in design, one must make several simple assumptions. In the case of cantilever walls, if the Rankine earth pressure theory is to be used for stability checks, we need to draw a vertical line $AB$ through point $A$, as shown in Figure 5.18a (which is located at the edge of

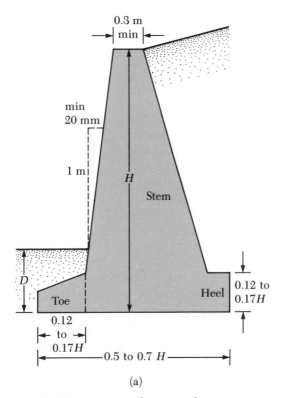

(a)

**Figure 5.17** Approximate dimensions for various components of retaining wall for initial stability checks: (a) gravity wall; (b) cantilever wall (see p. 238) (*note:* minimum dimension of $D$ is 0.6 m)

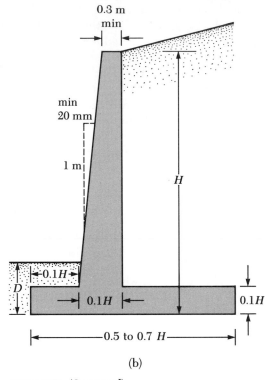

(b)

**Figure 5.17** (Continued)

the heel of the base slab). It is assumed that the Rankine active condition exists along the vertical plane $AB$. Rankine active earth pressure equations may then be used to calculate the lateral pressure on the face $AB$. In the analysis of stability for the wall, the force $P_{a(\text{Rankine})}$, the weight of soil above the heel ($W_s$), and the weight of the concrete ($W_c$) should all be taken into consideration. The assumption for the development of Rankine active pressure along the soil face $AB$ is theoretically correct if the shear zone bounded by the line $AC$ is not obstructed by the stem of the wall. The angle, $\eta$, that the line $AC$ makes with the vertical can be given by the equation

$$\eta = 45 + \frac{\alpha}{2} - \frac{\phi}{2} - \sin^{-1}\left(\frac{\sin \alpha}{\sin \phi}\right) \tag{5.49}$$

For gravity walls, a similar type of analysis can be used, as shown in Figure 5.18b. However, Coulomb's theory can also be used, as shown in Figure 5.18c. If *Coulomb's active pressure theory* is used, then the forces to be considered are only $P_{a(\text{Coulomb})}$ and the weight of the wall, $W_c$.

In the case of ordinary retaining walls, the problem of water tables and, hence, hydrostatic pressure are not encountered. Facilities for drainage from the soils retained are always provided.

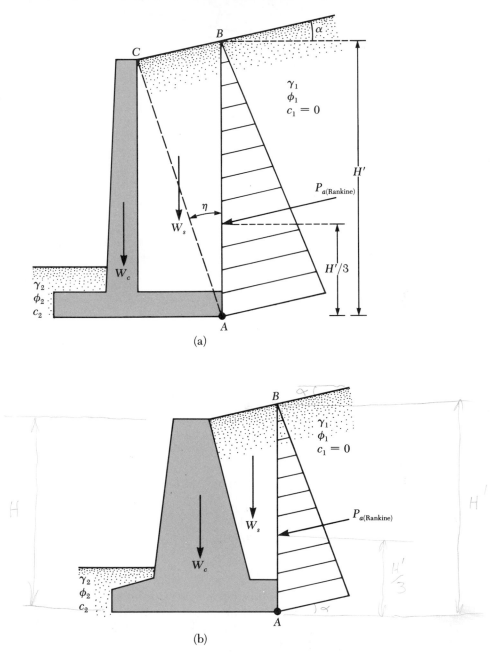

**Figure 5.18** Assumption for the determination of lateral earth pressure:
(a) cantilever wall; (b) and (c) gravity wall (see p. 240)

In several instances, for small retaining walls, *semiempirical charts* are used for evaluation of lateral earth pressure. Figures 5.19 and 5.20 show two semiempirical charts given by Terzaghi and Peck (1967). Figure 5.19 is for

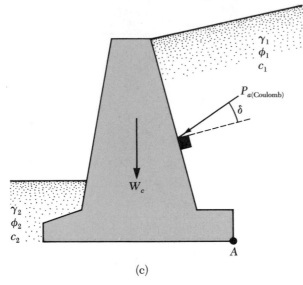

(c)

**Figure 5.18** (Continued)

backfills with plane surfaces, and Figure 5.20 is for backfills that slope upward from the crest of the wall for a limited distance and then become horizontal. Note that, in these figures, $\frac{1}{2}K_vH^2$ is the vertical component of the active force on the plane $AB$; similarly, $\frac{1}{2}K_hH'^2$ is the horizontal force. The numerals on the curves indicate the type of soil described in Table 5.8 on page 244.

## 5.14

### Stability Checks

To check the stability of a retaining wall, the following steps are necessary:

1. Check for *overturning* about its toe
2. Check for *sliding failure* along its base
3. Check for *bearing capacity failure* of the base
4. Check for *settlement*
5. Check for *overall stability*

This section will describe the procedure for checking for overturning and sliding and bearing capacity failure. The principles of investigation for settlement have been covered in Chapter 3 and will not be repeated here. Some problems regarding the overall stability of retaining walls have been discussed in Section 5.15.

### Check for Overturning

Figure 5.21 on page 245 shows the forces acting on a cantilever and gravity retaining wall with the assumption that the Rankine active pressure is acting along a vertical plane $AB$ drawn through the heel. $P_p$ is the Rankine passive pressure; its magnitude can be given as

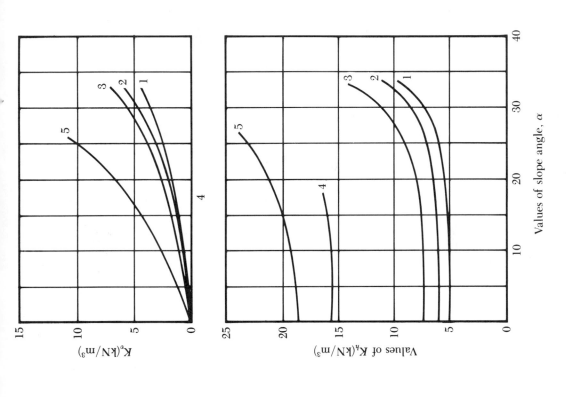

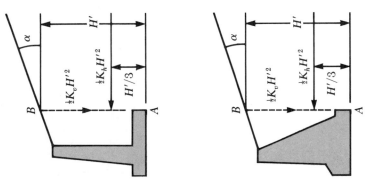

Note: Numerals on
curves indicate soil
types as described in
Table 5.8. For materials,
type-5 computations of
pressure may be based
on value of $H'$ 4 feet less
than actual value.

**Figure 5.19** Chart for estimating pressure of backfill against retaining
walls supporting backfills with plane surface (from *Soil Mechanics in
Engineering Practice*, 2nd ed., by K. Terzaghi and R. B. Peck. Copyright 1967 by John Wiley and Sons; reprinted by permission)

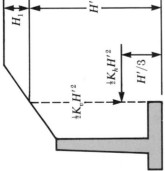

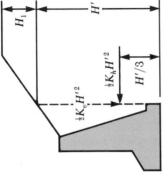

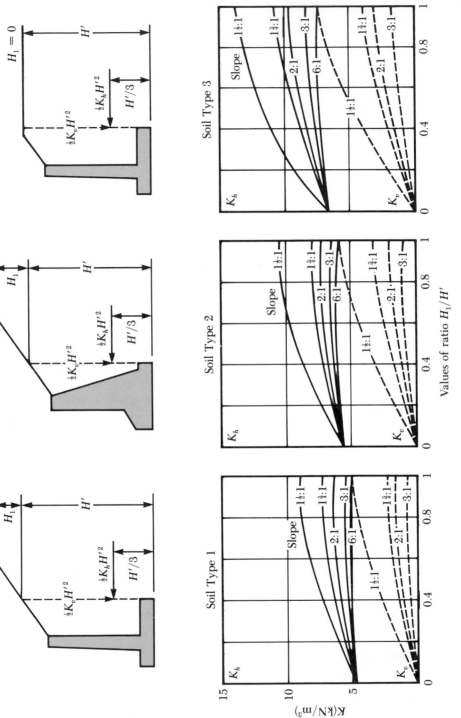

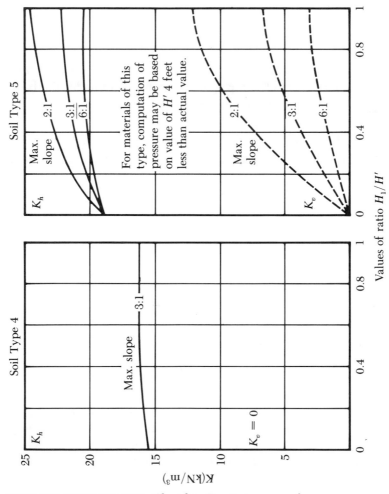

**Figure 5.20 (pages 242–243)**  Chart for estimating pressure of backfill against retaining walls supporting backfills with surface that slopes upward from crest of wall for limited distance and then becomes horizontal (from *Soil Mechanics in Engineering Practice*, 2nd ed., by K. Terzaghi and R. B. Peck. Copyright 1967 by John Wiley and Sons; reprinted by permission)

**Table 5.8**  Types of Backfill for Retaining Walls[a]

1. Coarse-grained soil without admixture of fine soil particles, very permeable (clean sand or gravel).
2. Coarse-grained soil of low permeability due to admixture of particles of silt size.
3. Residual soil with stones, fine silty sand, and granular materials with conspicuous clay content.
4. Very soft or soft clay, organic silts, or silty clays.
5. Medium or stiff clay, deposited in chunks and protected in such a way that a negligible amount of water enters the spaces between the chunks during floods or heavy rains. If this condition of protection cannot be satisfied, the clay should not be used as backfill material. With increasing stiffness of the clay, danger to the wall due to infiltration of water increases rapidly.

[a]From *Soil Mechanics in Engineering Practice*, 2nd ed., by K. Terzaghi and R. B. Peck. Copyright 1967 by John Wiley and Sons. Reprinted by permission.

$$P_p = \tfrac{1}{2} K_p \gamma_2 D^2 + 2c_2 \sqrt{K_p} D \tag{5.22}$$

where $\gamma_2$ = unit weight of soil in front of the heel and under the base slab
$\quad\quad K_p$ = Rankine passive earth pressure coefficient
$\quad\quad\quad$ = $\tan^2(45 + \phi_2/2)$
$\quad c_2, \phi_2$ = cohesion and soil friction angle, respectively

The factor of safety against overturning about the toe—that is, about point $C$ in Figure 5.21—can be expressed as

$$FS_{(\text{overturning})} = \frac{\Sigma \, M_R}{\Sigma \, M_O} \tag{5.50}$$

where $\Sigma \, M_O$ = sum of the moments of forces tending to overturn about point $C$
$\quad\quad \Sigma \, M_R$ = sum of the moments of forces tending to resist overturning about point $C$

The overturning moment can be given as

$$\Sigma \, M_O = P_h \left( \frac{H'}{3} \right) \tag{5.51}$$

where $P_h = P_a \cos \alpha$

For calculation of the resisting moment, $\Sigma \, M_R$ (neglecting $P_p$), a table (such as Table 5.9) can be prepared. The weight of the soil above the heel and the weight of the concrete (or masonry) are both forces that contribute to the resisting moment. Note that the force $P_v$ also contributes to the resisting moment. $P_v$ is the vertical component of the active force $P_a$, or

$$P_v = P_a \sin \alpha \tag{5.52}$$

The moment of the force $P_v$ about $C$ is equal to

$$M_v = P_v B = P_a \sin \alpha \cdot B \tag{5.53}$$

where $B$ = width of the base slab

Once $\Sigma \, M_R$ is known, the factor of safety can be calculated as

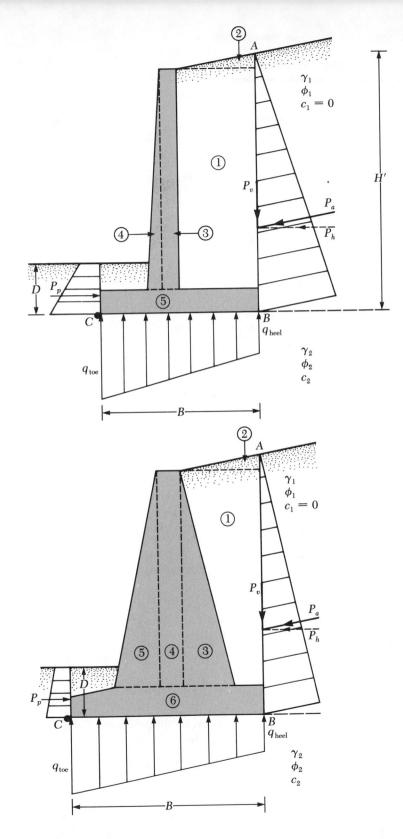

**Figure 5.21**
Check for
overturning;
assume Rankine
pressure is
valid

245

**Table 5.9** Procedure for Calculation of $\Sigma M_R$

| Section (1) | Area (2) | Weight/unit length of wall (3) | Moment arm measured from $C$ (4) | Moment about $C$ (5) |
|---|---|---|---|---|
| 1 | $A_1$ | $W_1 = \gamma_1 \times A_1$ | $X_1$ | $M_1$ |
| 2 | $A_2$ | $W_2 = \gamma_2 \times A_2$ | $X_2$ | $M_2$ |
| 3 | $A_3$ | $W_2 = \gamma_c \times A_3$ | $X_3$ | $M_3$ |
| 4 | $A_4$ | $W_2 = \gamma_c \times A_4$ | $X_4$ | $M_4$ |
| 5 | $A_5$ | $W_2 = \gamma_c \times A_5$ | $X_5$ | $M_5$ |
| 6 | $A_6$ | $W_2 = \gamma_c \times A_6$ | $X_6$ | $M_6$ |
| | | $P_v$ | $B$ | $M_v$ |
| | | $\Sigma V$ | | $\Sigma M_R$ |

Note: $\gamma_1$ = unit weight of backfill
$\gamma_c$ = unit weight of concrete

$$FS_{(overturning)} = \frac{M_1 + M_2 + M_3 + M_4 + M_5 + M_6 + M_v}{P_a \cos \alpha \dfrac{H'}{3}} \tag{5.54}$$

The usual minimum desirable value of the factor of safety with respect to overturning is 1.5 to 2.

Some designers prefer to use the following equation for determination of the factor of safety against overturning:

$$FS_{(overturning)} = \frac{M_1 + M_2 + M_3 + M_4 + M_5 + M_6}{(P_a \cos \alpha) \dfrac{H'}{3} - M_v} \tag{5.55}$$

## Check for Sliding Along the Base

The factor of safety against sliding may be expressed by the equation

$$FS_{(sliding)} = \frac{\Sigma F_{R'}}{\Sigma F_d} \tag{5.56}$$

where $\Sigma F_{R'}$ = sum of the horizontal resisting forces
$\Sigma F_d$ = sum of the horizontal driving forces

Referring to Figure 5.22, one sees that the shear strength of the soil below the base slab can be represented as

$$s = \sigma \tan \phi_2 + c_2$$

Thus, the maximum resisting force that can be derived from the soil per unit length of the wall along the bottom of the base slab is

$$R' = s \text{ (area of cross section)} = s(B \times 1) = B \sigma \tan \phi_2 + Bc_2$$

However

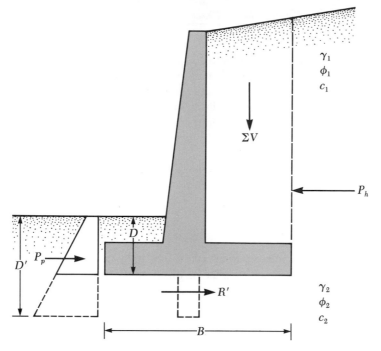

**Figure 5.22** Check for sliding along the base

$$B\sigma = \text{sum of the vertical force} = \Sigma \, V \text{ (see Table 5.9)}$$

So

$$R' = (\Sigma \, V) \tan \phi_2 + Bc_2$$

As can be seen in Figure 5.22, the passive force $P_p$ is also a horizontal resisting force. The expression for $P_p$ is given in Eq. (5.22). Hence

$$\Sigma \, F_{R'} = (\Sigma \, V) \tan \phi_2 + Bc_2 + P_p \tag{5.57}$$

The only horizontal force that will tend to cause sliding of the wall (*driving force*) is the horizontal component of the active force $P_a$, so

$$\Sigma \, F_d = P_a \cos \alpha \tag{5.58}$$

Combining Eqs. (5.56), (5.57), and (5.58)

$$FS_{(sliding)} = \frac{(\Sigma \, V) \tan \phi_2 + Bc_2 + P_p}{P_a \cos \alpha} \tag{5.59}$$

A minimum factor of safety of 1.5 against sliding is generally required.

In many cases, the passive force $P_p$ is ignored for calculation of the factor of safety with respect to sliding. The friction angle, $\phi_2$, is also reduced in several instances. The reduced soil friction angle may be in the order of one-half to two-thirds of the angle $\phi_2$. In a similar manner, the cohesion $c_2$ may be reduced to the value of 0.5 to 0.67 $c_2$. Thus

$$FS_{(\text{sliding})} = \frac{(\Sigma\ V)\tan(k_1\phi_2) + Bk_2c_2 + P_p}{P_a\ \cos\ \alpha}$$

(5.60)

where $k_1$ and $k_2$ are in the range of 1/2 to 2/3

In some instances, certain walls may not yield a desired factor of safety of 1.5. In order to improve their resistance to sliding, a base key may be used. Base keys are illustrated by broken lines in Figure 5.22. This figure indicates that the passive force at the toe *without the key* can be given as

$$P_p = \tfrac{1}{2}\gamma_2 D^2 K_p + 2c_2 D\sqrt{K_p}$$

However, if a key is included, the passive force per unit length of the wall will be

$$P_p = \tfrac{1}{2}\gamma_2 D'^2 K_p + 2c_2 D'\sqrt{K_p}$$

where $K_p = \tan^2(45 + \phi_2/2)$

Because $D' > D$, it is obvious that a key will help increase the passive resistance at the toe and, hence, the factor of safety against sliding. Usually the base key is constructed below the stem, and some main steel is run into the key.

### Check for Bearing Capacity Failure

The vertical pressure as transmitted to the soil by the base slab of the retaining wall should be checked against the ultimate bearing capacity of the soil. The nature of variation of the vertical pressure transmitted by the base slab into the soil is shown in Figure 5.23. Note that $q_{\text{toe}}$ and $q_{\text{heel}}$ are the *maximum* and the *minimum* pressures occurring at the ends of the toe and heel sections, respectively. The magnitudes of $q_{\text{toe}}$ and $q_{\text{heel}}$ can be determined in the following manner:

The sum of the vertical forces acting on the base slab is equal to $\Sigma\ V$ (see Col. 3, Table 5.9), and the horizontal force is $P_a\ \cos\ \alpha$. Let $R$ be the resultant force, or

$$\overrightarrow{R} = \overrightarrow{\Sigma\ V} + \overrightarrow{(P_a\ \cos\ \alpha)}$$

(5.61)

The net moment of these forces about the point $C$ (Figure 5.23) is equal to

$$M_{\text{net}} = \Sigma\ M_R - \Sigma\ M_O$$

(5.62)

Note that the values of $\Sigma\ M_R$ and $\Sigma\ M_O$ have been previously determined [see Col. 5, Table 5.9 and Eq. (5.51)]. Let the line of action of the resultant, $R$, intersect the base slab at $E$, as shown in Figure 5.23. The distance $CE$ can now be determined as

$$\overline{CE} = \overline{X} = \frac{M_{\text{net}}}{\Sigma\ V}$$

(5.63)

Hence, the eccentricity of the resultant, $R$, can be expressed as

$$e = \frac{B}{2} - \overline{CE}$$

(5.64)

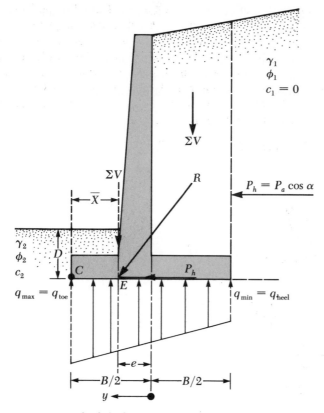

**Figure 5.23** Check for bearing capacity failure

The pressure distribution under the base slab can be determined by using the simple principles of mechanics of materials:

$$q = \frac{\Sigma V}{A} \pm \frac{M_{net} y}{I}$$

(5.65)

where $M_{net}$ = moment = $(\Sigma V)e$

$I$ = moment of inertia per unit length of the base section
$= 1/12\ (1)(B^3)$

For maximum and minimum pressures, the value of $y$ in Eq. (5.65) is equal to $B/2$. Substituting the preceding values into Eq. (5.65)

$$q_{max} = q_{toe} = \frac{\Sigma V}{(B)(1)} + \frac{e(\Sigma V)\dfrac{B}{2}}{\left(\dfrac{1}{12}\right)(B^3)} = \frac{\Sigma V}{B}\left(1 + \frac{6e}{B}\right)$$

(5.66)

Similarly

$$q_{min} = q_{heel} = \frac{\Sigma V}{B}\left(1 - \frac{6e}{B}\right)$$

(5.67)

Note that $\Sigma V$ includes the soil weight, as shown in Table 5.9, and that when the value of the eccentricity, $e$, becomes greater than $B/6$, $q_{min}$ becomes negative [Eq. (5.67)]. This indicates that there will be some tensile stress at the end of the heel section. This stress is not desirable because the tensile strength of soil is very small. If the analysis of a given design shows that $e > B/6$, then the design should be reproportioned, and calculations should be redone.

The relationships for the ultimate bearing capacity of a shallow foundation were discussed in Chapter 3 and can be expressed as

$$q_u = cN_cF_{cd}F_{ci} + qN_qF_{qd}F_{qi} + \tfrac{1}{2}\gamma_2 B'N_\gamma F_{\gamma d}F_{\gamma i} \tag{5.68}$$

where $q = \gamma_2 D$

$$B' = B - 2e$$

$$F_{cd} = 1 + 0.4\,\frac{D}{B'}$$

$$F_{qd} = 1 + 2\tan\phi_2(1 - \sin\phi_2)^2\,\frac{D}{B'}$$

$$F_{\gamma d} = 1$$

$$F_{ci} = F_{qi} = \left(1 - \frac{\psi^\circ}{90^\circ}\right)^2$$

$$F_{\gamma i} = \left(1 - \frac{\psi^\circ}{\phi^\circ}\right)^2$$

$$\psi^\circ = \tan^{-1}\left(\frac{P_a\cos\alpha}{\Sigma V}\right)$$

Note that the shape factors $F_{cs}$, $F_{qs}$, and $F_{ys}$, which were given in Eq. (3.16), are all equal to one, because they can be treated as a continuous foundation. For this reason, the shape factors are not shown in Eq. (5.68).

Once the ultimate bearing capacity of the soil has been calculated by using Eq. (5.68), the factor of safety against bearing capacity failure can be determined, or

$$FS_{(\text{bearing capacity})} = \frac{q_u}{q_{max}} \tag{5.69}$$

Generally, a factor of safety of 3 is required. In Chapter 3 we noted that the ultimate bearing capacity of shallow foundations occurs at a settlement of about 10% of the foundation width. In the case of retaining walls, the width $B$ is large. Hence, the ultimate load $q_u$ will occur at a fairly large foundation settlement. A factor of safety of 3 against bearing capacity failure may not insure, in all cases, that settlement of the structure will be within the tolerable limit. Thus this situation needs further investigation.

## Example 5.7

The cross section of a cantilever retaining wall is shown in Figure 5.24. Calculate the factors of safety with respect to overturning and sliding and bearing capacity.

## Solution

Referring to Figure 5.24

$$H' = H_1 + H_2 + H_3 = 2.6 \tan 10° + 6 + 0.7$$
$$= 0.458 + 6 + 0.7 = 7.158 \text{ m}$$

Rankine active force per unit length of wall $= P_a = \frac{1}{2}\gamma_1 H'^2 K_a$. For $\phi_1 = 30°$, $\alpha = 10°$, $K_a$ is equal to 0.350 (Table 5.6). Thus,

$$P_a = \frac{1}{2}(18)(7.158)^2(0.35) = 161.4 \text{ kN/m}$$
$$P_v = P_a \sin 10° = 161.4(\sin 10°) = 28.03 \text{ kN/m}$$
$$P_h = P_a \cos 10° = 161.4(\cos 10°) = 158.95 \text{ kN/m}$$

### Factor of Safety Against Overturning

The following table can now be prepared for determination of the resisting moment.

| Section* No. | Area (m²) | Weight/unit length (kN/m) | Moment arm from point C (m) | Moment (kN-m) |
|---|---|---|---|---|
| 1 | $6 \times 0.5 = 3$ | 70.74 | 1.15 | 81.35 |
| 2 | $\frac{1}{2}(0.2)6 = 0.6$ | 14.15 | 0.833 | 11.79 |
| 3 | $4 \times 0.7 = 2.8$ | 66.02 | 2.0 | 132.04 |
| 4 | $6 \times 2.6 = 15.6$ | 280.80 | 2.7 | 758.16 |
| 5 | $\frac{1}{2}(2.6)(0.458) = 0.595$ | 10.71 | 3.13 | 33.52 |
| | | $P_v = 28.03$ | 4.0 | 112.12 |
| | | $\Sigma V = 470.45$ | | $\Sigma\ 1128.98$ |
| | | | | $= \Sigma\ M_R$ |

*For section numbers, refer to Figure 5.24
$\gamma_{concrete} = 23.58 \text{ kN/m}^3$

The overturning moment, $M_O$

$$M_O = P_h\left(\frac{H'}{3}\right) = 158.95\left(\frac{7.158}{3}\right) = 379.25 \text{ kN-m}$$

$$FS_{(overturning)} = \frac{\Sigma\ M_R}{M_O} = \frac{1128.98}{379.25} = 2.98 > 2\text{—O.K.}$$

### Factor of Safety Against Sliding

From Eq. (5.60)

$$FS_{(sliding)} = \frac{(\Sigma\ V)\tan(k_1\phi_1) + Bk_2c_2 + P_p}{P_a \cos \alpha}$$

Let $k_1 = k_2 = 2/3$
Also

$$P_p = \frac{1}{2}K_p\gamma_2D^2 + 2c_2\sqrt{K_p}\,D$$

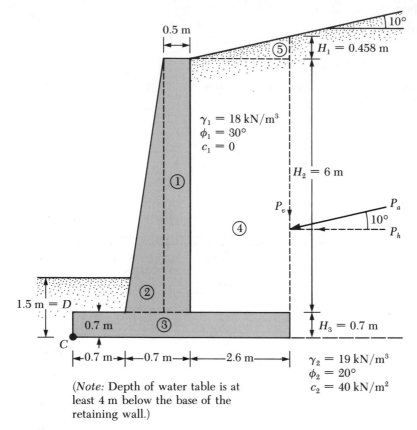

(*Note:* Depth of water table is at least 4 m below the base of the retaining wall.)

**Figure 5.24**

$$K_p = \tan^2\left(45 + \frac{\phi_2}{2}\right) = \tan^2(45 + 10) = 2.04$$

$$D = 1.5 \text{ m}$$

So

$$P_p = \tfrac{1}{2}(2.04)(19)(1.5)^2 + 2(40)(\sqrt{2.04})(1.5)$$
$$= 43.61 + 171.39 = 215 \text{ kN/m}$$

Hence

$$FS_{\text{(sliding)}} = \frac{(470.45)\tan\left(\dfrac{2 \times 20}{3}\right) + (4)\left(\dfrac{2}{3}\right)(40) + 215}{158.95}$$

$$= \frac{111.5 + 106.67 + 215}{158.95} = \underline{2.73 > 1.5\text{—O.K.}}$$

*Note:* For some designs, the depth $D$ for passive pressure calculation may be taken to be *equal to the thickness of the base slab*.

### Factor of Safety Against Bearing Capacity Failure

Combining Eqs. (5.62), (5.63), and (5.64)

$$e = \frac{B}{2} - \frac{\Sigma \, M_R - \Sigma \, M_O}{\Sigma \, V} = \frac{4}{2} - \frac{1128.98 - 379.25}{470.45}$$

$$= 0.406 \text{ m} < \frac{B}{6} = \frac{4}{6} = 0.666 \text{ m}$$

Again, from Eqs. (5.66) and (5.67)

$$q_{\substack{toe \\ heel}} = \frac{\Sigma \, V}{B}\left(1 \pm \frac{6e}{B}\right) = \frac{470.45}{4}\left(1 \pm \frac{6 \times 0.406}{4}\right) = 189.2 \text{ kN/m}^2 \text{ (toe)}$$

$$= 45.99 \text{ kN/m}^2 \text{ (heel)}$$

The ultimate bearing capacity of the soil can be determined from Eq. (5.68)

$$q_u = cN_cF_{cd}F_{ci} + qN_qF_{qd}F_{qi} + \tfrac{1}{2}\gamma_2 B\,'N_\gamma F_{\gamma d}F_{\gamma i}$$

For $\phi_2 = 20°$ (Table 3.2), $N_c = 14.83$, $N_q = 6.4$, and $N_\gamma = 5.39$. Also

$$q = \gamma_2 D = (19)(1.5) = 28.5 \text{ kN/m}^2$$

$$B' = B - 2e = 4 - 2(0.406) = 3.188 \text{ m}$$

$$F_{cd} = 1 + 0.4\left(\frac{D}{B'}\right) = 1 + 0.4\left(\frac{1.5}{3.188}\right) = 1.188$$

$$F_{qd} = 1 + 2\tan\phi_2(1 - \sin\phi_2)^2\left(\frac{D}{B'}\right) = 1 + 0.315\left(\frac{1.5}{3.188}\right) = 1.148$$

$$F_{\gamma d} = 1$$

$$F_{ci} = F_{qi} = \left(1 - \frac{\psi°}{90°}\right)^2$$

$$\psi = \tan^{-1}\left(\frac{P_a \cos\alpha}{\Sigma \, V}\right) = \tan^{-1}\left(\frac{158.95}{470.45}\right) = 18.67°$$

So

$$F_{ci} = F_{qi} = \left(1 - \frac{18.67}{90}\right)^2 = 0.628$$

$$F_{\gamma i} = \left(1 - \frac{\psi}{\phi}\right)^2 = \left(1 - \frac{18.67}{20}\right)^2 \approx 0$$

Hence

$$q_u = (40)(14.83)(1.188)(0.628) + (28.5)(6.4)(1.148)(0.628)$$
$$+ \tfrac{1}{2}(19)(5.93)(3.188)(1)(0)$$
$$= 442.57 + 131.50 + 0 = 574.07 \text{ kN/m}^2$$

$$FS_{(bearing\ capacity)} = \frac{q_u}{q_{toe}} = \frac{574.07}{189.2} = 3.03 > 3\text{—O.K.}$$

## Example
## 5.8

A gravity retaining wall is shown in Figure 5.25. Use $\delta = 2/3\phi_1$ and Coulomb's active earth pressure theory. Determine:

a. the factor of safety against overturning
b. the factor of safety against sliding
c. the pressure on the soil at the toe and heel

### Solution

$$H' = 5 + 1.5 = 6.5 \text{ m}$$

Coulomb's active force

$$P_a = \tfrac{1}{2}\gamma_1 H'^2 K_a$$

With $\alpha = 0°$, $\beta = 75°$, $\delta = 2/3\phi_1$, and $\phi_1 = 32°$, $K_a = 0.4023$ (Table 5.2). So

$$P_a = \tfrac{1}{2}(18.5)(6.5)^2(0.4023) = 157.22 \text{ kN/m}$$
$$P_h = P_a \cos(15 + \tfrac{2}{3}\phi_1) = 157.22 \cos 36.33 = 126.65 \text{ kN/m}$$
$$P_v = P_a \sin(15 + \tfrac{2}{3}\phi_1) = 157.22 \sin 36.33 = 93.14 \text{ kN/m}$$

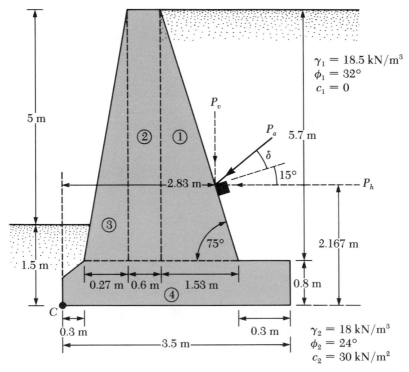

$\gamma_1 = 18.5 \text{ kN/m}^3$
$\phi_1 = 32°$
$c_1 = 0$

$\gamma_2 = 18 \text{ kN/m}^3$
$\phi_2 = 24°$
$c_2 = 30 \text{ kN/m}^2$

**Figure 5.25**

## Part a: Factor of Safety Against Overturning

Referring to Figure 5.25, one can prepare the following table:

| Area No. | Area (m²) | Weight* (kN) | Moment arm from $C$ (m) | Moment (kN-m) |
|---|---|---|---|---|
| 1 | $\frac{1}{2}(5.7)(1.53) = 4.36$ | 102.81 | 2.18 | 224.13 |
| 2 | $(0.6)(5.7) = 3.42$ | 80.64 | 1.37 | 110.48 |
| 3 | $\frac{1}{2}(0.27)(5.7) = 0.77$ | 18.16 | 0.98 | 17.80 |
| 4 | $\approx(3.5)(0.8) = 2.8$ | 66.02 | 1.75 | 115.54 |
| | | $P_v = 93.14$ | 2.83 | 263.59 |
| | | $\Sigma V = 360.77$ kN | | $\Sigma M_R = 731.54$ kN-m |

$*\gamma_{concrete} = 23.58$ kN/m³

*Note:* The weight of the soil above the back face of the wall is not taken into account in the preceding table. The reason for this has been explained in Figure 5.18c.

$$\text{Overturning moment} = M_O = P_h\left(\frac{H'}{3}\right) = 126.65(2.167) = 274.45 \text{ kN-m}$$

Hence

$$FS_{(overturning)} = \frac{\Sigma M_R}{\Sigma M_O} = \frac{731.54}{274.45} = 2.665 > 2\text{—O.K.}$$

## Part b: Factor of Safety Against Sliding

$$FS_{(sliding)} = \frac{(\Sigma V)\tan\left(\frac{2}{3}\phi_2\right) + \frac{2}{3}c_2B + P_p}{P_h}$$

$$P_p = \frac{1}{2}K_p\gamma_2D^2 + 2c_2\sqrt{K_p}D$$

$$K_p = \tan^2\left(45 + \frac{24}{2}\right) = 2.37$$

Hence

$$P_p = \frac{1}{2}(2.37)(18)(1.5)^2 + 2(30)(1.54)(1.5) = 186.59 \text{ kN/m}$$

So

$$FS_{(sliding)} = \frac{360.77\tan\left(\frac{2}{3} \times 24\right) + \frac{2}{3}(30)(3.5) + 186.59}{126.65}$$

$$= \frac{103.45 + 70 + 186.59}{126.65} = 2.84$$

If $P_p$ is ignored, the factor of safety would be 1.37.

Part c: Pressure on Soil at Toe and Heel

From Eqs. (5.62), (5.63), and (5.64)

$$e = \frac{B}{2} - \frac{\Sigma\,M_R - \Sigma\,M_O}{\Sigma\,V} = \frac{3.5}{2} - \frac{731.54 - 274.45}{360.77} = 0.483 < \frac{B}{6} = 0.583$$

$$q_{toe} = \frac{\Sigma\,V}{B}\left[1 + \frac{6e}{B}\right] = \frac{360.77}{3.5}\left[1 + \frac{(6)(0.483)}{3.5}\right] = 188.43 \text{ kN/m}^2$$

$$q_{heel} = \frac{V}{B}\left[1 - \frac{6e}{B}\right] = \frac{360.77}{3.5}\left[1 - \frac{(6)(0.483)}{3.5}\right] = 17.73 \text{ kN/m}^2$$

## 5.15
## Other Types of Possible Failure of Retaining Walls

In addition to the three types of possible failure mode for retaining walls discussed in the preceding section, two other types of failure could occur: *shallow shear failure* and *deep shear failure*.

*Shallow shear failure* in soil below the base of a retaining wall takes place along a cylindrical surface *abc* passing through the heel, as shown in Figure 5.26a. The center of the arc of the circle *abc* is located at *O*, which is found by trial and error (corresponds to the minimum factor of safety). This type of failure can occur as the result of excessive induced shear stress along the cylindrical surface in soil. In general, the factor of safety against horizontal sliding is lower than that obtained by shallow shear failure. So, if $FS_{(sliding)}$ is greater than about 1.5, shallow shear failure under the base may not occur.

*Deep shear failure* can occur along a cylindrical surface *abc*, as shown in Figure 5.26b, as the result of the existence of a weak layer of soil underneath the wall at a depth of about 1.5 times the height of the retaining wall. In such cases, the critical cylindrical failure surface *abc* has to be determined by trial and error with various centers, such as *O* (Figure 5.26b). The failure surface along which the minimum factor of safety is obtained is the *critical surface of sliding*. For the backfill slope with $\alpha$ less than about 10°, it has been observed that the critical failure circle passes through the edge of the heel slab (such as *def* in Figure 5.26b). In this situation, the minimum factor of safety also has to be determined by trial and error by changing the center of the trial circle.

An approximate procedure for determination of the factor of safety against a deep-seated shear failure for a gently sloping backfill ($\alpha < 10°$) is outlined as follows in a step-by-step manner (Teng, 1962) (refer to Figure 5.27):

**1.** Draw the retaining wall and the underlying soil layer to a convenient scale.

**2.** For a trial center *O*, draw an arc of a circle *abcd*. For all practical purposes, the weight of the soil in the area *abcde* may be taken to be symmetrical about a vertical line drawn through point *O*. Let the radius of the trial circle be equal to *r*.

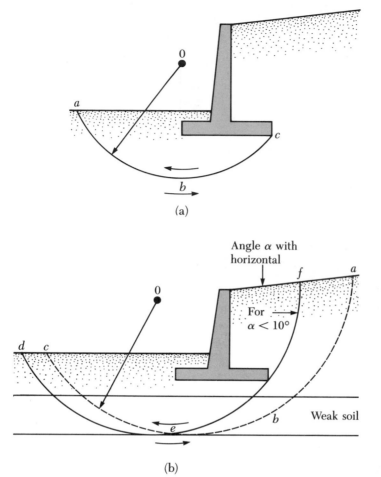

**Figure 5.26** (a) Shallow shear failure; (b) deep shear failure

**3.** To determine the driving force on the failure surface causing instability (Figure 5.27a), divide the area in the zone *efgh* into several slices. These slices can be treated as rectangles or triangles, as the case may be.

**4.** Determine the area of each of these slices, and then determine the weight $W$ of the soil (and/or concrete) contained inside each of these slices (per unit length of the wall).

**5.** Draw a vertical line through the centroid of each slice, and locate the point of intersection of each vertical line with the trial failure circle.

**6.** Join point $O$ (that is, the center of the trial circles) with the points of intersection as determined in Step 5.

**7.** Determine the angle, $\omega$, that each vertical line makes with the radial line.

**8.** Calculate $W \sin \omega$ for each slice.

**9.** Determine the active force $P_a$ on the face $df$, $\frac{1}{2}\gamma_1 H'^2 K_a$.

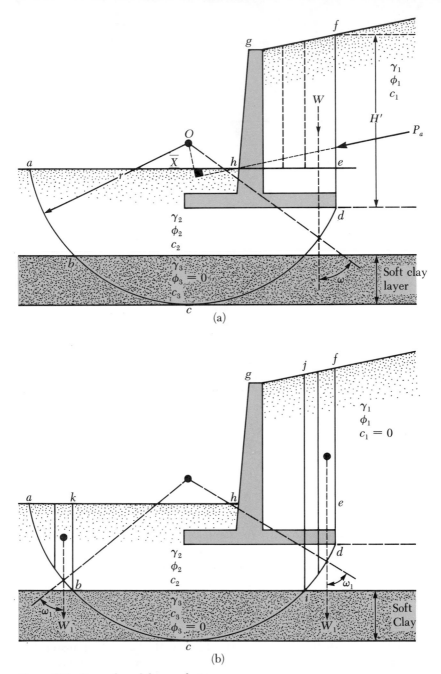

**Figure 5.27**   Deep shear failure analysis

**10.** The total driving force can now be calculated as

$$\sum (W \sin \omega) + \frac{P_a \overline{X}}{r} \tag{5.70}$$

where $\bar{X}$ = perpendicular distance between the line of action of $P_a$ and the center $O$

**11.** To determine the resisting force on the failure surface (Figure 5.27b), divide the area in the zones *abk* and *idefj* into several slices, and determine the weight of each slice, $W_1$ (per unit length of the wall). Note that the points *b* and *i* are on the top of the soft clay layer; the weight of each slice shown in Figure 5.27b is $W_1$ in contrast to the weight of each slice $W$, as shown in Figure 5.27a.

**12.** Draw a vertical line through the centroid of each slice, and locate the point of intersection of each line with the trial failure circle.

**13.** Join point $O$ with the points of intersection as determined in Step 12. Determine the angles, $\omega_1$, that the vertical lines make with the radial lines.

**14.** For each slice, obtain

$W_1 \tan \phi_2 \cos \omega_1$

**15.** Calculate

$c_2 l_1 + c_3 l_2 + c_2 l_3$

where $l_1$, $l_2$, and $l_3$ are the lengths of the arcs *ab*, *bi*, and *id*

**16.** The maximum resisting force that can be derived along the failure surface is

$$\sum (W_1 \tan \phi_2 \cos \omega_1) + c_2 l_1 + c_3 l_2 + c_2 l_4 \qquad (5.71)$$

**17.** To determine the factor of safety against deep shear failure for this trial failure surface:

$$FS_{\text{(deep shear failure)}} = \frac{\sum (W_1 \tan \phi_2 \cos \omega_1) + c_2 l_1 + c_3 l_2 + c_2 l_3}{\sum (W \sin \omega) + \dfrac{P_a \bar{X}}{r}} \qquad (5.72)$$

Several other trial failure surfaces may be drawn, and the factor of safety can be determined in a similar manner. The lowest value the factor of safety is the desired factor of safety.

## 5.16

### Comments Relating to Stability

When a weak soil layer is located at a shallow depth—that is, within a depth of about 1.5 times the height of the retaining wall—the bearing capacity of the weak layer should be carefully investigated. One should also look into the possibility of excessive settlement. In some cases, the use of lightweight backfill material behind the retaining wall may solve the problem.

In many instances, piles are used to transmit the foundation load to a firmer layer. However, often the thrust of the sliding wedge of soil, in the case of deep shear failure, may bend the piles and eventually cause them to fail. Careful attention should be given to this possibility when considering the option of pile foundations for retaining walls. (Pile foundations may sometimes be required for bridge abutments to avoid the problem of scouring.)

## 5.17

## Drainage from the Backfill of the Retaining Wall

As the result of rainfall or other wet conditions, the backfill material for a retaining wall may become saturated. Saturation will, in effect, increase the pressure on the wall and create an unstable condition. For this reason, adequate drainage must be provided by means of *weepholes* and/or *perforated drainage pipes* (see Figure 5.28).

The *weepholes,* if provided, should have a minimum diameter of about 0.1 m and be adequately spaced. Note that there is always a possibility that the backfill material may be washed into weepholes or drainage pipes. This will ultimately clog up the drainage facilities. Thus a filter material needs to be placed behind the weepholes or around the drainage pipes, as the case may be. Nowadays, filter cloth is used to serve the same purpose. Whenever granular soil is used as a filter, the principles outlined in Section 1.11 should be followed. Example Problem 5.9 gives the procedure for designing a filter.

## Example 5.9

Figure 5.29 shows the grain-size distribution of a backfill material. Using the conditions outlined in Section 1.11, determine the range of the grain-size distribution for the filter material.

### Solution

From the grain-size distribution curve given in Figure 5.29, the following values can be determined:

$$D_{15(B)} = 0.04 \text{ mm}$$
$$D_{85(B)} = 0.25 \text{ mm}$$
$$D_{50(B)} = 0.13 \text{ mm}$$

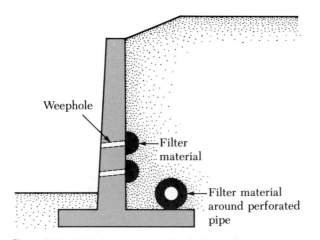

**Figure 5.28** Drainage provisions for the backfill of a retaining wall

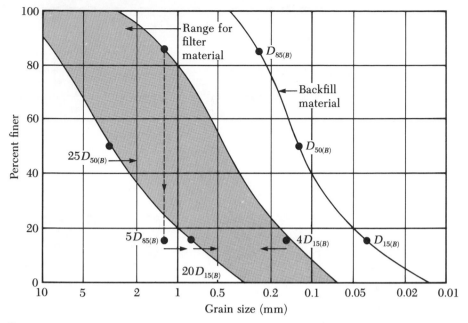

**Figure 5.29**

Conditions of filter

**1.** $D_{15(F)}$ should be less than $5D_{85(F)}$—that is, $5 \times 0.25 = 1.25$ mm.
**2.** $D_{15(F)}$ should be greater than $4D_{15(B)}$—that is, $4 \times 0.04 = 0.16$ mm.
**3.** $D_{50(F)}$ should be less than $25D_{50(B)}$—that is, $25 \times 0.13 = 3.25$ mm.
**4.** $D_{15(F)}$ should be less than $20D_{15(B)}$—that is, $20 \times 0.04 = 0.8$ mm.

These limiting points are plotted in Figure 5.29. Through these points, two curves can be drawn that are similar in nature to the grain-size distribution curve of the backfill material. These curves define the range for the filter material to be used.

# 5.18

## Provision of Joints in the Construction of Walls

A retaining wall may be constructed with all or any of the following joints:

**1.** *Construction joints (Figure 5.30a):* These are vertical and horizontal joints that are used between two successive pours of concrete. In order to increase the shear at the joints, keys may be used. If keys are not used, the surface of the first pour is cleaned and roughened before the next placement of concrete.

**2.** *Contraction joints (Figure 5.30b):* These are vertical joints (grooves) placed in the face of the wall (from the top of the base slab to the top of the wall) that allow the concrete to shrink without noticeable harm. The grooves may be about 6–8 mm wide and 12–16 mm deep.

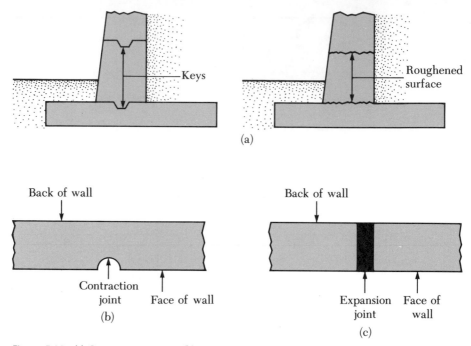

**Figure 5.30**  (a) Construction joints; (b) contraction joint; (c) expansion joint

**3.** *Expansion joints (Figure 5.30c):* To take into account the expansion of concrete caused by temperature changes, vertical expansion joints going from the base to the top of the wall may also be used. These joints may be filled with flexible joint fillers. In most cases, horizontal reinforcing steel bars running across the stem are continuous through all joints. The steel is greased to allow the concrete to expand.

## Problems

**5.1** Refer to Figure 5.3a on p. 211. Given: $H_1 = 10$ ft, $H_2 = 10$ ft, $q = 0$, $\gamma = 105$ lb/ft$^3$, $\gamma_{sat} = 122$ lb/ft$^3$, $c = 0$, and $\phi = 30°$. Determine the at-rest lateral earth force per foot length of the wall. Also find the location of the resultant. Use Eq. (5.3).

**5.2** Refer to Figure 5.3a. Given: $H_1 = 4.5$ m, $H_2 = 0$, $q = 0$, and $\gamma = 17$ kN/m$^2$. The backfill is an overconsolidated clay with a plasticity index of 23. If the overconsolidation ratio is 2.2, determine the at-rest lateral earth force per meter of the wall. Also find the location of the resultant. Use Eqs. (5.5) and (5.7).

**5.3** Redo Problem 5.2 assuming that the surcharge $q = 50$ kN/m$^2$.

**5.4** Refer to Figure 5.5a. Given: the height of the retaining wall, $H = 21$ ft the backfill is a saturated clay with $\phi = 0°$, $c = 630$ lb/ft$^2$, and $\gamma_{sat} = 113$ lb/ft$^3$.
a.  Determine the Rankine active pressure distribution diagram behind the wall.
b.  Determine the depth of the tensile crack, $z_c$.

c. Estimate the Rankine active force per foot of the wall before and after the occurrence of tensile crack.

**5.5** Refer to Problem 5.4.
   a. Draw the Rankine passive pressure distribution diagram behind the wall.
   b. Estimate the Rankine passive force per foot of the wall and also the location of the resultant.

**5.6** A vertical retaining wall (Figure 5.5a on p. 213) is 6.3 m high with a horizontal backfill. For the backfill, given $\gamma = 18.7$ kN/m³, $\phi = 24°$, and $c = 10$ kN/m². Determine the Rankine active force per unit length of the wall after the occurrence of the tensile crack.

**5.7** Repeat Problem 5.6 for the Rankine passive case.

**5.8** Refer to Figure P5.8. Given: $H_1 = 8.2$ ft, $H_2 = 14.8$ ft, $\gamma_1 = 107$ lb/ft³, $\phi_1 = 34°$, $c_1 = 0$, $\gamma_2 = 140$ lb/ft³, $\phi_2 = 25°$, and $c_2 = 10$ kN/m². Determine the Rankine active force per unit length of the wall.

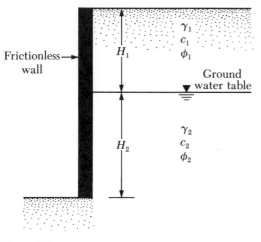

**Figure P5.8**

**5.9** Repeat Problem 5.8 for the Rankine passive case.

**5.10** Refer to Figure 5.8a on p. 220. Given: $H = 5$ m, $\gamma = 18.2$ kN/m³, $\phi = 30°$, $\delta = 20°$, $c = 0$, $\alpha = 10°$, and $\beta = 85°$. Determine the Coulomb's active force per meter of the wall and the location and direction of the resultant.

**5.11** For the retaining wall described in Problem 5.10, determine the Coulomb's passive force per meter of the wall and the location and direction of the resultant.

**5.12** Refer to Figure 5.12b on p. 229, which shows a vertical retaining wall with a horizontal backfill. Given: $H = 13.5$ ft, $\gamma = 105$ lb/ft³, $\phi = 35°$, and $\delta = 10°$. Based on Shields and Tolunay's work (Table 5.5), what would be the passive force per meter of the wall?

**5.13** Refer to Figure 5.13 on p. 230. For the retaining wall, given $H = 7.5$ m, $\phi = 32°$, $\alpha = 5°$, $\gamma = 18.2$ kN/m³, and $c = 0$.
   a. Determine the intensity of the Rankine active force at $z = 2, 4, 6$, and 7.5 m.
   b. Determine the Rankine active force per meter of the wall and also the location and direction of the resultant.

**5.14** Refer to Figure 5.14b on p. 233. Given: $H = 12$ ft, $a' = 3$ ft, $b' = 4.5$ ft, and $q = 525$ lb/ft².

   a. Determine the lateral force per unit length of the wall caused by surcharge loading only.

   b. Determine the location of the center of pressure, $\bar{z}$.

**5.15** For the cantilever retaining wall shown in Figure P5.15, given:

    Wall dimensions:  $H = 8$ m, $x_1 = 0.4$ m, $x_2 = 0.6$ m, $x_3 = 1.5$ m, $x_4 = 3.5$ m, $x_5 = 0.96$ m, $D = 1.75$ m, $\alpha = 10°$

    Soil properties:  $\gamma_1 = 16.8$ kN/m³, $\phi_1 = 32°$, $\gamma_2 = 17.6$ kN/m³, $\phi_2 = 28°$, $c_2 = 30$ kN/m²

Calculate the factor of safety with respect to overturning and sliding and bearing capacity.

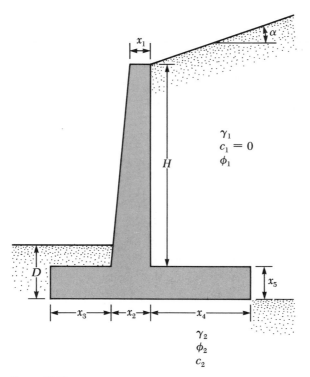

**Figure P5.15**

**5.16** Repeat Problem 5.15 with the following:

    Wall dimensions:  $H = 20$ ft, $x_1 = 12$ in., $x_2 = 27$ in., $x_3 = 4.5$ ft, $x_4 = 7.5$ ft, $x_5 = 2.75$ ft, $D = 4$ ft, $\alpha = 5°$

    Soil properties:  $\gamma_1 = 117$ lb/ft³, $\phi_1 = 34°$, $\gamma_2 = 107$ lb/ft³, $\phi_2 = 18°$, $c_2 = 1050$ lb/ft²

**5.17** Repeat Problem 5.15 with the following:

    Wall dimensions:  $H = 5.49$ m, $x_1 = 0.46$ m, $x_2 = 0.58$ m, $x_3 = 0.92$ m, $x_4 = 1.55$ m, $x_5 = 0.61$ m, $D = 1.22$ m, $\alpha = 0°$

    Soil properties:  $\gamma_1 = 18.08$ kN/m³, $\phi_1 = 36°$, $\gamma_2 = 19.65$ kN/m³, $\phi_2 = 15°$, $c_2 = 44$ kN/m²

**5.18** A gravity retaining wall is shown in Figure P5.18. Calculate the factor of safety with
respect to overturning and sliding and bearing capacity. Given:

Wall dimensions:   $H = 15$ ft, $x_1 = 1.5$ ft, $x_2 = 0.8$ ft, $x_3 = 5.25$ ft, $x_4 = 1.25$ ft,
                       $x_5 = 1.5$ ft, $x_6 = 2.5$ ft, $D = 4$ ft

Soil properties:   $\gamma_1 = 121$ lb/ft$^3$, $\phi_1 = 30°$, $\gamma_2 = 121$ lb/ft$^3$, $\phi_2 = 20°$,
                  $c_2 = 1000$ lb/ft$^2$

Use Rankine active pressure for calculation (see Figure 5.18b on p. 239).

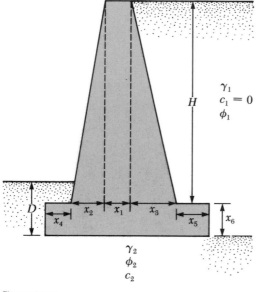

**Figure P5.18**

**5.19** Repeat Problem 5.18 using Coulomb's active pressure for calculation and $\delta = 2/3\phi_1$.

**5.20** Repeat Example Problem 5.8 using Rankine active pressure.

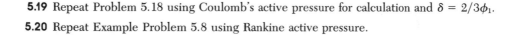

## References

Brooker, E. W., and Ireland, H. O. (1965). "Earth Pressure at Rest Related to Stress
History," *Canadian Geotechnical Journal*, Vol. 2, No. 1, pp. 1–15.

Caquot, A., and Kerisel, J. (1948). *Tables for Calculation of Passive Pressure, Active
Pressure, and Bearing Capacity of Foundations*, Gauthier-Villars, Paris, France.

Coulomb, C. A. (1776). *Essai sur une Application des Regles des Maximis et Minimum
à quelques Problemes de Statique Relatifs à l'Architecture*, Mem. Acad. Roy. des
Sciences, Paris, Vol. 3, p. 38.

Department of the Navy (1971). *Design Manual—Soil Mechanics, Foundations and
Earth Structures*, NAVFAC DM-7, Washington, D.C.

James, R. G., and Bransby, P. L. (1971). "A Velocity Field for Some Passive Earth
Pressure Problems," *Geotechnique*, London, England, Vol. 21, No. 1, pp. 61–84.

Jarquio, R. (1981). "Total Lateral Surcharge Pressure Due to Strip Load," *Journal of the
Geotechnical Engineering Division*, American Society of Civil Engineers, Vol. 107,
No. GT10, pp. 1424–1428.

Packshaw, S. (1969). "Earth Pressure and Earth Resistance," *A Century of Soil Mechanics*, The Institution of Engineers, London, England, pp. 409–435.

Shields, D. H., and Tolunay, A. Z. (1972). "Passive Pressure Coefficients for Sand by the Terzaghi and Peck Method," *Canadian Geotechnical Journal*, Vol. 9, No. 4.

Shields, D. H., and Tolunay, A. Z. (1973). "Passive Pressure by Method of Slices," *Journal of the Soil Mechanics and Foundations Division*, American Society of Civil Engineers, Vol. 99, No. SM12, pp. 1043–1053.

Teng, W. C. (1962). *Foundation Design*, Prentice-Hall, Englewood Cliffs, N.J.

Terzaghi, K., and Peck, R. B. (1967). *Soil Mechanics in Engineering Practice*, Wiley, New York.

C H A P T E R

# 6

# Sheet Pile Walls in Waterfront Structures

## 6.1

Introduction

Connected or semiconnected sheet piles are often used to build continuous walls for waterfront structures that may range from small waterfront pleasure-boat launching facilities to large dock facilities (Figure 6.1a). In contrast to the construction of other types of retaining wall, the building of sheet pile walls does not usually require dewatering of the site. Sheet piles are also used for some temporary structures, such as braced cuts (Figure 6.1b). The principles of sheet pile design used in braced cuts are given in Chapter 7.

There are several types of sheet pile that are in common use: (a) *wooden sheet piles*, (b) *precast concrete sheet piles*, and (c) *steel sheet piles*.

*Wooden sheet piles* are used only for temporary light structures located above the water table. The most common types are ordinary wooden planks and *Wakefield piles*. The wooden planks are about 50 mm × 300 mm in cross section and are driven edge to edge (Figure 6.2a). The Wakefield piles are made by nailing three planks together with the middle plank offset by 50–75 mm (Figure 6.2b). Wooden planks can also be milled to form *tongue-and-groove piles*. This is shown in Figure 6.2c. Figure 6.2d shows another type of wooden sheet pile that has grooves that are cut in the mills. Metal *splines* are driven into the grooves of the adjacent sheetings to hold them together after they are driven into the ground.

*Precast concrete sheet piles* are heavy and are designed with reinforcements to withstand the permanent stresses to which the structure will be

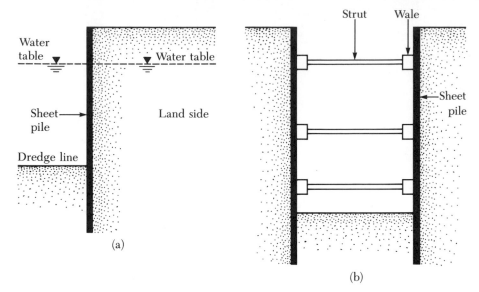

**Figure 6.1**  Examples of uses of sheet piles: (a) waterfront sheet pile wall; (b) braced cut

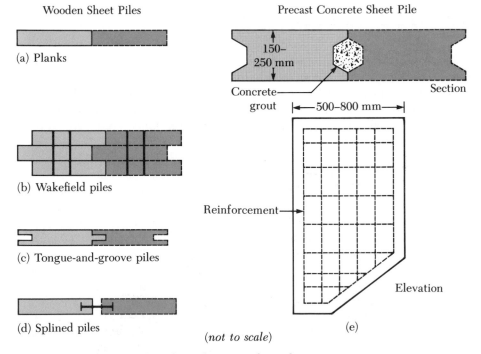

**Figure 6.2**  Various types of wooden and concrete sheet pile

subjected after construction and also to handle the stresses produced during construction. In cross section, these piles are about 500–800 mm wide and 150–250 mm thick. Figure 6.2e shows schematic diagrams of the elevation and the cross section of a reinforced concrete sheet pile.

The *steel sheet piles* in the United States are about 10–13 mm thick. European sections may be thinner and wider. Sheet pile sections may be Z, *deep arch, low arch,* or *straight web* sections. The interlocks of the sheet pile sections are shaped like a *thumb-and-finger* or a *ball-and-socket* type for water-tight connections. Figure 6.3a shows schematic diagrams of the thumb-and-finger type of interlocking for straight web sections. The ball-and-socket type of interlocking for Z section piles is shown in Figure 6.3b. Table 6.1 shows the properties of the sheet pile sections produced by the U.S. Steel Corporation. The allowable design flexural stress for the steel sheet piles is as follows:

| Type of steel | Allowable stress (MN/m²) |
|---|---|
| ASTM A-328 | 170 ($\approx 25{,}000$ lb/in.²) |
| ASTM A-572 | 210 ($\approx 30{,}000$ lb/in.²) |
| ASTM A-690 | 210 ($\approx 30{,}000$ lb/in.²) |

Steel sheet piles are convenient to use because of their resistance to high-driving stress developed when being driven into hard soils. They are also lightweight and reusable.

This chapter discusses the design principles of waterfront retaining structures built with sheet piles. The sheet pile walls may be divided into two basic categories: (a) *cantilever walls* and (b) *anchored walls*. In the construction of these walls, sheet piles may be driven into the ground and then the backfill placed on the land side; or, the sheet pile may first be driven into the ground and the soil in front of the sheet pile dredged. In any case, the soil used for backfill behind the sheet pile wall is usually granular. The soil below the dredge line may be sandy or clayey soil. The surface of soil on the water side is referred to as the *mud line*, or *dredge line*.

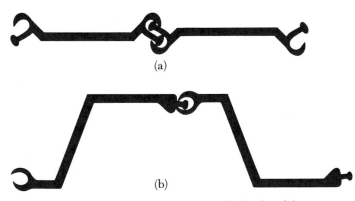

(a)

(b)

**Figure 6.3**  Nature of sheet pile connections: (a) thumb-and-finger type; (b) ball-and-socket type

**Table 6.1** Properties of Sheet Pile Sections (Produced by United States Steel Corporation)

| Section designation | Sketch of section | Driving distance (mm) | Section modulus | | Moment of inertia | | Comments |
|---|---|---|---|---|---|---|---|
| | | | m³/m of wall × 10⁵ | in.³/ft of wall | m⁴/m of wall × 10⁶ | in.⁴/ft of wall | |
| PZ-38 | 12.7 mm / 9.53 mm / 12.7 mm / 304.8 mm / Driving distance | 457.2 | 251.32 | 46.8 | 383.29 | 280.8 | Interlock with each other and also with PSA-23 or PSA-28 |
| PZ-32 | 12.7 mm / 9.53 mm / 12.7 mm / 292.1 mm | 533.4 | 205.67 | 38.3 | 300.85 | 220.4 | Interlock with each other and also with PSA-23 or PSA-28 |
| PZ-27 | 12.7 mm / 9.53 mm / 12.7 mm / 304.8 mm | 457.2 | 162.17 | 30.2 | 251.43 | 184.2 | Interlock with each other and also with PSA-23 or PSA-28 |
| PDA-27 | 9.53 mm / 12.7 mm | 406.4 | 57.46 | 10.7 | 54.33 | 39.8 | Interlock with each other |

**Table 6.1** (Continued)

| Section designation | Sketch of section | Driving distance (mm) | Section modulus m³/m of wall × 10⁵ | Section modulus in.³/ft of wall | Moment of inertia m⁴/m of wall × 10⁶ | Moment of inertia in.⁴/ft of wall | Comments |
|---|---|---|---|---|---|---|---|
| PMA-22 | 9.53 mm / 82.55 mm | 498.48 | 29.00 | 5.4 | 18.70 | 13.7 | Interlock with each other |
| PSA-28 | 12.7 mm / 34.13 mm | 406.4 | 13.43 | 2.5 | 6.14 | 4.5 | Interlock with each other |
| PSA-23 | 9.53 mm / 34.13 mm | 406.4 | 12.89 | 2.4 | 5.6 | 4.1 | Interlock with each other |
| PSX-32 | 11.51 mm | 419.1 | 12.89 | 2.4 | 5.05 | 3.7 | Interlock with each other |
| PS-32 | 12.7 mm | 381 | 10.20 | 1.9 | 3.96 | 2.9 | Interlock with each other |
| PS-28 | 9.53 mm | 381 | 10.20 | 1.9 | 3.82 | 2.8 | Interlock with each other |

271

## Cantilever Sheet Pile Wall

Cantilever sheet pile walls are usually recommended for walls of moderate height—about 10 m or less measured above the dredge line. In these walls, the sheet piles act like a wide cantilever beam above the dredge line. The basic principles for the estimation of net lateral pressure distribution on a cantilever sheet pile wall can be explained with the aid of Figure 6.4, which shows the nature of lateral yielding of a cantilever wall penetrating a sand layer below the dredge line. The wall rotates about a point $O$. Because the hydrostatic pressures at any depth from both sides of the wall will cancel each other, we will consider only the effective lateral soil pressures. In Zone A, the lateral pressure is only the active pressure from the land side. In Zone B, because of the nature of yielding of the wall, there will be active pressure from the land side and passive pressure from the water side. The condition is reversed in Zone C—that is, below the point of rotation, $O$. The net actual pressure distribution on the wall is like that shown in Figure 6.4b. However, for design purposes, Figure 6.4c shows a simplified version.

The following sections (Sections 6.2 and 6.3) present the mathematical formulation of the analysis of cantilever sheet pile walls. Note that, in some waterfront structures, the water level may fluctuate as the result of tidal effects. Care should be taken in determining the water level that will affect the net pressure diagram.

### 6.2

## Cantilever Sheet Piling Penetrating Sandy Soils

To develop the relationships for the proper depth of embedment of sheet piles driven into a granular soil, we refer to Figure 6.5a. The soil retained by

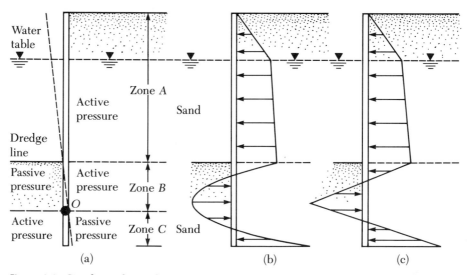

**Figure 6.4**   Cantilever sheet pile penetrating sand

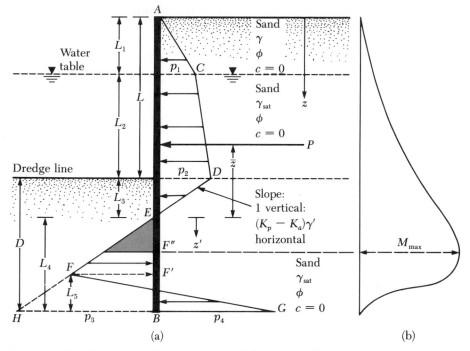

**Figure 6.5** Cantilever sheet pile penetrating sand: (a) variation of net pressure diagram; (b) variation of moment

the sheet piling above the dredge line is also sand. The water table is located at a depth of $L_1$ below the top of the wall. Let the angle of friction of the sand be $\phi$. The intensity of the active pressure at a depth $z = L_1$ can be given as

$$p_1 = \gamma L_1 K_a \qquad (6.1)$$

where $K_a$ = Rankine active pressure coefficient = $\tan^2(45 - \phi/2)$
$\quad\;\; \gamma$ = unit weight of soil above the water table

Similarly, the active pressure at a depth of $z = L_1 + L_2$ (that is, at the level of the dredge line) is equal to

$$p_2 = (\gamma L_1 + \gamma' L_2) K_a \qquad (6.2)$$

where $\gamma'$ = effective unit weight of soil = $\gamma_{sat} - \gamma_w$

Note that, at the level of the dredge line, the hydrostatic pressures from both sides of the wall are of the same magnitude and cancel each other.

In order to determine the net lateral pressure below the dredge line up to the point of rotation $O$, as shown in Figure 6.4a, one has to consider the passive pressure acting from the left side (water side) toward the right side (land side) and also the active pressure acting from the right side toward the left side of the wall. For such cases, ignoring the hydrostatic pressure from both sides of the wall, the active pressure at a depth $z$ can be given as

$$p_a = [\gamma L_1 + \gamma' L_2 + \gamma'(z - L_1 - L_2)] K_a \qquad (6.3)$$

Also, the passive pressure at that depth $z$ is equal to

$$p_p = \gamma'(z - L_1 - L_2)K_p \tag{6.4}$$

where $K_p$ = Rankine passive pressure coefficient = $\tan^2(45 + \phi/2)$

Hence, combining Eqs. (6.3) and (6.4), the net lateral pressure can be obtained as

$$
\begin{aligned}
p = p_a - p_p &= (\gamma L_1 + \gamma'L_2)K_a - \gamma'(z - L_1 - L_2)(K_p - K_a) \\
&= p_2 - \gamma'(z - L)(K_p - K_a)
\end{aligned}
\tag{6.5}
$$

where $L = L_1 + L_2$

The net pressure, $p$, becomes equal to zero at a depth $L_3$ below the dredge line; or

$$p_2 - \gamma'(z - L)(K_p - K_a) = 0$$

or

$$(z - L) = L_3 = \frac{p_2}{\gamma'(K_p - K_a)} \tag{6.6}$$

From the preceding equation, it is apparent that the slope of the net pressure distribution line $DEF$ is 1 vertical to $(K_p - K_a)\gamma'$ horizontal. So, in the pressure diagram

$$\overline{HB} = p_3 = L_4(K_p - K_a)\gamma' \tag{6.7}$$

At the bottom of the sheet pile, passive pressure $(p_p)$ acts from the right toward the left side, and active pressure acts from the left toward the right side of the sheet pile. So, at $z = L + D$

$$p_p = (\gamma L_1 + \gamma'L_2 + \gamma'D)K_p \tag{6.8}$$

At the same depth

$$p_a = \gamma'DK_a \tag{6.9}$$

Hence, the net lateral pressure at the bottom of the sheet pile is equal to

$$
\begin{aligned}
p_p - p_a = p_4 &= (\gamma L_1 + \gamma'L_2)K_p + \gamma'D(K_p - K_a) \\
&= (\gamma L_1 + \gamma'L_2)K_p + \gamma'L_3(K_p - K_a) + \gamma'L_4(K_p - K_a) \\
&= p_5 + \gamma'L_4(K_p - K_a)
\end{aligned}
\tag{6.10}
$$

where $p_5 = (\gamma L_1 + \gamma'L_2)K_p + \gamma'L_3(K_p - K_a)$ $\tag{6.11}$

$\qquad\quad D = L_3 + L_4$ $\tag{6.12}$

For the stability of the wall, the principles of statics can now be applied; or

$$\sum \text{ horizontal forces per unit length of wall} = 0$$

and

$$\sum \text{moment of the forces per unit length of wall about point } B = 0$$

For summation of the horizontal forces,

$$\text{area of the pressure diagram } ACDE - \text{area of } EFHB$$
$$+ \text{ area of } FHBG = 0$$

or

$$P - \tfrac{1}{2}p_3L_4 + \tfrac{1}{2}L_5(p_3 + p_4) = 0 \tag{6.13}$$

where $P$ = area of the pressure diagram $ACDE$

Summing the moment of all the forces about point $B$

$$P(L_4 + \bar{z}) - \left(\frac{1}{2}L_4 p_3\right)\left(\frac{L_4}{3}\right) + \frac{1}{2}L_5(p_3 + p_4)\left(\frac{L_5}{3}\right) = 0 \tag{6.14}$$

From Eq. (6.13)

$$L_5 = \frac{p_3 L_4 - 2P}{p_3 + p_4} \tag{6.15}$$

Combining Eqs. (6.7), (6.10), (6.14), and (6.15) and simplifying them further, one obtains the following fourth-degree equation in terms of $L_4$.

$$L_4^4 + A_1L_4^3 - A_2L_4^2 - A_3L_4 - A_4 = 0 \tag{6.16}$$

where $A_1 = \dfrac{p_5}{\gamma'(K_p - K_a)}$ (6.17)

$$A_2 = \frac{8P}{\gamma'(K_p - K_a)} \tag{6.18}$$

$$A_3 = \frac{6P[2\bar{z}\gamma'(K_p - K_a) + p_5]}{\gamma'^2(K_p - K_a)^2} \tag{6.19}$$

$$A_4 = \frac{P(6\bar{z}p_5 + 4P)}{\gamma'^2(K_p - K_a)^2} \tag{6.20}$$

## Step-By-Step Procedure for Obtaining the Pressure Diagram

Based on the preceding theory, the step-by-step procedure for obtaining the pressure diagram for a cantilever sheet pile wall penetrating into granular soil is as follows:

1. Calculate $K_a$ and $K_p$.
2. Calculate $p_1$ [Eq. (6.1)] and $p_2$ [Eq. (6.2)]. *Note:* $L_1$ and $L_2$ will be given.
3. Calculate $L_3$ [Eq. (6.6)].
4. Calculate $P$.
5. Calculate $\bar{z}$ (that is, the center of pressure for the area $ACDE$) by taking the moment about $E$.
6. Calculate $p_5$ [Eq. (6.11)].
7. Calculate $A_1$, $A_2$, $A_3$, and $A_4$ [Eqs. (6.17) to (6.20)].

**8.** Solve Eq. (6.16) by trial and error to determine $L_4$.
**9.** Calculate $p_4$ [Eq. (6.10)].
**10.** Calculate $p_3$ [Eq. (6.7)].
**11.** Obtain $L_5$ from Eq. (6.15).
**12.** Now the pressure distribution diagram as shown in Figure 6.5a can easily be drawn.
**13.** Obtain the theoretical depth [Eq. (6.12)] of penetration as $L_3 + L_4$. The actual depth of penetration is increased by about 20–30%.

*Note*: Some designers prefer to use a factor of safety on the passive earth pressure coefficient at the beginning. In that case, in Step 1

$$K_{p(design)} = \frac{K_p}{FS}$$

where $FS$ = factor of safety (usually between 1.5 to 2)

For this type of analysis, follow Steps 1 through 12 with the value of $K_a = \tan^2(45 - \phi/2)$ and $K_{p(design)}$ (instead of $K_p$). The actual depth of penetration can now be determined by adding $L_3$, obtained from Step 3, and $L_4$, obtained from Step 8.

## Calculation of Maximum Bending Moment

The nature of variation of the moment diagram for a cantilever sheet pile wall is shown in Figure 6.5b. The maximum moment will occur between the points $E$ and $F'$. To obtain the maximum moment ($M_{max}$) per unit length of the wall, one must determine the point of zero shear. Adopting a new axis $z'$ (with origin at point $E$) for zero shear

$$P = \tfrac{1}{2}(z')^2(K_p - K_a)\gamma'$$

or

$$z' = \sqrt{\frac{2P}{(K_p - K_a)\gamma'}} \tag{6.21}$$

Once the point of zero shear force is determined (point $F''$ in Figure 6.5a), the magnitude of the maximum moment can be obtained as

$$M_{max} = P(\bar{z} + z') - \tfrac{1}{2}\gamma'z'^2(K_p - K_a) \tag{6.22}$$

The sizing of the necessary profile of the sheet piling is then made according to the allowable flexural stress of the sheet pile material, or

$$S = \frac{M_{max}}{\sigma_{all}} \tag{6.23}$$

where $S$ = section modulus of the sheet pile required per unit length of the structure
$\sigma_{all}$ = allowable flexural stress of the sheet pile

## 6.3

## Cantilever Sheet Piling Penetrating Clay

In several cases, cantilever sheet piles must be driven into a clay layer possessing an undrained cohesion, $c$ ($\phi = 0$ concept). The net pressure diagram will be somewhat different than that shown in Figure 6.5a. Figure 6.6 shows a cantilever sheet pile wall driven into clay with a backfill of granular soil above the level of the dredge line. Let the water table be located at a depth of $L_1$ below the top of the wall. As before, using Eqs. (6.1) and (6.2), the intensity of the net pressures $p_1$ and $p_2$ can be calculated, and the diagram for pressure distribution above the level of the dredge line can be drawn.

The diagram for net pressure distribution below the dredge line can now be determined as follows.

At any depth $z$ greater than $L_1 + L_2$ and above the point of rotation (point $O$ in Figure 6.4a), the active pressure ($p_a$) from right to left can be expressed as

$$p_a = [\gamma L_1 + \gamma' L_2 + \gamma_{sat}(z - L_1 - L_2)]K_a - 2c\sqrt{K_a} \qquad (6.24)$$

where $K_a$ = Rankine active earth pressure coefficient; with $\phi = 0$, it is equal to one

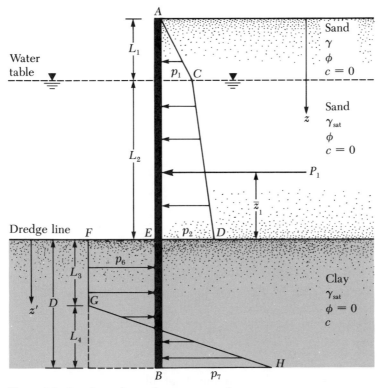

**Figure 6.6** Cantilever sheet pile penetrating clay

Similarly, the passive pressure ($p_p$) from left to right can be given as

$$p_p = \gamma_{sat}(z - L_1 - L_2)K_p + 2c\sqrt{K_p} \tag{6.25}$$

where $K_p$ = Rankine passive earth pressure coefficient; with $\phi = 0$, $K_p$ is equal to one

Thus, the net pressure

$$p_6 = p_p - p_a = [\gamma_{sat}(z - L_1 - L_2) + 2c]$$
$$- [\gamma L_1 + \gamma' L_2 + \gamma_{sat}(z - L_1 - L_2)] + 2c$$
$$= 4c - (\gamma L_1 + \gamma' L_2) \tag{6.26}$$

At the bottom of the sheet pile, the passive pressure from right to left is

$$p_p = (\gamma L_1 + \gamma' L_2 + \gamma_{sat}D) + 2c \tag{6.27}$$

Similarly, the active pressure from left to right is

$$p_a = \gamma_{sat}D - 2c \tag{6.28}$$

Hence, the net pressure

$$p_7 = p_p - p_a = 4c + (\gamma L_1 + \gamma' L_2) \tag{6.29}$$

For equilibrium analysis, $\Sigma F_H = 0$—that is, area of pressure diagram $ACDE$ − area of $EFIB$ + area of $GIH$ = 0, or

$$P_1 - [4c - (\gamma L_1 + \gamma' L_2)]D + \frac{1}{2}L_4[4c - (\gamma L_1 + \gamma' L_2)$$
$$+ 4c + (\gamma L_1 + \gamma' L_2)] = 0$$

where $P_1$ = area of the pressure diagram $ACDE$

Simplifying the preceding equation

$$L_4 = \frac{D[4c - (\gamma L_1 + \gamma' L_2)] - P_1}{4c} \tag{6.30}$$

Now, taking the moment about point $B$, $\Sigma M_B = 0$, or

$$P_1(D + \bar{z}_1) - [4c - (\gamma L_1 + \gamma' L_2)]\frac{D^2}{2} + \frac{1}{2}L_4(8c)\left(\frac{L_4}{3}\right) = 0 \tag{6.31}$$

where $\bar{z}_1$ = distance of the center of pressure of the pressure diagram $ACDE$ measured from the level of the dredge line

Combining Eqs. (6.30) and (6.31) yields

$$D^2[4c - (\gamma L_1 + \gamma' L_2)] - 2DP_1 - \frac{P_1(P_1 + 12c\bar{z}_1)}{(\gamma L_1 + \gamma' L_2) + 2c} = 0 \tag{6.32}$$

The preceding equation can now be solved to obtain $D$, the theoretical depth of penetration of the clay layer by the sheet pile.

## Step-by-Step Procedure to Obtain the Pressure Diagram

1. Calculate $K_a = \tan^2(45 - \phi/2)$ for the granular soil (backfill).
2. Obtain $p_1$ and $p_2$ [Eqs. (6.1) and (6.2)].
3. Calculate $P_1$ and $\bar{z}_1$.
4. Use Eq. (6.32) to obtain the theoretical value of $D$.
5. Using Eq. (6.30), calculate $L_4$.
6. Calculate $p_6$ and $p_7$ [Eqs. (6.26) and (6.29)].
7. Draw the pressure distribution diagram as shown in Figure 6.6.
8. The actual depth of penetration

$$D_{actual} = 1.4 \text{ to } 1.6 \, (D_{theoretical})$$

## Maximum Bending Moment

According to Figure 6.6, the maximum moment (that is, zero shear) will occur between $L_1 + L_2 < z < L_1 + L_2 + L_3$. Using a new coordinate system $z'$ ($z' = 0$ at dredge line) for zero shear

$$P_1 - p_6 z' = 0$$

or

$$z' = \frac{P_1}{p_6} \tag{6.33}$$

The magnitude of the maximum moment can now be obtained as

$$M_{max} = P_1(z' + \bar{z}_1) - \frac{p_6 z'^2}{2} \tag{6.34}$$

Knowing the maximum bending moment, one can determine the section modulus of the sheet pile section from Eq. (6.23).

## Example 6.1

Refer to Figure 6.5. For a cantilever sheet pile wall penetrating a granular soil, given: $L_1 = 2$ m, $L_2 = 3$ m. The granular soil has the following properties:

$$\phi = 32°$$

$$c = 0$$

$$\gamma = 15.9 \text{ kN/m}^3$$

$$\gamma_{sat} = 19.33 \text{ kN/m}^3$$

Make the necessary calculations to determine the theoretical and actual depth of penetration. Also determine the minimum size of sheet pile (section modulus) necessary.

### Solution

The step-by-step procedure given in Section 6.2 will be followed here.

Step 1

$$K_a = \tan^2\left(45 - \frac{\phi}{2}\right) = \tan^2\left(45 - \frac{32}{2}\right) = 0.307$$

$$K_p = \tan^2\left(45 + \frac{\phi}{2}\right) = 3.25$$

Step 2

$$p_1 = \gamma L_1 K_a = (15.9)(2)(0.307) = 9.763 \text{ kN/m}^2$$

$$p_2 = (\gamma L_1 + \gamma' L_2) K_a = [(15.9)(2) + (19.33 - 9.81)3]0.307$$

$$= 18.53 \text{ kN/m}^2$$

Step 3

$$L_3 = \frac{p_2}{\gamma'(K_p - K_a)} = \frac{18.53}{(19.33 - 9.81)(3.25 - 0.307)} = 0.66 \text{ m}$$

Step 4

$$P = \tfrac{1}{2} p_1 L_1 + p_1 L_2 + \tfrac{1}{2}(p_2 - p_1) L_2 + \tfrac{1}{2} p_2 L_3$$

$$= \tfrac{1}{2}(9.763)(2) + (9.763)(3) + \tfrac{1}{2}(18.53 - 9.763)3 + \tfrac{1}{2}(18.53)(0.66)$$

$$= 9.763 + 29.289 + 13.151 + 6.115 = 58.32 \text{ kN/m}$$

Step 5. Taking the moment about $E$

$$\bar{z} = \frac{1}{58.32}\left[9.763\left(0.66 + 3 + \frac{2}{3}\right) + 29.289\left(0.66 + \frac{3}{2}\right)\right.$$

$$\left. + 13.151\left(0.66 + \frac{3}{3}\right) + 6.115\left(0.66 \times \frac{2}{3}\right)\right] = 2.23 \text{ m}$$

Step 6

$$p_5 = (\gamma L_1 + \gamma' L_2) K_p + \gamma' L_3 (K_p - K_a)$$

$$= [(15.9)(2) + (19.33 - 9.81)3]3.25 + (19.33 - 9.81)(0.66)(3.25 - 0.307)$$

$$= 196.17 + 18.49 = 214.66 \text{ kN/m}^2$$

Step 7

$$A_1 = \frac{p_5}{\gamma'(K_p - K_a)} = \frac{214.66}{(9.52)(2.943)} = 7.66$$

$$A_2 = \frac{8P}{\gamma'(K_p - K_a)} = \frac{(8)(58.32)}{(9.52)(2.943)} = 16.65$$

$$A_3 = \frac{6P[2\bar{z}\gamma'(K_p - K_a) + p_5]}{\gamma'^2(K_p - K_a)^2}$$

$$= \frac{(6)(58.32)[(2)(2.23)(9.52)(2.943) + 214.66]}{(9.52)^2(2.943)^2} = 151.93$$

$$A_4 = \frac{P(6\bar{z}p_5 + 4P)}{\gamma'^2(K_p - K_a)^2}$$

$$= \frac{58.32[(6)(2.23)(214.66) + (4)(58.32)]}{(9.52)^2(2.943)^2} = 230.72$$

Step 8. From Eq. (6.16)

$$L_4^4 + 7.66L_4^3 - 16.65L_4^2 - 151.39L_4 - 230.72 = 0$$

The following table shows the solution of the preceding equation by trial and error.

| Assumed $L_4$ (m) | Left side of Eq. (6.16) |
|---|---|
| 4 | $-356.44$ |
| 5 | $+178.58$ |
| 4.8 | $+36.96$ |

So, $L_4 \approx 4.8$ m

Step 9

$$p_4 = p_5 + \gamma' L_4 (K_p - K_a)$$
$$= 214.66 + (9.52)(4.8)(2.943) = 349.14 \text{ kN/m}^2$$

Step 10

$$p_3 = \gamma'(K_p - K_a)L_4 = (9.52)(2.943)(4.8) = 134.48 \text{ kN/m}^2$$

Step 11

$$L_5 = \frac{p_3 L_4 - 2P}{p_3 + p_4} = \frac{(134.48)(4.8) - 2(58.32)}{134.48 + 349.14} = 1.09 \text{ m}$$

Step 12. The net pressure distribution diagram can now be drawn, as shown in Figure 6.5a.

Step 13. The actual depth of penetration $= 1.3(L_3 + L_4) = 1.3(0.66 + 4.8) = 7.1$ m. The theoretical depth of penetration $= 0.66 + 4.8 = 5.46$ m.

## Size of Sheet Piling

Using Eq. (6.21)

$$z' = \sqrt{\frac{2P}{\gamma'(K_p - K_a)}} = \sqrt{\frac{(2)(58.32)}{9.52(2.943)}} = 2.04$$

From Eq. (6.22)

$$M_{\max} = P(\bar{z} + z') - \tfrac{1}{2}\gamma'z'^2(K_p - K_a)$$
$$= (58.32)(2.23 + 2.04) - \tfrac{1}{2}(9.52)(2.04)^2(2.943)$$
$$= 249.03 - 58.3 = 190.73 \text{ kN-m}$$

The required section modulus of the sheet pile

$$S = \frac{M_{\max}}{\sigma_{\text{all}}}$$

With $\sigma_{all} = 172.5$ MN/m$^2$

$$S = \frac{190.73 \text{ kN-m}}{172.5 \times 10^3 \text{ kN/m}^2} = \underline{1.106 \times 10^{-3} \text{ m}^3/\text{m of wall}}$$

## Example 6.2

Redo Example Problem 6.1. Assume that the properties of the backfill material are the same. However, the soil below the dredge line is clay. The value of the uncon-solidated undrained shear strength of this clay is 47 kN/m$^2$.

### Solution

We will follow the step-by-step procedure given in Section 6.3; thus
Step 1

$$K_a = 0.307$$

Step 2

$$p_1 = 9.763 \text{ kN/m}^2$$
$$p_2 = 18.53 \text{ kN/m}^2$$

Step 3. Referring to the net pressure distribution diagram given in Figure 6.6

$$P_1 = \frac{1}{2}p_1L_1 + p_2L_2 + \frac{1}{2}(p_2 - p_1)L_2$$

$$= 9.763 + 29.289 + 13.151 = 52.2 \text{ kN/m}$$

$$\bar{z}_1 = \frac{1}{52.2}\left[9.763\left(3 + \frac{2}{3}\right) + 29.289\left(\frac{3}{2}\right) + 13.151\left(\frac{3}{3}\right)\right]$$

$$= 1.78 \text{ m}$$

Step 4. From Eq. (6.32)

$$D^2[4c - (\gamma L_1 + \gamma'L_2)] - 2DP_1 - \frac{P_1(P_1 + 12c\bar{z}_1)}{(\gamma L_1 + \gamma'L_2) + 2c} = 0$$

Substituting proper values

$$D^2\{(4)(47) - [(2)(15.9) + (19.33 - 9.81)3]\} - 2D(52.2)$$

$$- \frac{52.2[52.2 + (12)(47)(1.78)]}{[(15.9)(2) + (19.33 - 9.81)3] + (2)(47)} = 0$$

or

$$127.64 D^2 - 104.4 D - 357.15 = 0$$

Solving the preceding equation, $D = 2.13$ m.

Step 5. From Eq. (6.30)

$$L_4 = \frac{D[4c - (\gamma L_1 + \gamma'L_2)] - P_1}{4c}$$

$$4c - (\gamma L_1 + \gamma'L_2) = (4)(47) - [(15.9)(2) + (19.33 - 9.81)3]$$
$$= 127.64 \text{ kN/m}^2$$

So

$$L_4 = \frac{2.13(127.64) - 52.2}{(4)(47)} = 1.17 \text{ m}$$

Step 6

$$p_6 = 4c - (\gamma L_1 + \gamma'L_2) = 127.64 \text{ kN/m}^2$$
$$p_7 = 4c + (\gamma L_1 + \gamma'L_2) = 248.36 \text{ kN/m}^2$$

Step 7. The net pressure distribution diagram can now be drawn, as shown in Figure 6.6.

Step 8. $D_{\text{actual}} \approx 1.5 D_{\text{theoretical}} = 1.5(2.13) \approx \underline{3.2 \text{ m}}$

## Maximum Moment Calculation

From Eq. (6.33)

$$z' = \frac{P_1}{p_6} = \frac{52.2}{127.64} \approx 0.41 \text{ m}$$

Again, from Eq. (6.34)

$$M_{\text{max}} = P_1(z' + \bar{z}_1) - \frac{p_6 z'^2}{2}$$

So

$$M_{\text{max}} = 52.2(0.41 + 1.78) - \frac{127.64(0.41)^2}{2}$$
$$= 114.32 - 10.73 = 103.59 \text{ kN-m}$$

The minimum required section modulus (assuming $\sigma_{\text{all}} = 172.5 \text{ MN/m}^2$)

$$S = \frac{103.59 \text{ kN-m}}{172.5 \times 10^3 \text{ kN/m}^2} = \underline{0.6 \times 10^{-3} \text{ m}^3/\text{m of the wall}}$$

---

## Anchored Sheet Pile Wall

When the height of the backfill material behind a cantilever sheet pile wall exceeds about 10 m, it becomes more economical to tie the sheet pile wall near the top to anchor plates, anchor walls, or anchor piles. This is referred to as *anchored sheet piling*, or *anchored bulkhead*. Anchors minimize the depth of required penetration by the sheet piles and also reduce the cross section area and weight of the sheet piles needed for construction. However, the tie rods and anchors must be carefully designed.

There are two basic methods of constructing anchored sheet pile walls: (a) the *free earth support* method and (b) the *fixed earth support* method. Figure 6.7 shows the nature of deflection of the sheet piles designed by the two methods.

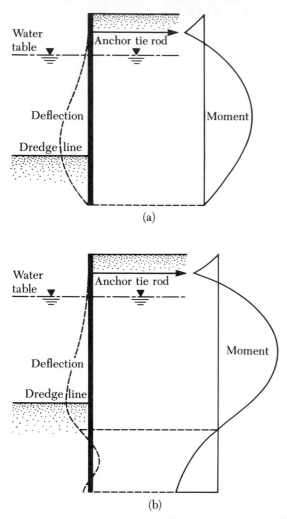

**Figure 6.7**    Nature of variation of deflection and moment for anchored sheet piles: (a) free earth support method; (b) fixed earth support method

The free earth support method is the method of minimum penetration depth. Below the dredge line, no pivot point exists for the statical system. The nature of variation of the bending moment with depth for both of the methods is also shown in Figure 6.7.

## 6.4

### Free Earth Support Method for Penetration of Sandy Soil

Figure 6.8 shows an anchor sheet pile wall with a granular soil backfill; the wall has been driven into a granular soil. The tie rod connecting the sheet pile and the anchor is located at a depth of $l_1$ below the top of the sheet pile wall.

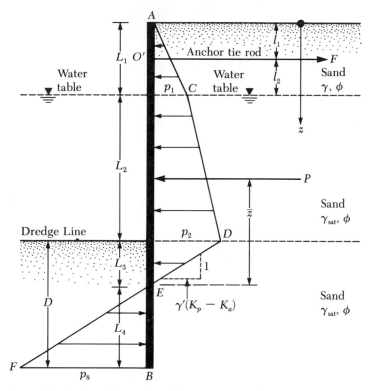

**Figure 6.8** Anchored sheet pile wall penetrating sand

The diagram of net pressure distribution above the dredge line will be similar to that shown in Figure 6.5. At a depth $z = L_1$, $p_1 = \gamma L_1 K_a$; and, at $z = L_1 + L_2$, $p_2 = (\gamma L_1 + \gamma' L_2)K_a$. Below the dredge line, the net pressure will be equal to zero at a depth of $z = (L_1 + L_2 + L_3)$. The relation for $L_3$ can be given by Eq. (6.6), or

$$L_3 = \frac{p_2}{\gamma'(K_p - K_a)}$$

At a depth $z = (L_1 + L_2 + L_3 + L_4)$, the net pressure can be given by

$$p_8 = \gamma'(K_p - K_a)L_4 \tag{6.35}$$

Note that the slope of the line $DEF$ is 1 vertical to $\gamma'(K_p - K_a)$ horizontal.

For equilibrium of the sheet pile, $\Sigma$ horizontal forces $= 0$, and $\Sigma$ moment about $O' = 0$. (*Note:* Point $O'$ is located at the level of the tie rod.)

Summing the forces in the horizontal direction (per unit length of the wall)

area of the pressure diagram $ACDE$ − area of $EBF$ − $F = 0$

where $F =$ tension in the tie rod/unit length of the wall, or

$$P - \tfrac{1}{2}p_8 L_4 - F = 0$$

or

$$F = P - \tfrac{1}{2}[(\gamma'(K_p - K_a)]L_4^2 \tag{6.36}$$

where $P$ = area of the pressure diagram $ACDE$

Now, taking the moment about point $O'$

$$-P[(L_1 + L_2 + L_3) - (\bar{z} + l_1)] + \tfrac{1}{2}[\gamma'(K_p - K_a)] \cdot L_4^2$$
$$\cdot (l_2 + L_2 + L_3 + \tfrac{2}{3}L_4) = 0$$

or

$$L_4^3 + 1.5 L_4^2(l_2 + L_2 + L_3) - \frac{3P[(L_1 + L_2 + L_3) - (\bar{z} + l_1)]}{\gamma'(K_p - K_a)} = 0 \tag{6.37}$$

The preceding equation can be solved by trial and error to determine the theoretical depth, $L_4$. Thus, the theoretical depth of penetration is equal to

$$D_{\text{theoretical}} = L_3 + L_4$$

The theoretical depth is increased by about 30–40% in the actual construction work.

$$D_{\text{actual}} = 1.3 \text{ to } 1.4 \, D_{\text{theoretical}} \tag{6.38}$$

The step-by-step procedure in Section 6.2 pointed out that a factor of safety can be applied to $K_p$ at the beginning [that is, $K_{p(\text{design})} = K_p/FS$]. If this is done, there is no need to increase the theoretical depth by 30–40%.

The maximum theoretical moment to which the sheet pile will be subjected will occur at a depth between $z = L_1$ to $z = L_1 + L_2$. The depth, $z$, for zero shear and, hence, maximum moment can be evaluated from the equation

$$\tfrac{1}{2}p_1L_1 - F + p_1(z - L_1) + \tfrac{1}{2}K_a\gamma'(z - L_1)^2 = 0 \tag{6.39}$$

Once the value of $z$ is determined, the magnitude of the maximum moment can be obtained easily. The procedure of determination of the holding capacity of anchors is treated in Section 6.7.

## 6.5

### Free Earth Support Method for Penetration of Clay

Figure 6.9 shows an anchor sheet pile wall that has penetrated a clay soil and that has a granular soil backfill. The diagram of pressure distribution above the dredge line is similar to that shown in Figure 6.6. The net pressure distribution below the dredge line (from $z = L_1 + L_2$ to $z = L_1 + L_2 + D$) can be given as [Eq. (6.26)]

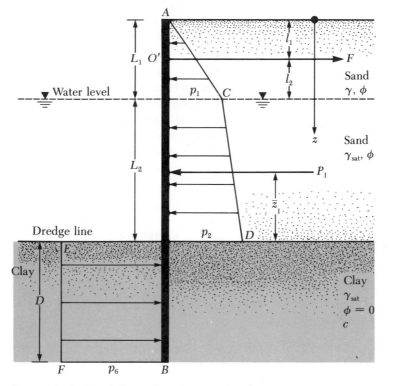

**Figure 6.9** Anchored sheet pile wall penetrating clay

$$p_6 = 4c - (\gamma L_1 + \gamma' L_2)$$

For static equilibrium, summing the forces in the horizontal direction

$$P_1 - p_6 D = F \qquad (6.40)$$

where $P_1$ = area of the pressure diagram $ACD$
$F$ = anchor force per unit length of the sheet pile wall

Again, taking the moment about $O'$

$$P_1(L_1 + L_2 - l_1 - \overline{z}_1) - p_6 D\left(l_2 + L_2 + \frac{D}{2}\right) = 0$$

Simplification of the preceding equation yields

$$p_6 D^2 + 2p_6 D(L_1 + L_2 - l_1) - 2P_1(L_1 + L_2 - l_1 - \overline{z}_1) = 0 \qquad (6.41)$$

The theoretical depth of penetration, $D$, can be determined from the preceding equation.

As in Section 6.4, the maximum moment in this case will occur at a depth $L_1 < z < L_1 + L_2$. The depth of zero shear (and thus the maximum moment) can be determined by using Eq. (6.39).

## 6.6

### Moment Reduction for Anchored Sheet Pile Walls

Sheet piles are flexible. Because of this flexibility, sheet pile walls will yield. Their yielding results in a redistribution of lateral earth pressure. This change will tend to reduce the maximum bending moment, $M_{max}$, as calculated by the procedure outlined in Sections 6.4 and 6.5. For that reason, Rowe (1952, 1957) suggested a procedure to reduce the maximum design moment on the sheet pile walls as obtained via the free-earth support method. This section discusses the procedure of moment reduction as proposed by Rowe.

In Figure 6.10, which is valid for the case of a sheet pile penetrating sand, the following notations have been used:

1. $H'$ = total height of pile drive (that is, $L_1 + L_2 + D_{actual}$)

2. Relative flexibility of pile = $\rho = 10.91 \times 10^{-7} \left( \dfrac{H'^4}{EI} \right)$     (6.42)

where $H$ is in meters

        $E$ = Young's modulus of the pile material ($MN/m^2$)

        $I$ = moment of inertia of the pile section per foot of the wall ($m^4$/m of wall)

3. $M_d$ = design moment

4. $M_{max}$ = maximum theoretical moment

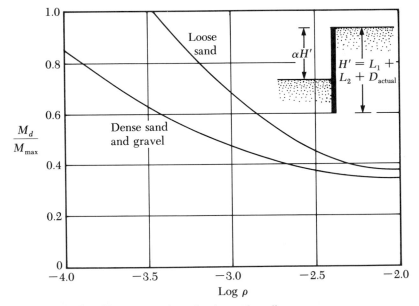

**Figure 6.10**   Plot of log $\rho$ vs. $M_d/M_{max}$ for sheet pile walls penetrating sand (after Rowe, 1952)

The procedure for the use of the moment reduction diagram (Figure 6.10) is as follows:

**Step 1.** Choose a sheet pile section (such as those given in Table 6.1).
**Step 2.** Find the section modulus ($S$) of the selected section (Step 1) per unit length of the wall.
**Step 3.** Determine the moment of inertia of the section (Step 1) per unit length of the wall.
**Step 4.** Obtain $H'$ and calculate $\rho$ [Eq. (6.42)].
**Step 5.** Find log $\rho$.
**Step 6.** Find the moment capacity of the pile section chosen in Step 1 as $M_d = \sigma_{all} \cdot S$.
**Step 7.** Determine $M_d/M_{max}$. Note that $M_{max}$ is the maximum theoretical moment determined before.
**Step 8.** Plot log $\rho$ (Step 5) and $M_d/M_{max}$ in Figure 6.10.
**Step 9.** Repeat Steps 1 through 8 for several sections. The points that fall above the curve (loose sand or dense sand, as the case may be) are all *safe sections*. Those points that fall below the curve are the *unsafe sections*. The cheapest section can now be chosen from those points that fall above the proper curve. Note that the section chosen will have an $M_d < M_{max}$.

For piles penetrating clay soils, the notations in Figure 6.11 are as follows:

**1.** Stability number =

$$S_n = 1.25\frac{c}{(\gamma L_1 + \gamma' L_2)} \tag{6.43}$$

where $c$ = undrained cohesion ($\phi = 0$ condition)

For the definition of $\gamma$, $\gamma'$, $L_1$, and $L_2$, see Figure 6.9.

**2.**

$$\alpha = \frac{L_1 + L_2}{L_1 + L_2 + D_{actual}} \tag{6.44}$$

**3.** Flexibility number, $\rho$ [see Eq. (6.42)]
**4.** $M_d$ = design moment
   $M_{max}$ = maximum theoretical moment

The procedure for moment reduction using Figure 6.11 is as follows:

**Step 1.** Obtain $H'$.
**Step 2.** Determine $\alpha = (L_1 + L_2)/H'$.
**Step 3.** Determine $S_n$ [Eq. (6.43)].
**Step 4.** For a given value of $\alpha$ and $S_n$ (Steps 2 and 3), determine $M_d/M_{max}$ for various values of log $\rho$ from Figure 6.11 and plot a graph of $M_d/M_{max}$ vs. log $\rho$.
**Step 5.** Follow Steps 1 through 9 as outlined for the case of moment reduction of sheet pile walls penetrating granular soil.

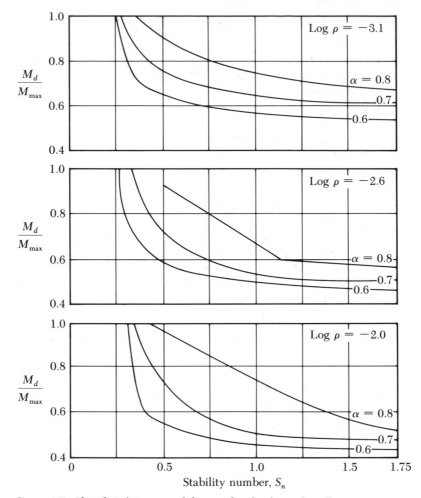

**Figure 6.11**    Plot of $M_d/M_{max}$ vs. stability number for sheet pile wall
penetrating clay (after Rowe, 1957)

## Example 6.3

Redo Example Problem 6.1 for an anchored sheet pile wall to:

a. Determine the theoretical and actual depth of penetration
b. Find the anchor force per unit length of the wall
c. Determine $M_{max}$
d. Find the most appropriate design section using Rowe's moment reduction diagram

The anchor line is located at a depth of 1 m below the top of the wall.

### Solution

Refer to Figure 6.8 for the nature of variation of the net lateral pressure diagram.

From Example Problem 6.1

$$p_1 = 9.763 \text{ kN/m}^2$$
$$p_2 = 18.53 \text{ kN/m}^2$$
$$L_3 = 0.66 \text{ m}$$
$$P = 58.32 \text{ kN/m}$$
$$\bar{z} = 2.23 \text{ m}$$

## Part a: Depth of Penetration

Referring to Eq. (6.37)

$$L_3^4 + 1.5L_4^2(l_2 + L_2 + L_3) - \frac{3P[(L_1 + L_2 + L_3) - (\bar{z} + l_1)]}{\gamma'(K_p - K_a)} = 0$$

Noting that

$$l_1 = 1 \text{ m} \qquad K_p = 3.25$$
$$l_2 = 1 \text{ m} \qquad K_a = 0.307$$

$$L_4^3 + 1.5\, L_4^2(1 + 3 + 0.66) - \frac{3(58.32)[(2 + 3 + 0.66) - (2.33 + 1)]}{9.52(3.25 - 0.307)} = 0$$

or

$$L_4^3 + 6.99\, L_4^2 - 14.55 = 0 \qquad\qquad\qquad (a)$$

The value of $L_4$ can be obtained by trial and error, as shown in the following table.

| Assumed $L_4$ (m) | Left side of Eq. (a) |
|---|---|
| 2.0 | +21.41 |
| 1.5 | +3.55 |
| 1.4 | +2.89 |
| 1.3 | −0.54 |

The preceding table shows that $L_4$ is approximately equal to 1.4 m. Hence,

$$D_{\text{theoretical}} = L_3 + L_4 = 0.66 + 1.4 = 2.06 \text{ m}$$
$$D_{\text{actual}} \approx 1.4\, D_{\text{theory}} = (1.4)(2.06) = 2.88 \text{ m (rounded off to 2.9 m)}$$

## Part b: Anchor Force

From Eq. (6.36)

$$F = P - \tfrac{1}{2}[\gamma'(K_p - K_a)]L_4^2$$
$$= 58.32 - \tfrac{1}{2}[9.52(3.25 - 0.307)](1.4)^2 = 30.86 \text{ kN/m}$$

## Part c: Maximum Moment ($M_{\text{max}}$)

Referring to Eq. (6.39), for zero shear

$$\tfrac{1}{2}p_1L_1 - F + p_1(z - L_1) + \tfrac{1}{2}K_a\gamma'(z - L_1)^2 = 0$$

or

$$(\tfrac{1}{2})(9.763)(2) - 30.86 + (9.763)(z - 2) + \tfrac{1}{2}(0.307)(9.52)(z - 2)^2 = 0$$

Let $z - 2 = x$. So

$$9.763 - 30.86 + 9.763x + 1.461x^2 = 0$$

$$x^2 + 6.682x - 14.44 = 0$$

$$x = 1.72 \text{ m}$$

or

$$z = x + 2 = 1.72 + 2 = 3.72 \text{ m}$$

$$(L_1 + L_2 < z < L_1 \text{—checks})$$

Taking the moment about the point of zero shear force (that is, $z = 3.72$ m or $x = 1.72$ m)

$$M_{max} = -\left(\frac{1}{2}p_1 L_1\right)\left[\left(x + \left(\frac{1}{3}\right)(2)\right)\right] + F(x + 1) - (p_1 x)\left(\frac{x}{2}\right) - \frac{1}{2}K_a \gamma'(x)^2\left(\frac{x}{3}\right)$$

or

$$M_{max} = -(9.763)(2.387) + (30.86)(2.72) - \frac{9.763(1.72)^2}{2} - \frac{(0.307)(9.52)(1.72)^3}{6}$$

$$= -23.3 + 83.94 - 14.44 - 2.48 = \underline{43.72 \text{ kN-m/m}}$$

Compare this moment with that obtained in Example Problem 6.1.

## Part d: Moment Reduction for Design of Section

$$H' = L_1 + L_2 + D_{actual} = 2 + 3 + 2.9 = 7.9 \text{ m}$$

For moment reduction, the following table can be prepared.

| Section (1) | $I$ (m$^4$/m) (2) | $H'$ (m) (3) | $\rho = \dfrac{10.91 \times 10^{-7}H'^4}{EI}$ (4) | $\log \rho$ (5) | $S$ (m$^3$/m) (6) | $M_d = S \cdot \sigma_{all}$ (kN-m) (7) | $\dfrac{M_d}{M_{max}}$ (8) |
|---|---|---|---|---|---|---|---|
| PDA-27 | $54.33 \times 10^{-6}$ | 7.9 | 0.000378 | $-3.42$ | $57 \times 10^{-5}$ | 98.32 | 2.250 |
| PMA-22 | $18.7 \times 10^{-6}$ | 7.9 | 0.001098 | $-2.96$ | $29 \times 10^{-5}$ | 50.02 | 1.140 |
| PSX-32 | $5.05 \times 10^{-6}$ | 7.9 | 0.004080 | $-2.39$ | $12.89 \times 10^{-5}$ | 22.24 | 0.508 |
| PS-28 | $3.82 \times 10^{-6}$ | 7.9 | 0.005370 | $-2.27$ | $10.2 \times 10^{-5}$ | 17.60 | 0.403 |

(1), (2), (6) from Table 6.1
(4) Eq. (6.42), $E = 207 \times 10^3$ MN/m$^2$
(7) $\sigma_{all} = 172,500$ kN/m$^2$
(8) $M_{max} = 43.72$ kN-m/m

Figure 6.12 shows the calculated values of $\log \rho$ and the corresponding values of $M_d/M_{max}$ (it is assumed that the soil behaves like loose sand). Also shown in Figure 6.12 is the design curve of $\log \rho$ vs. $M_d/M_{max}$, as given by Rowe (from Figure 6.10). Note that all points plot above the curve, so they are all safe for design use. The point corresponding to Section PS-28 is closest to the curve: this is the section to be adopted.

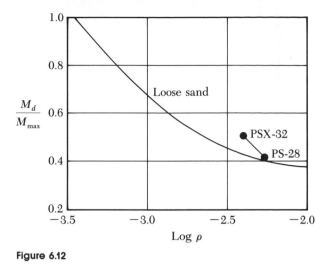

**Figure 6.12**

## 6.7

Fixed Earth Support Method for Penetration of Sandy Soil

When using the fixed earth support method, one assumes that the toe of the pile is restrained from rotating, as shown in Figure 6.13a. The net lateral pressure distribution diagram for this condition is also shown in Figure 6.13a. In the fixed earth support solution, the lower portion of the pressure distribution diagram—that is, $HFH'GB$—is replaced by a concentrated force, $P'$. In order to calculate $L_4$, a simplified solution called the *equivalent beam solution* is generally used. In order to understand the equivalent beam solution, refer to point $I$, which is the point of inflection on the deflected shape of the sheet pile. At this point, the pile can be assumed to be hinged and the bending moment to be equal to zero (Figure 6.13b). The vertical distance between the point $I$ and the dredge line is equal to $L_5$. Blum (1931) has given a mathematical solution between $L_5$ and $L_1 + L_2$. Figure 6.13d is a plot of $L_5/(L_1 + L_2)$ vs. the soil friction angle, $\phi$.

With the known values of $\phi$ and $L_1 + L_2$, the magnitude of $L_5$ can now be obtained. The portion of sheet pile (Figure 6.13c) above the point $I$ can now be treated as a beam that resists the net lateral earth pressure via the anchor force $F$ (kN/m) and the shear $P''$ (kN/m). The shear force $P''$ can be calculated by taking the moment about $O'$ (that is, at the anchor level).

Once the value of $P''$ is known, the length $L_4$ can be obtained by taking the moment about the point $H$ (see the bottom diagram of Figure 6.13c). The depth of penetration $D$ can now be given as 1.2 to 1.4 $(L_3 + L_4)$.

### Step-by-Step Procedure for Obtaining $D$

The following step-by-step procedure for calculating the depth of embedment of sheet piles is based on the procedure just described.

**Step 1.** Obtain $K_a$ and $K_p$.

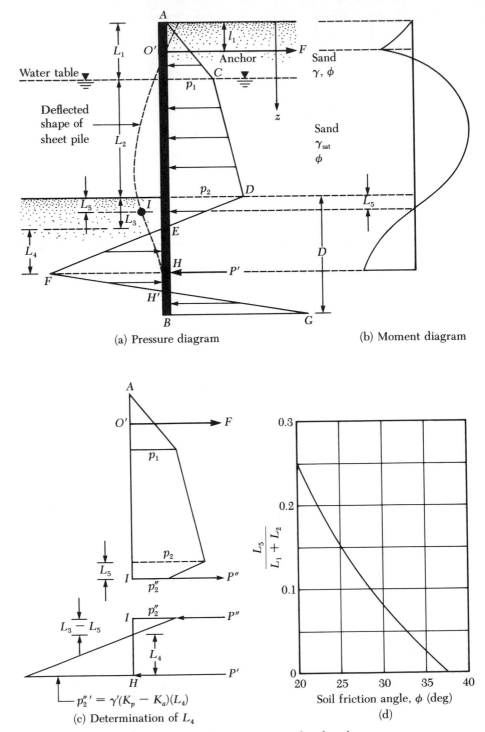

(a) Pressure diagram

(b) Moment diagram

(c) Determination of $L_4$

(d)

**Figure 6.13**   Fixed earth support method for penetration of sandy soil

**Step 2.** Calculate $p_1$ and $p_2$ from Eqs. (6.1) and (6.2), respectively.
**Step 3.** Calculate $L_3$ using Eq. (6.6).
**Step 4.** Determine $L_5$ from Figure 6.13d.
**Step 5.** Calculate $p_2''$ (Figure 6.13c)

$$p_2'' = \frac{p_2(L_3 - L_5)}{L_3} \tag{6.45}$$

**Step 6.** Draw the pressure distribution for the portion of the sheet pile located above $I$, as shown in Figure 6.13c.
**Step 7.** For the diagram drawn in Step 6, take the moment about $O'$ to calculate $P''$.
**Step 8.** Knowing $P''$, draw the pressure distribution diagram for the portion of the sheet pile located between points $I$ and $H$, as shown in Figure 6.13c. Note that in this diagram $p_2'''$ is equal to $\gamma'(K_p - K_a)(L_4)$.
**Step 9.** For the diagram drawn in Step 8, take the moment about $H$ to calculate $L_4$.
**Step 10.** Calculate $D = 1.2$ to $1.4(L_3 + L_4)$.

## 6.8

## Anchors

Sections 6.5 through 6.7 presented the analysis of anchored sheet pile walls. These sections also discussed the expression for obtaining the force, $F$, per unit length of the sheet pile wall that has to be taken by the anchors. This section covers in more detail the various types of anchor generally used and the procedures for evaluating their ultimate resistive capacity.

The general types of anchor used in sheet pile walls are as follows:

**1.** Anchor plates and beams (deadman)
**2.** Tie backs
**3.** Vertical anchor piles
**4.** Anchor beams supported by batter (compression and tension) piles

*Anchor plates and beams* are generally made of cast-concrete blocks (Figure 6.14a). The anchors are attached to the sheet pile by *tie rods*. A *wale* is placed at the front or back face of a sheet pile for the purpose of conveniently attaching the tie rod to the wall. To protect the tie rod from corrosion, it is generally coated with paint or asphaltic materials.

In the construction of *tie backs,* bars or cables are placed in predrilled holes (Figure 6.14b) with concrete grout (cables are commonly high-strength, pre-stressed steel tendons). Figures 6.14c and 6.14d show a vertical anchor pile and an anchor beam with batter piles.

### Placement of Anchors

The resistance offered by anchor plates and beams is primarily derived from the passive force of the soil located in front of them. To learn how to determine the best location for an anchor plate (for maximum efficiency), refer

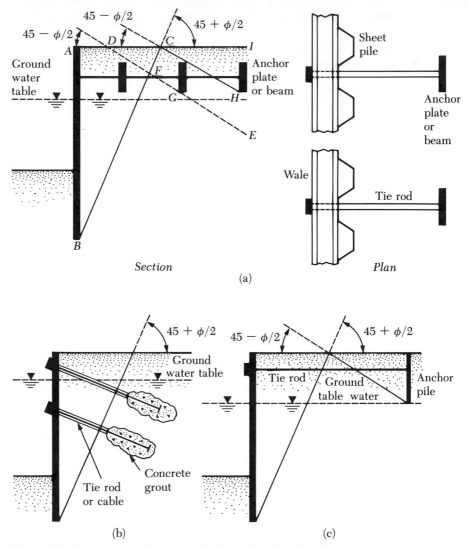

**Figure 6.14**   Various types of anchoring for sheet pile walls: (a) anchor plate or beam; (b) tie back; (c) vertical anchor pile; (d) anchor beam with batter piles

to Figure 6.14a, in which $AB$ is the sheet pile wall. If the anchor is placed inside the wedge $ABC$, which is the Rankine active zone, it would not provide any resistance to failure. As an alternative, one could place the anchor in the zone $CFEH$. Note that line $DFG$ is the slip line for the Rankine passive pressure. If part of the passive wedge is located inside the active wedge $ABC$, full passive resistance of the anchor cannot be realized upon failure of the sheet pile wall. However, if the anchor is placed in zone $ICH$, the Rankine passive zone in front of the anchor slab or plate is located completely outside the Rankine active zone $ABC$. In this case, full passive resistance from the anchor can be realized.

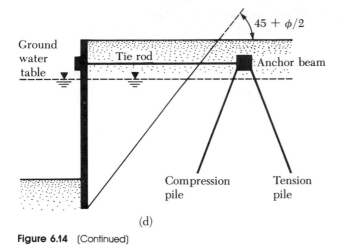

Figure 6.14 (Continued)

Figures 6.14b, 6.14c, and 6.14d also show the proper locations for placement of tie backs, vertical anchor piles, and anchor beams supported by batter piles.

### Calculation of the Ultimate Resistance Offered by Anchor Plates and Beams in Sand

Teng (1962) has proposed the following equation to determine the ultimate resistance of anchor plates or walls in granular soils located at or near the ground surface ($H/h \leq 1.5$ to 2 in Figure 6.15).

$$P_u = B(P_p - P_a) \quad \text{(for continuous plates or beams—that is, } B/h \approx \infty)$$

$$(6.46)$$

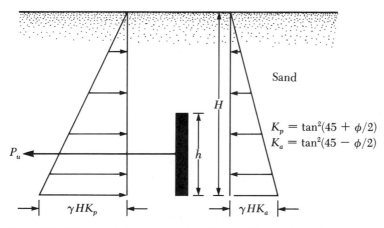

Figure 6.15 Ultimate resistance of anchor plates and beams in sand— Eqs. (6.46) and (6.49)

where        $P_u$ = ultimate resistance of anchor

$B$ = length of anchor at right angle to the cross section shown

$P_p$ and $P_a$ = Rankine passive and active force per unit length of anchor

Note that $P_p$ acts in front of the anchor, as shown in Figure 6.15. Also

$$P_p = \frac{1}{2}\gamma H^2 \tan^2\left(45 + \frac{\phi}{2}\right) \tag{6.47}$$

and

$$P_a = \frac{1}{2}\gamma H^2 \tan^2\left(45 - \frac{\phi}{2}\right) \tag{6.48}$$

Equation (6.46) is valid for the plane-strain type of condition. For all practical cases, $B/h > 5$ may be considered as plane-strain condition.

For $B/h <$ about 5, considering the three-dimensional failure surface (that is, accounting for the friction resistance developed at the two ends of an anchor), Teng (1962) has given the following relation for the ultimate anchor resistance:

$$P_u = B(P_p - P_a) + \frac{1}{3}K_O\gamma(\sqrt{K_p} + \sqrt{K_a})H^3\tan\phi$$

$$\left(\text{for }\frac{H}{h} \le 1.5 \text{ to } 2\right) \tag{6.49}$$

where $K_O$ = earth pressure coefficient at rest $\approx 0.4$

More recently, Neeley, Stuart, and Graham (1973) studied the problem of anchor pullout resistance in sand using the *equivalent free surface method*. The failure surface in soil assumed in their work is shown in the insert of Figure 6.16. At failure, the shearing stress mobilized along $OX$ in Figure 6.16 can be defined as

$$m = \frac{s_o}{\sigma_o \tan\phi} \tag{6.50}$$

where    $m$ = shear stress mobilization factor

$s_o, \sigma_o$ = shear stress at failure and normal stress (on $OX$), respectively

For a conservative design, $m$ can be taken as zero. According to the method of Neeley et al.,

$$P_u = M_{\gamma q}(\gamma h^2)BS' \quad \text{(for } H/h < \text{ about 5)} \tag{6.51}$$

where    $P_u$ = ultimate resistance of anchor

$M_{\gamma q}$ = force coefficient

$S'$ = shape factor

The variation of $M_{\gamma q}$ for various values of soil friction angle, $\phi$, and $H/h$ is given in Figure 6.16 (for $m = 0$). Figure 6.17a gives the experimental values of the shape factors ($S'$) for various values of $B/h$ and $H/h$. Note that in certain

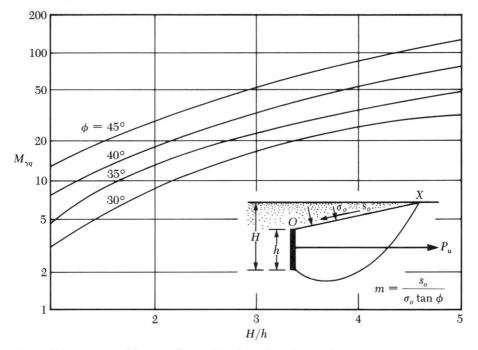

**Figure 6.16** Variation of force coefficient, $M_{\gamma q}$, for anchor plates and beams in sand (redrawn after Neeley et al., 1973)

circumstances, such as in the bearing capacity problem, the lateral displacement of anchors may need to be limited. Relatively few studies have been conducted so far to determine the relation of load to displacement of anchors. Figure 6.17b shows an example of nondimensional displacement of anchors for various values of $B/h$ and $H/h$ as obtained by Neeley et al. experimentally in medium to dense sand. Das (1975) and Das and Seeley (1975) also found essentially similar relations for anchors tested in loose sand. Based on the experimental results, Das and Seeley (1975) presented the following load-displacement relationship for anchors

$$\overline{P} = \frac{\overline{\Delta}}{0.15 + 0.85\,\overline{\Delta}} \qquad (6.52)$$

where $\overline{P} = \dfrac{\text{load on anchor at horizontal displacement, }\Delta}{\text{ultimate load at horizontal displacement, }\Delta_u} \qquad (6.53)$

$$\overline{\Delta} = \frac{\Delta}{\Delta_u} \qquad (6.54)$$

The relationship given by Eq. (6.52) was found to be valid for $B/h$ varying from 1 to 5 and $H/h$ varying from 1 to 5. Example Problem 6.4 shows the procedure for the use of Figure 6.17b and Eq. (6.52) to estimate the anchor load for a given maximum displacement.

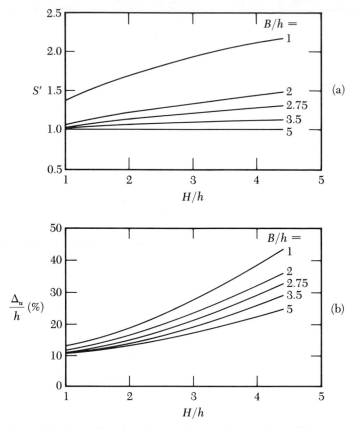

**Figure 6.17**   Vertical anchor plates or beams: (a) shape factor; (b) horizontal displacement at ultimate load (redrawn after Neeley et al., 1973)

## Ultimate Resistance of Anchor Plates and Beams in Clay ($\phi = 0$ condition)

For anchor plates and beams in clay, Mackenzie (1955) has experimentally determined the variation of $P_u$ with $H/h$. This is shown in Figure 6.18. Note that for $H/h > 12$, $(P_u)/(hBc)$ is approximately constant and is equal to about 8.5 ($c$ = undrained cohesion for $\phi = 0$ condition).

### Factor of Safety for Anchor Plates and Beams

The allowable resistance per anchor plate, $P_{all}$, can be given as

$$P_{all} = \frac{P_u}{FS} \tag{6.55}$$

where $FS$ = factor of safety

Generally, a factor of safety of 2 is suggested.

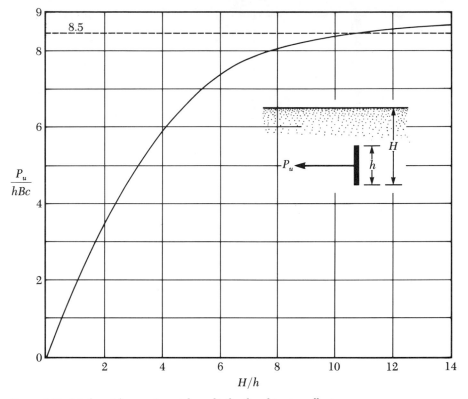

**Figure 6.18** Mackenzie's experimental results for the ultimate pullout capacity of anchors in clay (from *Foundations, Retaining and Earth Structures*, 2nd ed., by G. P. Tschebotarioff. Copyright © 1973 by McGraw-Hill; used with the permission of the McGraw-Hill Book Company)

## Spacing of Anchor Plates

The center-to-center spacing of anchors, $s_p$, can be obtained thus:

$$s_p = \frac{P_{\text{all}}}{F} \tag{6.56}$$

where $F$ = force per unit length of the sheet pile

## Ultimate Resistance of Tie Backs

According to Figure 6.19, the ultimate resistance offered by a tie back in sand can be given as follows:

$$P_u = \pi d l \bar{\sigma}_v' K \tan \phi \tag{6.57}$$

where $P_u$ = ultimate resistance
$\phi$ = angle of friction of soil
$\bar{\sigma}_v'$ = average effective vertical stress (= $\gamma x$ in dry sand)
$K$ = earth pressure coefficient

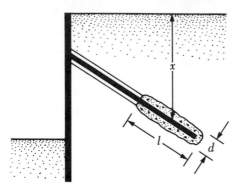

**Figure 6.19**  Parameters for defining the ultimate resistance of tie backs

The value of $K$ can be taken to be equal to the earth pressure coefficient at rest ($K_o$) if the concrete grout is placed under pressure (Littlejohn, 1970). The lower limit of $K$ can be taken to be equal to the Rankine active earth pressure coefficient.

In clays, the ultimate resistance of tie backs can be approximated as

$$P_u = \pi dl c_a \qquad (6.58)$$

where $c_a$ = adhesion

The value of $c_a$ can be approximated as $2/3\ c$ (where $c$ = undrained cohesion).

A factor of safety of 1.5–2 may be used over the ultimate resistance to obtain the allowable resistance offered by each tie back.

## Example 6.4

Refer to Figure 6.16. For a square anchor plate 0.4 m × 0.4 m ($B \times h$) in size in a sandy soil with $H = 1$ m, determine the following:

a. The ultimate resistance, $P_u$
b. The anchor resistance for a maximum horizontal displacement $\Delta = 40$ mm

Given: $\gamma = 16.51$ kN/m$^3$ and $\phi = 35°$. Use the method presented by Neeley et al.

**Solution**
**Part a**

From Eq. (6.51)

$$P_u = M_{\gamma q}(\gamma h^2) BS'$$

With $H/h = 1/0.4 = 2.5$, $B/h = 0.4/0.4 = 1$, and $\phi = 35°$, we refer to Figure 6.16 to find $M_{\gamma q} = 18$. Again, with $H/h = 2.5$ and $B/H = 1$, $S' = 1.75$ (Figure 6.17a). So

$$P_u = (18)(16.51)(0.4)^2(0.4)(1.75) = \underline{33.28\ \text{kN}}$$

**Part b**

For $H/h = 2.5$, $B/h = 1$, $\Delta_u/h = 0.24$ (Figure 6.17b). Hence at ultimate load, $P_u$, the displacement of the anchor will be

$$\Delta_u = 0.24h = 0.24 \times 0.4 = 0.096 \text{ m} = 96 \text{ mm}$$

If $\Delta = 40$ mm, $\bar{\Delta} = \Delta/\Delta_u = 40/96 = 0.417$. Referring to Eq. (6.52)

$$\bar{P} = \frac{\bar{\Delta}}{0.15 + 0.85 \, \bar{\Delta}}$$

Hence

$$\bar{P} = \frac{0.417}{0.15 + (0.85)(0.417)} = 0.827$$

Thus

$$P_{\text{at } \Delta} = 40 \text{ mm} = 0.827 \times P_u = \underline{27.52 \text{ kN}}$$

## Problems

**6.1** Figure P6.1 shows a cantilever sheet pile wall penetrating a granular soil. Given: $L_1 = 8$ ft, $L_2 = 15$ ft, $\gamma = 100$ lb/ft³, $\gamma_{\text{sat}} = 110$ lb/ft³, $\phi = 35°$.
a. What is the theoretical depth of embedment, $D$?
b. Draw the net pressure distribution diagram.
c. Assuming a 30% increase of $D$, what should be the total length of sheet piles to be used?

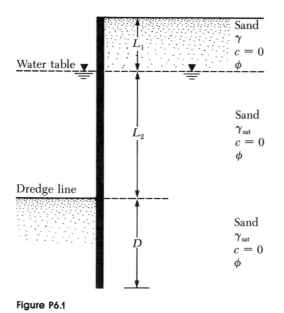

**Figure P6.1**

**6.2** Redo Problem 6.1a, b, and c with the following changes: $L_1 = 3$ m, $L_2 = 6$ m, $\gamma = 17.3$ kN/m³, $\gamma_{\text{sat}} = 19.4$ kN/m³, $\phi = 30°$.

**6.3** Redo Problem 6.1a, b, and c with the following changes: $L_1 = 10$ ft, $L_2 = 15$ ft, $\gamma = 105$ lb/ft³, $\gamma_{sat} = 124$ lb/ft³, $\phi = 32°$.

**6.4** For Problem 6.1, Part b, determine the maximum moment for the sheet pile section and then choose the most appropriate sheet pile section.

**6.5** For Problem 6.2, determine the theoretical maximum moment for the sheet pile section and then choose one.

**6.6** For Problem 6.3, calculate the theoretical maximum moment and choose a sheet pile section.

**6.7** Refer to Figure P6.7. Given: $L_1 = 2.4$ m, $L_2 = 4.6$ m, $\gamma = 15.7$ kN/m³, $\gamma_{sat} = 17.3$ kN/m³, $c = 29$ kN/m², $\phi = 30°$.
  a. Find the theoretical depth of penetration, $D$.
  b. Draw the pressure distribution diagram with all dimensions.
  c. Increase $D$ by 40%: What is the length of sheet piles needed for this work?

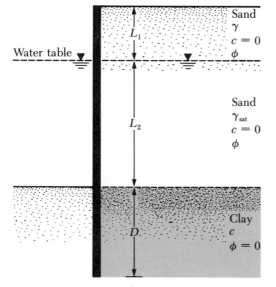

**Figure P6.7**

**6.8** Redo Problem 6.7a, b, and c with the following: $L_1 = 5$ ft, $L_2 = 20$ ft, $\gamma = 108$ lb/ft³, $\gamma_{sat} = 122.4$ lb/ft³, $\phi = 36°$, $c = 800$ lb/ft².

**6.9** For Problem 6.7, determine the maximum moment and choose a sheet pile section.

**6.10** For Problem 6.8, determine the maximum moment and the appropriate section.

**6.11** An anchored sheet pile bulkhead is shown in Figure P6.11. Given: $L_1 = 3.05$ m, $L_2 = 6.1$ m, $l_1 = 1.53$ m, $\gamma = 16$ kN/m³, $\gamma_{sat} = 19.5$ kN/m³, $\phi = 30°$ (use the free earth support method).
  a. Calculate the theoretical value of the depth of embedment, $D$.
  b. Draw the pressure distribution diagram.
  c. Determine the anchor force per unit length of the wall.

**6.12** Refer to Problem 6.11. Assume that $D_{actual} = 1.3\,D_{theory}$.
  a. Determine the theoretical maximum moment.
  b. Choose a sheet pile section using Rowe's moment reduction technique.

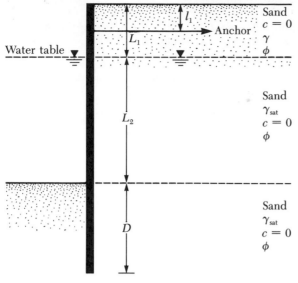

**Figure P6.11**

**6.13** Redo Problem 6.11 with the following: $L_1 = 9$ ft, $L_2 = 26$ ft, $l_1 = 5$ ft, $\gamma = 108.5$ lb/ft³, $\gamma_{sat} = 128.5$ lb/ft³, $\phi = 35°$. Use the free earth support method.

**6.14** Refer to Problem 6.13. Assume $D_{actual} = 1.4 D_{theory}$. Using the Rowe's moment reduction method, choose a sheet pile section.

**6.15** An anchored sheet pile bulkhead is shown in Figure P6.15. Given: $L_1 = 2$ m, $L_2 = 6$ m, $l_1 = 1$ m, $\gamma = 16$ kN/m³, $\gamma_{sat} = 18.86$ kN/m³, $\phi = 32°$, and $c = 27$ kN/m².
a. Determine the theoretical depth of embedment, $D$.
b. Calculate the anchor force per unit length of the sheet pile wall.
Use the free earth support method.

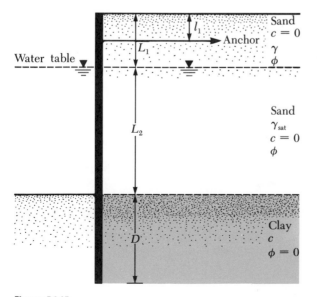

**Figure P6.15**

**6.16** Solve Problem 6.15 with the following: $L_1 = 10.8$ ft, $L_2 = 21.6$ ft, $l_1 = 5.4$ ft, $\gamma = 108$ lb/ft$^3$, $\gamma_{sat} = 127.2$ lb/ft$^3$, $\phi = 35°$, and $c = 850$ lb/ft$^2$.

**6.17** An anchor slab is shown in Figure P6.17. Given: $H = 0.9$ m, $h = 0.3$ m, $\gamma = 17$ kN/m$^3$, and $\phi = 32°$. Calculate the ultimate holding capacity of the anchor slab if the width $B$ is (a) 0.3 m, (b) 0.6 m, and (c) 0.9 m.

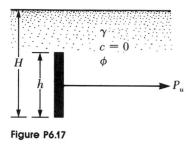

**Figure P6.17**

**6.18** Refer to Problem 6.17. Estimate the probable horizontal displacement of the anchors at ultimate load. Also estimate the resistance of anchors if the allowable displacement is 1 in.

## References

Blum, H. (1931). *Einspannungsverhältnisse bei Bohlwerken*, W. Ernst und Sohn, Berlin, Germany.

Das, B. M. (1975). "Pullout Resistance of Vertical Anchors," *Journal of the Geotechnical Engineering Division*, American Society of Civil Engineers, Vol. 101, No. GT1, pp. 87–91.

Das, B. M., and Seeley, G. R. (1975). "Load-Displacement Relationships for Vertical Anchor Plates," *Journal of the Geotechnical Engineering Division*, American Society of Civil Engineers, Vol. 101, No. GT7, pp. 711–715.

Littlejohn, G. S. (1970). "Soil Anchors," *Proceedings*, Conference on Ground Engineering, Institute of Civil Engineers, London, pp. 33–44.

Mackenzie, T. R. (1955). *Strength of Deadman Anchors in Clay*, M.S. Thesis, Princeton University, Princeton, N. J.

Neeley, W. J., Stuart, J. G., and Graham, J. (1973). "Failure Loads of Vertical Anchor Plates in Sand," *Journal of the Soil Mechanics and Foundations Division*, American Society of Civil Engineers, Vol. 99, No. SM9, pp. 669–685.

Rowe, P. W. (1952). "Anchored Sheet Pile Walls," *Proceedings*, Institute of Civil Engineers, Vol. 1, Part 1, pp. 27–70.

Rowe, P. W. (1957). "Sheet Pile Walls in Clay," *Proceedings*, Institute of Civil Engineers, Vol. 7, pp. 629–654.

Teng, W. C. (1962). *Foundation Design*, Prentice-Hall, Englewood Cliffs, N. J.

Tschebotarioff, G. P. (1973). *Foundations, Retaining and Earth Structures*, 2nd ed., McGraw-Hill, New York.

# Braced Cuts

## 7.1

### Introduction

Sometimes construction work requires ground excavations with vertical or near-vertical faces—for example, basements of buildings in developed areas or underground transportation facilities at shallow depths below the ground surface (cut-and-cover type of construction). The vertical faces of the cuts need to be protected by temporary bracing systems to avoid failure that may be accompanied by considerable settlement or by bearing capacity failure of nearby foundation(s).

Figure 7.1 shows two types of braced cut commonly used in construction work. One type uses the *soldier beam* (Figure 7.1a), which is driven into the ground before excavation. Soldier beams are vertical steel or timber beams. After the excavation is started, *laggings* are placed between the soldier beams as the excavation proceeds. Laggings are horizontal timber planks. When the excavation reaches the desired depth, *wales* and *struts* are properly installed (wales and struts are horizontal steel beams). The struts act like horizontal columns. Figure 7.1b shows another type of braced excavation. In this case, interlocking *sheet piles* are driven into the soil before excavation. As the excavation proceeds, wales and struts are inserted immediately after reaching the appropriate depth.

To design braced excavations (that is, to select wales, struts, sheet piles, and soldier beams), one must know the lateral earth pressure to which the braced cuts will be subjected. This topic is discussed in Section 7.2; subsequent sections cover the procedures of analysis and design of braced cuts.

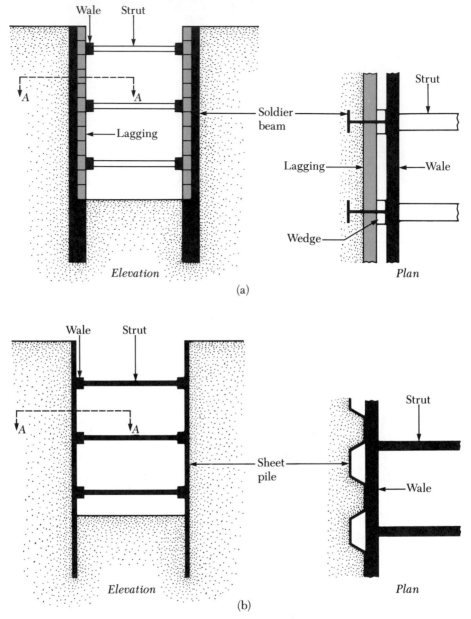

**Figure 7.1**  Types of braced cut: (a) use of soldier beams; (b) use of sheet piles

## 7.2

### Lateral Earth Pressure in Braced Cuts

In Chapter 5 we learned that a retaining wall rotates about its bottom (Figure 7.2a). With sufficient yielding of the wall, the lateral earth pressure can be approximated to be equal to that obtained by Rankine's or Coulomb's theory.

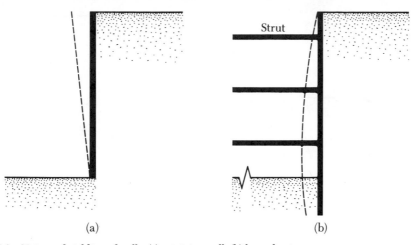

**Figure 7.2**   Nature of yielding of walls: (a) retaining wall; (b) braced cut

In contrast to retaining walls, braced cuts show a different type of wall yielding (see Figure 7.2b). In this case, the deformation of the wall gradually increases with the depth of excavation. The variation of the amount of deformation will depend on several factors, such as the type of soil, the depth of excavation, and the workmanship. However, one can easily visualize that, with very little wall yielding at the top of the cut, the lateral earth pressure will be close to the at-rest pressure. At the bottom of the wall, with a much larger degree of yielding, the lateral earth pressure will be substantially lower than the Rankine active earth pressure. As a result, the distribution of lateral earth pressure will vary substantially in comparison to the linear distribution assumed in the case of retaining walls.

A theoretical evaluation of the total lateral force, $P$, imposed on a wall can be made by using Terzaghi's general wedge theory (1943a) (Figure 7.3a), in which the failure surface is assumed to be the arc of a logarithmic spiral, defined by the equation

$$r = r_o e^{\theta \tan \phi} \tag{7.1}$$

where $\phi$ = angle of friction of soil

A detailed outline for the evaluation of $P$ is beyond the scope of this text; readers should check a soil mechanics text for more information (for example, Das, 1979). However, a comparison of the lateral earth pressure for braced cuts in sand (with angle of wall friction $\delta = 0$) with that for a retaining wall ($\delta = 0$) is shown in Figure 7.3b. If $\delta = 0$, a retaining wall of height $H$ will be subjected to a Rankine active earth pressure, and the resultant active force will intersect the wall at a distance of $nH$ measured from the bottom of the wall. For this case, $n = 1/3$. In contrast, the value of $n$ for a braced cut may vary from 0.33 to 0.5 or 0.6. The general wedge theory can also be used to analyze braced cuts in saturated clay (for example, see Das and Seeley, 1975).

In any event, when choosing a lateral soil pressure distribution for design of braced cuts, one should keep in mind that the nature of failure in braced cuts

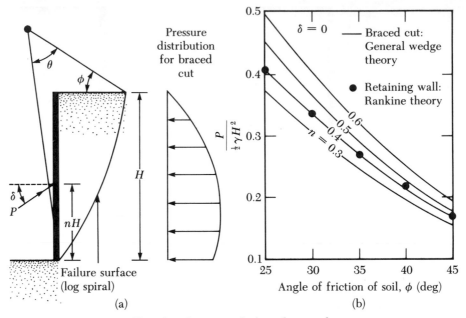

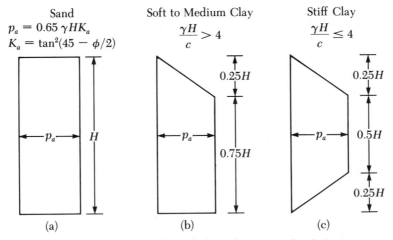

**Figure 7.3**  Comparison of lateral earth pressure for braced cuts and retaining walls in sand ($\delta = 0$)

is much different from that in retaining walls. After observation of several braced cuts, Peck (1969) suggested using *design pressure envelopes* for braced cuts in sand and clay. Figure 7.4 shows Peck's pressure envelopes, to which the following guidelines apply:

    **1.** Figure 7.4a is for braced cuts constructed in dry or moist sand. Note $K_a$ is the Rankine active earth pressure coefficient.

**Figure 7.4**  Peck's pressure envelopes for braced cuts in sand and clay [*note:* in Figure (b), $p_a = \gamma H[1 - (4c/\gamma H)]$ or about $0.3\gamma H$, whichever is higher; in Figure (c), $p_a = 0.2\gamma H$ to $0.4\gamma H$, with an average $0.3\gamma H$]

**2.** For cuts in clay, first calculate the value of $\gamma H/c$ (where $c$ = undrained cohesion of the clay located on the sides of the cuts; $\phi = 0$ concept). If $\gamma H/c$ is less than or equal to 4, the pressure envelope shown in Figure 7.4c should be used. The value of $p_a$ varies between $0.2\gamma H$ and $0.4\gamma H$, with an average of $0.3\gamma H$. If $\gamma H/c$ is greater than 4, the pressure envelope shown in Figure 7.4b should be used. In this case, $p_a$ may be equal to $\gamma H[1 - (4c/\gamma H)]$ or $0.3\gamma H$, whichever is greater. Peck's pressure envelopes are sometimes referred to as *apparent pressure envelopes*.

Sometimes one encounters layers of both sand and clay when constructing a braced cut. In this case, Peck (1943) proposed that an equivalent value of cohesion ($\phi = 0$ concept) should be determined in the following manner (refer to Figure 7.5a):

$$c_{av} = \frac{1}{2H}[\gamma_s K_s H_s^2 \tan \phi_s + (H - H_s)n'q_u] \tag{7.2}$$

where $H$ = total height of the cut
$\quad \gamma_s$ = unit weight of sand
$\quad H_s$ = height of the sand layer
$\quad K_s$ = a lateral earth pressure coefficient for the sand layer ($\approx 1$)
$\quad \phi_s$ = angle of friction of sand
$\quad q_u$ = unconfined compression strength of clay
$\quad n'$ = a coefficient of progressive failure (ranges from 0.5 to one; average value 0.75)

The average unit weight, $\gamma_a$, of the layers can be expressed as

$$\gamma_a = \frac{1}{H}[\gamma_s H_s + (H - H_s)\gamma_c] \tag{7.3}$$

where $\gamma_c$ = saturated unit weight of clay layer

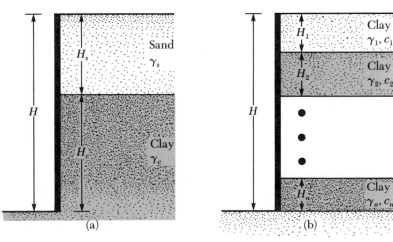

**Figure 7.5** Layered soils in braced cuts

Once the average values of cohesion and unit weight are determined, the pressure envelopes in clay (Figure 7.4b and c) can be used to design the cuts.

In a similar manner, when a number of clay layers are encountered in the cut (Figure 7.5b), the average undrained cohesion can be expressed by the equation

$$c_{av} = \frac{1}{H}(c_1 H_1 + c_2 H_2 + \cdots + c_n H_n) \tag{7.4}$$

where $c_1, c_2, \ldots, c_n$ = undrained cohesion in layers 1, 2, $\ldots$, $n$
$H_1, H_2, \ldots, H_n$ = thicknesses of layers 1, 3, $\ldots$, $n$

The average unit weight, $\gamma_a$, can be given as

$$\gamma_a = \frac{1}{H}(\gamma_1 H_1 + \gamma_2 H_2 + \gamma_3 H_3 + \cdots + \gamma_n H_n) \tag{7.5}$$

## 7.3

### Design of Various Components of a Braced Cut

#### Struts

In construction work, the struts should have a minimum vertical spacing of about 2.75 m or more. The struts are actually horizontal columns subject to bending. The load-carrying capacity of columns will depend on the *slenderness ratio*, $l/r$. The slenderness ratio can be reduced by providing vertical and horizontal supports at intermediate points. For cuts with large widths, it may be necessary to splice the struts. In the case of braced cuts in clayey soils, the depth of the first strut below the ground surface should be less than the depth of tensile crack, $z_c$. From Eq. (5.11)

$$\sigma_a = \gamma z K_a - 2c\sqrt{K_a}$$

where $K_a$ = coefficient of Rankine active pressure

For determination of the depth of tensile crack

$$\sigma_a = 0 = \gamma z_c K_a - 2c\sqrt{K_a}$$

or

$$z_c = \frac{2c}{\sqrt{K_a}\,\gamma}$$

With $\phi = 0$, $K_a = \tan^2(45 - \phi/2) = 1$. So

$$z_c = \frac{2c}{\gamma}$$

A simplified conservative procedure can be used to determine the strut loads. This procedure will vary depending on the engineers involved in the project. Following is a step-by-step outline of it (refer to Figure 7.6).

**1.** Draw the pressure envelope for the braced cut (see Figure 7.4). Also show the proposed strut levels. Figure 7.6a shows a pressure envelope for a

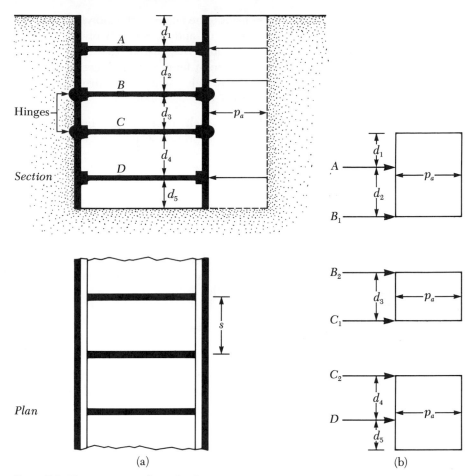

**Figure 7.6** Determination of strut loads

sandy soil; however, it could also be for a clay. Also, in this figure, the strut
levels are marked $A$, $B_1$, $C$, and $D$. The sheet piles (or soldier beams) can be
assumed to be hinged at the strut levels, except for the top and bottom ones.
In Figure 7.6a, the hinges are at the level of struts $B$ and $C$. (Many designers
also assume the sheet piles, or soldier beams, to be hinged at all strut levels,
except for the top.)

  **2.** Determine the reactions for the two simple cantilever beams (top and
bottom) and all the simple beams in between. In Figure 7.6b, these reactions
are $A$, $B_1$, $B_2$, $C_1$, $C_2$, and $D$.

  **3.** The strut loads in Figure 7.6 can now be calculated as follows:

$$P_A = (A)(s)$$
$$P_B = (B_1 + B_2)(s)$$
$$P_C = (C_1 + C_2)(s)$$
$$P_D = (D)(s)$$

(7.6)

where $\quad$ $P_A$, $P_B$, $P_C$, $P_D$ = loads to be taken by the individual struts at levels
$A$, $B$, $C$, and $D$, respectively

$A$, $B_1$, $B_2$, $C_1$, $C_2$, $D$ = reactions calculated in Step 2 (note unit: force/unit length of the braced cut)

$s$ = horizontal spacing of the struts (see plan in Figure 7.6a)

**4.** Knowing the strut loads at each level and the intermediate bracing conditions, one can now select the proper sections by using the steel construction manual.

## Sheet Piles

In order to design the sheet piles, perform the following steps:

**1.** For each of the sections shown in Figure 7.6b, determine the maximum bending moment.

**2.** Determine the maximum value of the maximum bending moment ($M_{max}$) obtained in Step 1. Note that the unit of this moment will be, for example, kN-m/meter length of the wall.

**3.** Obtain the section modulus of the sheet piles:

$$S = \frac{M_{max}}{\sigma_{all}} \tag{7.7}$$

where $\sigma_{all}$ = allowable flexural stress of the sheet pile material

**4.** The sheet pile section can now be chosen from a table such as Table 6.1.

## Wales

**1.** Wales can be treated as continuous horizontal members if they are spliced properly. Conservatively, they may also be treated as though they are pinned at the struts. For the section shown in Figure 7.6a, the maximum moments for the wales (assuming that they are pinned at the struts) are as follows:

at level $A$, $M_{max} = \dfrac{(A)(s^2)}{8}$

at level $B$, $M_{max} = \dfrac{(B_1 + B_2)s^2}{8}$

at level $C$, $M_{max} = \dfrac{(C_1 + C_2)s^2}{8}$

at level $D$, $M_{max} = \dfrac{(D)(s^2)}{8}$

where $A$, $B_1$, $B_2$, $C_1$, $C_2$, and $D$ are the reactions under the struts per unit length of the wall (Step 2 of strut design).

**2.** Determine the section modulus of the wales:

$$S = \frac{M_{max}}{\sigma_{all}}$$

The wales are sometimes fastened to the sheet piles at points that satisfy the lateral support requirements.

## Example
## 7.1

The cross section of a long braced cut is shown in Figure 7.7a.

a. Draw the earth pressure envelope.
b. Determine the strut loads at levels $A$, $B$, and $C$.
c. Determine the section of the struts subjected to the largest load.
d. Determine the sheet pile section required.
e. Determine a design section for the wales at level $B$.

*Note:* The struts are placed at 3 m center-to-center in the plan.

**Solution**
Part a

Given: $\gamma = 18$ kN/m$^3$, $c = 35$ kN/m$^2$, and $H = 6$ m.

$$\frac{\gamma H}{c} = \frac{(18)(7)}{35} = 3.6 < 4$$

So, the pressure envelope will be like the one in Figure 7.4c. This is plotted in Figure 7.7a with maximum pressure intensity, $p_a$, equal to $0.3\gamma H = 0.3(18)(7) = 37.8$ kN/m$^2$.

Part b

For determination of the strut loads, refer to Figure 7.7b. Taking the moment about $B_1$, $\Sigma M_{B_1} = 0$

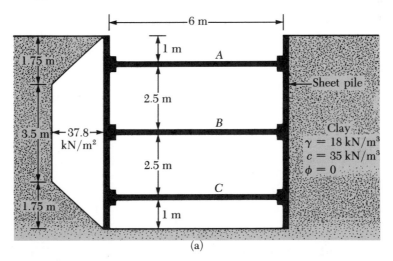

(a)

**Figure 7.7**

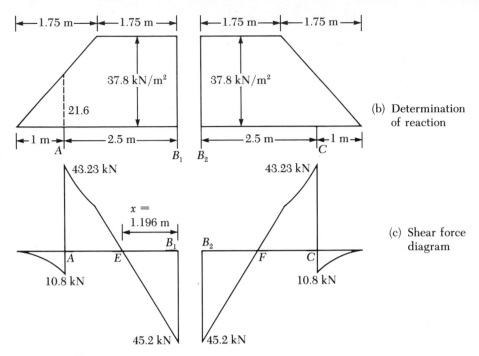

(b) Determination of reaction

(c) Shear force diagram

**Figure 7.7**  (Continued)

$$A(2.5) = \left(\frac{1}{2}\right)(37.8)(1.75)\left(1.75 + \frac{1.75}{3}\right) - (1.75)(37.8)\left(\frac{1.75}{2}\right) = 0$$

or

$A = 54.02$ kN/m

Also, $\Sigma$ vertical forces $= 0$. Thus

$\frac{1}{2}(1.75)(37.8) + (37.8)(1.75) = A + B_1$

$33.08 + 66.15 - A = B_1$

So

$B_1 = 45.2$ kN/m

Due to symmetry

$B_2 = 45.2$ kN/m

$C = 54.02$ kN/m

Strut load at level

$P_a = 54.02 \times$ horizontal spacing, $s = 54.02 \times 3 = 162.06$ kN

$P_B = (B_1 + B_2)3 = (45.2 + 45.2)3 = 271.2$ kN

$P_C = 54.02 \times 3 = 162.06$ kN

## Part c

The struts at level $B$ are subjected to the largest load—that is, $P_B = 271.2$ kN. For the struts, effective length ($KL$: refer to the American Institute of Steel Construction, *Manual of Steel Construction*, 1980, pp. 3–29) with respect to $x$ and $y$ axes is 6 m. Accordingly, the section W 250 mm $\times$ 49 kg/m (in English units, it is section W 10 $\times$ 33) will be more than sufficient. (*Note: $F_y = 248.4$ MN/m².*)

## Part d

Refer to the left side of Figure 7.7b. For the maximum moment, the shear force should be zero. The nature of variation of the shear force is shown in Figure 7.7c. The location of point $E$ can be given as

$$x = \frac{\text{reaction at } B_1}{37.8} = \frac{45.2}{37.8} = 1.196 \text{ m}$$

$$\text{The magnitude of moment at } A = \frac{1}{2}(1)\left(\frac{37.8}{1.75} \times 1\right)\left(\frac{1}{3}\right)$$

$$= 3.6 \text{ kN-m/meter of wall}$$

$$\text{The magnitude of moment at } E = (45.2 \times 1.196) - (37.8 \times 1.196)\left(\frac{1.196}{2}\right)$$

$$= 54.06 - 27.03 = 27.03 \text{ kN-m/meter of wall}$$

Because the loading on the left and right sections of Figure 7.7b are the same, the magnitude of moments at $F$ and $C$ (Figure 7.7c) will be the same as $E$ and $A$, respectively. Hence, the maximum moment = 27.03 kN-m/meter of wall.

The section modulus of the sheet piles,

$$S = \frac{M_{max}}{\sigma_{all}} = \frac{27.03 \text{ kN-m}}{170 \times 10^3 \text{ kN/m}^2} = 15.9 \times 10^{-5} \text{ m}^3/\text{m of the wall}$$

According to Table 6.1, section PMA-22 can be used.

## Part e

The reaction at level $B$ has been calculated in Part b. Hence

$$M_{max} = \frac{(B_1 + B_2)s^2}{8} = \frac{(45.2 + 45.2)3^2}{8} = 101.7 \text{ kN-m}$$

$$\text{Section modulus, } S_x = \frac{101.7}{\sigma_{all}} = \frac{101.7}{0.6F_y} = \frac{101.7}{0.6(248.4 \times 1000)}$$

$$= 0.682 \times 10^{-3} \text{ m}^3$$

So, a section of W 310 mm $\times$ 54 kg/m (in English units, W 12 $\times$ 36) with an $S_x = 0.754 \times 10^{-3}$ m³ can be used. (*Note: $L_c = 2.1$ m and $L_u = 4.085$ m—AISC Manual.*)

# 7.4

## Stability of Braced Cuts

### Heave of the Bottom of the Cut in Clay

Braced cuts in clay may become unstable as a result of the heaving of the bottom of the excavation. Terzaghi (1943b) has analyzed the factor of safety of braced excavations against bottom heave. The failure surface for such a case is shown in Figure 7.8. The vertical load per unit length of the cut at the level of the bottom of the cut along the line $bd$ and $af$ is equal to

$$Q = \gamma H B_1 - cH \tag{7.8}$$

where $B_1 = 0.7B$

$c$ = cohesion ($\phi = 0$ concept)

This load $Q$ can be treated like a load per unit length on a continuous foundation at the level of $bd$ (and $af$) having a width of $B_1 = 0.7B$. Based on Terzaghi's bearing capacity theory, the net ultimate load-carrying capacity per unit length of this foundation can be given by the equation [Chapter 3; see Eqs. (3.3) and (3.34)]

$$Q_u = cN_cB_1 = 5.7cB_1$$

Hence, from Eq. (7.8), the factor of safety against bottom heave can be given as

$$FS = \frac{Q_u}{Q} = \frac{5.7cB_1}{\gamma H B_1 - cH} = \frac{1}{H}\left(\frac{5.7c}{\gamma - \dfrac{c}{0.7B}}\right) \tag{7.9}$$

The preceding factor of safety [Eq. (7.9)] has been derived based on the assumption that the clay layer is homogeneous, at least up to a depth of $0.7B$ below the bottom of the cut. However, if a hard layer of rock or rock-like material is located at a depth $D < 0.7B$, the failure surface will be modified to some extent. In such a case, the factor of safety can be modified to the form

$$FS = \frac{1}{H}\left(\frac{5.7c}{\gamma - c/D}\right) \tag{7.10}$$

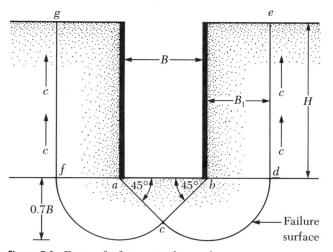

*Note: cd and cf are arcs of circles with centers at $b$ and $a$, respectively*

**Figure 7.8**   Factor of safety against bottom heave

Bjerrum and Eide (1956) also studied the problem of bottom heave for braced cuts in clay, and they proposed the following equation for the factor of safety

$$FS = \frac{cN_c}{\gamma H} \tag{7.11}$$

The bearing capacity factor $N_c$ varies with the ratio of $H/B$ and also $L/B$ (where $L$ = length of the cut). For infinitely long cuts $(B/L = 0)$, $N_c = 5.14$ at $H/B = 0$ and increases to a value of $N_c = 7.6$ at $H/B = 4$. Beyond that—that is, for $H/B > 4$—the value of $N_c$ remains constant. For cuts square in plan $(B/L = 1)$, $N_c = 6.3$ at $H/B = 0$, and $N_c = 9$ for $H/B \geq 4$. In general, at any given $H/B$

$$N_{c(\text{rectangle})} = N_{c(\text{square})}\left(0.84 + 0.16\frac{B}{L}\right) \tag{7.12}$$

Figure 7.9 shows the variation of the value of $N_c$ for $L/B = 1, 2, 3,$ and $\infty$. In any case, a factor of safety of 1.25 to 1.5 is desired.

### Stability of the Bottom of the Cut in Sand

The bottom of a cut in sand is generally stable. When the ground water table is encountered, the bottom of the cut is stable as long as the water level inside the excavation is higher than the ground water level. If the water level inside the cut is lowered below the ground water level by pumping, instability may be created as a result of the upward seepage of water into the cut. Section 7.5 discusses this problem in more detail.

### Lateral Yielding of Sheet Piles

In braced cuts, some lateral movement of sheet pile walls may be expected (Figure 7.10). Of course, the lateral yield will depend on several factors, the

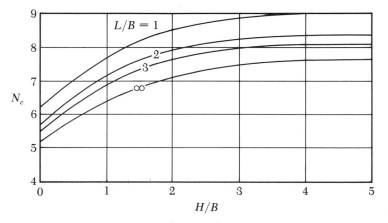

**Figure 7.9** Variation of $N_c$ with $L/B$ and $H/B$—based on Bjerrum and Eide's equation [Eq. (7.12)]

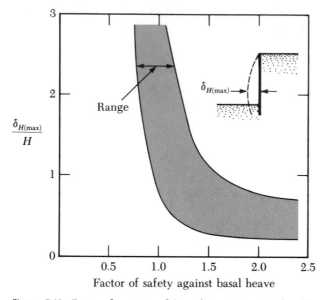

**Figure 7.10**  Range of variation of $\delta_{H(max)}/H$ vs. FS against basal heave from field observation (redrawn after Mana and Clough, 1981)

most important of which is time elapsed after excavation that is required for the placement of wales and struts. Mana and Clough (1981) analyzed the field records of several braced cuts in clay from the San Francisco, Oslo (Norway), Boston, Chicago, and Bowline Point (New York) areas. Under ordinary construction conditions, it was found that the maximum lateral wall yield $[\delta_{H(max)}]$ has a definite relationship with the factor of safety against heave. This is shown in Figure 7.10. Note that the factor of safety against heave as plotted in Figure 7.10 has been calculated by using Eqs. (7.9) and (7.10).

In several instances, the sheet piles (or the soldier piles as the case may be) are driven to a certain depth below the bottom of the excavation. This is done to reduce the lateral yielding of the walls during the last stages of excavation. The lateral yielding of the walls will cause settlement of the ground surface surrounding the cut. The degree of lateral yielding, however, depends mostly on the soil type below the bottom of the cut. If clay below the cut extends to a great depth and $\gamma H/c$ is less than about 6, extension of the sheet piles or soldier piles below the bottom of the cut will help considerably in reducing the lateral yield of the walls. However, under similar circumstances, if $\gamma H/c$ is about 8, the extension of sheet piles into the clay below the cut does not help to a great extent. In such circumstances, one may expect a great degree of wall yielding that may result in the total collapse of the bracing systems. If a hard soil layer is located below a clay layer at the bottom of the cut, the piles should be embedded in the stiffer layer. This will have a great effect in reducing the lateral yield.

## Ground Settlement

The lateral yielding of walls will generally induce ground settlement ($\delta_V$)

around a braced cut. This is generally referred to as *ground loss*. Based on several field observations, Peck (1969) has provided curves for prediction of ground settlement in various types of soil (see Figure 7.11). The magnitude of ground loss varies extensively; however, Figure 7.11 can be used as a general guide.

Based on the field data obtained from various cuts in the areas of San Francisco, Oslo, and Chicago, Mana and Clough (1981) have provided a correlation between the maximum lateral yield of sheet piles $[\delta_{H(max)}]$ and the maximum ground settlement $[\delta_{V(max)}]$. This is shown in Figure 7.12. It can be seen that

$$\delta_{V(max)} \approx 0.5 \text{ to } 1 \ \delta_{H(max)} \tag{7.13}$$

## Example
## 7.2

Refer to Example Problem 7.1. Determine the factor of safety against bottom heave using Eqs. (7.9) and (7.11).

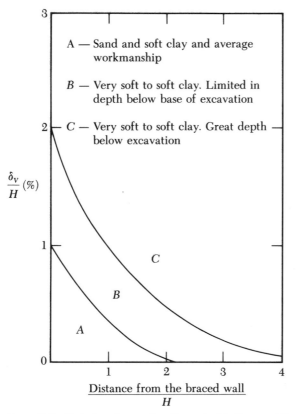

Figure 7.11   Variation of ground settlement with distance (after Peck, 1969)

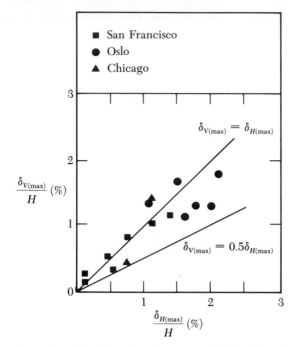

**Figure 7.12** Variation of maximum lateral yield with maximum ground settlement (after Mana and Clough, 1981)

**Solution**

In Example Problem 7.1, $\gamma = 18$ kN/m³, $c = 35$ kN/m², and $H = 7$ m.

Factor of Safety from Eq. (7.9)

$$FS = \frac{1}{H}\left(\frac{5.7c}{\gamma - \dfrac{c}{0.7B}}\right) = \frac{1}{7}\left[\frac{(5.7)(35)}{18 - \dfrac{35}{(0.7)(6)}}\right] = \underline{2.95}$$

Factor of Safety from Eq. (7.11)

$$FS = \frac{cN_c}{\gamma H}$$

According to Figure 7.9, for $H/B = 7/6 = 1.16$ and $B/L \approx 0$, the value of $N_c$ is equal to 6.46. Thus

$$FS = \frac{(35)(6.46)}{(18)(7)} = \underline{1.79}$$

**7.5**

## Failure of Single Wall Cofferdams by Piping

Sheet piles are sometimes driven for excavations that need dewatering (Figure 7.13). In such cases, the factor of safety against piping should be

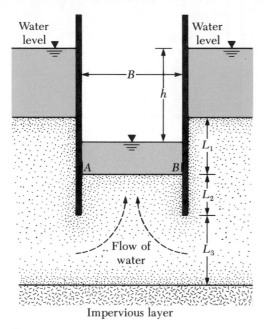

**Figure 7.13**

checked. [*Piping* is another term for failure by heave, as defined in Section 1.12; see Eq. (1.48).] Piping may occur when high hydraulic gradient is set up as a result of the flow of water into the excavation. One can check the factor of safety by drawing flow nets and determining the maximum exit gradient [$i_{max(exit)}$] that will occur at points $A$ and $B$. Figure 7.14 shows a flow net to illustrate the problem. The maximum exit gradient for this flow net can be calculated as

$$i_{max(exit)} = \frac{\dfrac{h}{N_d}}{a} \tag{7.14}$$

where $a$ = length of the flow element at $A$ (or $B$)
$N_d$ = number of drops (*Note:* In Figure 7.14, $N_d = 8$.)

The factor of safety against piping can be expressed as

$$FS = \frac{i_{cr}}{i_{max(exit)}} \tag{7.15}$$

where $i_{cr}$ = critical hydraulic gradient

The relationship for $i_{cr}$ has been given in Chapter 1 [Eq. (1.48)] as

$$i_{cr} = \frac{G_s - 1}{e + 1}$$

The value of $i_{cr}$ varies between 0.9 and 1.1 in most soils with an average of about one. A factor of safety of about 1.5 is desirable.

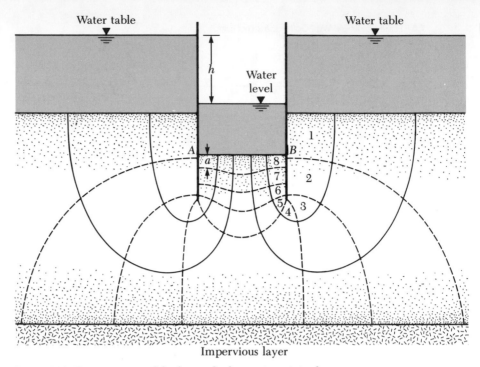

**Figure 7.14** Determination of the factor of safety against piping by drawing flow net

Marsland (1958) has suggested the following values of $L_2$ (minimum penetration) for a factor of safety of 1.5 against piping for excavations in sand.

Cuts in Loose Sand ($L_3 = \infty$)

| Width of excavation, $\dfrac{B}{L_1}$ | $\dfrac{L_2}{L_1}$ |
|:---:|:---:|
| 0.25 | 1.4 |
| 0.5 | 1.2 |
| 1.0 | 1.0 |
| 2.0 | 0.9 |
| 4.0 | 0.8 |
| 8.0 | 0.7 |

Cuts in Dense Sand ($L_3 = \infty$)

| Width of excavation, $\dfrac{B}{L_1}$ | $\dfrac{L_2}{L_1}$ |
|:---:|:---:|
| 0.25 | 1.3 |
| 0.5 | 1.0 |
| 1.0 | 0.8 |
| 2.0 | 0.6 |
| 4.0 | 0.5 |
| 8.0 | 0.4 |

The maximum exit gradient for sheeted excavations in sands with $L_3 = \infty$ can also be theoretically evaluated (Harr, 1962). (Only the results of these mathematical derivations will be presented here. For further details, refer to the original work.) To calculate the maximum exit gradient, refer to Figures 7.15 and 7.16 and perform the following steps:

1. Determine the modulus, $m$, from Figure 7.15 by obtaining $2L_2/B$ (or $B/2L_2$) and $2L_1/B$.
2. With the known modulus and $2L_1/B$, refer to Figure 7.16 and determine $L_2 i_{exit(max)}/h$. Because $L_2$ and $h$ will be known, $i_{exit(max)}$ can be calculated.
3. The factor of safety against piping can be evaluated by using Eq. (7.15).

Another method to prevent piping and increase the factor of safety is to lower the ground water level, thereby reducing the head $h$ shown in Figure 7.13. This can be done by pumping from well points or deep wells placed below the level of the bottom of the sheet piles.

## Example 7.3

Refer to Figure 7.13. Given: $h = 4.5$ m, $L_1 = 5$ m, $L_2 = 4$ m, $B = 5$ m, and $L_3 = \infty$. Determine the factor of safety against piping.

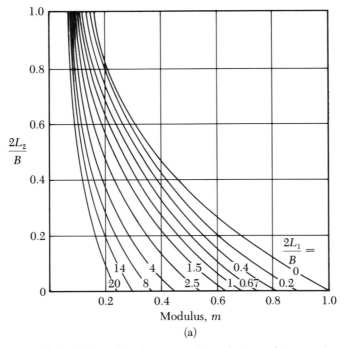

**Figure 7.15** Variation of modulus (from *Groundwater and Seepage*, by M. E. Harr. Copyright © 1962 by McGraw-Hill. Used with the permission of the McGraw-Hill Book Company)

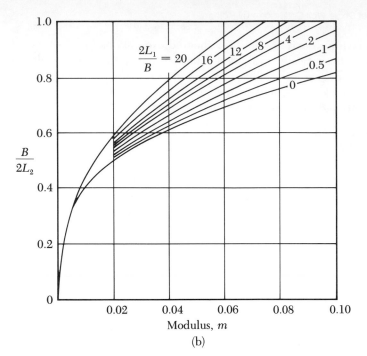

$\frac{2L_1}{B} = 20$ 16 12 8 4 2 1 0.5 0

Modulus, $m$

(b)

**Figure 7.15** (Continued)

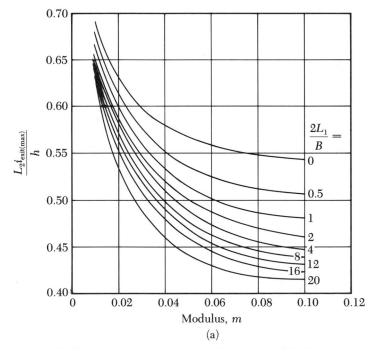

$\frac{2L_1}{B} =$

0

0.5

1

2

4

8

12

16

20

Modulus, $m$

(a)

**Figure 7.16** Variation of maximum exit gradient with modulus (from *Groundwater and Seepage*, by M. E. Harr, Copyright © 1962 by McGraw-Hill. Used with the permission of the McGraw-Hill Book Company)

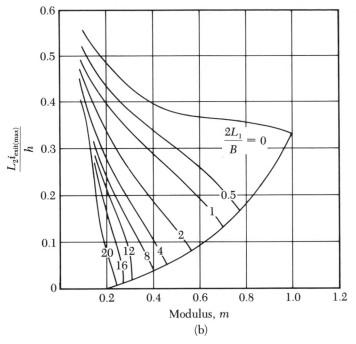

**Figure 7.16** (Continued)

## Solution

$$\frac{2L_1}{B} = \frac{2(5)}{5} = 2$$

$$\frac{B}{2L_2} = \frac{5}{2(4)} = 0.625$$

According to Figure 7.15b, for $2L_1/B = 2$ and $B/2L_2 = 0.625$, $m \approx 0.033$. Again, referring to Figure 7.16a, for $m = 0.033$ and $2L_1/B = 2$, $L_2 i_{exit(max)}/h = 0.55$. Hence

$$i_{exit(max)} = \frac{0.55(h)}{L_2} = 0.55(4.5)/4 = 0.619$$

$$FS = \frac{i_{cr}}{i_{max(exit)}} = \frac{1}{0.619} = \underline{1.616}$$

## Problems

**7.1** Refer to the braced cut shown in Figure P7.1. Given: $\gamma = 112$ lb/ft³, $\phi = 32°$, and $c = 0$. The struts are located at 12 ft center-to-center in the plan. Draw the earth pressure envelope and determine the strut loads at levels $A$, $B$, and $C$.

**7.2** For the braced cut described in Problem 7.1, determine:
a. the sheet pile section;
b. the section of the strut subjected to the largest load;
c. the section of the wales at level $A$.

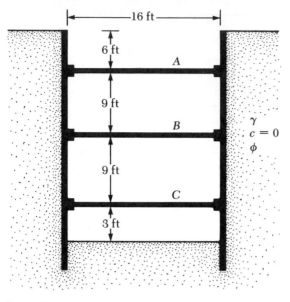

**Figure P7.1**

**7.3** Redo Problem 7.1 with $\gamma = 116$ lb/ft³, $\phi = 35°$, $c = 0$, and center-to-center strut spacing in plan $= 10$ ft.

**7.4** Determine the sheet pile section required for the braced cut described in Problem 7.3.

**7.5** Refer to Figure 7.5a. For the braced cut, given: $H = 6$ m; $H_s = 2$ m; $\gamma_s = 16.2$ kN/m³; angle of friction of sand, $\phi_s = 34°$; $H_c = 4$ m, $\gamma_c = 17.5$ kN/m³; and unconfined compression strength of clay layer, $q_u = 68$ kN/m².
  a. Estimate the average cohesion $(c_{av})$ and average unit weight $(\gamma_{av})$ for the construction of the earth pressure envelope.
  b. Plot the earth pressure envelope.

**7.6** Refer to Figure 7.5b on p. 311, which shows a braced cut in clay. Given: $H = 22$ ft, $H_1 = 6$ ft, $c_1 = 2125$ lb/ft², $\gamma_1 = 111$ lb/ft³, $H_2 = 8$ ft, $c_2 = 1565$ lb/ft², $\gamma_2 = 107$ lb/ft³, $H_3 = 8$ ft, $c_3 = 1670$ lb/ft², $\gamma_3 = 109$ lb/ft³.
  a. Determine the average cohesion $(c_{av})$ and average unit weight $(\gamma_{av})$ for calculation of the earth pressure envelope.
  b. Plot the earth pressure envelope.

**7.7** Refer to Figure P7.7. Given: $\gamma = 17.5$ kN/m³, $c = 30$ kN/m², and center-to-center spacing of struts $= 5$ m. Draw the earth pressure envelope and determine the strut loads at levels $A$, $B$, and $C$.

**7.8** For the braced cut described in Problem 7.7, determine:
  a. the sheet pile section;
  b. the section of the strut subjected to the largest load.

**7.9** Redo Problem 7.7 assuming $c = 60$ kN/m².

**7.10** Determine the factor of safety against bottom heave for the braced cut described in Problem 7.7. Use Eqs. (7.9) and (7.11). For Eq. (7.11), assume the length of the cut, $L = 18$ m.

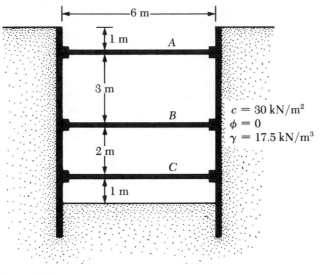

**7.11** Determine the factor of safety against bottom heave for the braced cut described in Problem 7.9. Use Eq. (7.11). Length of the cut is 12.5 m.

## References

American Institute of Steel Construction (1980). *Manual of Steel Construction*, Eighth Edition, Chicago.

Bjerrum, L., and Eide, O. (1956). "Stability of Strutted Excavation in Clay," *Geotechnique*, Vol. 6., No. 1, pp. 32–47.

Das, B. M. (1979). *Introduction to Soil Mechanics*, Iowa State University Press, Ames, Iowa.

Das, B. M., and Seeley, G. R. (1975). "Active Thrust on Braced Cut in Clay," *Journal of the Construction Division*, American Society of Civil Engineers, Vol. 101, No. C04, pp. 945–949.

Harr, M. E. (1962). *Ground Water and Seepage*, McGraw-Hill, New York.

Mana, A. I., and Clough, G. W. (1981). "Prediction of Movements for Braced Cuts in Clay," *Journal of the Geotechnical Engineering Division*, American Society of Civil Engineers, Vol. 107, No. GT8, pp. 759–777.

Marsland, A. (1958). "Model Experiments to Study the Influence of Seepage on the Stability of a Sheeted Excavation in Sand," *Geotechnique*, Vol. 3, p. 223.

Peck, R. B. (1943). "Earth Pressure Measurements in Open Cuts, Chicago (Ill.) Subway," *Transactions*, American Society of Civil Engineers, Vol. 108, pp. 1008–1058.

Peck, R. B. (1969). "Deep Excavation and Tunneling in Soft Ground," *Proceedings*, Seventh International Conference on Soil Mechanics and Foundation Engineering, Mexico City, State-of-the-Art Volume, pp. 225–290.

Terzaghi, K. (1943a). "General Wedge Theory of Earth Pressure," *Transactions*, American Society of Civil Engineers, Vol. 106, pp. 68–97.

Terzaghi, K. (1943b). *Theoretical Soil Mechanics*, Wiley, New York.

# Pile Foundations

## 8.1

### Introduction

Piles are structural members that are made of steel, concrete, and/or timber. They are used to build pile foundations, which are deep and which cost more than shallow foundations (Chapter 3). Despite the cost, the use of piles often becomes necessary to ensure that the structure under consideration is safe. Following is a list of some of the conditions that require pile foundations:

**1.** When the upper soil layer(s) is highly compressible and too weak to support the load transmitted by the superstructure, piles are used to transmit the load to underlying bedrock. This is shown in Figure 8.1a. When bedrock is not located at a reasonable depth below the ground surface, piles are used to gradually transmit the structural load to the soil. The resistance to the applied structural load is derived mainly from the frictional resistance developed at the soil-pile interface (Figure 8.1b).

**2.** When subjected to horizontal forces (see Figure 8.1c), pile foundations can resist by bending, while still supporting the vertical load transmitted by the superstructure. This type of situation is generally encountered in the design and construction of earth-retaining structures and foundations of tall structures that are subjected to high wind and/or earthquake forces.

**3.** In many cases, expansive and collapsible soils (Chapter 10) may be encountered at the site of a proposed structure. These soils may extend to a great depth below the ground surface. Expansive soils swell and shrink depending on the increase or decrease of moisture content. The swelling

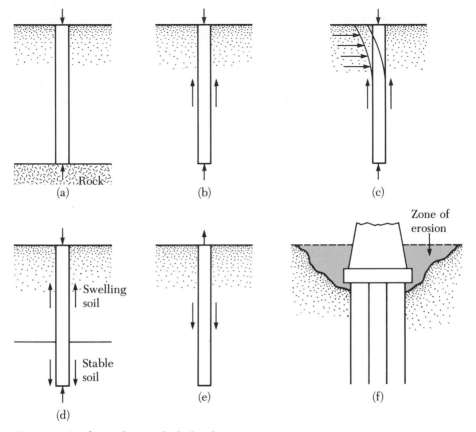

**Figure 8.1** Conditions for use of pile foundations

pressure of such soils can be considerably high. If shallow foundations are used in such circumstances, the structure may encounter considerable damage. However, pile foundations may be considered as an alternative when piles are extended beyond the active zone, which undergoes swelling and shrinking (Figure 8.1d).

Soils such as loess are collapsible in nature. When these soils undergo an increase of moisture content, their structures may break down. A sudden decrease of void ratio of soil induces large settlements of structures supported by shallow foundations. In such cases, pile foundations may be used in which piles are extended into stable soil layers beyond the zone of possible moisture change.

**4.** Foundations of some structures, such as transmission towers, offshore platforms, and basement mats below the water table, are subjected to uplifting forces. Piles are sometimes used for these foundations to resist the uplifting force (Figure 8.1e).

**5.** Bridge abutments and piers are occasionally constructed over pile foundations to avoid the possible loss of bearing capacity that a shallow foundation might suffer because of soil erosion at the ground surface (Figure 8.1f).

Although several investigations—both theoretical and experimental—have been conducted in the past to predict the behavior and the load-bearing capacity of piles in granular and cohesive soils, the dynamics are not yet entirely understood and may never be. The design of pile foundations may be considered somewhat of an "art" as a result of the uncertainties involved in working with some subsoil conditions. This chapter discusses the present state-of-the-art for design and analysis of pile foundations.

## 8.2
## Types of Pile in Use

Different types of pile are used in construction work depending on the type of load to be carried, the subsoil conditions, and the ground water table. Piles can be divided into the following categories: (a) *steel piles*, (b) *concrete piles*, (c) *wooden (timber) piles*, and (d) *composite piles*. This section discusses each of these types in detail.

### Steel Piles

The steel piles generally used are either *pipe piles* or *rolled steel H-section piles*. Pipe piles can be driven into the ground with their ends open or closed. Wide-flange and I-section steel beams can also be used as piles. However, H-section piles are usually preferred because their web and flange thicknesses are equal. In wide-flange and I-section beams, the web thicknesses are smaller than the thicknesses of the flange. Table 8.1 gives the dimensions of some standard H-section steel piles used in the United States. Table 8.2 shows a list of selected pipe sections frequently used for piling purposes. In many cases, the pipe piles are filled with concrete after driving.

The allowable design load for steel piles may be given as

$$Q_{all} = A_s \sigma_{all} \tag{8.1}$$

where $A_s$ = cross-sectional area of the steel
$\sigma_{all}$ = allowable stress of steel

Based on geotechnical considerations, once the design load for a pile is fixed, it is always advisable to check if $Q_{(design)}$ is within the allowable range as defined by Eq. (8.1).

Steel piles, when necessary, are spliced by welding or by riveting. Figure 8.2a on page 334 shows a typical condition of splicing by welding for an H-pile. A typical case of splicing by welding for a pipe pile is shown in Figure 8.2b. Figure 8.2c shows a diagram of splicing an H-pile by rivets and bolts.

When hard driving conditions are expected, such as driving through dense gravel, shale, and soft rock, the steel piles can be fitted with driving points or shoes. Figures 8.2d and e are diagrams of two types of shoe used for pipe piles.

Steel piles may be subject to corrosion. For example, swamps, peats, and other organic soils are corrosive. Soils that have a pH greater than 7 are not so corrosive. In order to take into account the effect of corrosion, an additional thickness of steel (over the actual design cross-sectional area) is generally recom-

**Table 8.1**  Common H-Pile Sections Used in the United States

| Designation, size (mm) × weight (kN/M) | Depth $d_1$ (mm) | Section area (m² × 10⁻³) | Flange and web thickness $w$ (mm) | Flange width $d_2$ (mm) | Moment of inertia (m⁴ × 10⁻⁶) $I_{xx}$ | $I_{yy}$ |
|---|---|---|---|---|---|---|
| HP 200 × 0.52 | 204 | 6.84 | 11.3 | 207 | 49.5 | 16.8 |
| HP 250 × 0.834 | 254 | 10.8 | 14.4 | 260 | 123 | 42 |
| × 0.608 | 246 | 8.0 | 10.6 | 256 | 87.5 | 24 |
| HP 310 × 1.226 | 312 | 15.9 | 17.5 | 312 | 271 | 89 |
| × 1.079 | 308 | 14.1 | 15.49 | 310 | 237 | 77.5 |
| × 0.912 | 303 | 11.9 | 13.1 | 308 | 197 | 63.7 |
| × 0.775 | 299 | 10.0 | 11.05 | 306 | 164 | 62.9 |
| HP 330 × 1.462 | 334 | 19.0 | 19.45 | 335 | 370 | 123 |
| × 1.264 | 329 | 16.5 | 16.9 | 333 | 314 | 104 |
| × 1.069 | 324 | 13.9 | 14.5 | 330 | 263 | 86 |
| × 0.873 | 319 | 11.3 | 11.7 | 328 | 210 | 69 |
| HP 360 × 1.707 | 361 | 22.2 | 20.45 | 378 | 508 | 184 |
| × 1.491 | 356 | 19.4 | 17.91 | 376 | 437 | 158 |
| × 1.295 | 351 | 16.8 | 15.62 | 373 | 374 | 136 |
| × 1.060 | 346 | 13.8 | 12.82 | 371 | 303 | 109 |

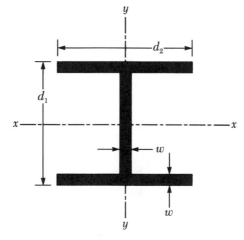

mended. In many circumstances, factory-applied epoxy coatings on piles work satisfactorily against corrosion. These coatings are not very easily damaged by pile driving. Concrete encasement of steel piles in most corrosive zones is also practiced as a protection against corrosion.

## Concrete Piles

Concrete piles can be divided into two basic categories: (a) *precast piles* and (b) *cast-in-situ piles*. *Precast piles* can be prepared by using ordinary reinforcement, and they can be square or octagonal in cross section (Figure 8.3). The

**Table 8.2**   Selected Pipe Pile Sections

| Outside diameter (mm) | Wall thickness (mm) | Area of steel (cm²) |
|---|---|---|
| 219 | 3.17 | 21.5 |
|  | 4.78 | 32.1 |
|  | 5.56 | 37.3 |
|  | 7.92 | 52.7 |
| 254 | 4.78 | 37.5 |
|  | 5.56 | 43.6 |
|  | 6.35 | 49.4 |
| 305 | 4.78 | 44.9 |
|  | 5.56 | 52.3 |
|  | 6.35 | 59.7 |
| 406 | 4.78 | 60.3 |
|  | 5.56 | 70.1 |
|  | 6.35 | 79.8 |
| 457 | 5.56 | 80 |
|  | 6.35 | 90 |
|  | 7.92 | 112 |
| 508 | 5.56 | 88 |
|  | 6.35 | 100 |
|  | 7.92 | 125 |
| 610 | 6.35 | 121 |
|  | 7.92 | 150 |
|  | 9.53 | 179 |
|  | 12.70 | 238 |

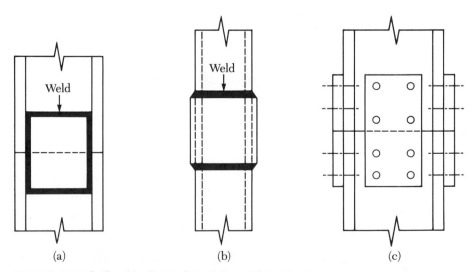

**Figure 8.2**   Steel piles: (a) splicing of H-pile by welding; (b) splicing of pipe pile by welding; (c) splicing of H-pile by rivets and bolts; (d) flat driving point of pipe pile; (e) conical driving point of pipe pile

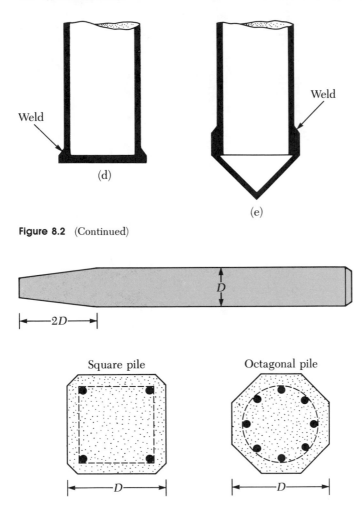

(d)

(e)

Weld

Weld

**Figure 8.2** (Continued)

$D$

$2D$

Square pile

Octagonal pile

$D$

$D$

**Figure 8.3** Precast piles with ordinary reinforcement

reinforcement is provided to enable the pile to resist the bending moment developed during pickup and transportation, the vertical load, and the bending moment caused by lateral load. The piles are cast to desired lengths and cured before being transported to the work sites.

Precast piles can also be prestressed by the use of high-strength steel prestressing cables. The ultimate strength of these steel cables is about 1800 $MN/m^2$ ( $\approx$ 261 ksi). In casting the piles, the cables are pretensioned up to about 900–1300 $MN/m^2$ ( $\approx$ 130–188 ksi), and concrete is poured around them. After proper curing, the cables are cut and thus produce a compressive force on the pile section. Table 8.3 gives additional information about prestressed concrete piles with square and octagonal cross sections.

*Cast-in-situ,* or *cast-in-place, piles* are built by making a hole in the ground and then filling it with concrete. Various types of cast-in-place concrete pile are used in construction nowadays, and most of them have been patented by their

**Table 8.3** Typical Prestressed Concrete Piles In Use

| Pile Shape* | D (mm) | Area of cross section (cm²) | Perimeter (mm) | Number of Strands 12.7 mm diameter | Number of Strands 11.1 mm diameter | Minimum effective prestress force (kN) | Section modulus (m³ × 10⁻³) | Design bearing capacity (kN) Concrete strength (MN/m²) 34.5 | Design bearing capacity (kN) Concrete strength (MN/m²) 41.4 |
|---|---|---|---|---|---|---|---|---|---|
| S | 254 | 645 | 1016 | 4 | 4 | 312 | 2.737 | 556 | 778 |
| O | 254 | 536 | 838 | 4 | 4 | 258 | 1.786 | 462 | 555 |
| S | 305 | 929 | 1219 | 5 | 6 | 449 | 4.719 | 801 | 962 |
| O | 305 | 768 | 1016 | 4 | 5 | 369 | 3.097 | 662 | 795 |
| S | 356 | 1265 | 1422 | 6 | 8 | 610 | 7.489 | 1091 | 1310 |
| O | 356 | 1045 | 1168 | 5 | 7 | 503 | 4.916 | 901 | 1082 |
| S | 406 | 1652 | 1626 | 8 | 11 | 796 | 11.192 | 1425 | 1710 |
| O | 406 | 1368 | 1346 | 7 | 9 | 658 | 7.341 | 1180 | 1416 |
| S | 457 | 2090 | 1829 | 10 | 13 | 1010 | 15.928 | 1803 | 2163 |
| O | 457 | 1729 | 1524 | 8 | 11 | 836 | 10.455 | 1491 | 1790 |
| S | 508 | 2581 | 2032 | 12 | 16 | 1245 | 21.844 | 2226 | 2672 |
| O | 508 | 2136 | 1077 | 10 | 14 | 1032 | 14.355 | 1842 | 2239 |
| S | 559 | 3123 | 2235 | 15 | 20 | 1508 | 29.087 | 2694 | 3232 |
| O | 559 | 2587 | 1854 | 12 | 16 | 1250 | 19.107 | 2231 | 2678 |
| S | 610 | 3658 | 2438 | 18 | 23 | 1793 | 37.756 | 3155 | 3786 |
| O | 610 | 3078 | 2032 | 15 | 19 | 1486 | 34.794 | 2655 | 3186 |

*S = square section; O = octagonal section

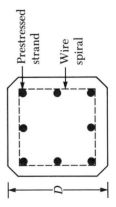

Wire spiral
Prestressed strand

Prestressed strand
Wire spiral

manufacturers. These piles can be divided into two broad categories: (a) *cased* and (b) *uncased*. Both types may have a *pedestal* at the bottom.

Cased piles are made by driving a steel casing into the ground with the help of a mandrel placed inside the casing. When the pile reaches the proper depth, the mandrel is withdrawn and the casing is filled with concrete. Figures 8.4a, b, c, and d show some examples of cased piles without a pedestal. Table 8.4 gives additional information about these cased piles. Figure 8.4e shows a cased pile with a pedestal. The pedestal is an expanded concrete bulb that is formed by dropping a hammer on fresh concrete.

Figures 8.4f and g are two types of uncased pile, one with a pedestal and the other without. The uncased piles are made by first driving the casing to the desired depth and then filling it with fresh concrete. The casing is gradually withdrawn in steps.

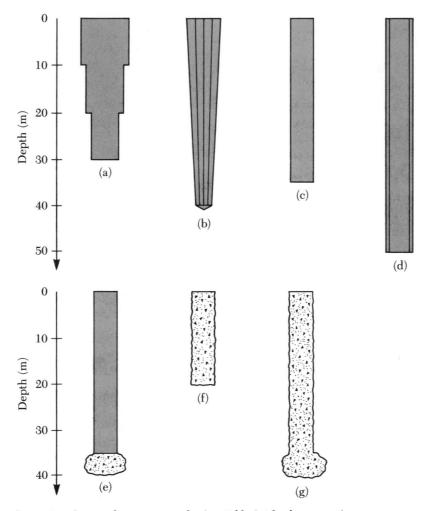

**Figure 8.4** Cast-in-place concrete piles (see Table 8.4 for descriptions)

**Table 8.4**  Descriptions of the Cast-In-Place Piles Shown in Figure 8.4

| Part in Figure 8.4 | Name of pile | Type of casing | Maximum usual depth of pile (m) |
|---|---|---|---|
| a | Raymond Step-Taper | Corrugated, thin cylindrical casing | 30 |
| b | Monotube or Union Metal | Thin fluted, tapered steel casing driven without mandrel | 40 |
| c | Western cased | Thin sheet casing | 30–40 |
| d | Seamless pipe or Armco | Straight steel pipe casing | 50 |
| e | Franki cased pedestal | Thin sheet casing | 30–40 |
| f | Western uncased without pedestal | ---- | 15–20 |
| g | Franki uncased pedestal | ---- | 30–40 |

The allowable load for cast-in-place concrete piles may be given by the following equations:

*Cased pile*

$$Q_{all} = A_s f_s + A_c f_c \qquad (8.2a)$$

where $A_s$ = area of cross section of steel
$A_c$ = area of cross section of concrete
$f_s$ = allowable stress of steel
$f_c$ = allowable stress of concrete

*Uncased pile*

$$Q_{all} = A_c f_c \qquad (8.2b)$$

## Timber Piles

Timber piles are tree trunks that have had their branches carefully trimmed off. The maximum length of most timber piles is 10–20 m. In order to qualify for use as a pile, the timber should be straight, sound, and without any defects. The American Society of Civil Engineers' *Manual of Practice*, No. 17 (1959), classifies timber piles into three categories:

**1.** *Class A Piles*: These piles carry heavy loads. Minimum diameter of the butt should be 356 mm.

**2.** *Class B Piles*: These are used to carry medium loads. Minimum butt diameter should be 305–330 mm.

**3.** *Class C Piles*: For use in temporary construction work. They can be used for structures on a permanent basis when the entire pile is below the water table. Minimum butt diameter should be 305 mm.

In any case, pile tips should not have a diameter less than 150 mm.

Timber piles cannot withstand hard driving stress; therefore, the pile capacity is generally limited to about 220–270 kN (25–30 tons). Steel shoes may be used to avoid damage at the pile tip (bottom). The tops of timber piles may also be damaged during the driving operation. The crushing of the wooden fibers caused by the impact of the hammer is referred to as *brooming*. To avoid damage to the pile top, a metal band or a cap may be used.

Splicing of timber piles should be avoided, particularly when they are expected to carry tensile load or lateral load. However, if splicing is necessary, it can be done by using either *pipe sleeves* (Figure 8.5a) or *metal straps and bolts* (Figure 8.5b). The length of the pipe sleeve should be at least five times the diameter of the pile. The butting ends should be cut square so that full contact can be maintained. The spliced portions should be carefully trimmed so that they fit tightly to the inside of the pipe sleeve. In the case of metal straps and bolts, the butting ends should also be cut square. Also, the sides of the spliced portion should be trimmed plane for putting the straps on.

Timber piles can stay undamaged indefinitely if they are surrounded by saturated soil. However, in a marine environment timber piles are subject to attack by various organisms and can be extensively damaged in a few months. When located above the water table, the piles are subject to attack by insects. The life of the piles may be increased by treating them with preservatives such as creosote oil.

The allowable load-carrying capacity of wooden piles may be given as

$$Q_{\text{all}} = A_p f_w \tag{8.3}$$

where $A_p$ = average area of cross section of the pile
       $f_w$ = allowable stress for the timber

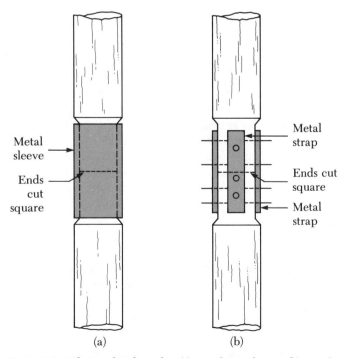

(a)                              (b)

**Figure 8.5**  Splicing of timber piles: (a) use of pipe sleeves; (b) use of metal straps and bolts

## Composite Piles

The upper and lower portions of composite piles are made of different materials. For example, composite piles may be made of *steel and concrete* or *timber and concrete*. Steel and concrete piles consist of a lower portion of steel and an upper portion of cast-in-place concrete. This type of pile is the one used when the length of the pile required for adequate bearing exceeds the capacity of simple cast-in-place concrete piles. Timber and concrete piles usually consist of a lower portion of timber pile below the permanent water table and an upper portion of concrete. In any case, it is difficult to form proper joints between two dissimilar materials, and, for that reason, composite piles are not widely used.

## Comparison of Pile Types

Several factors affect the selection of piles for a given structure at a given site. Table 8.5 gives a brief comparison of the advantages and disadvantages of the various types of pile based on the pile material.

## 8.3

## Estimation of Pile Length

Selecting the type of pile to be used and estimating its necessary length are fairly difficult tasks that require good judgment. In addition to the classification given in Section 8.2, piles can be divided into three major categories depending on their lengths and the mechanisms of load transfer to the soil. These categories are (a) *point bearing piles*, (b) *friction piles*, and (c) *compaction piles*.

## Point Bearing Piles

If bedrock or rock-like material is located at a given site within a reasonable depth that has been well established by soil-boring records, piles can be extended to the rock layer (Figure 8.6a, p. 344). In this case, the ultimate capacity of the piles entirely depends on the load-bearing capacity of the firm material, thus the piles are called *point bearing piles*. In most of these cases, the necessary length of the pile can be fairly well established.

Instead of bedrock, if a fairly compact and hard stratum of soil is located at a reasonable depth, piles can be extended a few meters into the hard stratum (Figure 8.6b). Piles with pedestals can be constructed on the bed of the hard stratum. For these types of pile, the ultimate pile load can be expressed as

$$Q_u = Q_p + Q_s \tag{8.4}$$

where $Q_p$ = load carried at the pile point
$Q_s$ = load carried by skin friction developed at the side of the pile (caused by shearing resistance between the soil and the pile)

If $Q_s$ is very small, then

$$Q_u \approx Q_p \tag{8.5}$$

In this case, necessary pile length can also be accurately estimated if proper subsoil exploration records are available.

## Friction Piles

When there is no layer of rock or rock-like material located at a reasonable depth at a given site, point bearing piles become very long and uneconomical. For this type of subsoil condition, piles are driven through the softer material to specified depths (Figure 8.6c). The ultimate load of these piles can be expressed by Eq. (8.4). However, if the value of $Q_p$ is relatively small

$$Q_u \approx Q_s \tag{8.6}$$

These piles are called *friction piles* because most of the resistance is derived from skin friction. However, the term *friction pile*, although used often in literature, is a misnomer: in clayey soils, the resistance to applied load is also caused by *adhesion*.

The length of friction piles depends on the shear strength of the soil, the applied load, and the pile size. To determine the necessary lengths of these piles, one needs a good understanding of the soil-pile interaction, good judgment, and experience. Theoretical procedures for the calculation of load-bearing capacity of piles are given in Section 8.6.

## Compaction Piles

Under certain circumstances, piles are driven in granular soils to achieve proper compaction of soil close to the ground surface. These piles are called *compaction piles*. The length of compaction piles depends on factors such as (a) relative density of the soil before compaction, (b) desired relative density of the soil after compaction, and (c) required depth of compaction. These piles are generally short; however, some field tests are necessary to arrive at a reasonable figure.

## 8.4

## Installation of Piles

Most piles are driven into the ground by means of *hammers* or *vibratory drivers*. In special circumstances, piles can also be inserted by *jetting* or *partial augering*. The hammers used for pile driving can be of several types, such as (a) drop hammer, (b) single-acting air or steam hammer, (c) double-acting and differential air or steam hammer, and (d) diesel hammer. In the driving operation, a cap is attached to the top of the pile. A cushion may be used between the pile and the cap. This cushion has the effect of evening out the hammer impulses; however, its use is optional. A hammer cushion is placed on the pile cap. The hammer drops on the cushion.

Figure 8.7 (pp. 344–345) illustrates pile driving using various hammers. A drop hammer (Figure 8.7a) is raised by a winch and allowed to drop from a certain height $H$. This is the oldest type of hammer used for pile driving. The main disadvantage with the drop hammer is the slow rate of hammer blows. The principle of the single-acting air or steam hammer is shown in Figure 8.7b. In this case, the striking part, or ram, is raised by air or steam pressure, and then it drops by gravity. Figure 8.7c shows the operation of the double-acting and differential air or steam hammer. For these hammers, air or steam is used both

**Table 8.5** Comparisons of Piles Made of Different Materials

| Pile type | Usual length of piles (m) | Maximum length of pile (m) | Usual load (kN) | Approximate maximum load (kN) | Comments |
|---|---|---|---|---|---|
| Steel | 15–60 | Practically unlimited | 300–1200 | Eq. (8.1) | *Advantages*<br>a. Easy to handle with respect to cutoff and extension to the desired length<br>b. Can stand high driving stresses<br>c. Can penetrate hard layers such as dense gravel, soft rock<br>d. High load-carrying capacity<br>*Disadvantages*<br>a. Relatively costly material<br>b. High level of noise during pile driving<br>c. Subject to corrosion<br>d. H-piles may be damaged or deflected from the vertical during driving through hard layers or past major obstructions |
| Precast concrete | *Precast: 10–15*<br>*Prestressed: 10–35* | *Precast: 30*<br>*Prestressed: 60* | 300–3000 | *Precast: 800–900*<br>*Prestressed: 7500–8500* | *Advantages*<br>a. Can be subjected to hard driving<br>b. Corrosion resistant<br>c. Can be easily combined with concrete superstructure<br>*Disadvantages*<br>a. Difficult to achieve proper cutoff<br>b. Difficult to transport |
| Cased cast-in-place concrete | 5–15 | 15–40 | 200–500 | 800 | *Advantages*<br>a. Relatively cheap<br>b. Possibility of inspection before pouring concrete<br>c. Easy to extend<br>*Disadvantages*<br>a. Difficult to splice after concreting<br>b. Thin casings may be damaged during driving |

342

**Table 8.5** (Continued)

| Pile type | Usual length of piles (m) | Maximum length of pile (m) | Usual load (kN) | Approximate maximum load (kN) | Comments |
|---|---|---|---|---|---|
| Uncased cast-in-place concrete | 5–15 | 30–40 | 300–500 | 700 | *Advantages*<br>a. Initially economical<br>b. Can be finished at any elevation<br>*Disadvantages*<br>a. Voids may be created if concrete is placed rapidly<br>b. Difficult to splice after concreting<br>c. In soft soils, the sides of the hole may cave in, thus squeezing the concrete |
| Wood | 10–15 | 30 | 100–200 | 270 | *Advantages*<br>a. Economical<br>b. Easy to handle<br>c. Permanently submerged piles are fairly resistant to decay<br>*Disadvantages*<br>a. Decay above water table<br>b. Can be damaged in hard driving<br>c. Low load-bearing capacity<br>d. Low resistance to tensile load when spliced |

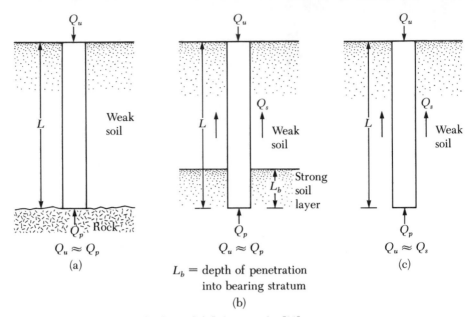

**Figure 8.6**  Point bearing piles [(a) and (b)]; friction piles [(c)]

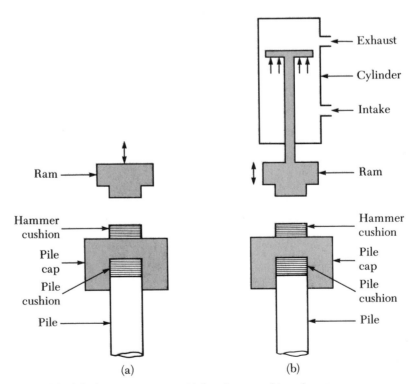

**Figure 8.7**  Pile-driving equipment: (a) drop hammer; (b) single-acting air or steam hammer; (c) double-acting and differential air or steam hammer; (d) diesel hammer; (e) vibratory pile driver

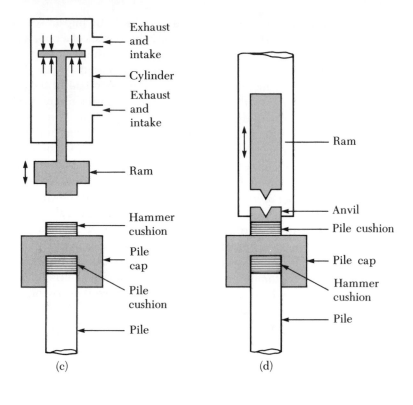

(c)

(d)

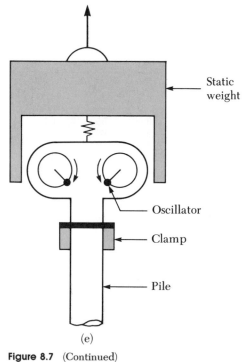

(e)

**Figure 8.7** (Continued)

to raise the ram and to push it downward. This increases the impact velocity of the ram. The diesel hammer (Figure 8.7d) essentially consists of a ram, an anvil block, and a fuel-injection system. During the operation, the ram is first raised and fuel is injected near the anvil. Then the ram is released. When the ram drops, it compresses the air-fuel mixture. This compression ignites the air-fuel mixture. This, in effect, pushes the pile downward and raises the ram. Diesel hammers work well under hard driving conditions. In soft soils, the downward movement of the pile is rather large, and the upward movement of the ram is small. This may not be sufficient to ignite the air-fuel system, so the ram may have to be lifted manually. Tables 8.6 and 8.7 give partial lists of the commercially available diesel, single-acting, double-acting, and differential hammers.

The principles of operation of a vibratory pile driver are shown in Figure 8.7e. This driver essentially consists of two counter-rotating weights. The horizontal components of the centrifugal force generated as a result of rotating masses cancel each other. As a result, this produces a sinusoidal dynamic vertical force on the pile and helps the pile to be driven downward.

*Jetting* is a technique sometimes used in pile driving, when the pile needs to penetrate a thin layer of hard soil (such as sand and gravel) overlying a softer soil layer. In this technique, water is discharged at the pile point by means of a pipe 50–75 mm in diameter to wash and loosen the sand and gravel.

Piles are sometimes driven at an angle to the horizontal. These are referred to as *batter piles*. Batter piles are used in group piles when higher lateral load-bearing capacity is required. Piles can also be advanced by partial au-

**Table 8.6**   Partial List of Typical Air and Steam Hammers

| Maker of hammer* | Model No. | Type of hammer | Rated energy | | Blows per minute | Ram weight | |
|---|---|---|---|---|---|---|---|
| | | | kN-m | kips-ft | | kN | kips |
| V | 3100 | Single acting | 406.8 | 300 | 58 | 448.8 | 100 |
| V | 540 | Single acting | 271.2 | 200 | 48 | 181.9 | 40.9 |
| V | 060 | Single acting | 244.1 | 180 | 62 | 266.9 | 60 |
| MKT | OS-60 | Single acting | 244.1 | 180 | 55 | 266.9 | 60 |
| V | 040 | Single acting | 162.7 | 120 | 60 | 177.9 | 40 |
| V | 400C | Differential | 153.9 | 113.5 | 100 | 177.9 | 40 |
| R | 8/0 | Single acting | 110.2 | 81.25 | 35 | 111.2 | 25 |
| MKT | S-20 | Single acting | 81.4 | 60 | 60 | 89 | 20 |
| R | 5/0 | Single acting | 77.2 | 56.9 | 44 | 77.8 | 17.5 |
| V | 200-C | Differential | 68.1 | 50.2 | 98 | 89 | 20 |
| R | 150-C | Differential | 66.1 | 48.75 | 95–105 | 66.7 | 15 |
| MKT | S-14 | Single acting | 50.9 | 37.5 | 60 | 62.3 | 14 |
| V | 140C | Differential | 48.8 | 36 | 103 | 62.3 | 14 |
| V | 08 | Single acting | 35.3 | 26 | 50 | 35.6 | 8 |
| MKT | S-8 | Single acting | 35.3 | 26 | 55 | 35.6 | 8 |
| MKT | 11B3 | Double acting | 26.1 | 19.2 | 95 | 22.2 | 5 |
| MKT | C-5 | Double acting | 21.7 | 16.0 | 110 | 22.2 | 5 |
| V | 30-C | Double acting | 9.9 | 7.3 | 133 | 13.3 | 3 |

*V—Vulcan Iron Works, Florida
MKT—McKiernan-Terry, New Jersey
R—Raymond International, Inc., Texas

**Table 8.7** Partial List of Typical Diesel Hammers

| Maker of hammer* | Model No. | Rated energy | | Blows per minute | Piston weight | |
|---|---|---|---|---|---|---|
| | | kN-M | kips-ft | | kN | kips |
| K | K150 | 379.7 | 280 | 45–60 | 147.2 | 33.1 |
| M | MB70 | 191.2–86 | 141–63.4 | 38–60 | 70.5 | 15.84 |
| K | K-60 | 143.2 | 105.6 | 42–60 | 58.7 | 13.2 |
| K | K-45 | 123.5 | 91.1 | 39–60 | 44.0 | 9.9 |
| M | M-43 | 113.9–51.3 | 84–37.8 | 40–60 | 42.1 | 9.46 |
| K | K-35 | 96 | 70.8 | 39–60 | 34.3 | 7.7 |
| MKT | DE70B | 85.4–57 | 63–42 | 40–50 | 31.1 | 7.0 |
| K | K-25 | 68.8 | 50.7 | 39–60 | 24.5 | 5.51 |
| V | N-46 | 44.1 | 32.55 | 50–60 | 17.6 | 3.96 |
| L | 520 | 35.7 | 26.3 | 80–84 | 22.6 | 5.07 |
| M | M-14S | 35.3–16.1 | 26–11.88 | 42–60 | 13.2 | 2.97 |
| V | N-33 | 33.4 | 24.6 | 50–60 | 13.3 | 3.0 |
| L | 440 | 24.7 | 18.2 | 86–90 | 17.8 | 4.0 |
| MKT | DE20 | 24.4–16.3 | 18.0–12.0 | 40–50 | 8.9 | 2.0 |
| MKT | DE-10 | 11.9 | 8.8 | 40–50 | 4.9 | 1.1 |
| L | 180 | 11.0 | 8.1 | 90–95 | 7.7 | 1.73 |

*V—Vulcan Iron Works, Florida
M—Mitsubishi International Corporation
MKT—McKiernan-Terry, New Jersey
L—Link, Belt, Cedar Rapids, Iowa
K—Kobe Diesel

gering. This means that power augers (Chapter 2) can be used to predrill holes part of the way. The piles can then be inserted into the holes and driven to the desired depth.

Based on the nature of their placement, piles can be divided into two categories: *displacement piles* and *nondisplacement piles*. Driven piles are displacement piles, because they move some soil laterally, and, hence, there is a tendency for densification of soil surrounding them. Concrete piles and closed-ended pipe piles are high displacement piles. However, steel H-piles displace less soil laterally during driving, and so they are low displacement piles. In contrast, bored piles are nondisplacement type. Their placement causes very little change in the state of stress in the soil.

# 8.5

## Load Transfer Mechanism

The load transfer mechanism from a pile to the soil is highly complicated. In order to understand it, consider a pile, $L$, as shown in Figure 8.8a. Let the load on the pile be gradually increased from zero to $Q_{(z\,=\,0)}$ at the ground surface. Part of this load will be resisted by the side friction developed along the shaft $(Q_1)$ and part by the soil below the tip of the pile $(Q_2)$. Now, how are $Q_1$ and $Q_2$ related to the total load? If measurements are made to obtain the load carried by the pile shaft $[Q_{(z)}]$ at any depth $z$, the nature of variation will be like that shown in Curve 1 of Figure 8.8b. The *frictional resistance per unit area* $[f_{(z)}]$

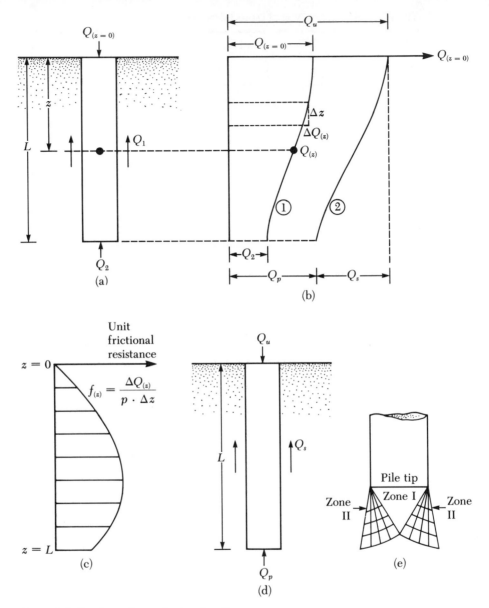

**Figure 8.8**  Load transfer mechanism for piles

at any given depth $z$ can be determined as

$$f_{(z)} = \frac{\Delta Q_{(z)}}{(p)(\Delta z)}$$  (8.7)

where $p$ = perimeter of pile cross section

The nature of variation of $f_{(z)}$ with depth is shown in Figure 8.8c.
If the load $Q$ at the ground surface is gradually increased, maximum fric-

tional resistance along the pile shaft will be fully mobilized when the relative displacement between the soil and the pile is about 5–10 mm irrespective of pile size and length $L$. However, the maximum point resistance $Q_2 = Q_p$ will not be mobilized until the pile tip has gone through a movement of about 10–25% of the pile width (or diameter). The lower limit applies to driven piles, and the upper limit is for bored piles. At ultimate load (Figure 8.8d and Curve 2 in Figure 8.8b), $Q_{(z = 0)} = Q_u$.

$$Q_1 = Q_s$$

and

$$Q_2 = Q_p$$

The preceding explanation indicates that $Q_s$ (or the unit skin friction, $f$, along the pile shaft) is developed at a *much smaller pile displacement compared to the point resistance, $Q_p$*. This can be seen from Vesic's (1970) pile load test results in granular soil, as given in Figure 8.9. Note that these results are for *pipe piles in dense sand*.

At ultimate load, the failure surface in the soil at the pile tip (bearing capacity failure caused by $Q_p$) is of the nature shown in Figure 8.8e. Note that pile foundations are deep foundations, and the soil fails mostly in a *punching mode*, as illustrated in Figure 3.1c and Figure 3.2 (pp. 102 and 103). This means that a *triangular zone*, I, is developed at the pile tip, which is pushed downward without producing any other visible slip surface. In dense sands and stiff clayey soils, a *radical shear zone*, II, may partially develop. Hence, the load displacement curves of piles will resemble those shown in Figure 3.1c.

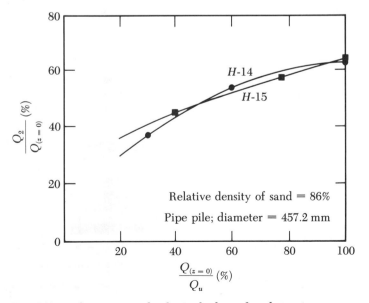

**Figure 8.9** Relative magnitude of point load transferred at various stages of pile loading (redrawn after Vesic, 1970)

## 8.6

### Equations for Estimation of Pile Capacity

The ultimate load-carrying capacity of a pile can be given by a simple equation as the sum of the load carried at the pile point plus the total frictional resistance (skin friction) derived from the soil-pile interface (Figure 8.10a), or

$$Q_u = Q_p + Q_s \tag{8.8}$$

where $Q_u$ = ultimate pile capacity
$Q_p$ = load-carrying capacity of the pile point
$Q_s$ = frictional resistance

Numerous published studies relate to the determination of the values of $Q_p$ and $Q_s$. Excellent reviews of most of the recent investigations have been provided by Vesic (1977), Meyerhof (1976), and Coyle and Castello (1981). These studies provide insight into the problem of determination of ultimate pile capacity.

### Point Bearing Capacity, $Q_p$

The ultimate bearing capacity of shallow foundations has been discussed in Chapter 3. According to Terzaghi's equations

$$q_u = 1.3\, cN_c + qN_q + 0.4\gamma\, BN_\gamma \qquad \text{(for shallow square foundations)}$$

and

$$q_u = 1.3\, cN_c + qN_q + 0.3\gamma\, BN_\gamma \qquad \text{(for shallow circular foundations)}$$

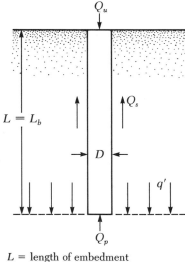

$L$ = length of embedment
$L_b$ = length of embedment in bearing
stratum

(a)

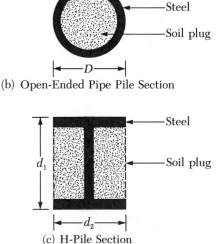

(b) Open-Ended Pipe Pile Section

(c) H-Pile Section

(*Note*: $A_p$ = area of steel + soil plug)

**Figure 8.10**

In a similar manner, the general bearing capacity equation for shallow foundations has been given in Chapter 3 (for vertical loading) as

$$q_u = cN_cF_{cs}F_{cd} + qN_qF_{qs}F_{qd} + \tfrac{1}{2}\,\gamma\,BN_\gamma F_{\gamma s}F_{\gamma d}$$

Hence, in general, the ultimate load bearing capacity can be expressed as

$$q_u = cN_c^* + qN_q^* + \gamma BN_\gamma^* \tag{8.9a}$$

where $N_c^*$, $N_q^*$, and $N_\gamma^*$ are the bearing capacity factors that include the necessary shape and depth factors.

Pile foundations are deep. However, the ultimate resistance per unit area
developed at the pile tip ($q_p$) can be expressed by an equation similar in form
to that shown in Eq. (8.9a), although the values of $N_c^*$, $N_q^*$, and $N_\gamma^*$ will change.
The notation used in this chapter for the width of pile is $D$. Hence, substitution
of $D$ for $B$ in Eq. (8.9a) results in the form

$$q_u = q_p = cN_c^* + qN_q^* + \gamma DN_\gamma^* \tag{8.9b}$$

Because the width $D$ of a pile is relatively small, the term $\gamma DN_\gamma^*$ can be dropped
from the right side of the preceding equation without making a serious error,
or

$$q_p = cN_c^* + q'N_q^* \tag{8.10}$$

Note that the term $q$ has been replaced by $q'$ in Eq. (8.10) to signify effective
vertical stress. Hence, the point bearing of piles can be expressed as

$$Q_p = A_p q_p = A_p(cN_c^* + q'N_q^*) \tag{8.11a}$$

where $A_p$ = area of pile tip
      $c$ = cohesion of the soil supporting the pile tip
     $q_p$ = unit point resistance
     $q'$ = effective vertical stress at the level of the pile tip
 $N_c^*, N_q^*$ = the bearing capacity factors

There are several methods for the determination of the bearing capacity
factors $N_c^*$ and $N_q^*$, including Meyerhof's method and Vesic's method, which we
will now discuss.

*Meyerhof's Method:* The point bearing capacity ($q_p$) of a pile in sand generally increases with the depth of embedment in the bearing stratum to the width
of pile ratio ($L_b/D$) and reaches a maximum value at an embedment ratio of
$L_b/D = (L_b/D)_{cr}$. Note that in a homogeneous soil $L_b$ is equal to the actual
embedment length of the pile, $L$ (see Figure 8.10a). However, in Figure 8.6b,
where a pile has penetrated into a bearing stratum, $L_b < L$. Beyond the critical
embedment ratio ($L_b/D)_{cr}$, the value of $q_p$ remains constant (that is, $q_p = q_l$).
This fact is shown in Figure 8.11 for the case of a homogeneous soil—that is,
$L = L_b$. The variation of $(L_b/D)_{cr}$ with the soil friction angle is shown in Figure
8.12. Based on the given variation of $(L_b/D)_{cr}$, Meyerhof (1976) has recommended the following procedure for estimation of the point bearing capacity of
a pile in granular soil.

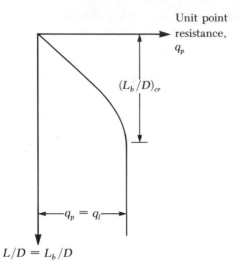

$$L/D = L_b/D$$

**Figure 8.11**   Nature of variation of unit point resistance in a homogeneous sand

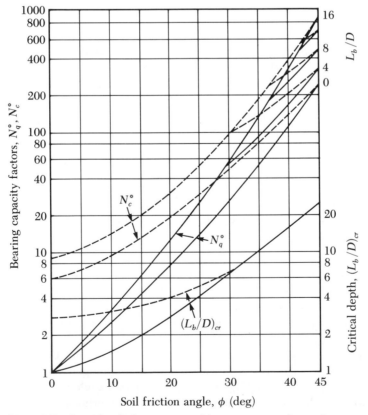

**Figure 8.12**   Critical embedment ratio and bearing capacity factors for various soil friction angles (after Meyerhof, 1976)

**1.** For sand, since $c = 0$, Eq. (8.11a) becomes equal to

$$Q_p = A_p q_p = A_p q' N_q^*$$                                (8.11b)

**2.** Determine the soil friction angle, $\phi$.

**3.** Determine the $L_b/D$ ratio for the pile.

**4.** Obtain $(L_b/D)_{cr}$ from Figure 8.12.

**5.** Determine the appropriate value of $N_q^*$ from Figure 8.12 corresponding to the given $L_b/D$ obtained in Step 3. Note that the value of $N_q^*$ increases linearly with $L_b/D$ and reaches a maximum at $L_b/D \approx (L_b/D)_{cr}/2$.

**6.** Use the $N_q^*$ value calculated in Step 5 to obtain $Q_p$ as

$$Q_p = A_p q' N_q^* \leq A_p q_l$$                                (8.12)

The limiting point resistance can be given as

$$q_l(kN/m^2) = 50 N_q^* \tan \phi$$                                (8.13)

where $\phi$ = soil friction angle in the bearing stratum

Based on field observations, Meyerhof (1976) also suggested that the ultimate point resistance, $q_p$, in a homogeneous granular soil ($L = L_b$) can be obtained from standard penetration numbers as

$$q_p(kN/m^2) = 40 NL/D \leq 400\, N$$                                (8.14)

where $N$ = average standard penetration number near the pile point
(about $10D$ above and $4D$ below the pile point)

In many circumstances, a pile may initially penetrate a weak sand layer and then a dense layer, as shown in Figure 8.13. For these piles

$$q_p = q_{l(l)} + \frac{[q_{l(d)} - q_{l(l)}]L_b}{10D} \leq q_{l(d)}$$                                (8.15)

where $q_{l(l)}$ = limiting unit point resistance in the loose sand determined from Eq. (8.13) using the maximum value of $N_q^*$ and $\phi$ value of the loose sand

$q_{l(d)}$ = limiting unit point resistance in the dense sand determined from Eq. (8.13) using the maximum value of $N_q^*$ and $\phi$ value of the dense sand

$L_b$ = depth of penetration into the dense sand layer

For piles in saturated clays in undrained conditions ($\phi = 0$)

$$Q_p = N_c^* c_u A_p = 9 c_u A_p$$                                (8.16)

where $c_u$ = undrained cohesion of the soil below the pile tip

For clays with $c$ and $\phi$ parameters present (effective stress basis), the ultimate point load can be given by the same relation presented in Eq. (8.11a). In most design problems, the assumed value of $\phi$ is less than about 30°. For $\phi$ less than 30°, use the following procedure for obtaining $N_c^*$ and $N_q^*$ from Figure 8.12.

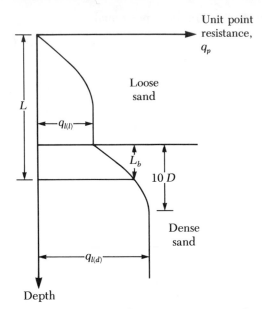

**Figure 8.13**   Variation of unit point resistance in layered soil

1. Obtain $(L_b/D)_{cr}$ for the given value of $\phi$ from Figure 8.12.
2. Calculate $L_b/D$.
3. If $L_b/D \geq (L_b/D)_{cr}/2$, take the maximum values of $N_c^*$ and $N_q^*$ from Figure 8.12.
4. If $L_b/D < (L_b/D)_{cr}/2$, then

$$N_c^* = N_{c\,(\text{at } L_b/D = 0)}^* + [N_{c\,(\text{max})}^* - N_{c\,(\text{at } L_b/D = 0)}^*] \left[ \frac{\left(\dfrac{L_b}{D}\right)}{0.5\left(\dfrac{L_b}{D}\right)_{cr}} \right] \tag{8.17}$$

$$N_q^* = N_{q\,(\text{at } L_b/D = 0)}^* + [N_{q\,(\text{max})}^* - N_{q\,(\text{at } L_b/D = 0)}^*] \left[ \frac{\left(\dfrac{L_b}{D}\right)}{0.5\left(\dfrac{L_b}{D}\right)_{cr}} \right] \tag{8.18}$$

*Vesic's Method*: Vesic (1977) proposed a method for estimating the pile point bearing capacity based on the theory of *expansion of cavities*. According to this theory, based on effective stress parameters,

$$Q_p = A_p q_p = A_p(cN_c^* + \sigma_o' N_\sigma^*) \tag{8.19}$$

where $\sigma_o'$ = mean normal ground stress (effective) at the level of the pile point

$$= \left(\frac{1 + 2K_o}{3}\right) q' \tag{8.20}$$

$$K_o = \text{earth pressure coefficient at rest} = 1 - \sin \phi \qquad (8.21)$$

$N_c^*, N_\sigma^* = $ bearing capacity factors

Note that Eq. (8.19) is a modification of Eq. (8.11a) with

$$N_\sigma^* = \frac{3N_q^*}{(1 + 2K_o)} \qquad (8.22)$$

The relation for $N_c^*$ given in Eq. (8.19) can be expressed as

$$N_c^* = (N_q^* - 1) \cot \phi \qquad (8.23)$$

According to the theory of Vesic

$$N_q^* = f(I_{rr}) \qquad (8.24)$$

where $I_{rr} = $ reduced rigidity index for the soil

However

$$I_{rr} = \frac{I_r}{1 + I_r \Delta} \qquad (8.25)$$

where $I_r = $ rigidity index $= \dfrac{E_s}{2(1 + \mu_s)(c + q' \tan \phi)} = \dfrac{G_s}{c + q' \tan \phi}$ \quad (8.26)

$E_s = $ Young's modulus of soil
$\mu_s = $ Poisson's ratio of soil
$G_s = $ shear modulus of soil
$\Delta = $ average volumatic strain in the plastic zone below the pile point

For conditions of no volume change (that is, dense sand or saturated clay), $\Delta = 0$. So

$$I_r = I_{rr} \qquad (8.27)$$

Table 8.8 gives the values of $N_c^*$ and $N_\sigma^*$ for various values of the soil friction angle ($\phi$) and $I_{rr}$. For $\phi = 0$ (that is, undrained condition)

$$N_c^* = \frac{4}{3}(lnI_{rr} + 1) + \frac{\pi}{2} + 1 \qquad (8.28)$$

The values of $I_r$ can be estimated from laboratory consolidation and triaxial tests corresponding to the proper stress levels. However, for preliminary use the following values are recommended:

| Soil type | $I_r$ |
|---|---|
| Sand | 70–150 |
| Silts and clays (drained condition) | 50–100 |
| Clays (undrained condition) | 100–200 |

Irrespective of the theoretical procedure adopted in calculating $Q_p$, it should be kept in mind that full value cannot be realized until the pile tip has

**Table 8.8**  Bearing Capacity Factors for Deep Foundations, $N_c^*$ and $N_\sigma^*$

| $\phi$ | 10 | 20 | 40 | 60 | 80 | 100 | 200 | 300 | 400 | 500 |
|---|---|---|---|---|---|---|---|---|---|---|
| 0 | 6.97 | 7.90 | 8.82 | 9.36 | 9.75 | 10.04 | 10.97 | 11.51 | 11.89 | 12.19 |
|   | 1.00 | 1.00 | 1.00 | 1.00 | 1.00 | 1.00 | 1.00 | 1.00 | 1.00 | 1.00 |
| 1 | 7.34 | 8.37 | 9.42 | 10.04 | 10.49 | 10.83 | 11.92 | 12.57 | 13.03 | 13.39 |
|   | 1.13 | 1.15 | 1.16 | 1.18 | 1.18 | 1.19 | 1.21 | 1.22 | 1.23 | 1.23 |
| 2 | 7.72 | 8.87 | 10.06 | 10.77 | 11.28 | 11.69 | 12.96 | 13.73 | 14.28 | 14.71 |
|   | 1.27 | 1.31 | 1.35 | 1.38 | 1.39 | 1.41 | 1.45 | 1.48 | 1.50 | 1.51 |
| 3 | 8.12 | 9.40 | 10.74 | 11.55 | 12.14 | 12.61 | 14.10 | 15.00 | 15.66 | 16.18 |
|   | 1.43 | 1.49 | 1.56 | 1.61 | 1.64 | 1.66 | 1.74 | 1.79 | 1.82 | 1.85 |
| 4 | 8.54 | 9.96 | 11.47 | 12.40 | 13.07 | 13.61 | 15.34 | 16.40 | 17.18 | 17.80 |
|   | 1.60 | 1.70 | 1.80 | 1.87 | 1.91 | 1.95 | 2.07 | 2.15 | 2.20 | 2.24 |
| 5 | 8.99 | 10.56 | 12.25 | 13.30 | 14.07 | 14.69 | 16.69 | 17.94 | 18.86 | 19.59 |
|   | 1.79 | 1.92 | 2.07 | 2.16 | 2.23 | 2.28 | 2.46 | 2.57 | 2.65 | 2.71 |
| 6 | 9.45 | 11.19 | 13.08 | 14.26 | 15.14 | 15.85 | 18.17 | 19.62 | 20.70 | 21.56 |
|   | 1.99 | 2.18 | 2.37 | 2.50 | 2.59 | 2.67 | 2.91 | 3.06 | 3.18 | 3.27 |
| 7 | 9.94 | 11.85 | 13.96 | 15.30 | 16.30 | 17.10 | 19.77 | 12.46 | 22.71 | 23.73 |
|   | 2.22 | 2.46 | 2.71 | 2.88 | 3.00 | 3.10 | 3.43 | 3.63 | 3.79 | 3.91 |
| 8 | 10.45 | 12.55 | 14.90 | 16.41 | 17.54 | 18.45 | 21.51 | 23.46 | 24.93 | 26.11 |
|   | 2.47 | 2.76 | 3.09 | 3.31 | 3.46 | 3.59 | 4.02 | 4.30 | 4.50 | 4.67 |
| 9 | 10.99 | 13.29 | 15.91 | 17.59 | 18.87 | 19.90 | 23.39 | 25.64 | 27.35 | 28.73 |
|   | 2.74 | 3.11 | 3.52 | 3.79 | 3.99 | 4.15 | 4.70 | 5.06 | 5.33 | 5.55 |
| 10 | 11.55 | 14.08 | 16.97 | 18.86 | 20.29 | 21.46 | 25.43 | 28.02 | 29.99 | 31.59 |
|    | 3.04 | 3.48 | 3.99 | 4.32 | 4.58 | 4.78 | 5.48 | 5.94 | 6.29 | 6.57 |
| 11 | 12.14 | 14.90 | 18.10 | 20.20 | 21.81 | 23.13 | 27.64 | 30.61 | 32.87 | 34.73 |
|    | 3.36 | 3.90 | 4.52 | 4.93 | 5.24 | 5.50 | 6.37 | 6.95 | 7.39 | 7.75 |
| 12 | 12.76 | 15.77 | 19.30 | 21.64 | 23.44 | 24.92 | 30.03 | 33.41 | 36.02 | 38.16 |
|    | 3.71 | 4.35 | 5.10 | 5.60 | 5.98 | 6.30 | 7.38 | 8.10 | 8.66 | 9.11 |
| 13 | 13.41 | 16.69 | 20.57 | 23.17 | 25.18 | 26.84 | 32.60 | 36.46 | 39.44 | 41.89 |
|    | 4.09 | 4.85 | 5.75 | 6.35 | 6.81 | 7.20 | 8.53 | 9.42 | 10.10 | 10.67 |
| 14 | 14.08 | 17.65 | 21.92 | 24.80 | 27.04 | 28.89 | 35.38 | 39.75 | 43.15 | 45.96 |
|    | 4.51 | 5.40 | 6.47 | 7.18 | 7.74 | 8.20 | 9.82 | 10.91 | 11.76 | 12.46 |
| 15 | 14.79 | 18.66 | 23.35 | 26.53 | 29.02 | 31.08 | 38.37 | 43.32 | 47.18 | 50.39 |
|    | 4.96 | 6.00 | 7.26 | 8.11 | 8.78 | 9.33 | 11.28 | 12.61 | 13.64 | 14.50 |
| 16 | 15.53 | 19.73 | 24.86 | 28.37 | 31.13 | 33.43 | 41.58 | 47.17 | 51.55 | 55.20 |
|    | 5.45 | 6.66 | 8.13 | 9.14 | 9.93 | 10.58 | 12.92 | 14.53 | 15.78 | 16.83 |
| 17 | 16.30 | 20.85 | 26.46 | 30.33 | 33.37 | 35.92 | 45.04 | 51.32 | 56.27 | 60.42 |
|    | 5.98 | 7.37 | 9.09 | 10.27 | 11.20 | 11.98 | 14.77 | 16.69 | 18.20 | 19.47 |
| 18 | 17.11 | 22.03 | 28.15 | 32.40 | 35.76 | 38.59 | 48.74 | 55.80 | 61.38 | 66.07 |
|    | 6.56 | 8.16 | 10.15 | 11.53 | 12.62 | 13.54 | 16.84 | 19.13 | 20.94 | 22.47 |
| 19 | 17.95 | 23.26 | 29.93 | 34.59 | 38.30 | 41.42 | 52.71 | 60.61 | 66.89 | 72.18 |
|    | 7.18 | 9.01 | 11.31 | 12.91 | 14.19 | 15.26 | 19.15 | 21.87 | 24.03 | 25.85 |
| 20 | 18.83 | 24.56 | 31.81 | 36.92 | 40.99 | 44.43 | 56.97 | 65.79 | 72.82 | 78.78 |
|    | 7.85 | 9.94 | 12.58 | 14.44 | 15.92 | 17.17 | 21.73 | 24.94 | 27.51 | 29.67 |
| 21 | 19.75 | 25.92 | 33.80 | 39.38 | 43.85 | 47.64 | 61.51 | 71.34 | 79.22 | 85.90 |
|    | 8.58 | 10.95 | 13.97 | 16.12 | 17.83 | 19.29 | 24.61 | 28.39 | 31.41 | 33.97 |
| 22 | 20.71 | 27.35 | 35.89 | 41.98 | 46.88 | 51.04 | 66.37 | 77.30 | 86.09 | 93.57 |
|    | 9.37 | 12.05 | 15.50 | 17.96 | 19.94 | 21.62 | 27.82 | 32.23 | 35.78 | 38.81 |
| 23 | 21.71 | 28.84 | 38.09 | 44.73 | 50.08 | 54.66 | 71.56 | 83.68 | 93.47 | 101.83 |
|    | 10.21 | 13.24 | 17.17 | 19.99 | 22.26 | 24.20 | 31.37 | 36.52 | 40.68 | 44.22 |
| 24 | 22.75 | 30.41 | 40.41 | 47.63 | 53.48 | 58.49 | 77.09 | 90.51 | 101.39 | 110.70 |
|    | 11.13 | 14.54 | 18.99 | 22.21 | 24.81 | 27.04 | 35.32 | 41.30 | 46.14 | 50.29 |
| 25 | 23.84 | 32.05 | 42.85 | 50.69 | 57.07 | 62.54 | 82.98 | 97.81 | 109.88 | 120.23 |
|    | 12.12 | 15.95 | 20.98 | 24.64 | 27.61 | 30.16 | 39.70 | 46.61 | 52.24 | 57.06 |

**Table 8.8** (Continued)

| $\phi$ | 10 | 20 | 40 | 60 | 80 | 100 | 200 | 300 | 400 | 500 |
|---|---|---|---|---|---|---|---|---|---|---|
| 26 | 24.98 | 33.77 | 45.42 | 53.93 | 60.87 | 66.84 | 89.25 | 105.61 | 118.96 | 130.44 |
|  | 13.18 | 17.47 | 23.15 | 27.30 | 30.69 | 33.60 | 44.53 | 52.51 | 59.02 | 64.62 |
| 27 | 26.16 | 35.57 | 48.13 | 57.34 | 64.88 | 71.39 | 95.02 | 113.92 | 128.67 | 141.39 |
|  | 14.33 | 19.12 | 25.52 | 30.21 | 34.06 | 37.37 | 49.88 | 59.05 | 66.56 | 73.04 |
| 28 | 27.40 | 37.45 | 50.96 | 60.93 | 69.12 | 76.20 | 103.01 | 122.79 | 139.04 | 153.10 |
|  | 15.57 | 20.91 | 28.10 | 33.40 | 37.75 | 41.51 | 55.77 | 66.29 | 74.93 | 82.40 |
| 29 | 28.69 | 39.42 | 53.95 | 64.71 | 73.58 | 81.28 | 110.54 | 132.23 | 150.11 | 165.61 |
|  | 16.90 | 22.85 | 30.90 | 36.87 | 41.79 | 46.05 | 62.27 | 74.30 | 84.21 | 92.80 |
| 30 | 30.03 | 41.49 | 57.08 | 68.69 | 78.30 | 86.64 | 118.53 | 142.27 | 161.91 | 178.98 |
|  | 18.24 | 24.95 | 33.95 | 40.66 | 46.21 | 51.02 | 69.43 | 83.14 | 94.48 | 104.33 |
| 31 | 31.43 | 43.64 | 60.37 | 72.88 | 83.27 | 92.31 | 126.99 | 152.95 | 174.49 | 193.23 |
|  | 19.88 | 27.22 | 37.27 | 44.79 | 51.03 | 56.46 | 77.31 | 92.90 | 105.84 | 117.11 |
| 32 | 32.89 | 45.90 | 63.82 | 77.29 | 88.50 | 98.28 | 135.96 | 164.29 | 187.87 | 208.43 |
|  | 21.55 | 29.68 | 40.88 | 49.30 | 56.30 | 62.41 | 85.96 | 103.66 | 118.39 | 131.24 |
| 33 | 34.41 | 48.26 | 67.44 | 81.92 | 94.01 | 104.58 | 145.46 | 176.33 | 202.09 | 224.62 |
|  | 23.34 | 32.34 | 44.80 | 54.20 | 62.05 | 68.92 | 95.46 | 115.51 | 132.24 | 146.87 |
| 34 | 35.99 | 50.72 | 71.24 | 86.80 | 99.82 | 111.22 | 155.51 | 189.11 | 217.21 | 241.84 |
|  | 25.28 | 35.21 | 49.05 | 59.54 | 68.33 | 76.02 | 105.90 | 128.55 | 147.51 | 164.12 |
| 35 | 37.65 | 53.30 | 75.22 | 91.91 | 105.92 | 118.22 | 166.14 | 202.64 | 233.27 | 260.15 |
|  | 27.36 | 38.32 | 53.67 | 65.36 | 75.17 | 83.78 | 117.33 | 142.89 | 164.33 | 183.16 |
| 36 | 39.37 | 55.99 | 79.39 | 97.29 | 112.34 | 125.59 | 177.38 | 216.98 | 250.30 | 279.60 |
|  | 29.60 | 41.68 | 58.68 | 71.69 | 82.62 | 92.24 | 129.87 | 158.65 | 182.85 | 204.14 |
| 37 | 41.17 | 58.81 | 83.77 | 102.94 | 119.10 | 133.34 | 189.25 | 232.17 | 268.36 | 300.26 |
|  | 32.02 | 45.31 | 64.13 | 78.57 | 90.75 | 101.48 | 143.61 | 175.95 | 203.23 | 227.26 |
| 38 | 43.04 | 61.75 | 88.36 | 108.86 | 126.20 | 141.50 | 201.78 | 248.23 | 287.50 | 322.17 |
|  | 34.63 | 49.24 | 70.03 | 86.05 | 99.60 | 111.56 | 158.65 | 194.94 | 225.62 | 252.71 |
| 39 | 44.99 | 64.83 | 93.17 | 115.09 | 133.66 | 150.09 | 215.01 | 265.23 | 307.78 | 345.41 |
|  | 37.44 | 53.50 | 76.45 | 94.20 | 109.24 | 122.54 | 175.11 | 215.78 | 250.23 | 280.71 |
| 40 | 47.03 | 68.04 | 98.21 | 121.62 | 141.51 | 159.13 | 228.97 | 283.19 | 329.24 | 370.04 |
|  | 40.47 | 58.10 | 83.40 | 103.05 | 119.74 | 134.52 | 193.13 | 238.62 | 277.26 | 311.50 |
| 41 | 49.16 | 71.41 | 103.49 | 128.48 | 149.75 | 168.63 | 243.69 | 302.17 | 351.95 | 396.12 |
|  | 43.74 | 63.07 | 90.96 | 112.68 | 131.18 | 147.59 | 212.84 | 263.67 | 306.94 | 345.34 |
| 42 | 51.38 | 74.92 | 109.02 | 135.68 | 158.41 | 178.62 | 259.22 | 322.22 | 375.97 | 423.74 |
|  | 47.27 | 68.46 | 99.16 | 123.16 | 143.64 | 161.83 | 234.40 | 291.13 | 339.52 | 382.53 |
| 43 | 53.70 | 78.60 | 114.82 | 143.23 | 167.51 | 189.13 | 275.59 | 343.40 | 401.36 | 452.96 |
|  | 51.08 | 74.30 | 108.08 | 134.56 | 157.21 | 177.36 | 257.99 | 321.22 | 375.28 | 423.39 |
| 44 | 56.13 | 82.45 | 120.91 | 151.16 | 177.07 | 200.17 | 292.85 | 365.75 | 428.21 | 483.88 |
|  | 55.20 | 80.62 | 117.76 | 146.97 | 172.00 | 194.31 | 283.80 | 354.20 | 414.51 | 468.28 |
| 45 | 58.66 | 86.48 | 127.28 | 159.48 | 187.12 | 211.79 | 311.04 | 389.35 | 456.57 | 516.58 |
|  | 59.66 | 87.48 | 128.28 | 160.48 | 188.12 | 212.79 | 312.03 | 390.35 | 457.57 | 517.58 |
| 46 | 61.30 | 90.70 | 133.97 | 168.22 | 197.67 | 224.00 | 330.20 | 414.26 | 486.54 | 551.16 |
|  | 64.48 | 94.92 | 139.73 | 175.20 | 205.70 | 232.96 | 342.94 | 429.98 | 504.82 | 571.74 |
| 47 | 64.07 | 95.12 | 140.99 | 177.40 | 208.77 | 236.85 | 350.41 | 440.54 | 518.20 | 587.72 |
|  | 69.71 | 103.00 | 152.19 | 191.24 | 224.88 | 254.99 | 376.77 | 473.42 | 556.70 | 631.25 |
| 48 | 66.97 | 99.75 | 148.35 | 187.04 | 220.43 | 250.36 | 371.70 | 468.28 | 551.64 | 626.36 |
|  | 75.38 | 111.78 | 165.76 | 208.73 | 245.81 | 279.06 | 413.82 | 521.08 | 613.65 | 696.64 |
| 49 | 70.01 | 104.60 | 156.09 | 197.17 | 232.70 | 264.58 | 394.15 | 497.56 | 586.96 | 667.21 |
|  | 81.54 | 121.33 | 180.56 | 227.82 | 268.69 | 305.37 | 454.42 | 573.38 | 676.22 | 768.53 |
| 50 | 73.19 | 109.70 | 164.21 | 207.83 | 245.60 | 279.55 | 417.82 | 528.46 | 624.28 | 710.39 |
|  | 88.23 | 131.73 | 196.70 | 248.68 | 293.70 | 334.15 | 498.94 | 630.80 | 744.99 | 847.61 |

From "Design of Pile Foundations," by A. S. Vesic, in NCHRB *Synthesis of Highway Practice 42,* Transportation Research Board, 1977. Reprinted by permission.
*Note:* Upper number $N_c^*$, lower number $N_\sigma^*$

gone through a settlement of 10–25% of the width of the pile. This is critical in the case of sand.

### Frictional Resistance, $Q_s$

The frictional or skin resistance of a pile can be written as

$$Q_s = \Sigma \, p \, \Delta L f \tag{8.29}$$

where $p$ = perimeter of the pile section
   $\Delta L$ = pile length (Figure 8.14a)
   $f$ = unit friction resistance at any given depth $z$

*Frictional Resistance in Sand:* The unit frictional resistance at a given depth for a pile can be expressed as

$$f = K\sigma_v' \, \tan \delta \tag{8.30}$$

where $K$ = earth pressure coefficient
   $\sigma_v'$ = effective vertical stress at the depth under consideration
   $\delta$ = soil-pile friction angle

In reality, the value of $K$ varies with depth. It is approximately equal to the Rankine passive earth pressure coefficient $(K_p)$ at the top of the pile and may be less than the at-rest earth pressure coefficient $(K_o)$ at the pile tip. It will also depend on the nature of pile installation. Based on presently available results, the following average values of $K$ are recommended for use in Eq. (8.30).

For bored or jetter piles

$K = K_o = 1 - \sin \phi$

For low-displacement driven piles

$K = K_o$—lower limit
$K = 1.4K_o$—upper limit

For high-displacement driven piles

$K = K_o$—lower limit
$K = 1.8K_o$—upper limit

As we have seen, the effective vertical stress, $\sigma_v'$, for use in Eq. (8.30) increases with pile depth up to a maximum limit at a depth of 15–20 pile diameters and remains constant thereafter. This is shown in Figure 8.14b. This critical depth, $L'$, depends on several factors, such as soil friction angle and compressibility and relative density. A conservative estimate would be to assume

$$L' = 15D \tag{8.31}$$

The values of $\delta$ from various investigations appear to be in the range of $0.5\phi$ to $0.8\phi$. Proper judgment should be used in choosing the value of $\delta$.

Meyerhof (1976) has also indicated that the average unit frictional resistance $(f_{av})$ for driven high-displacement piles can be obtained from average

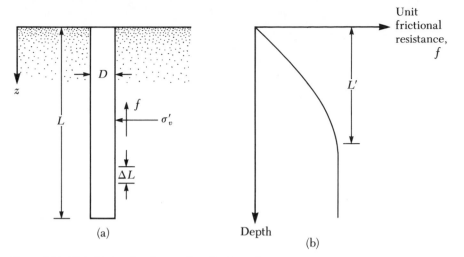

**Figure 8.14**  Unit frictional resistance for piles in sand

standard penetration resistance values as

$$f_{av} \, (kN/m^2) = 2\overline{N} \tag{8.32}$$

where $\overline{N}$ = average value of standard penetration resistance

For driven low-displacement piles

$$f_{av} \, (kN/m^2) = \overline{N} \tag{8.33}$$

Thus

$$Q_s = pLf_{av} \tag{8.34}$$

*Frictional (or Skin) Resistance in Clay:* There are several methods presently available to obtain unit frictional (or skin) resistance of piles in clay. Some of the presently accepted procedures are briefly described below.

$\lambda$ *Method:* This method was proposed by Vijayvergiya and Focht (1972). This method assumes that the displacement of soil caused by pile driving results in a passive lateral pressure at any depth, and the average unit skin resistance can be given as

$$f_{av} = \lambda(\overline{\sigma}_v' + 2c_u) \tag{8.35}$$

where $\overline{\sigma}_v'$ = mean effective vertical stress for the entire embedment length
$c_u$ = mean undrained shear strength ($\phi = 0$ concept)

The value of $\lambda$ will change with the depth of pile penetration (see Figure 8.15). Thus, the total frictional resistance can now be calculated as

$$Q_s = pLf_{av}$$

Care should be taken in obtaining the values of $\overline{\sigma}_v'$ and $c_u$ in layered soil. This can be explained with the aid of Figure 8.16. According to Figure 8.16b,

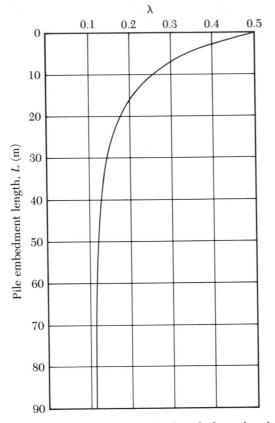

**Figure 8.15** Variation of $\lambda$ with pile embedment length (redrawn after McClelland, 1974)

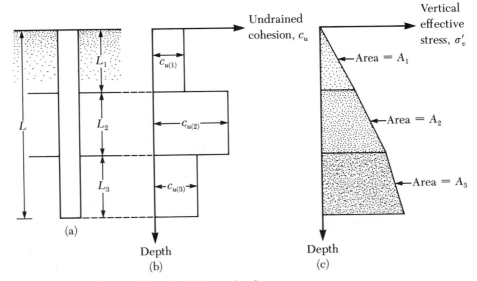

**Figure 8.16** Application of $\lambda$ method in layered soil

the mean value of $c_u$ is $(c_{u(1)}L_1 + c_{u(2)}L_2 + \ldots)/L$. In a similar manner, Figure 8.16c shows the plot of the variation of effective stress with depth. The mean effective stress

$$\bar{\sigma}'_v = \frac{A_1 + A_2 + A_3 + \ldots}{L} \tag{8.36}$$

where $A_1$, $A_2$, $A_3$, $\ldots$ = areas of the vertical effective stress diagrams

*α Method:* According to the α method, the unit skin resistance in clayey soils can be represented by the equations

$$f = \alpha c_u \tag{8.37a}$$

where $\alpha$ = empirical adhesion factor

The approximate variation of the value of α is shown in Figure 8.17. Note that for normally consolidated clays with $c_u \le$ about 50 kN/m², the value of α is equal to one. Thus

$$Q_s = \Sigma fp \, \Delta L = \Sigma \, \alpha c_u p \, \Delta L \tag{8.37b}$$

*β Method:* When piles are driven into saturated clays, the pore water pressure in the soil around the piles increases. This excess pore water pressure in normally consolidated clays may be of the magnitude of 4–6 times $c_u$. However, within a month or so, this pressure gradually dissipates. Hence, the unit frictional resistance for the pile can be determined on the basis of the effective

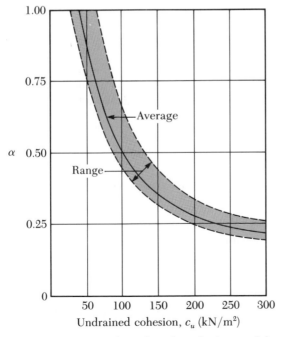

**Figure 8.17** Variation of α with undrained cohesion of clay

stress parameters of the clay in a remolded state (that is, $c = 0$). Thus, at any given depth

$$f = \beta \sigma_v'$$ (8.38)

where $\sigma_v' =$ vertical effective stress at any depth

$$\beta = K \tan \phi_R$$ (8.39)

$\phi_R =$ drained friction angle of remolded clay
$K =$ earth pressure coefficient

The value of $K$ can be conservatively taken as the earth pressure coefficient at rest, or

$$K = 1 - \sin \phi_R \text{ (for normally consolidated clays)}$$ (8.40a)

and

$$K = (1 - \sin \phi_R) \sqrt{OCR} \text{ (for overconsolidated clays)}$$ (8.40b)

where $OCR =$ overconsolidation ratio

Combining Eqs. (8.38), (8.39), and (8.40), for normally consolidated clays

$$f = (1 - \sin \phi_R) \tan \phi_R \sigma_v'$$ (8.41a)

and, for overconsolidated clays

$$f = (1 - \sin \phi_R) \tan \phi_R \sqrt{OCR} \; \sigma_v'$$ (8.41b)

Once the value of $f$ is determined, the total frictional resistance can be evaluated as

$$Q_s = \Sigma \, fp \, \Delta L$$

## Allowable Pile Capacity

Once the total ultimate load-carrying capacity of a pile is determined by summing the point bearing capacity and the frictional (or skin) resistance, a reasonable factor of safety should be used to obtain the total allowable load for each pile, or

$$Q_{all} = \frac{Q_u}{FS}$$ (8.42)

where $Q_{all} =$ allowable load-carrying capacity for each pile
$FS =$ factor of safety

The factor of safety generally used ranges from 2.5–4, depending on the uncertainties of ultimate load calculation.

## General Comments

Although calculations for the ultimate load-carrying capacity of piles are made by using Eqs. (8.9a)–(8.41b), one needs to keep the following points in mind:

**1.** For a given initial value of the soil friction angle ($\phi$), driven piles in sand may show about 50–100% higher unit point resistance compared to bored piles. This is a result of the densification of soil during pile driving.

**2.** In sandy soils, cast-in-place piles with pedestals may show about 50–100% higher unit point resistance compared to cast-in-place piles without pedestals. The high-impact energy of the hammer building the pedestal causes substantial soil compaction and thus an increase of the soil friction angle.

**3.** In the calculation of the area of cross section ($A_p$) and the perimeter ($p$) of piles with developed profiles, such as H-piles and open-ended pipe piles, the effect of soil plug should be considered. According to Figure 8.10b and c, for pipe piles

$$A_p = \left(\frac{\pi}{4}\right)D^2$$

and

$$p = 2\pi D$$

Similarly, for H-piles

$$A_p = d_1 \cdot d_2$$
$$p = 2(d_1 + d_2)$$

Also, note that for H-piles, because $d_2 > d_1$, $D = d_1$.

**4.** The ultimate point load relations given in Eqs. (8.11) and (8.19) are for the gross ultimate point load; that is, they include the weight of the pile. So, the net ultimate point load can approximately be given as

$$Q_{p(net)} = Q_{p(gross)} - q'$$

In practice, for soils with $\phi > 0$, this is not done, and $Q_{p(net)}$ is assumed to be equal to $Q_{p(gross)}$.

In cohesive soils with $\phi = 0$, the value of $N_q^*$ is equal to one (Figure 8.12). Hence, from Eq. (8.11a)

$$Q_{p(gross)} = c_u N_c^* + q'$$

So

$$Q_{p(net)} = (c_u N_c^* + q') - q' = c_u N_c^* = 9c_u = Q_p$$

This is the relation given in Eq. (8.16).

## 8.7
## Coyle and Castello Design Correlations

Coyle and Castello (1981) have analyzed several large-scale field load tests of driven piles in sand. For sands, the ultimate load can be given by the equation

$$Q_u = Q_p + Q_s = q'N_q^*A_p + f_{av}pL \tag{8.43}$$

where $q'$ = effective vertical stress at the pile tip

$f_{av}$ = average frictional resistance for the entire pile length and can be given by the relation

$$f_{av} = K\overline{\sigma}_v' \tan \delta \qquad (8.44)$$

where $K$ = lateral earth pressure coefficient

$\overline{\sigma}_v'$ = average effective overburden pressure

$\delta$ = soil-pile friction angle

Based on this study, the calculated values of the bearing capacity factor $(N_q^*)$ have been correlated with the embedment ratio $L/D$. Figure 8.18 shows the values of $N_q^*$ for various embedment ratios and soil friction angles. Note that $N_q^*$ gradually increases with $L/D$ to a maximum value and decreases thereafter.

In a similar manner, the values of deduced $K$ for various values of $\phi$ and $L/D$ ratios are given in Figure 8.19. It can be seen that, for any given soil friction angle, $K$ decreases in a linear manner with embedment ratio. Note that, in Figure 8.19, it is assumed that

$$\delta = 0.8\phi \qquad (8.45)$$

Hence, combining Eqs. (8.43), (8.44), and (8.45)

$$Q_u = q'N_q^*A_p + pLK\overline{\sigma}_v' \tan (0.8\phi) \qquad (8.46)$$

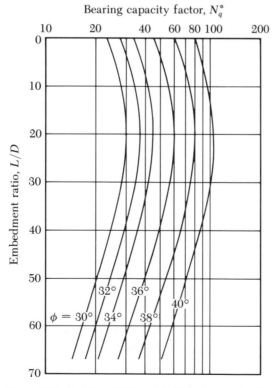

**Figure 8.18** Variation of $N_q^*$ with $L/D$ (redrawn after Coyle and Castello, 1981)

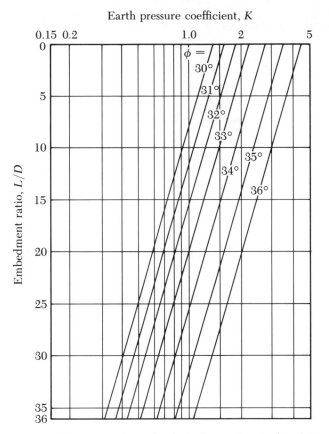

**Figure 8.19** Variation of $K$ with $L/D$ (redrawn after Coyle and Castello, 1981)

Using the results of 24 pile load tests, Coyle and Castello have shown that Eq. (8.46) can predict the ultimate load with an error band of ±30%, with a majority falling within an error band of ±20%.

## Example 8.1

A fully embedded precast, prestressed concrete pile is 12 m long and is driven into a homogeneous sand layer ($c = 0$). The pile is square in cross section with sides measuring 305 mm. The dry unit weight of sand ($\gamma_d$) is 16.8 kN/m³, and the average soil friction angle is 35°. The average standard penetration resistance near the vicinity of the pile tip is 16. Calculate the ultimate point load on the pile by the following methods:

   a. Coyle and Castello's method [Eq. (8.43) and Figure 8.18];
   b. Meyerhof's method [Eq. (8.11b) and Figure 8.12];
   c. Vesic's method—use $I_r = 90 = I_{rr}$ [Eq. (8.19)];
   d. Eq. (8.14).
   e. Compare the results of parts a through d and estimate a design value.

### Solution

**Part a: Coyle and Castello's Method**

$$Q_p = q'N_q^* A_p$$

From Table 8.3, $A_p = 929 \text{ cm}^2 = 0.0929 \text{ m}^2$

$$q' = \gamma_d L = (16.8)(12) = 201.6 \text{ kN/m}^2$$

Now, $L/D = 12/0.305 = 39.34$. For $L/D = 39.34$ and $\phi = 35°$, Figure 8.18 gives $N_q^* \approx 45$. So

$$Q_p = (201.6)(45)(0.0929) = \underline{842.8 \text{ kN}}$$

### Part b: Meyerhof's Method

Because it is a homogeneous soil, $L_b = L$. For $\phi = 35°$, $(L_b/D)_{cr} = (L/D)_{cr} \approx 10$ (from Figure 8.12). So, for this pile, $L_b/D = 39.34 > (L_b/D)_{cr}$. Hence, from Figure 8.12, $N_q^* \approx 120$.

$$Q_p = A_p q' N_q^* = (0.0929)(201.6)(120) = 2247.4 \text{ kN}$$

However, from Eq. (8.13)

$$q_l = 50 N_q^* \tan \phi = 50(120) \tan 35° = 4201.25 \text{ kN/m}^2$$

So

$$Q_p = A_p q_l = (0.0929)(4201) = 390.3 \text{ kN} < A_p q' N_q^*$$

So

$$Q_p \approx \underline{390 \text{ kN}}$$

### Part c: Vesic's Method

Given $I_{rr} \approx 90$. With $\phi = 35°$, Table 8.8 gives $N_\sigma^* \approx 79.5$. From Eq. (8.19)

$$Q_p = A_p \sigma_o' N_\sigma^*$$

$$\sigma_o' = \frac{1 + 2K_o}{3} q'$$

$$K_o = 1 - \sin \phi = 1 - \sin 35° = 0.43$$

So

$$\sigma_o' = \left[\frac{1 + 2(0.43)}{3}\right](201.6) \approx 125 \text{ kN/m}^2$$

So

$$Q_p = (0.0929)(125)(79.5) \approx \underline{923 \text{ kN}}$$

### Part d: Eq. (8.14)

Given: the average standard penetration resistance near the pile tip = 16. So, from Eq. (8.14)

$$q_p = 40 N \frac{L}{D} \leq 400 N$$

$$Q_p = A_p q_p = (0.0929)(40)(16)39.34 = 2339 \text{ kN}$$

However, the limiting value is

$$Q_p = A_p 400N = (0.0929)(400)(16) = 594.6 \text{ kN} \approx 595 \text{ kN}$$

### Part e: Estimation of Design Value

For this problem, Vesic's equation gives a much higher value (923 kN). The next highest value is obtained from the equation given by Coyle and Castello (842.8 kN). For a conservative estimation, one may take

$$Q_p = \frac{390.3 + 595}{2} \approx 493 \text{ kN}$$

## Example 8.2

Refer to the pile given in Example Problem 8.1. (a) Determine the total frictional resistance using $K = 1.4$ and $\delta = 0.6\phi$ [use Eqs. (8.29), (8.30), and (8.31)]. (b) Determine the total frictional resistance by using Coyle and Castello's method.

### Solution
### Part a:

The unit skin friction at any depth is given by Eq. (8.30) as

$$f = K\sigma_v' \tan \delta$$

Also, from Eq. (8.31)

$$L' = 15D$$

So, for depth $z = 0-15D$, $\sigma_v' = \gamma z = 16.8 \times z$ (kN/m²) and beyond that—that is, $z \geq 15D$, $\sigma_v' = \gamma(15D) = (16.8)(15 \times 0.305) = 76.86$ kN/m². This is shown in Figure 8.20.

*Frictional resistance from $z = 0 - 15D$:*

$$Q_s = pL'f_{av} = [(4)(0.305)][15D]\left[\frac{(1.4)(76.86)\tan(0.6 \times 35)}{2}\right]$$

$$= (1.22)(4.575)(20.65) = 115.26 \text{ kN}$$

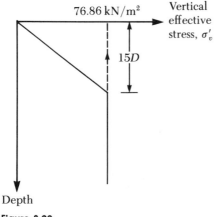

76.86 kN/m²

Vertical effective stress, $\sigma_v'$

15D

Depth

**Figure 8.20**

*Frictional resistance from* $z = 15D$ *to* 12 m

$$Q_s = p(L - L')f_{av} = [(4)(0.305)][12 - 4.575][(1.4)(76.86)\tan(0.6 \times 35)]$$
$$= (1.22)(7.425)(41.3) = 374.1 \text{ kN}$$

So, the total frictional resistance is equal to $115.26 + 374.1 = 489.35 \text{ kN} \approx 490 \text{ kN}$.

### Part b: Coyle and Castello's Method

From Eqs. (8.44) and (8.45)

$$f_{av} = K\bar{\sigma}'_v \tan \delta$$
$$\delta = 0.8\phi$$
$$Q_s = f_{av}pL = [K\bar{\sigma}'_v \tan(0.8\phi)]pL$$

For this pile, $L/D = 39.34$. According to Figure 8.19 for determination of $K$, this is out of the range of the graph. By interpolation, for $L/D = 39.34$ and $\phi = 35°$, $K \approx 0.7$. Now

$$\bar{\sigma}'_v = \frac{\gamma L}{2} = \frac{(16.8)(12)}{2} = 100.8 \text{ kN/m}^2$$

So

$$Q_s = [(0.7)(100.8)\tan(0.8 \times 35)][4 \times 0.305][12] = 549.3 \text{ kN} \approx 550 \text{ kN}$$

## Example 8.3

Refer to Example Problems 8.1 and 8.2. Using a factor of safety of 3, estimate the allowable load for the piles.

### Solution

$$Q_u = Q_p + Q_s$$

From Example 8.1, $Q_p = 490$ kN. Also, from Example 8.2, $Q_s = 490$ kN to 550 kN. So, use $Q_s = (490 + 550)/2 = 1040/2 = 520$ kN. So

$$Q_{all} = \frac{Q_s}{FS} = \frac{490 + 520}{3} = 336.7 \approx 337 \text{ kN}$$

Referring to Table 8.3, the design bearing capacity of this pile is 801 kN, which is greater than 337 kN. So, $Q_{all} = 337$ kN.

## Example 8.4

An HP 310 × 1.079 steel pile is driven into sand, as shown in Figure 8.21a.

a. Calculate the ultimate point load by (1) Meyerhof's procedure, (2) Vesic's procedure ($I_r = 150 = I_{rr}$), (3) using standard penetration resistance equations. (Given: average value of $N$ in the vicinity of pile point is 45.)
b. Estimate the value of the ultimate point load from the calculations in part a.
c. Calculate the ultimate frictional resistance, $Q_s$. Use Eqs. (8.29) to (8.31), $K = 1.4$ and $\delta = 0.6\phi$.
d. Calculate the allowable pile load. Use $FS = 4$.

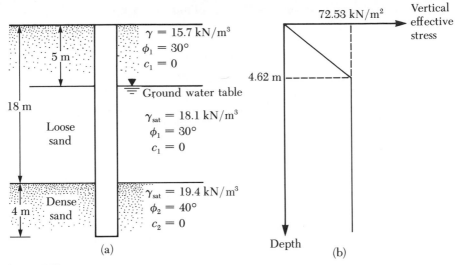

**Figure 8.21**

Also check the allowable load-bearing capacity of the steel section of the pile. Use $\sigma_{\text{all}}$ for steel to be 62,000 $kN/m^2$.

### Solution

In Table 8.1, depth of the pile section, $d_1 = 308$ mm and flange width $= 310$ mm. Area of the pile section, $A_p$, for capacity calculation $= 0.308 \times 0.310 = 0.0955$ $m^2$.

### Part a: Calculation of Ultimate Point Load

*Meyerhof's procedure*: The variation of the unit point resistance will be like that in Figure 8.13. The depth of penetration of the lower dense sand layer, $L_b$, is 4 m. So $L_b/D = 4/0.308 = 12.99 > 10$. Hence, referring to Eq. (8.15)

$$q_p = q_{l(d)} = 50 N_q^* \tan \phi_2$$

For $\phi_2 = 40°$, $N_q^* \approx 350$ (Figure 8.12). Thus

$$q_p = (50)(350)(\tan 40°) \approx 14,684 \text{ } kN/m^2$$

So

$$Q_p = (14684)(0.0955) = \underline{1402 \text{ kN}}$$

Also, check Eq. (8.11). From Eq. (8.11), with $c = 0$, $Q_p = A_p q' N_q^*$.

$$q' = 5(15.7) + 13(18.1 - 9.81) + 4(19.4 - 9.81)$$

$$= 78.5 + 107.77 + 38.36 = 224.63 \text{ } kN/m^2$$

So

$$Q_p = (0.0955)(224.63)(350) = 7508 \text{ kN}$$

Because $Q_p = 1402$ kN $< 7508$ kN, Eq. (8.15) controls. Thus, $Q_p = \underline{1402 \text{ kN}}$.

*Vesic's procedure*: Given: $I_r = 150 = I_{rr}$. From Eq. (8.19)

$$Q_p = A_p \sigma'_o N^*_\sigma$$

$$K_o = 1 - \sin \phi = 1 - \sin 40° = 0.357$$

$$\sigma'_o = \frac{1 + 2K_o}{3} q' = \left[\frac{1 + (2)(0.357)}{3}\right] (224.63) = 128.34 \text{ kN/m}^2$$

From Table 8.8, for $\phi = 40°$ and $I_{rr} = 150$, the value of $N^*_\sigma \approx (134.52 + 193.13)/2 = 163.8$. So

$$Q_p = (0.0955)(128.34)(163.8) = \underline{2008 \text{ kN}}$$

*Standard Penetration Resistance Equations*: Given: average penetration per resistance, N, near the pile point is about 45. From Eq. (8.14)

$$q_p = 40N\frac{L}{D} \leq 400N$$

So

$$q_p = 40(45)\left(\frac{22}{0.308}\right) = 128571 \text{ kN/m}^2$$

or

$$q_p = (400)(N) = (400)(45) = 18000 \text{ kN/m}^2$$

Hence, $q_p = 18,000 \text{ kN/m}^2$ controls. So

$$Q_p = A_p q_p = (0.0955)(18000) = \underline{1719 \text{ kN}}$$

## Part b: Estimation of Value for $Q_p$

Considering all three results, we might use

$$Q_p = \frac{1402 + 2008 + 1719}{3} \approx \underline{1709 \text{ kN}}$$

## Part c: Calculation of Ultimate Frictional Resistance

According to Eq. (8.31)

$$L' = 15D = 15(0.308) = 4.62 \text{ m}$$

For ultimate frictional resistance, $\sigma'_v$ will remain constant for $z > 4.62$ m. The assumed variation of $\sigma'_v$ with depth is shown in Figure 8.21b.

Frictional resistance from $z = 0$ to $4.62$ m $=$

$$pLf_{av} = 2(0.308 + 0.310)(4.62)\left(\frac{K\sigma'_v \tan \delta}{2}\right)$$

$$= 5.71\left[\frac{(1.4)(72.53) \tan (0.6 \times 30)}{2}\right] = \underline{94.2 \text{ kN}}$$

Frictional resistance from $z = 4.62$ m to $22$ m $=$

$$pLf_{av} = 2(0.308 + 0.310)(22 - 4.62)(K\sigma'_v \tan \delta)$$
$$= 21.48[(1.4)(72.53) \tan \delta]$$

As an approximation, the value of $\delta$ may be taken as $0.6\phi_1 = (0.6)(30) = 18°$ for the entire length. Thus

$$Q_{s(z\,=\,4.62-22\text{ m})} = (21.48)(1.4)(72.53) \tan 18° = 708.7 \text{ kN}$$

So, the total frictional resistance =

$$Q_s = Q_{s(z\,=\,0-4.62\text{ m})} + Q_{s(z\,=\,4.62-22\text{ m})}$$

$$= 94.2 + 708.7 = 802.9 \text{ kN} \approx 803 \text{ kN}$$

## Part d: Calculation of Allowable Load

$Q_u = Q_p + Q_s$. From Part b, $Q_p = 1709$ kN. From Part c, $Q_s = 802.9$ kN. So $Q_u \approx 1709 + 803 = 2512$ kN.

$$Q_{\text{all}} = \frac{Q_u}{FS} = \frac{2512}{4} = 628 \text{ kN}$$

We also need to check the allowable load-bearing capacity for the steel pile section. Table 8.1 shows that the area of steel section for the pile is $14.1 \times 10^{-3}$ m².

$$Q_{\text{all}} = (\sigma_{\text{all}})14.1 \times 10^{-3}$$

$$\sigma_{\text{all}} = 62,000 \text{ kN/m}^2$$

So

$$Q_{\text{all}} = (62,000)(14.1 \times 10^{-3}) = 874.2 \text{ kN}$$

Hence, the allowable pile load is 628 kN ($<$874.2 kN).

## Example 8.5

A driven pipe pile in clay is shown in Figure 8.22a. The pipe has an outside diameter of 406 mm, and wall thickness is 6.35 mm.

a. Calculate the net point bearing capacity. Use. Eq. (8.16).
b. Calculate the skin resistance (1) by using Eq. (8.37) ($\alpha$ method), (2) by using Eq. (8.35) ($\lambda$ method), and (3) by using Eq. (8.38) ($\beta$ method). Given: $\phi_R = 30°$ for all clay layers. The top 10 m of clay is normally consolidated. The bottom clay layer has an OCR of 2.
c. Estimate the net allowable pile capacity. Use $FS = 4$.

## Solution

Area of cross section of pile including the soil inside the pile =

$$A_p = \frac{\pi}{4}D^2 = \frac{\pi}{4}(0.406)^2 = 0.1295 \text{ m}^2$$

## Part a: Calculation of Net Point Bearing Capacity

From Eq. (8.16)

$$Q_p = A_p q_p = A_p N_c^* c_{u(2)} = (0.1295)(9)(100) = 116.55 \text{ kN}$$

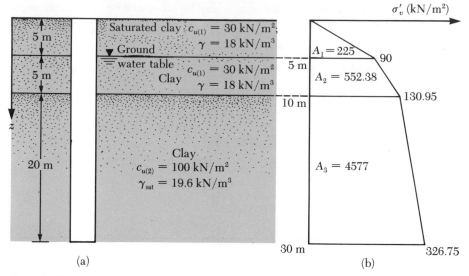

(a)                                                    (b)

**Figure 8.22**

## Part b: Calculation of Skin Resistance

(1) Use of Eq. (8.37): From Eq. (8.37b)

$$Q_s = \Sigma \, \alpha c_u p \, \Delta L$$

For the top soil layer, $c_{u(1)} = 30 \text{ kN/m}^2$. According to the average plot of Figure 8.17, $\alpha_1 = 1.0$. Similarly, for the bottom soil layer, $c_{u(2)} = 100 \text{ kN/m}^2$; $\alpha_2 = 0.5$. Thus

$$\begin{aligned}
Q_s &= \alpha_1 c_{u(1)}[(\pi)(0.406)] \, 10 + \alpha_2 c_{u(2)}[(\pi)(0.406)] \, 20 \\
&= (1)(30) \, [(\pi)(0.406)] \, 10 + (0.5)(100) \, [(\pi)(0.406)] \, 20 \\
&= 382.7 + 1275.5 = \underline{1658.2 \text{ kN}}
\end{aligned}$$

(2) Use of Eq. (8.35): $f_{av} = \lambda(\overline{\sigma}'_v + 2c_u)$

The average value of $c_u$ is equal to

$$\frac{c_{u(1)}(10) + c_{u(2)}(20)}{30} = \frac{(30)(10) + (100)(20)}{30} = 76.7 \text{ kN/m}^2$$

To obtain the average value of $\overline{\sigma}'_v$, the diagram for vertical effective stress variation with depth is plotted in Figure 8.22b. From Eq. (8.36)

$$\overline{\sigma}'_v = \frac{A_1 + A_2 + A_3}{L} = \frac{225 + 552.38 + 4577}{30} = 178.48 \text{ kN/m}^2$$

The value of $\lambda$ can be obtained from Figure 8.15 as 0.14. So

$$f_{av} = 0.14[178.48 + (2)(76.7)] = 46.46 \text{ kN/m}^2$$

Hence

$$Q_s = pLf_{av} = \pi(0.406)(30)(46.46) = \underline{1777.8 \text{ kN}}$$

(3) Use of Eq. (8.38): The top clay layer (10 m) is normally consolidated. $\phi_R = 30°$.
For $z = 0–5$ m [Eq. (8.41a)]:

$$f_{av(1)} = (1 - \sin \phi_R) \tan \phi_R \sigma'_{v(av)}$$

$$= (1 - \sin 30°)(\tan 30°)\left(\frac{0 + 90}{2}\right) = 13.0 \text{ kN/m}^2$$

Similarly, for $z = 5–10$ m:

$$f_{av(2)} = (1 - \sin 30°)(\tan 30°)\left(\frac{90 + 130.95}{2}\right) = 31.9 \text{ kN/m}^2$$

For $z = 10–30$ m:

$$f_{av} = (1 - \sin \phi_R) \tan \phi_R \sqrt{OCR} \, \sigma'_{v(av)}$$

Given $OCR = 2$, so

$$f_{av(3)} = (1 - \sin 30°)(\tan 30°) \sqrt{2} \left(\frac{130.95 + 326.75}{2}\right) = 93.43 \text{ kN/m}^2$$

So

$$Q_s = p[f_{av(1)}(5) + f_{av(2)}(5) + f_{av(3)}(20)]$$
$$= (\pi)(0.406)[(13)(5) + (31.9)(5) + (93.43)(20)] = \underline{2669.7 \text{ kN}}$$

## Part c: Calculation of Net Ultimate Capacity, $Q_u$

If we compare the three values just given, we see that the $\alpha$ and $\lambda$ methods give close results. So use

$$Q_s = \frac{1658.1 + 1777.8}{2} \approx 1718 \text{ kN}$$

Thus

$$Q_u = Q_p + Q_s = 116.46 + 1718 = 1834.46 \text{ kN}$$

$$Q_{all} = \frac{Q_u}{FS} = \frac{1834.46}{4} = \underline{458.6 \text{ kN}}$$

## 8.8
## Point Bearing Capacity of Piles Resting on Rocks

Sometimes piles are driven to the underlying rock layers. In such cases, one must evaluate the bearing capacity of the rocks. The ultimate unit point resistance in rocks can be approximately given as (Goodman, 1980)

$$q_p = q_u(N_\phi + 1) \tag{8.47}$$

where $N_\phi = \tan^2(45 + \phi/2)$
$q_u$ = unconfined compression strength of rock
$\phi$ = drained angle of friction

The unconfined compression strength of rocks can be determined by laboratory tests on rock specimens collected during field investigation. However, extreme caution should be used in obtaining the proper value of $q_u$, because laboratory specimens are usually small in diameter. As the diameter of the specimen increases, the unconfined compression strength decreases. This is referred to as the *scale effect*. For specimens larger than about 1 meter in diameter, the value of $q_u$ remains approximately constant. There appears to be a 4- to 5-fold reduction of the magnitude of $q_u$ in this process. The scale effect in rocks is primarily caused by randomly distributed large and small fractures and also by progressive ruptures along the slip lines. Hence, it is always recommended that

$$q_{u(\text{design})} = \frac{q_{u(\text{lab})}}{5} \tag{8.48}$$

Table 8.9 lists some representative values of (laboratory) unconfined compression strengths of rocks. Representative values of the rock friction angle, $\phi$, are given in Table 8.10.

A factor of safety of *at least 3* should be used to determine the allowable point bearing capacity of piles. Thus

$$Q_{p(\text{all})} = \frac{[q_u(N_\phi + 1)]A_p}{FS}$$

**Table 8.9**  Unconfined Compressive Strengths of Specimens of Representative Rocks

|  |  |  |
|---|---|---|
|  | $q_u$ | |
| Description | MPa | psi |
| Berea sandstone | 73.8 | 10,700 |
| Navajo sandstone | 214.0 | 31,030 |
| Tensleep sandstone | 72.4 | 10,500 |
| Hackensack siltstone | 122.7 | 17,800 |
| Monticello Dam s.s. (greywacke) | 79.3 | 11,500 |
| Solenhofen limestone | 245.0 | 35,500 |
| Bedford limestone | 51.0 | 7,400 |
| Tavernalle limestone | 97.9 | 14,200 |
| Oneota dolomite | 86.9 | 12,600 |
| Lockport dolomite | 90.3 | 13,100 |
| Flaming Gorge shale | 35.2 | 5,100 |
| Micaceous shale | 75.2 | 10,900 |
| Dworshak Dam gneiss |  |  |
| 45° to foliation | 162.0 | 23,500 |
| Quartz mica schist ⊥ schistocity | 55.2 | 8,000 |
| Baraboo quartzite | 320.0 | 46,400 |
| Taconic marble | 62.0 | 8,990 |
| Cherokee marble | 66.9 | 9,700 |
| Nevada Test Site granite | 141.1 | 20,500 |
| Pikes Peak granite | 226.0 | 32,800 |
| Cedar City tonalite | 101.5 | 14,700 |
| Palisades diabase | 241.0 | 34,950 |
| Nevada Test Site basalt | 148.0 | 21,500 |
| John Day basalt | 355.0 | 51,500 |
| Nevada Test Site tuff | 11.3 | 1,639 |

From *Introduction to Rock Mechanics*, by R. E. Goodman. Copyright 1980 by John Wiley and Sons. Reprinted by permission.

**Table 8.10** Representative Values for angle of Internal Friction ($\phi$) for Selected Rocks

| Description | Porosity (%) | $\phi$ (deg) |
|---|---|---|
| Berea sandstone | 18.2 | 27.8 |
| Bartlesville sandstone | | 37.2 |
| Pottsville sandstone | 14.0 | 45.2 |
| Repetto siltstone | 5.6 | 32.1 |
| Muddy shale | 4.7 | 14.4 |
| Stockton shale | | 22.0 |
| Edmonton bentonitic shale | | |
| (water content 30%) | 44.0 | 7.5 |
| Sioux quartzite | | 48.0 |
| Texas slate; loaded | | |
| 30° to cleavage | | 21.0 |
| 90° to cleavage | | 26.9 |
| Georgia marble | 0.3 | 25.3 |
| Wolf Camp limestone | | 34.8 |
| Indiana limestone | 19.4 | 42.0 |
| Chalk | 40.0 | 31.5 |
| Hasmark dolomite | 3.5 | 35.5 |
| Blaine anhydrite | | 29.4 |
| Inada biotite granite | 0.4 | 47.7 |
| Stone Mountain granite | 0.2 | 51.0 |
| Nevada Test Site basalt | 4.6 | 31.0 |
| Schistose gneiss | | |
| 90° to schistocity | 0.5 | 28.0 |
| 30° to schistocity | 1.9 | 27.6 |

From *Introduction to Rock Mechanics*, by R. E. Goodman. Copyright 1980 by John Wiley and Sons. Reprinted by permission.

## 8.9
## Elastic Settlement of Piles

The settlement of a pile under a vertical working load ($Q_w$) is caused by three factors:

$$s = s_1 + s_2 + s_3 \tag{8.49}$$

where $s$ = total pile settlement
   $s_1$ = settlement of pile shaft
   $s_2$ = settlement of pile caused by the load at the pile point
   $s_3$ = settlement of pile caused by the load transmitted along the pile shaft

The procedures for estimating the preceding three elements of pile settlement are as follows:

### Determination of $s_1$

If one assumes the pile material to be elastic, then the deformation of the pile shaft can be evaluated using the fundamental principles of mechanics of materials:

$$s_1 = \frac{(Q_{wp} + \xi Q_{ws})L}{A_p E_p} \tag{8.50}$$

where $Q_{wp}$ = load carried at the pile point under working load condition
$\quad\quad Q_{ws}$ = load carried by frictional (skin) resistance under working load condition
$\quad\quad A_p$ = area of pile cross section
$\quad\quad L$ = length of pile
$\quad\quad E_p$ = Young's modulus of the pile material

The magnitude of $\xi$ will depend on the nature of unit frictional (skin) resistance distribution along the pile shaft. If the distribution of $f$ is uniform or parabolic in nature, as shown in Figure 8.23a and b, $\xi$ is equal to 0.5. However, for triangular distribution of $f$ (Figure 8.23c), the value of $\xi$ is about 0.67 (Vesic, 1977).

### Determination of $s_2$

The settlement of a pile caused by the load carried at the pile point can be expressed in a form similar to that given for shallow foundations [Eq. (3.63)]:

$$s_2 = \frac{q_{wp} D}{E_s} (1 - \mu_s^2) I_{wp} \tag{8.51}$$

where $D$ = width or diameter of pile
$\quad\quad q_{wp}$ = point load per unit area at the pile point = $Q_{wp}/A_p$
$\quad\quad E_s$ = Young's modulus of soil
$\quad\quad \mu_s$ = Poisson's ratio of soil
$\quad\quad I_{wp}$ = influence factor

For all practical purposes, $I_{wp}$ may be taken to be equal to $\alpha_r$ as given in Eq. (3.63) and can be evaluated from Figure 3.13 (p. 125). In the absence of any experimental results, representative values of Poisson's ratio can be obtained from Table 3.4 (p. 129).

Vesic (1977) has also proposed a semiempirical method to obtain the mag-

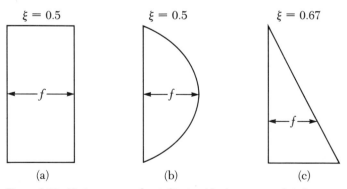

$\xi = 0.5$ $\quad\quad\quad\quad$ $\xi = 0.5$ $\quad\quad\quad\quad$ $\xi = 0.67$

$\quad\quad$ (a) $\quad\quad\quad\quad\quad\quad$ (b) $\quad\quad\quad\quad\quad\quad$ (c)

**Figure 8.23**  Various types of unit friction (skin) resistance distribution along the pile shaft

nitude of the settlement, $s_2$. This can be expressed by the following equation:

$$s_2 = \frac{Q_{wp}C_p}{Dq_p}$$ (8.52)

where $q_p$ = ultimate point resistance of the pile
$C_p$ = an empirical coefficient

Representative values of $C_p$ for various soils are given in Table 8.11.

**Table 8.11** Typical Values of $C_p$ [Eq. (8.52)]

| Soil type | Driven pile | Bored pile |
|---|---|---|
| Sand (dense to loose) | 0.02–0.04 | 0.09–0.18 |
| Clay (stiff to soft) | 0.02–0.03 | 0.03–0.06 |
| Silt (dense to loose) | 0.03–0.05 | 0.09–0.12 |

From "Design of Pile Foundations," by A. S. Vesic, in NCHRB *Synthesis of Highway Practice 42*, Transportation Research Board, 1977. Reprinted by permission.

### Determination of $s_3$

The settlement of a pile caused by the load carried by the pile shaft can be given by a relation similar to Eq. (8.51), or

$$s_3 = \left(\frac{Q_{ws}}{pL}\right)\frac{D}{E_s}(1 - \mu_s^2)I_{ws}$$ (8.53)

where $p$ = perimeter of the pile
$L$ = embedded length of pile
$I_{ws}$ = influence factor

Note that the term $Q_{ws}/pL$ in the preceding equation is the average value of $f$ along the pile shaft. The influence factor $I_{ws}$ can be expressed by a simple empirical relation as (Vesic, 1977)

$$I_{ws} = 2 + 0.35 \sqrt{\frac{L}{D}}$$ (8.54)

Vesic (1977) has also proposed a simple empirical relation similar to Eq. (8.52) for obtaining $s_3$ as

$$s_3 = \frac{Q_{ws}C_s}{Lq_p}$$ (8.55a)

where $C_s$ = an empirical constant = $(0.93 + 0.16 \sqrt{L/D})C_p$ (8.55b)

The values of $C_p$ for use in Eq. (8.55a) can be estimated from Table 8.11.

### Example 8.6

Consider the pile described in Example Problems 8.1 to 8.3. The allowable working load is 337 kN. If 240 kN is contributed by the frictional resistance and 97 kN is from the point load, determine the elastic settlement of the pile. Use $E_p = 21 \times 10^6$ kN/m$^2$, $E_s = 30,000$ kN/m$^2$, and $\mu_s = 0.3$.

## Solution

We will use Eq. (8.49)

$$s = s_1 + s_2 + s_3$$

From Eq. (8.50)

$$s_1 = \frac{(Q_{wp} + \xi Q_{ws})L}{A_p E_p}$$

Let $\xi = 0.6$ and $E_p = 21 \times 10^6$ kN/m$^2$. So

$$s_1 = \frac{[97 + (0.6)(240)]12}{(0.305)^2(21 \times 10^6)} = 0.00148 \text{ m} = 1.48 \text{ mm}$$

From Eq. (8.51)

$$s_2 = \frac{q_{wp}D}{E_s}(1 - \mu_s^2)I_{wp}$$

From Figure 3.13, $I_{wp} = 0.82$

$$q_{wp} = \frac{Q_{wp}}{A_p} = \frac{97}{(0.305)^2} = 1042.7 \text{ kN/m}^2$$

So

$$s_2 = \left[\frac{(1042.7)(0.305)}{30,000}\right](1 - 0.3^2)(0.82) = 0.0079 \text{ m} = 7.9 \text{ mm}$$

Again, from Eq. (8.53)

$$s_3 = \left(\frac{Q_{wp}}{pL}\right)\frac{D}{E_s}(1 - \mu_s^2)I_{ws}$$

$$I_{ws} = 2 + 0.35\sqrt{\frac{L}{D}} = 2 + 0.35\sqrt{\frac{12}{0.305}} = 4.2$$

So

$$s_3 = \frac{240}{(\pi \times 0.305)(12)}\left(\frac{0.305}{30,000}\right)(1 - 0.3^2)(4.2) = 0.00081 \text{ m} = 0.81 \text{ mm}$$

Hence, the total settlement

$$s = 1.48 + 7.9 + 0.81 = 10.19 \text{ mm}$$

## 8.10
## Pullout Resistance of Piles

Section 8.1 noted that, under certain construction conditions, piles are subjected to uplifting forces. The ultimate resistance of piles subjected to such force has not received much attention among researchers until recently. The gross ultimate resistance of a pile subjected to uplifting force can be written as (Figure 8.24)

$$T_{ug} = T_{un} + W \tag{8.56}$$

where $T_{ug}$ = gross uplift capacity
$T_{un}$ = net uplift capacity
$W$ = effective weight of the pile

The net ultimate uplift capacity of piles embedded in saturated clays has been studied by Das and Seeley (1982). According to this study

$$T_{un} = Lp\alpha'c_u \tag{8.57}$$

where $L$ = length of the pile
$p$ = perimeter of pile section
$\alpha'$ = adhesion coefficient at soil-pile interface
$c_u$ = undrained cohesion of clay

For cast-*in-situ* concrete piles

$$\alpha' = 0.9 - 0.00625c_u \text{ (for } c_u \leq 80 \text{ kN/m}^2) \tag{8.58a}$$

and

$$\alpha' = 0.4 \text{ (for } c_u > 80 \text{ kN/m}^2) \tag{8.58b}$$

Similarly, for pipe piles

$$\alpha' = 0.715 - 0.0191c_u \text{ (for } c_u \leq 27 \text{ kN/m}^2) \tag{8.59a}$$

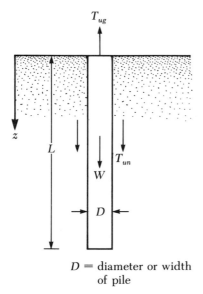

$D$ = diameter or width of pile

**Figure 8.24** Uplift capacity of piles

and

$$\alpha' = 0.2 \text{ (for } c_u > 27 \text{ kN/m}^2) \tag{8.59b}$$

When piles are embedded in granular soils ($c = 0$), the net ultimate uplift capacity can be expressed as (Das and Seeley, 1975)

$$T_{un} = \int_0^L (f_u p)dz \tag{8.60a}$$

where $f_u$ = unit skin friction during uplift
$p$ = perimeter of pile cross section

The unit skin friction during uplift ($f_u$) usually varies in the manner shown in Figure 8.25a. It increases linearly up to a depth of $z = L_{cr}$; beyond that it remains constant. For $z \leq L_{cr}$

$$f_u = K_u \sigma_v' \tan \delta \tag{8.60b}$$

where $K_u$ = uplift coefficient
$\sigma_v'$ = effective vertical stress at a depth of $z$
$\delta$ = soil-pile friction angle

The variation of the uplift coefficient with soil friction angle $\phi$ is given in Figure 8.25b. Based on the experience of the author, it appears that the values of $L_{cr}$ and $\delta$ are dependent on the relative density of soil. Figure 8.25c shows the approximate nature of these variations with the relative density of soil. For calculation of the net ultimate uplift capacity of piles, the following procedure is suggested.

**1.** Determine the relative density of the soil and, using Figure 8.25c, obtain the value of $L_{cr}$.
**2.** If the length of the pile $L$ is less than or equal to $L_{cr}$

$$T_{un} = p\int_0^L f_u \, dz = p\int_0^L (\sigma_v' K_u \tan \delta)dz \tag{8.61a}$$

In dry soils, $\sigma_v' = \gamma z$ (where $\gamma$ = unit weight of soil). So

$$T_{un} = p\int_0^L (\sigma_v' K_u \tan \delta)dz = p\int_0^L \gamma z K_u \tan \delta \, dz = \tfrac{1}{2}p\gamma L^2 K_u \tan \delta \tag{8.61b}$$

The values of $K_u$ and $\delta$ can be obtained from Figure 8.25b and c.
**3.** For the case in which $L > L_{cr}$

$$T_{un} = p\int_0^L f_u \, dz = p\left[\int_0^{L_{cr}} f_u \, dz + \int_{L_{cr}}^L f_u \, dz\right]$$

$$= p\left\{\int_0^{L_{cr}} [\sigma_v' K_u \tan \delta]dz + \int_{L_{cr}}^L \left[\sigma_{v(\text{at } z = L_{cr})}' K_u \tan \delta\right]dz\right\} \tag{8.62a}$$

In dry soils, Eq. (8.62a) changes to the following simplified form

$$T_{un} = \tfrac{1}{2}p\gamma L_{cr}^2 K_u \tan \delta + p\gamma L_{cr} K_u \tan \delta (L - L_{cr}) \tag{8.62b}$$

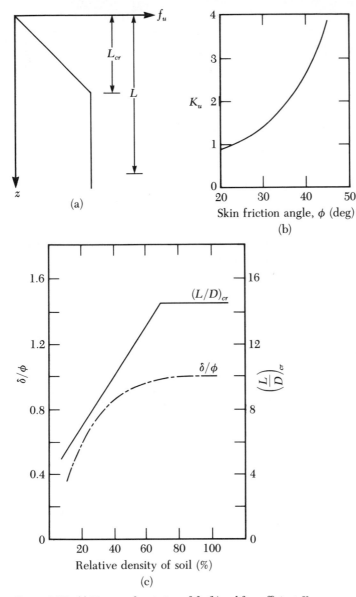

**Figure 8.25** (a) Nature of variation of $f_u$; (b) uplift coefficient $K_u$; (c) variation of $\delta/\phi$ and $(L/D)_{cr}$ with relative density of sand

The values of $K_u$ and $\delta$ can be determined from Figure 8.25b and c.

For estimation of the net allowable uplift capacity, a factor of safety of 2–3 is recommended. Thus

$$T_{u(all)} = \frac{T_{ug}}{FS}$$

where $T_{u(all)}$ = allowable uplift capacity

## Example
## 8.7

For the pipe pile given in Example Problem 8.5, determine the net ultimate uplift capacity.

### Solution

We will use Eq. (8.57) for this problem. Because the top and bottom layers have clays with $c_u > 27$ kN/m², the value of $\alpha'$ for both layers is 0.2. So, from Eq. (8.57)

$$T_{un} = p\alpha' \ \Sigma \ c_u \ \Delta L = \pi(0.406)(0.2)[(30)(10) + (100)(20)]$$

$$= 586.7 \ \text{kN}$$

## Example
## 8.8

Refer to Example Problem 8.1. For the concrete pile, determine the net ultimate pullout capacity. Assume the relative density of soil to be 60%.

### Solution

From Figure 8.25c, for a relative density of 60%, $(L/D)_{cr} \approx 12.7$. So

$$L_{cr} = (12.7)(0.305) = 3.87 \ \text{m}$$

So, $L = 12$ m $> L_{cr}$. For this case, Eq. (8.62b) is to be used. Or

$$T_{un} = \tfrac{1}{2}p\gamma L_{cr}^2 K_u \tan \delta + p\gamma L_{cr} K_u \tan \delta(L - L_{cr})$$

From Figure 8.25b, for $\phi = 35°$, $K_u = 1.9$. Similarly, from Figure 8.25c, for a relative density $= 60\%$, $(\delta/\phi) \approx 0.97$. So, $\delta = (0.97)(35) = 33.95°$. Substituting these values into Eq. (8.62b)

$$T_{un} = (\tfrac{1}{2})(4 \times 0.305)(16.8)(3.87)^2(1.9)\tan(33.95)$$

$$+ (4 \times 0.305)(16.8)(3.87)(1.9)\tan(33.95)(12 - 3.87)$$

$$= 1021.2 \ \text{kN}$$

## 8.11
## Laterally Loaded Vertical Piles

### Granular Soils

A general solution for determination of moments and displacements of a vertical pile subjected to lateral load and moment at the ground surface has been given by Matlock and Reese (1960). Consider a pile of length $L$ subjected to a lateral force $Q_g$ and a moment $M_g$ at the ground surface (that is, at $z = 0$), as shown in Figure 8.26a. Figure 8.26b shows the general nature of the deflected shape of the pile and the soil resistance caused by the applied load and the moment.

According to a simpler Winkler's model, an elastic medium (which is soil in this case) can be replaced by a series of infinitely close independent elastic

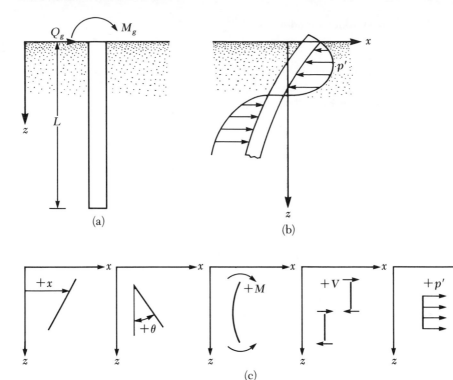

**Figure 8.26** (a) Laterally loaded pile; (b) soil resistance on pile caused by lateral load; (c) sign conventions for displacement, slope, moment, shear, and soil reaction

springs. With this assumption, one can write that

$$k = \frac{p' \ (kN/m)}{x \ (m)} \tag{8.63}$$

where $k$ = modulus of subgrade reaction
$\quad p'$ = pressure on soil
$\quad x$ = deflection

The subgrade modulus for granular soils at a depth $z$ can be defined as

$$k_z = n_h z \tag{8.64}$$

where $n_h$ = constant of modulus of horizontal subgrade reaction

Referring to Figure 8.26b and using the theory of beams on an elastic foundation, one can write

$$E_p I_p \frac{d^4 x}{dz^4} = p' \tag{8.65}$$

where $E_p$ = Young's modulus of the pile material
$\quad I_p$ = moment of inertia of the pile section

Based on Winkler's model

$$p' = -kx \tag{8.66}$$

The sign in the preceding equation is negative because the soil reaction is in the opposite direction to the pile deflection.

Combining Eqs. (8.65) and (8.66)

$$E_p I_p \frac{d^4 x}{dz^4} + kx = 0 \tag{8.67}$$

Solution of the preceding equation results in the following expressions:
Pile deflection at any depth $[x_z(z)]$:

$$x_z(z) = A_x \frac{Q_g T^3}{E_p I_p} + B_x \frac{M_g T^2}{E_p I_p} \tag{8.68}$$

Slope of pile at any depth $[\theta_z(z)]$:

$$\theta_z(z) = A_\theta \frac{Q_g T^2}{E_p I_p} + B_\theta \frac{M_g T}{E_p I_p} \tag{8.69}$$

Moment of pile at any depth $[M_z(z)]$:

$$M_z(z) = A_m Q_g T + B_m M_g \tag{8.70}$$

Shear force on pile at any depth $[V_z(z)]$:

$$V_z(z) = A_v Q_g + B_v \frac{M_g}{T} \tag{8.71}$$

Soil reaction at any depth $[p_z'(z)]$:

$$p_z'(z) = A_{p'} \frac{Q_g}{T} + B_{p'} \frac{M_g}{T^2} \tag{8.72}$$

where $A_x$, $B_x$, $A_\theta$, $B_\theta$, $A_m$, $B_m$, $A_v$, $B_v$, $A_{p'}$, and $B_{p'}$ are coefficients and

$T$ = characteristic length of the soil-pile system

$$= \sqrt[5]{\frac{E_p I_p}{n_h}} \tag{8.73}$$

$n_h$ has been defined in Eq. (8.64).

When the pile length, $L \geq 5T$, it is considered to be a *long pile*. For $L \leq 2T$, the pile is considered to be a *rigid pile*. Table 8.12 gives the values of the coefficients for long piles ($L/T \geq 5$) in Eqs. (8.68) to (8.72). Note that, in the first column of Table 8.12, $Z$ is the nondimensional depth, or

$$Z = \frac{z}{T} \tag{8.74}$$

The positive sign conventions for $x_z(z)$, $\theta_z(z)$, $M_z(z)$, $V_z(z)$, and $p_z'(z)$ assumed in the derivations in Table 8.12 are shown in Figure 8.26c. Also, Figure 8.27 shows the variation of $A_x$, $B_x$, $A_m$, and $B_m$ for various values of $L/T = Z_{max}$. These

**Table 8.12** Coefficients for Long Piles, $k_z = n_h z$

| $Z$ | $A_x$ | $A_\theta$ | $A_m$ | $A_v$ | $A_p{}'$ | $B_x$ | $B_\theta$ | $B_m$ | $B_v$ | $B_p{}'$ |
|---|---|---|---|---|---|---|---|---|---|---|
| 0.0 | 2.435 | -1.623 | 0.000 | 1.000 | 0.000 | 1.623 | -1.750 | 1.000 | 0.000 | 0.000 |
| 0.1 | 2.273 | -1.618 | 0.100 | 0.989 | -0.227 | 1.453 | -1.650 | 1.000 | -0.007 | -0.145 |
| 0.2 | 2.112 | -1.603 | 0.198 | 0.956 | -0.422 | 1.293 | -1.550 | 0.999 | -0.028 | -0.259 |
| 0.3 | 1.952 | -1.578 | 0.291 | 0.906 | -0.586 | 1.143 | -1.450 | 0.994 | -0.058 | -0.343 |
| 0.4 | 1.796 | -1.545 | 0.379 | 0.840 | -0.718 | 1.003 | -1.351 | 0.987 | -0.095 | -0.401 |
| 0.5 | 1.644 | -1.503 | 0.459 | 0.764 | -0.822 | 0.873 | -1.253 | 0.976 | -0.137 | -0.436 |
| 0.6 | 1.496 | -1.454 | 0.532 | 0.677 | -0.897 | 0.752 | -1.156 | 0.960 | -0.181 | -0.451 |
| 0.7 | 1.353 | -1.397 | 0.595 | 0.585 | -0.947 | 0.642 | -1.061 | 0.939 | -0.226 | -0.449 |
| 0.8 | 1.216 | -1.335 | 0.649 | 0.489 | -0.973 | 0.540 | -0.968 | 0.914 | -0.270 | -0.432 |
| 0.9 | 1.086 | -1.268 | 0.693 | 0.392 | -0.977 | 0.448 | -0.878 | 0.885 | -0.312 | -0.403 |
| 1.0 | 0.962 | -1.197 | 0.727 | 0.295 | -0.962 | 0.364 | -0.792 | 0.852 | -0.350 | -0.364 |
| 1.2 | 0.738 | -1.047 | 0.767 | 0.109 | -0.885 | 0.223 | -0.629 | 0.775 | -0.414 | -0.268 |
| 1.4 | 0.544 | -0.893 | 0.772 | -0.056 | -0.761 | 0.112 | -0.482 | 0.688 | -0.456 | -0.157 |
| 1.6 | 0.381 | -0.741 | 0.746 | -0.193 | -0.609 | 0.029 | -0.354 | 0.594 | -0.477 | -0.047 |
| 1.8 | 0.247 | -0.596 | 0.696 | -0.298 | -0.445 | -0.030 | -0.245 | 0.498 | -0.476 | 0.054 |
| 2.0 | 0.142 | -0.464 | 0.628 | -0.371 | -0.283 | -0.070 | -0.155 | 0.404 | -0.456 | 0.140 |
| 3.0 | -0.075 | -0.040 | 0.225 | -0.349 | 0.226 | -0.089 | 0.057 | 0.059 | -0.213 | 0.268 |
| 4.0 | -0.050 | 0.052 | 0.000 | -0.106 | 0.201 | -0.028 | 0.049 | -0.042 | 0.017 | 0.112 |
| 5.0 | -0.009 | 0.025 | -0.033 | 0.015 | 0.046 | 0.000 | -0.011 | -0.026 | 0.029 | -0.002 |

From *Drilled Pier Foundations*, by R. J. Woodward, W. S. Gardner, and D. M. Greer. Copyright 1972 by McGraw-Hill Book Company. Used with the permission of McGraw-Hill Book Company.

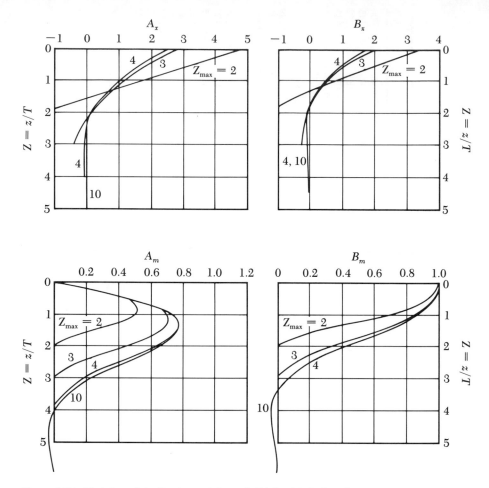

**Figure 8.27**  Variation of $A_x$, $B_x$, $A_m$, and $B_m$ with Z (after Matlock and Reese, 1960)

figures indicate that when $L/T$ is greater than about 5, the coefficients do not change. This is true of long piles only.

To calculate the characteristic length $T$ for the pile, one needs to assume a proper value of $n_h$. Some representative values of $n_h$ are given in Table 8.13.

## Cohesive Soils

Solutions similar to those given in Eqs. (8.68) to (8.72) have been given by Davisson and Gill (1963) for the case of piles embedded in clay. According to these solutions

$$x_z(z) = A'_x \frac{Q_g R^3}{E_p I_p} + B'_x \frac{M_g R^2}{E_p I_p} \qquad (8.75)$$

and

$$M_z(z) = A'_m Q_g R + B'_m M_g \qquad (8.76)$$

**Table 8.13** Representative Values of $n_h$

| Soil type | $n_h (kN/m^3)$ |
|---|---|
| Dry or moist sand | Loose: 1800–2200<br>Medium: 5500–7000<br>Dense: 15000–18000 |
| Submerged sand | Loose: 1000–1400<br>Medium: 3500–4500<br>Dense: 9000–12000 |

*Note:* $1kN/m^3 = 6.36 \ lb/ft^3$

where $A'_x$, $B'_x$, $A'_m$, and $B'_m$ are coefficients, and

$$R = \sqrt[4]{\frac{E_p I_p}{k}} \tag{8.77}$$

The values of the $A$ and $B$ coefficients are given in Figure 8.28. Note that, in this figure,

$$Z' = \frac{z}{R} \tag{8.78}$$

and

$$Z'_{max} = \frac{L}{R} \tag{8.79}$$

To use Eqs. (8.75) and (8.76), one must know the magnitude of the characteristic length, $R$. This can be calculated from Eq. (8.77) provided the coefficient of the subgrade reaction is known. For sands, the coefficient of

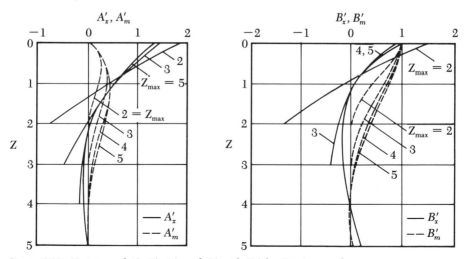

**Figure 8.28** Variation of $A'_x$, $B'_x$, $A'_m$, and $B'_m$ with $Z$ (after Davisson and Gill, 1963)

subgrade reaction was given by Eq. (8.64), which showed a linear variation with depth. However, in cohesive soils, the subgrade reaction can be assumed to be approximately constant with depth. Vesic (1961) has proposed the following equation to estimate the value of $k$.

$$k = 0.65 \sqrt[12]{\frac{E_s D^4}{E_p I_p}} \frac{E_s}{1 - \mu_s^2} \qquad (8.80)$$

where $E_s$ = Young's modulus of soil
$D$ = pile width (or diameter)
$\mu_s$ = Poisson's ratio of the soil

The Young's modulus of clay, $E_s$, can be obtained from laboratory consolidation of the soil as

$$E_s = \frac{3(1 - \mu_s)}{m_v} \qquad (8.81)$$

where $m_v$ = volume coefficient of compressibility—see Eq. (1.65);

$$m_v = \frac{\Delta e}{\Delta p(1 + e_{av})}$$

The value of $\mu_s$ can be assumed to vary between 0.3–0.4.

## Example 8.9

Consider a steel H-pile (HP 250 × 0.834) 25 m long embedded fully in a granular soil. Assume $n_h$ = 12000 kN/m³. The allowable displacement at the top of the pile is 8 mm. Determine the allowable lateral load, $Q_g$. Assume $M_g$ is equal to zero.

### Solution

From Table 8.1, for a HP 250 × 0.834 pile,

$$I_p = 123 \times 10^{-6}\,\text{m}^4 \text{ (about the strong axis)}$$
$$E_p = 207 \times 10^6\,\text{kN/m}^2$$

From Eq. (8.73)

$$T = \sqrt[5]{\frac{E_p I_p}{n_h}} = \sqrt[5]{\frac{(207 \times 10^6)(123 \times 10^{-6})}{12{,}000}} = 1.16\text{ m}$$

$L/T = 25/1.16 = 21.55 > 5$. So, this is a long pile. Because $M_g = 0$, Eq. (8.68) takes the form

$$x_z(z) = A_x \frac{Q_g T^3}{E_p I_p}$$

So

$$Q_g = \frac{x_z(z) E_p I_p}{A_x T^3}$$

Given $x_z(z = 0) = 8$ mm $= 0.008$ m. At $z = 0$, $A_x = 2.435$ (Table 8.12). So,

$$Q_g = \frac{(0.008)(207 \times 10^6)(123 \times 10^{-6})}{(2.435)(1.16^3)} = 53.59 \text{ kN}$$

The preceding value of $Q_g = 53.59$ kN has been determined based on the *limiting displacement condition only*. However, the value of $Q_g$ based on the *moment capacity* of the pile also needs to be determined. For that, referring to Eq. (8.70) (with $M_g = 0$),

$$M_z(z) = A_m Q_g T$$

According to Table 8.12, the maximum value of $A_m$ at any depth is equal to 0.772. The maximum allowable moment that the pile can carry is equal to

$$M_{z(\max)} = \sigma_{\text{all}} \frac{I_p}{\dfrac{d_1}{2}}$$

Let $\sigma_{\text{all}} = 125000$ kN/m². From Table 8.1, $I_p = 123 \times 10^{-6}$ m⁴ and $d_1 = 0.254$ m. So

$$\frac{I_p}{\left(\dfrac{d_1}{2}\right)} = \frac{123 \times 10^{-6}}{\left(\dfrac{0.254}{2}\right)} = 968.5 \times 10^{-6} \text{ m}^3$$

Now

$$Q_g = \frac{M_{z(\max)}}{A_m T} = \frac{(968.5 \times 10^{-6})(125,000)}{(0.772)(1.16)} = \underline{135.2 \text{ kN}}$$

This value of $Q_g = 135.2$ kN is greater than 53.59 kN. So the deflection criteria apply. Hence, $Q_g = \underline{53.59 \text{ kN}}$.

This is only the first approximation. The validity of the assumption of $n_h = 12,000$ kN/m³ may now be checked using $Q_g = 53.59$ kN.

## 8.12

## Pile Driving Formulas

To develop the desired load-carrying capacity, a point-bearing pile must sufficiently penetrate the dense soil layer or have sufficient contact with a layer of rock. This requirement cannot always be satisfied by driving a pile to a predetermined depth, because soil profiles vary. For that reason, several equations have been developed to calculate the ultimate capacity of a pile during driving. These dynamic equations are widely used in the field to determine if the pile has reached satisfactory bearing value at the predetermined depth. One of the earliest of these dynamic equations—commonly referred to as the *Engineering News Record (ENR) formula*—is derived on the basis of the work-energy theory. This means that

energy imparted by the hammer per blow =
(pile resistance) × (penetration per hammer blow)

According to the ENR formula, the pile resistance is the ultimate load $Q_u$ and can be expressed as

$$Q_u = \frac{W_R h}{S + C}$$  (8.82)

where $W_R$ = weight of the ram (for example, see Table 8.6)
 $h$ = height of fall of the ram
 $S$ = penetration of pile per hammer blow
 $C$ = a constant

The pile penetration, $S$, is usually based on the average value obtained from the last few driving blows. In the equation's original form, the following values of $C$ were recommended:

For drop hammers:

---

$C$ = 2.54 cm (if the units of $S$ and $h$ are in centimeters)
$C$ = 1 in. (if the units of $S$ and $h$ are in inches)

For steam hammers:

---

$C$ = 0.254 cm (if the units of $S$ and $h$ are in centimeters)
$C$ = 0.1 in. (if the units of $S$ and $h$ are in inches)

Also, a factor of safety, $FS$ = 6, was recommended to estimate the allowable pile capacity. Note that, for single- and double-acting hammers, the term $W_R h$ can be replaced by $EH_E$ (where $E$ = hammer efficiency and $H_E$ = rated energy of hammer). Thus

$$Q_u = \frac{EH_E}{S + C}$$  (8.83)

The ENR pile driving formula has gone through several revisions over the years. A recent form—the *modified ENR formula*—can be given as

$$Q_u = \frac{EW_R h}{S + C} \cdot \frac{W_R + n^2 W_p}{W_R + W_p}$$  (8.84)

where $E$ = hammer efficiency
 $C$ = 0.254 cm or 0.1 in., depending on the units of $S$ and $h$
 $W_p$ = weight of the pile
 $n$ = coefficient of restitution between the ram and the pile cap

The efficiencies of various pile driving hammers, $E$, are in the following ranges

| Hammer type | Efficiency, $E$ |
| --- | --- |
| Single- and double-acting hammers | 0.7 to 0.85 |
| Diesel hammers | 0.8 to 0.9 |
| Drop hammers | 0.7 to 0.9 |

Representative values of the coefficient of restitution, $n$, are given in the following table.

| Pile material | Coefficient of restitution, $n$ |
|---|---|
| Cast iron hammer and concrete piles (without cap) | 0.4 to 0.5 |
| Wood cushion on steel piles | 0.3 to 0.4 |
| Wooden piles | 0.25 to 0.3 |

A factor of safety ($FS$) of 4–6 may be used in Eq. (8.84) to obtain the allowable load-bearing capacity of a pile.

The Michigan State Highway Commission (1965) undertook a study for obtaining a rational pile driving equation. At three diverse sites, a total of 88 piles were driven. Based on these tests, Michigan adopted a modified ENR formula:

$$Q_u = \frac{2.5H_E}{S + C} \frac{W_R + n^2 W_p}{W_R + W_p} \tag{8.85}$$

where $H_E$ = manufacturer's maximum rated hammer energy

$C$ = 0.254 or 0.1 in. (depending on the units of $S$ and $H_E$)

A factor of safety of 6 is recommended.

Another equation referred to as the *Danish formula* also yields results as reliable as any other equation's, and it is expressed as

$$Q_u = \frac{EH_E}{S + \sqrt{\dfrac{EH_E L}{2A_p E_p}}} \tag{8.86}$$

where $E$ = hammer efficiency

$H_E$ = rated hammer energy

$E_p$ = Young's modulus of the pile material

$L$ = length of pile

$A_p$ = area of the pile cross section

(*Note:* The area of the soil plug as shown in Figure 8.10b and c should not be included.)

One has to use units consistently in the preceding equation. A factor of safety varying from 3 to 6 is recommended to estimate the allowable load-bearing capacity of piles.

Other frequently used equations for pile driving are given by the Pacific Coast Uniform Building Code (International Conference of Building Officials, 1982) and by Janbu (1953). These relations are as follows.

*Pacific Coast Uniform Building Code Formula:*

$$Q_u = \frac{(EH_E)\left(\dfrac{W_R + nW_p}{W_R + W_p}\right)}{S + \dfrac{Q_u L}{AE}} \tag{8.87}$$

The value of $n$ in the preceding equation should be taken to be 0.25 for steel piles and 0.1 for all other piles. A factor of safety of 4 is generally recommended.

*Janbu's Formula:*

$$Q_u = \frac{EH_E}{K_u S} \tag{8.88}$$

where
$$K_u = C_d(1 + \sqrt{1 + \lambda/C_d}) \tag{8.89}$$
$$C_d = 0.75 + 0.15(W_p/W_R) \tag{8.90}$$
$$\lambda = (EH_E L/A_p E_p S^2) \tag{8.91}$$

A factor of safety of about 4–5 is generally recommended.

## Example 8.10

A precast concrete pile 0.305 m × 0.305 m in cross section is driven by a Vulcan hammer (Model No. 08). Given the following:

Maximum rated hammer energy = 35.3 kN-m (Table 8.6)
Weight of ram = 35.6 kN (Table 8.6)
Total length of pile = 20 m
Hammer efficiency = 0.8
Coefficient of restitution = 0.45
Weight of pile cap = 3.2 kN
Number of blows for last 25.4 mm of penetration = 5

Estimate the allowable pile capacity by using:

a. Eq. (8.83)  (use *FS* = 6)
b. Eq. (8.84)  (use *FS* = 5)
c. Eq. (8.86)  (use *FS* = 4)

**Solution**

Part a

Use of Eq. (8.83):

$$Q_u = \frac{EH_E}{S + C}$$

Given: $E = 0.8$, $H_E = 35.3$ kN-m.

$$S = \frac{25.4}{5} = 5.08 \text{ mm} = 0.508 \text{ cm}$$

So

$$Q_u = \frac{(0.8)(35.3)(100)}{0.508 + 0.254} = 3706 \text{ kN}$$

Hence

$$Q_{\text{all}} = \frac{Q_u}{FS} = \frac{3706}{6} = \underline{617.7 \text{ kN}}$$

## Part b

Use of Eq. (8.84):

$$Q_u = \frac{EW_R h}{S + C} \frac{W_R + n^2 W_p}{W_R + W_p}$$

Weight of pile $= LA_p \gamma_c = (20)(0.305)^2(23.58) = 43.87$ kN

$W_p$ = weight of pile + weight of cap = $43.87 + 3.2 = 47.07$ kN

So

$$Q_u = \left[\frac{(0.8)(35.3)(100)}{0.508 + 0.254}\right]\left[\frac{35.6 + (0.45)^2(47.07)}{35.6 + 47.07}\right]$$

$$= (3706)(0.546) \approx 2024 \text{ kN}$$

$$Q_{all} = \frac{Q_u}{FS} = \frac{2024}{5} = 404.8 \text{ kN} \approx 405.\text{kN}$$

## Part c

Use of Eq. (8.86):

$$Q_u = \frac{EH_E}{S + \sqrt{\dfrac{EH_E L}{2A_p E_p}}}$$

$E_p \approx 20.7 \times 10^6$ kN/m$^2$. So

$$\sqrt{\frac{EH_E L}{2A_p E_p}} = \sqrt{\frac{(0.8)(35.3)(20)}{(2)(0.305)^2(20.7 \times 10^6)}} = 0.0121 \text{ m} = 1.21 \text{ cm}$$

Hence

$$Q_u = \frac{(0.8)(35.3)(100)}{0.508 + 1.21} = 1644 \text{ kN}$$

$$Q_{all} = \frac{Q_u}{FS} = \frac{1644}{4} = 411 \text{ kN}$$

## 8.13
## Stress on Pile During Pile Driving

The maximum stress developed on a pile during the driving operation can be estimated from the pile-driving formulas presented in the preceding section. This can be shown, for example, by taking the modified ENR formula given in Eq. (8.84)

$$Q_u = \frac{EW_R h}{S + C} \frac{W_R + n^2 W_p}{W_R + W_p}$$

Note that, in the preceding equation, $S$ is equal to the average penetration per hammer blow. This can also be expressed as

$$S = \frac{2.54}{N} \tag{8.92}$$

where $S$ is in centimeters and $N$ = number of hammer blows per 2.54 cm of penetration. Thus

$$Q_u = \frac{EW_Rh}{\frac{2.54}{N} + 0.254} \cdot \frac{W_R + n^2W_p}{W_R + W_p} \tag{8.93}$$

One can now assume different values of $N$ for a given hammer and pile, and calculate $Q_u$. The driving stress can then be calculated for each value of $N$ as $Q_u/A_p$. This procedure can be demonstrated with a set of numerical values. Assume that a prestressed concrete pile 25 m in length has to be driven by an 11B3 (MKT) hammer. The pile sides measure 254 mm. From Table 8.3, for this pile

$$A_p = 645 \text{ cm}^2$$

The weight of the pile, $W_p = (A_pL)\gamma_c = (645/10,000)(25)(23.58) \approx 38$ kN. Let the weight of cap = 3 kN. So, $W_p = 38 + 3 = 41$ kN. Again, from Table 8.6, for an 11B3 hammer,

$$\text{rated energy} = 26.1 \text{ kN-m} = 26.1 \times 100 \text{ kN-cm} = H_E = W_Rh$$

Weight of ram = 22.2 kN
Assume that the hammer efficiency = $E = 0.85$, and $n = 0.35$. Substituting these values in Eq. (8.93)

$$Q_u = \left[\frac{0.85(26.1 \times 100)}{\frac{25.4}{N} + 0.254}\right]\left[\frac{22.2 + 0.35^2(41)}{22.2 + 41}\right] = \frac{955.6}{\frac{25.4}{N} + 0.254}$$

Now the following table can be prepared.

| $N$ | $Q_u$ (kN) | $A_p$ (m²) | $Q_u/A_p$ (MN/m²) |
|---|---|---|---|
| 0 | 0 | $645 \times 10^{-4}$ | 0 |
| 2 | 627 | $645 \times 10^{-4}$ | 9.72 |
| 4 | 1075 | $645 \times 10^{-4}$ | 16.67 |
| 6 | 1410 | $645 \times 10^{-4}$ | 21.87 |
| 8 | 1672 | $645 \times 10^{-4}$ | 25.92 |
| 10 | 1881 | $645 \times 10^{-4}$ | 29.16 |
| 12 | 2052 | $645 \times 10^{-4}$ | 31.82 |
| 20 | 2508 | $645 \times 10^{-4}$ | 38.88 |

Both the number of hammer blows per 2.54 cm and the stress can now be plotted in a graph, as shown in Figure 8.29. If such a curve is prepared, the number of blows per 2.54 cm of pile penetration corresponding to the allowable pile-driving stress can be easily determined.

In practice, the driving stresses in wooden piles are limited to about $0.7f_u$.

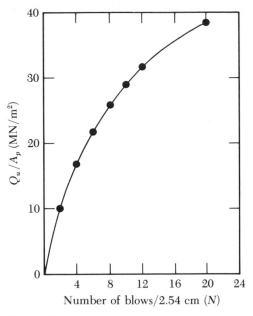

Figure 8.29

Similarly, for concrete and steel piles the driving stresses are limited to about $0.6f_c'$ and $0.85\sigma_y$, respectively.

In most cases, wooden piles are driven with a hammer energy of less than 60 kN-m. The driving resistances are mostly limited to 4–5 blows per 2.54 cm of pile penetration. For concrete and steel piles, the usual $N$ values adopted are 6–8 and 12–14, respectively.

## 8.14

### Pile Load Tests

In most large projects, a specific number of load tests must be conducted on piles. This is primarily because of the unreliability of prediction methods. Vertical and lateral load-bearing capacity of a given pile can be tested in the field. Figure 8.30a shows a schematic diagram of the pile load test arrangement in the field. This arrangement is for testing in *axial compression*. The load to the pile is applied by a hydraulic jack. Step loads are applied to the pile, and sufficient time is allowed to elapse after each step load so that the settlement rate reaches a small value. The settlement of the pile is recorded by means of dial gauges. The amount of load to be applied for each step will vary depending on local building codes. Most building codes require that each step load be about one-fourth of the proposed working load and should be carried out to at least a total load of two times the proposed working load. After reaching the desired pile load, the pile is gradually unloaded.

Load tests on piles in sand can be carried out immediately after the piles are driven. However, care should be taken in deciding the time lapse between

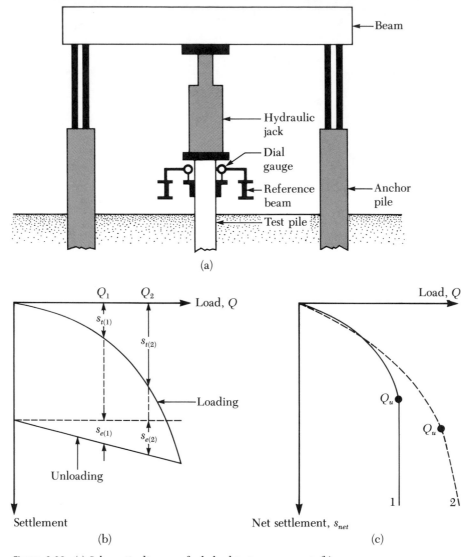

**Figure 8.30** (a) Schematic diagram of pile load test arrangement; (b) plot of load vs. total settlement; (c) plot of load vs. net settlement

driving and starting the load test when piles are embedded in clay. This time lapse can range from 30–60 days or more, because the soil requires some time to gain its *thixotropic strength*.

Figure 8.30b shows the nature of the load settlement diagram for loading and unloading that is obtained from the field. For any given load ($Q$), the net pile settlement can be calculated as follows: When $Q = Q_1$

net settlement, $s_{\text{net}(1)} = s_{t(1)} - s_{e(1)}$

When $Q = Q_2$

net settlement, $s_{net(2)} = s_{t(2)} - s_{e(2)}$

.
.
.

where $s_{net}$ = net settlement
$s_e$ = elastic settlement of the pile itself
$s_t$ = total settlement

These values of $Q$ can be plotted in a graph against the corresponding net settlement ($s_{net}$). This is shown in Figure 8.30c. The ultimate load of the pile can be determined from this graph. The settlement of the pile may increase with load up to a certain point beyond which the load-settlement curve becomes vertical. The load corresponding to the point where the $Q$ vs. $s_{net}$ curve becomes vertical is the ultimate load ($Q_u$) for the pile. This is shown by Curve 1 in Figure 8.30c. In many cases, the latter stage of the load-settlement curve is almost linear, showing a large degree of settlement for a small increment of load. This is shown by Curve 2 in Figure 8.30c. The ultimate load ($Q_u$) for such a case is determined from the point of the $Q$ vs. $s_{net}$ curve where this steep linear portion starts.

The load test procedure just described requires application of step loads on the piles and measurement of settlement. This is a *load-controlled* mode of test. Another technique used for pile load test is the *constant-rate-of-penetration* test. In this type of test, the load on the pile is continuously increased to maintain a constant *rate-of-penetration* that can vary in the range of 0.25–2.5 mm/min. This test gives a type of load-settlement plot similar to that obtained from the load-controlled test. Other modes of pile load tests include cyclic loading, in which an incremental load is repeatedly applied and removed.

## 8.15
### Group Piles—Efficiency

In most cases, piles are used in groups, as shown in Figure 8.31, to transmit the structural load to the soil. A *pile cap* is constructed over *group piles*. The pile cap can be in contact with the ground, as in most of the cases (Figure 8.31a), or it may be well above the ground, as in the case of construction of offshore platforms (Figure 8.31b).

The preceding sections have discussed the load-bearing capacity of single piles. Determination of the load-bearing capacity of group piles is an extremely complicated problem and has not yet been fully resolved. When the piles are placed close to each other, it is reasonable to assume that the stresses transmitted by the piles to the soil medium will overlap (Figure 8.31c), and this may reduce the load-bearing capacity of the piles. Ideally, the piles in a group should be spaced in such a way that the load-bearing capacity of the group should not be less than the sum of the bearing capacity of the individual piles. In practice,

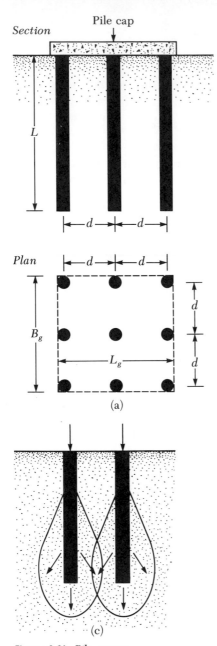

*Section*

Pile cap

$L$

$\leftarrow d \rightarrow\!\leftarrow d \rightarrow$

*Plan* $\quad \leftarrow d \rightarrow\!\leftarrow d \rightarrow$

$B_g$

$L_g$

$d$

$d$

(a)

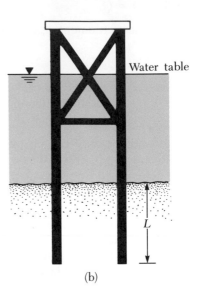

Water table

$L$

(b)

Number of piles in group $= n_1 \times n_2$
*Note:* $L_g \geq B_g$
$L_g = (n_1 - 1)d + 2(D/2)$
$B_g = (n_2 - 1)d + 2(D/2)$

(c)

**Figure 8.31** Pile groups

the center-to-center pile spacings $(d)$ are kept to a minimum of $2.5D$. However, in ordinary situations, they are kept at about $3$–$3.5D$.

The efficiency of the load-bearing capacity of a group pile may be defined as

$$\eta = \frac{Q_{g(u)}}{\Sigma Q_u} \tag{8.94}$$

where $\eta$ = group efficiency

$Q_{g(u)}$ = ultimate load-bearing capacity of the group pile

$Q_u$ = ultimate load-bearing capacity of each pile without the group effect

## Piles in Sand

Many structural engineers use a simplified analysis to obtain the group efficiency for friction piles in sand. This can be explained with the aid of Figure 8.31a. Depending on their spacing within the group, the piles may act in one of the two following ways: (1) as a *block* with dimensions $L_g \times B_g \times L$, or (2) as *individual piles*. If the piles act as a block, the frictional capacity can be given as $f_{av}p_gL \approx Q_{g(u)}$. [Note: $p_g$ = perimeter of the cross section of block = $2(n_1 + n_2 - 2)d + 4D$, and $f_{av}$ = average unit frictional resistance.] Similarly, for each pile acting individually, $Q_u \approx pLf_{av}$. (Note: $p$ = perimeter of the cross section of each pile.) Thus

$$\eta = \frac{Q_{g(u)}}{\Sigma Q_u} = \frac{f_{av}[2(n_1 + n_2 - 2)d + 4D]L}{n_1 n_2 pLf_{av}}$$

$$= \frac{2(n_1 + n_2 - 2)d + 4D}{pn_1 n_2} \tag{8.95}$$

Hence

$$Q_{g(u)} = \left[\frac{2(n_1 + n_2 - 2)d + 4D}{pn_1 n_2}\right]\Sigma Q_u \tag{8.96}$$

From Eq. (8.96), if the center-to-center spacings $(d)$ are large, one may obtain $\eta > 1$. In that case, the piles will behave as individual piles. Thus, in practice, if $\eta < 1$

$$Q_{g(u)} = \eta\Sigma Q_u$$

and, if $\eta \geq 1$

$$Q_{g(u)} = \Sigma Q_u$$

Another equation that is quoted often among design engineers is the *Converse-Labarre equation*, which can be stated as

$$\eta = 1 - \left[\frac{(n_1 - 1)n_2 + (n_2 - 1)n_1}{90n_1 n_2}\right]\theta \tag{8.97}$$

where $\theta$ (deg) = $\tan^{-1}(D/d)$ $\tag{8.98}$

Figure 8.32 shows a set of the author's laboratory model test results for round piles driven into dense sand. Note that the group efficiency, in reality, can be larger than one. This is because of the soil compaction zones created around the piles during driving. Based on the experimental observations for the

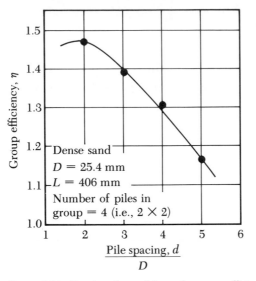

**Figure 8.32**  Results of a model test for group efficiency of piles in dense sand

behavior of group piles in sand made so far, the following general conclusions can be drawn:

**1.** For *driven* group piles in *sand* with $d \geq 3D$, $Q_{g(u)}$ may be taken to be equal to $\Sigma\, Q_u$. This includes the frictional and the point bearing capacity of individual piles.
**2.** For *bored* group piles in *sand* at conventional spacings (that is, $d \approx 3D$), $Q_{g(u)}$ may be taken to be equal to 2/3 to 3/4 times $\Sigma\, Q_u$ (frictional and point bearing capacity of individual piles).

### Piles in Clay

The ultimate load-bearing capacity of group piles in clay can be estimated in the following manner:

**1.** Determine $\Sigma\, Q_u = n_1 n_2 (Q_p + Q_s)$. From Eq. (8.16)

$$Q_p = A_p [9c_{u(p)}]$$

where $c_{u(p)}$ = undrained cohesion of the clay at the pile tip

Also, from Eq. (8.37b)

$$Q_s = \Sigma\, \alpha p c_u\, \Delta L$$

So

$$\Sigma\, Q_u = n_1 n_2 [9A_p c_{u(p)} + \Sigma\, \alpha p c_u\, \Delta L] \qquad (8.99)$$

**2.** Determine the ultimate capacity assuming that the piles in the group act as a block with dimensions of $L_g \times B_g \times L$. The skin resistance of the block =

$$\Sigma \, p_g c_u \, \Delta L = \Sigma \, 2(L_g + B_g)c_u \, \Delta L$$

The point bearing capacity can be calculated =

$$A_p q_p = A_p c_{u(p)} N_c^* = (L_g B_g)c_{u(p)} N_c^*$$

The value of the bearing capacity factor ($N_c^*$) may be taken from Figure 7.9 on p. 319. Note the change of notations: Along the abscissa, the term $H/B$ in Figure 7.9 is equivalent to $L/B_g$ for this problem; Also, $L/B$ in Figure 7.9 is equivalent to $L_g/B_g$. Thus, the ultimate load is equal to

$$\Sigma \, Q_u = L_g B_g c_{u(p)} N_c^* + \Sigma \, 2(L_g + B_g)c_u \, \Delta L \tag{8.100}$$

**3.** Compare the two values of Eqs. (8.99) and (8.100). The *lower* of the two values is equal to $Q_{g(u)}$.

### Piles in Rock

For point bearing piles resting on rocks, most building codes specify that $Q_{g(u)} = \Sigma \, Q_u$, provided that the minimum center-to-center spacing of piles is equal to $D + 300$ mm. For H-piles and piles with square cross sections, the value of $D$ is equal to the diagonal dimension of the pile cross section.

### General Comments

A pile cap resting on soil, as shown in Figure 8.31a, will contribute to the load-bearing capacity of a pile group. However, this contribution may be neglected for design purposes, because the support may be lost as a result of soil erosion or excavation during the lifetime of the project.

### Example 8.11

Refer to Figure 8.31a. Given: $n_1 = 4$, $n_2 = 3$, $D = 305$ mm, $d = 2.5D$. The piles are square in cross section and are embedded in sand. Use Eq. (8.95) to obtain the group efficiency.

#### Solution

From Eq. (8.95)

$$\eta = \frac{2(n_1 + n_2 - 2)d + 4D}{pn_1 n_2}$$

$$d = 2.5D = (2.5)(305) = 762.5 \text{ mm}$$

$$p = 4D = (4)(305) = 1220 \text{ mm}$$

So

$$\eta = \frac{2(4 + 3 - 2)762.5 + 1220}{(1220)(4)(3)} = 0.604 = \underline{60.4\%}$$

## Example 8.12

Redo Example Problem 8.11 using Eq. (8.97).

**Solution**

According to Eq. (8.97)

$$\eta = 1 - \left[\frac{(n_1 - 1)n_2 + (n_2 - 1)n_1}{90n_1 n_2}\right] \tan^{-1}\left(\frac{D}{d}\right)$$

However,

$$\tan^{-1}\left(\frac{D}{d}\right) = \tan^{-1}\left(\frac{1}{2.5}\right) = 21.8°$$

So

$$\eta = 1 - \left[\frac{(3)(3) + (2)(4)}{(90)(3)(4)}\right](21.8°) = 0.657 = 65.7\%$$

## Example 8.13

Refer to Figure 8.31a. Given: $n_1 = 4$, $n_2 = 3$, $D = 305$ mm, $d = 1220$ mm, and $L = 15$ m. The piles are square in cross section and embedded in a homogeneous clay with $c_u = 70$ kN/m². Using a factor of safety equal to 4, determine the allowable load-bearing capacity of the group pile.

**Solution**

From Eq. (8.99)

$$\Sigma Q_u = n_1 n_2 [9A_p c_{u(p)} + \Sigma \alpha p c_u \, \Delta L]$$
$$A_p = (0.305)(0.305) = 0.093 \text{ m}^2$$
$$p = (4)(0.305) = 1.22 \text{ m}$$

Given: $c_u = 70$ kN/m². From Figure 8.17, for $c_u = 70$ kN/m², $\alpha = 0.63$. So

$$\Sigma Q_u = (4)(3)[(9)(0.093)(70) + (0.63)(1.22)(70)(15)]$$
$$= 12(58.59 + 807.03) \approx 10,387 \text{ kN}$$

Again, from Eq. (8.100), the ultimate block capacity is
$L_g B_g c_{u(p)} N_c^* + \Sigma 2(L_g + B_g)c_u \, \Delta L$.

$$L_g = (n_1 - 1)d + 2\left(\frac{D}{2}\right) = (4 - 1)(1.22) + 0.305 = 3.965 \text{ m}$$

$$B_g = (n_2 - 1)d + 2\left(\frac{D}{2}\right) = (3 - 1)(1.22) + 0.305 = 2.745 \text{ m}$$

$$\frac{L}{B_g} = \frac{15}{2.745} = 5.46$$

$$\frac{L_g}{B_g} = \frac{3.965}{2.745} = 1.44$$

From Figure 7.9, $N_c^* \approx 8.6$. So

$$\text{block capacity} = (3.965)(2.745)(70)(8.6) + 2(3.965 + 2.745)(70)(15)$$
$$= 6552 + 14091 = 20{,}643 \text{ kN}$$

So

$$Q_{g(u)} = 10{,}387 \text{ kN} < 20{,}643 \text{ kN}$$

$$Q_{g(\text{all})} = \frac{Q_{g(u)}}{FS} = \frac{10{,}387}{4} \approx \underline{2597 \text{ kN}}$$

## 8.16
## Consolidation Settlement of Group Piles

The consolidation settlement of a group pile in clay can be approximately estimated by using the 2 : 1 stress distribution method shown in Figure 3.27 on p. 146. The procedure of calculation uses the following steps (refer to Figure 8.33):

**1.** Let the depth of embedment of the piles be equal to $L$. The group is subjected to a total load of $Q_g$. If the pile cap is below the original ground surface, $Q_g$ is equal to the total load of the superstructure on the piles minus the effective weight of soil above the pile group removed by excavation.

**2.** Assume that the load $Q_g$ is transmitted to the soil beginning at a depth of $2L/3$ from the top of the pile, as shown in Figure 8.33. This is depth $z = 0$ in the figure. The load $Q_g$ spreads out along 2 vertical : 1 horizontal lines from this depth. Lines $aa'$ and $bb'$ are the two 2 : 1 lines.

**3.** Calculate the stress increase caused at the middle of each soil layer by the load $Q_g$:

$$\Delta p_i = \frac{Q_g}{(B_g + z_i)(L_g + z_i)} \tag{8.101}$$

where $\Delta p_i$ = stress increase at the middle of layer $i$
$L_g, B_g$ = length and width of the plan of pile group, respectively
$z_i$ = distance from $z = 0$ to the middle of the clay layer, $i$

For example, in Figure 8.33 for Layer No. 2, $z_i = L_1/2$. Similarly, for Layer No. 3, $z_i = L_1 + L_2/2$; and, for Layer No. 4, $z_i = L_1 + L_2 + L_3/2$. Note, however, that there will be no stress increase in clay layer No. 1, because it is above the horizontal plane $(z = 0)$ from which the stress distribution to the soil starts.

**4.** Calculate the settlement of each layer caused by the increased stress:

$$\Delta s_i = \left[ \frac{\Delta e_{(i)}}{1 + e_{o(i)}} \right] H_i \tag{8.102}$$

where $\Delta s_i$ = consolidation settlement of layer $i$
$\Delta e_{(i)}$ = change of void ratio caused by the stress increase in layer $i$
$e_o$ = initial void ratio of layer $i$ (before construction)

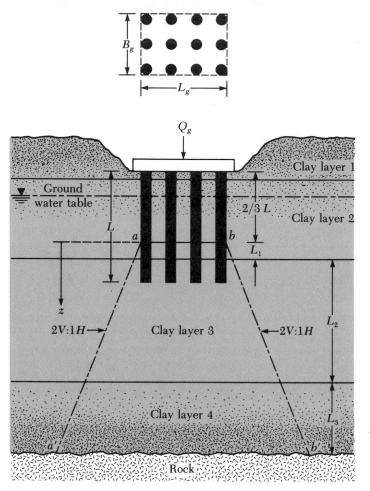

**Figure 8.33**  Consolidation settlement of group piles

$H_i$ = thickness of layer $i$ (*Note:* In Figure 8.33, the value of $H_i$ for Layer No. 2 is equal to $L_1$. For layer 3, $H_i = L_2$; and for layer 4, $H_i = L_3$.)

Relations for $\Delta e_{(i)}$ are given in Chapter 1.

**5.** Total consolidation settlement of the pile group is then

$$\Delta s_{g(c)} = \Sigma \Delta s_i \tag{8.103}$$

Note that consolidation settlement of piles may also be initiated by fills placed nearby, adjacent floor loads, and lowering of water tables.

## Example 8.14

A group pile in clay is shown in Figure 8.34. Determine the consolidation settlement of the pile groups. All clays are normally consolidated.

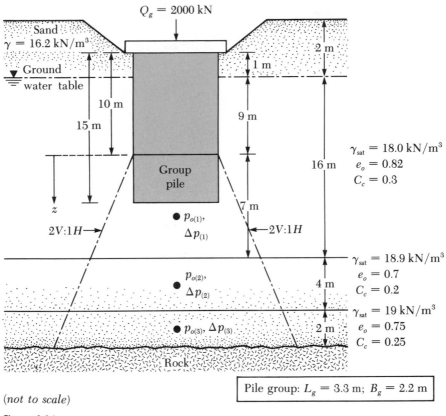

$Q_g = 2000$ kN

Sand
$\gamma = 16.2$ kN/m³

Ground water table

2 m

1 m

10 m

15 m

9 m

Group pile

16 m

$\gamma_{sat} = 18.0$ kN/m³
$e_o = 0.82$
$C_c = 0.3$

7 m

$z$

2V:1H →

$p_{o(1)}$,
$\Delta p_{(1)}$

← 2V:1H

$\gamma_{sat} = 18.9$ kN/m³
$e_o = 0.7$
$C_c = 0.2$

$p_{o(2)}$,
$\Delta p_{(2)}$

4 m

$\gamma_{sat} = 19$ kN/m³
$e_o = 0.75$
$C_c = 0.25$

$p_{o(3)}$, $\Delta p_{(3)}$

2 m

Rock

Pile group: $L_g = 3.3$ m; $B_g = 2.2$ m

(*not to scale*)

**Figure 8.34**

## Solution

Because the lengths of the piles are 15 m each, the stress distribution starts at a depth of 10 m below the top of the pile. Given: $Q_g = 2000$ kN.

### Calculation of Settlement of Clay Layer 1

For normally consolidated clays

$$\Delta s_1 = \left[\frac{C_{c(1)}H_1}{1 + e_{o(1)}}\right]\log\left[\frac{p_{o(1)} + \Delta p_{(1)}}{p_{o(1)}}\right]$$

$$\Delta p_{(1)} = \frac{Q_g}{(L_g + z_1)(B_g + z_1)} = \frac{2000}{(3.3 + 3.5)(2.2 + 3.5)} = 51.6 \text{ kN/m}^2$$

$$p_{o(1)} = 2(16.2) + 12.5(18.0 - 9.81) = 134.8 \text{ kN/m}^2$$

So

$$\Delta s_1 = \frac{(0.3)(7)}{1 + 0.82}\log\left[\frac{134.8 + 51.6}{134.8}\right] = 0.1624 \text{ m} = \underline{162.4 \text{ mm}}$$

### Settlement of Layer 2

$$\Delta s_2 = \frac{C_{c(2)} H_2}{1 + e_{o(2)}} \log\left[\frac{p_{o(2)} + \Delta p_{(2)}}{p_{o(2)}}\right]$$

$$p_{o(2)} = 2(16.2) + 16(18.0 - 9.81) + 2(18.9 - 9.81) = 181.62 \text{ kN/m}^2$$

$$\Delta p_{(2)} = \frac{2000}{(3.3 + 9)(2.2 + 9)} = 14.52 \text{ kN/m}^2$$

Hence

$$\Delta s_2 = \frac{(0.2)(4)}{1 + 0.7} \log\left[\frac{181.62 + 14.52}{181.62}\right] = 0.0157 \text{ m} = \underline{15.7 \text{ mm}}$$

### Settlement of Layer 3

$$p_{o(3)} = 181.62 + 2(18.9 - 9.81) + 1(19 - 9.81) = 208.99 \text{ kN/m}^2$$

$$\Delta p_{(3)} = \frac{2000}{(3.3 + 12)(2.2 + 12)} = 9.2 \text{ kN/m}^2$$

$$\Delta s_3 = \frac{(0.25)(2)}{1 + 0.75} \log\left[\frac{208.99 + 9.2}{208.99}\right] = 0.0054 \text{ m} = \underline{5.4 \text{ mm}}$$

Hence, total settlement =

$$\Delta s_g = 162.4 + 15.7 + 5.4 = \underline{183.5 \text{ mm}}$$

## 8.17
## Elastic Settlement of Pile Groups

Several investigations relating to the settlement of pile groups have been reported in the literature with widely varying results. The simplest relation for the settlement of group piles has been given by Vesic (1969) as

$$s_{g(e)} = \sqrt{\frac{B_g}{Ds}} \tag{8.104}$$

where $s_{g(e)}$ = elastic settlement of group piles
$B_g$ = width of pile group section (see Figure 8.31a)
$D$ = width or diameter of each pile in the group
$s$ = elastic settlement of each pile at comparable working load (see Section 8.9)

For pile groups in sand and gravel, Meyerhof (1976) suggested the following empirical relation for elastic settlement.

$$s_{g(e)} \text{ (mm)} = \frac{0.92q \sqrt{B_g} I}{N_{cor}} \tag{8.105}$$

where $q \text{ (kN/m}^2) = Q_g/(L_g B_g)$               (8.106)
$L_g$ and $B_g$ = length and width of the pile group section (in m)
$N_{cor}$ = average corrected standard penetration number within seat of settlement ( $\approx B_g$ deep below the tip of the piles)

$$I = \text{influence factor} = 1 - L/8B_g \geq 0.5 \qquad (8.107)$$
$$L = \text{length of embedment of piles}$$

In a similar manner, the pile group settlement can be related to the cone penetration resistance as

$$s_{g(e)} = \frac{qB_gI}{2q_c} \qquad (8.108)$$

where $q_c$ = average cone penetration resistance within the seat of settlement

In Eq. (8.108), all symbols are in consistent units.

## 8.18

## Negative Skin Friction

Negative skin friction is a downward drag force on the pile by the soil surrounding it. This can happen under several conditions, such as the following:

**1.** If a fill of clay soil is placed over a granular soil layer into which a pile is driven, the fill will gradually consolidate. This consolidation process will provide a downward drag force on the pile (Figure 8.35a) during the period of consolidation.

**2.** If a fill of granular soil is placed over a layer of soft clay, as shown in Figure 8.35b, it will induce the process of consolidation in the clay layer and thus provide a downward drag on the pile.

**3.** Lowering of the ground water table will increase the vertical effective stress on the soil at any depth. This will induce consolidation settlement in clay. If a pile is located in the clay layer, it will be subjected to a downward drag force.

In some cases, the downward drag force could be excessive and cause foundation failure. This section outlines two tentative methods for the calculation of negative skin friction.

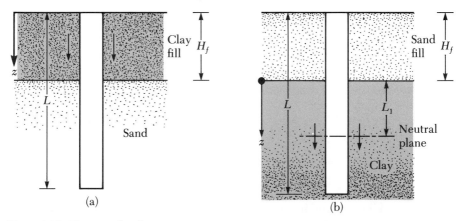

**Figure 8.35** Negative skin friction

## Clay Fill Over Granular Soil (Figure 8.35a)

Similar to the $\beta$ method presented in Section 8.6, the negative (downward) skin stress on the pile can be given by the relation

$$f_n = K'\sigma_v' \tan \delta \tag{8.109}$$

where $K' =$ earth pressure coefficient $= K_o = 1 - \sin \phi$
$\sigma_v' =$ vertical effective stress at any depth $z = \gamma_f' z$
$\gamma_f' =$ effective unit weight of fill
$\delta =$ soil-pile friction angle $\approx 0.5-0.7\phi$

Hence, the total downward drag force, $Q_n$, on a pile is

$$Q_n = \int_0^{H_f} (pK'\gamma_f' \tan \delta)z \, dz = \frac{pK'\gamma_f' H_f^2 \tan \delta}{2} \tag{8.110}$$

where $H_f =$ height of the fill

If the fill is above the water table, the effective unit weight $\gamma_f'$ should be replaced by the moist unit weight.

## Granular Soil Fill Over Clay (Figure 8.35b)

In this case, there is enough evidence to show that the negative skin stress on the pile may exist from $z = 0$ to $z = L_1$, which is referred to as the *neutral depth* (see Vesic, 1977, pp. 25–26 for discussion). The neutral depth may be given as (Bowles, 1982)

$$L_1 = \frac{(L - H_f)}{L_1} \left[ \frac{L - H_f}{2} + \frac{\gamma_f' H_f}{\gamma'} \right] - \frac{2\gamma_f' H_f}{\gamma'} \tag{8.111}$$

where $\gamma_f'$ and $\gamma' =$ effective unit weights of the fill and the underlying clay layer, respectively

Once the value of $L_1$ is determined, the downward drag force can be obtained in the following manner: The unit negative skin friction at any depth from $z = 0$ to $z = L_1$ is

$$f_n = K'\sigma_v' \tan \delta \tag{8.112}$$

where $K' = K_o = 1 - \sin \phi$
$\sigma_v' = \gamma_f' H_f + \gamma' z$
$\delta = 0.5-0.7\phi$

Hence, the total drag force

$$Q_n = \int_0^{L_1} pf_n \, dz = \int_0^{L_1} pK'(\gamma_f' H_f + \gamma' z)\tan \delta \cdot dz$$

$$= (pK'\gamma_f' H_f \tan \delta)L_1 + \frac{L_1^2 pK'\gamma' \tan \delta}{2} \tag{8.113}$$

If the soil and the fill are above the water table, the effective unit weights should be replaced by moist unit weights. In some cases, the piles can be coated with asphalt in the down-drag zone to avoid this problem.

## Example 8.15

Refer to Figure 8.35a. $H_f = 2$ m. The pile is circular in cross section with a diameter of 0.305 m. For the fill that is above the ground water table, $\gamma_f = 16$ kN/m$^3$ and $\phi = 32°$. Determine the total downward drag force.

**Solution**

From Eq. (8.110)

$$Q_n = \frac{pK'\gamma_f H_f^2 \tan \delta}{2}$$

$$p = \pi(0.305) = 0.958 \text{ m}$$

$$K' = 1 - \sin \phi = 1 - \sin 32° = 0.47$$

$$\delta = (0.6)(32) = 19.2°$$

$$Q_n = \frac{(0.958)(0.47)(16)(2)^2 \tan 19.2°}{2} = \underline{5.02 \text{ kN}}$$

## Example 8.16

Refer to Figure 8.35b. $H_f = 2$ m, pile diameter $= 0.305$ m, $\gamma_f = 16.5$ kN/m$^3$, $\phi_{clay} = 34°$, $\gamma_{sat(clay)} = 17.2$ kN/m$^3$, and $L = 20$ m. The water table coincides with the top of the clay layer. Determine the down-drag force.

**Solution**

The depth to the neutral plane is given by Eq. (8.111) as

$$L_1 = \frac{L - H_f}{L_1} \left( \frac{L - H_f}{2} + \frac{\gamma_f H_f}{\gamma'} \right) - \frac{2\gamma_f H_f}{\gamma'}$$

Note that $\gamma_f'$ in Eq. (8.111) has been replaced by $\gamma_f$, because the fill is above the water table. So

$$L_1 = \frac{(20 - 2)}{L_1} \left[ \frac{(20 - 2)}{2} + \frac{(16.5)(2)}{(17.2 - 9.81)} \right] - \frac{(2)(16.5)(2)}{(17.2 - 9.81)}$$

$$L_1 = \frac{242.4}{L_1} - 8.93$$

or

$$L_1 = 11.75 \text{ m}$$

Now, referring to Eq. (8.113)

$$Q_n = (pK'\gamma_f H_f \tan \delta)L_1 + \frac{L_1^2 pK'\gamma' \tan \delta}{2}$$

$$p = \pi(0.305) = 0.958 \text{ m}$$

$$K' = 1 - \sin 34° = 0.44$$

$$Q_n = (0.958)(0.44)(16.5)(2)[\tan(0.6 \times 34)](11.75)$$

$$+ \frac{(11.75)^2(0.958)(0.44)(17.2 - 9.81)[\tan(0.6 \times 34)]}{2}$$

$$= 60.78 + 79.97 = \underline{40.75 \text{ kN}}$$

## Problems

**8.1** A concrete pile is 50 ft long and 16 in. × 16 in. in cross section. The pile is fully embedded in sand. Given: for sand $\gamma = 110$ lb/ft³ and $\phi = 30°$. Calculate:
a. the ultimate point load, $Q_p$, by Meyerhof's method [Eqs. (8.11), (8.12), and (8.13)];
b. the total frictional resistance [Eqs. (8.30) and (8.31)]. Use $K = 1.3$ and $\delta = 0.08\phi$.

**8.2** Solve Problem 8.1 by Coyle and Castello's method (Figures 8.18 and 8.19).

**8.3** Solve Problem 8.1, Part a, by Vesic's method [Eq. (8.19)]. Use $I_r = I_{rr} = 50$.

**8.4** Redo Problem 8.1 with the following changed soil properties: $\gamma = 18.4$ kN/m³ and $\phi = 37°$.

**8.5** Solve Problem 8.4 by Coyle and Castello's method (Figures 8.18 and 8.19).

**8.6** A pipe pile (closed end) that has been driven into a layered sand is shown in Figure P8.6. Calculate:
a. the ultimate point load by Meyerhof's procedure;
b. the ultimate point load by Vesic's procedure ($I_r = I_{rr} = 50$);
c. an approximate ultimate point load on the basis of Parts a and b;
d. the ultimate frictional resistance, $Q_s$—use Eqs. (8.29) to (8.31), $K = 1.4$, $\delta = 0.7\phi$;
e. the allowable load of the pile—use $FS = 4$.

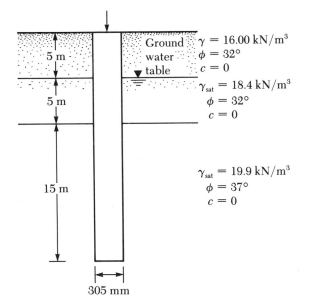

**Figure P8.6**

**8.7** A concrete pile 16 in. × 16 in. in cross section is shown in Figure P8.7. Determine the allowable load that the pile can carry ($FS = 4$). Use the $\alpha$ method for determination of the skin resistance.

**8.8** For the pile shown in Figure P8.7, determine the ultimate skin resistance by using the $\lambda$ method.

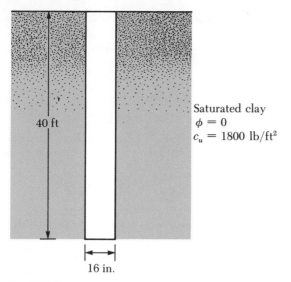

40 ft

Saturated clay
$\phi = 0$
$c_u = 1800 \text{ lb/ft}^2$

16 in.

**Figure P8.7**

**8.9** A concrete pile 405 mm × 405 mm in cross section is shown in Figure P8.9. Calculate the ultimate skin resistance by using:
a. the $\alpha$ method;
b. the $\lambda$ method;
c. the $\beta$ method.
Use $\phi_R = 25°$ for all clays. All clays are normally consolidated.

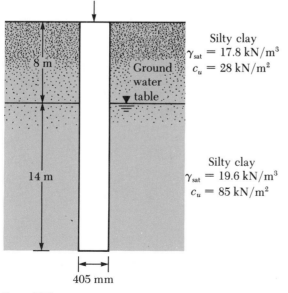

8 m

Ground
water
▼ table

Silty clay
$\gamma_{sat} = 17.8 \text{ kN/m}^3$
$c_u = 28 \text{ kN/m}^2$

14 m

Silty clay
$\gamma_{sat} = 19.6 \text{ kN/m}^3$
$c_u = 85 \text{ kN/m}^2$

405 mm

**Figure P8.9**

**8.10** The allowable working load on a prestressed concrete pile 21 m long that has been driven into sand is 502 kN. The pile is octagonal in shape with $D = 356$ mm (see Table 8.3). 350 kN of the allowable load is carried by the skin resistance and the rest by point bearing. Given:

$$E_p = 21 \times 10^6 \text{ kN/m}^2 \qquad \mu_s = 0.35$$
$$E_s = 25 \times 10^3 \text{ kN/m}^2 \qquad \zeta = 0.55 \text{ [Eq. (8.50)]}$$

Determine the elastic settlement of the pile. [Use 0.8 for $I_{wp}$ in Eq. (8.51).]

**8.11** A steel pile (H-section; HP 12 × 84; see Table B.1 in Appendix B) is driven to a layer of sandstone. The length of the pile is 60 ft. Following are the properties of the sandstone:

$$\text{Unconfirmed compression strength} = q_{u(\text{lab})} = 10,500 \text{ lb/in.}^2$$
$$\text{Angle of friction} = 37°$$

Using a factor of safety of 4, estimate the allowable point load that can be carried by the pile.

**8.12** A precast concrete pile with a cross section of 406 mm × 406 mm is embedded in sand. The length of the pile is 10.4 m. Assume $\gamma_{\text{sand}} = 15.8 \text{ kN/m}^3$, $\phi_{\text{sand}} = 30°$, relative density of sand = 70%. Estimate the allowable pullout capacity of the pile ($FS = 4$).

**8.13** Redo Problem 8.12 with the following changes: $\gamma_{\text{sand}} = 118 \text{ lb/ft}^3$, $\phi_{\text{sand}} = 37°$, and relative density of compaction of sand = 80%.

**8.14** A concrete pile 50 ft long is embedded in a saturated clay with $c_u = 850 \text{ lb/ft}^2$. The pile is 12 in. × 12 in. in cross section. Using $FS = 4$, determine the allowable pullout capacity of the pile. Use Eq. (8.58).

**8.15** Redo Problem 8.14 with the following changes: The top 5 m of the clay has a $c_u = 25 \text{ kN/m}^2$, and below that $c_u = 55 \text{ kN/m}^2$. Use Eq. (8.58).

**8.16** A steel H-pile (Section: HP 14 × 102; see Table B.1 in Appendix B) is driven by a MKT S-20 hammer (see Table 8.6). Given:

Length of pile = 80 ft
Coefficient of restitution = 0.35
Weight of pile cap = 17.8 kip
Hammer efficiency = 0.84
Number of blows for last 1 in. of penetration = 8

Estimate the ultimate pile capacity by using Eqs. (8.83), (8.84), and (8.86). For the pile, $E_p = 30 \times 10^6 \text{ lb/in.}^2$

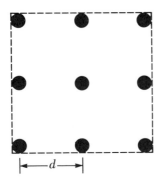

**Figure P8.17**

**8.17** The plan of a group pile (friction pile) in sand is shown in Figure P8.17. The piles are circular in cross section and have an outside diameter of 18 in. The center-to-center spacings of the piles ($d$) are 36 in. Using Eq. (8.95), find the efficiency of the pile group.

**8.18** Refer to Problem 8.17. If the center-to-center pile spacing is increased to 1050 mm, what will be the group efficiency?

**8.19** Solve Problem 8.17 using the Converse-Labarre equation.

**8.20** Solve Problem 8.18 using the Converse-Labarre equation.

**8.21** The plan of a group pile is shown in Figure P8.17. Assume that the piles are embedded in a saturated homogeneous clay having a $c_u$ = 95.8 kN/m². Given:

Diameter of piles ($D$) = 406 mm
Center-to-center spacing of piles = 850 mm
Length of piles = 18.5 m

Find the allowable load-carrying capacity of the pile group. Use $FS$ = 3.

**8.22** The section of a group pile (3 × 4) in a layered saturated clay is shown in Figure P8.22. The piles are square in cross section (14 in. × 14 in.). The center-to-center spacing ($d$) of the piles is 35 in. Determine the allowable load-bearing capacity of the pile group. Use $FS$ = 4.

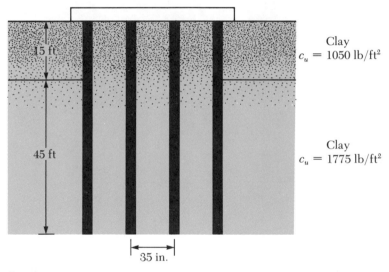

**Figure P8.22**

**8.23** Figure P8.23 on page 414 shows a group pile in clay. Determine the consolidation settlement of the group.

**8.24** Refer to Example Problem 8.14. If the pile group has $L_g$ = 3.5 m and $B_g$ = 3 m; and the clay layer No. 1 is preconsolidated ($p_c$ = 155 kN/m², $C_s = \frac{1}{4}C_c$), determine the settlement of the pile group.

**8.25** Figure 8.35a shows a pile. Let $L$ = 50 ft; diameter of pile = 18 in.; height of clay fill, $H_f$ = 11.5; unit weight of fill, $\gamma_f$ = 112 lb/ft³; and $\phi_{fill}$ = 28°. Determine the total down-drag force on the pile. Assume that the fill is located above the ground water table. Assume $\delta = 0.6\phi_{fill}$.

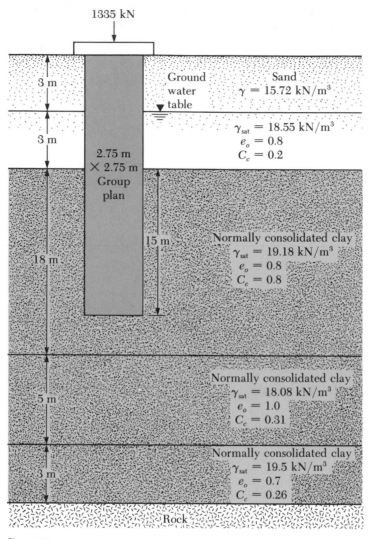

1335 kN

2.75 m
X 2.75 m
Group
plan

3 m

3 m

18 m

15 m

5 m

3 m

Ground
water
table

Sand
$\gamma = 15.72$ kN/m³

$\gamma_{sat} = 18.55$ kN/m³
$e_o = 0.8$
$C_c = 0.2$

Normally consolidated clay
$\gamma_{sat} = 19.18$ kN/m³
$e_o = 0.8$
$C_c = 0.8$

Normally consolidated clay
$\gamma_{sat} = 18.08$ kN/m³
$e_o = 1.0$
$C_c = 0.31$

Normally consolidated clay
$\gamma_{sat} = 19.5$ kN/m³
$e_o = 0.7$
$C_c = 0.26$

Rock

**Figure P8.23**

**8.26** Redo Problem 8.25 assuming that the ground water table coincides with the top of the fill and $\gamma_{sat(fill)} = 124.5$ lb/ft³. Other quantities remaining the same, what would be the down-drag force on the pile? Assume $\delta = 0.6\phi_{fill}$.

**8.27** Refer to Figure 8.35b. Let $L = 19$ m, $\gamma_{fill} = 15.2$ kN/m³, $\gamma_{sat(clay)} = 19.5$ kN/m³, $\phi_{clay} = 30°$, $H_f = 3.2$ m, and pile diameter $= 0.460$ m. The ground water table coincides with the top of the clay layer. Determine the total down-drag force on the pile. Assume $\delta = 0.5\phi_{clay}$.

# References

American Society of Civil Engineers (1959). "Timber Piles and Construction Timbers," *Manual of Practice*, No. 17.

Bowles, J.E. (1982). *Foundation Design and Analysis*, McGraw-Hill, New York.

Coyle, H.M., and Castello, R.R. (1981). "New Design Correlations for Piles in Sand," *Journal of the Geotechnical Engineering Division*, American Society of Civil Engineers, Vol. 107, No. GT7, pp. 965–986.

Das, B.M., and Seeley, G.R. (1975). "Uplift Capacity of Buried Model Piles in Sand," *Journal of the Geotechnical Engineering Division*, American Society of Civil Engineers, Vol. 101, No. GT10, pp. 1091–1094.

Das, B.M., and Seeley, G.R. (1982). "Uplift Capacity of Pipe Piles in Saturated Clay," *Soils and Foundations*, The Japanese Society of Soil Mechanics and Foundation Engineering, Vol. 22, No. 1, pp. 91–94.

Davisson, M.T., and Gill, H.L. (1963). "Laterally Loaded Piles in a Layered Soil System," *Journal of the Soil Mechanics and Foundations Division*, American Society of Civil Engineers, Vol. 89, No. SM3, pp. 63–94.

Goodman, R.E. (1980). *Introduction to Rock Mechanics*, Wiley, New York.

International Conference of Building Officials (1982). "Uniform Building Code," Whittier, Calif.

Janbu, N. (1953). *An Energy Analysis of Pile Driving with the use of Dimensionless Parameters*, Norwegian Geotechnical Institute, Oslo, Publication No. 3.

Matlock, H., and Reese, L.C. (1960). "Generalized Solution for Laterally Loaded Piles," *Journal of the Soil Mechanics and Foundations Division*, American Society of Civil Engineers, Vol. 86, No. SM5, Part I, pp. 63–91.

McClelland, B. (1974). "Design of Deep Penetration Piles for Ocean Structures," *Journal of the Geotechnical Engineering Division*, American Society of Civil Engineers, Vol. 100, No. GT7, pp. 709–747.

Meyerhof, G.G. (1976). "Bearing Capacity and Settlement of Pile Foundations," *Journal of the Geotechnical Engineering Division*, American Society of Civil Engineers, Vol. 102, No. GT3, pp. 197–228.

Michigan State Highway Commission (1965). *A Performance Investigation of Pile Driving Hammers and Piles*, Lansing, Michigan, 338 pp.

Vesic, A.S. (1961). "Bending of Beams Resting on Isotropic Elastic Solids," *Journal of the Engineering Mechanics Division*, American Society of Civil Engineers, Vol. 87, No. EM2, pp. 35–53.

Vesic, A.S. (1969). "Experiments with Instrumented Pile Groups in Sand," American Society for Testing and Materials; Special Technical Publication, No. 444, pp. 177–222.

Vesic, A.S. (1970). "Tests on Instrumented Piles—Ogeechee River Site," *Journal of the Soil Mechanics and Foundations Division*, American Society of Civil Engineers, Vol. 96, No. SM2, pp. 561–584.

Vesic, A.S. (1977). *Design of Pile Foundations*, National Cooperative Highway Research Program Synthesis of Practice No. 42, Transportation Research Board, Washington, D.C.

Vijayvergiya, V.N., and Focht, J.A., Jr. (1972). *A New Way to Predict Capacity of Piles in Clay*, Offshore Technology Conference Paper 1718, Fourth Offshore Technology Conference, Houston, Texas.

Woodward, R.J., Gardner, W.S., and Greer, D.M. (1972). *Drilled Pier Foundations*, McGraw-Hill, New York.

# Drilled-Pier and Caisson Foundations

## 9.1

### Introduction

The terms *caisson, pier,* and *drilled pier* are often used interchangeably in foundation engineering; they all refer to a *cast-in-place pile having a diameter of about 750 mm* or more, with or without steel reinforcements and with or without an enlarged bottom. In order to avoid confusion, this text will use the following definitions:

The term *drilled pier* is used when a hole is drilled or excavated to the bottom of a structure's foundation and then filled with concrete. Depending on the soil conditions, casings or *laggings* (boards or sheet piles) may be used to prevent the soil around the hole from caving in during construction. The diameter of the pier shaft is usually large enough for a person to enter for inspection.

The use of drilled-pier foundations has several advantages. For example:

**1.** A single drilled pier can be used to replace a group of piles and the pile cap.

**2.** It is easier to construct drilled piers in deposits of dense sand and gravel than it is to drive piles.

**3.** Construction of drilled piers can be completed before the completion of grading operations.

**4.** When piles are driven by a hammer, the gound vibration may cause damage to the nearby structures. The use of drilled piers does not present such hazards.

**5.** Piles driven into clay soils may produce ground heaving and may also cause previously driven piles to move laterally. Such conditions do not exist in the construction of drilled piers.

**6.** There is no hammer noise during the construction of drilled piers, as there is during pile driving.

**7.** Because the base of a drilled pier can be enlarged, it provides great resistance to the uplifting load (Section 9.8).

**8.** The surface over which the base of the drilled pier is constructed can be visually inspected.

**9.** Construction of drilled piers generally requires light, mobile equipment. Under proper soil conditions, they may prove to be more economical than pile foundations.

**10.** Drilled piers have high resistance to lateral loads.

There are also several drawbacks to the use of drilled-pier construction. The concreting operation always needs very close supervision. This can be delayed to some extent during bad weather. A more detailed soil exploration is generally required before making decisions about drilled-pier construction than is the case in other types of foundation construction. Also, as in the case of braced cuts, deep excavations for drilled piers may induce substantial ground loss and damage to closely located structures.

The term *caisson* refers to a substructure element used at wet construction sites, such as rivers, lakes, and docks. For the construction of caissons, a hollow shaft or a box is sunk into position to rest on firm ground. The lower part of the shaft or the box is provided with a cutting edge to help it penetrate the soft soil layers below the water level and come to rest on a load-bearing stratum. The material inside the shaft or box is dredged through the openings at the top, and then concrete is poured in. Bridge abutments, quay walls, and structures for shore protection can be built over caissons.

## Drilled Piers

### 9.2

### Types of Drilled Pier

Drilled piers can be classified according to the manner in which they are designed to transfer the structural load to the substratum. Figure 9.1a shows a drilled pier that has a *straight shaft*. It extends through the upper layer(s) of poor soil, and its tip rests on a strong load-bearing soil layer or rock. The shaft can be cased with steel shell or pipe when required (as in the case of cased, cast-in-place concrete piles; Figure 8.4). For these piers, the resistance to the applied load may develop from end bearing and also from side friction at the pier perimeter and soil interface.

A *belled pier* (Figure 9.1b and c) consists of a straight shaft with a bell at the bottom. The bell rests on good bearing soil. The bell can be constructed in the shape of a dome (Figure 9.1b), or it can be angled (Figure 9.1c). For angled bells, the underreaming tools commercially available are such that the sides can make 30° to 45° angles with the vertical. For the majority of drilled piers

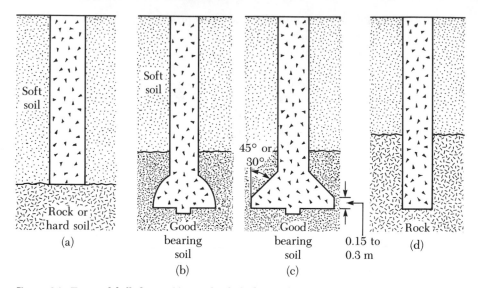

**Figure 9.1**  Types of drilled pier: (a) straight shafted pier; (b) and
(c) belled pier; (d) straight shafted pier socketed into rock

constructed in the United States, the entire load-carrying capacity is assigned
to the end bearing only. However, in certain circumstances, the end-bearing
capacity and the side friction are taken into account. In Europe, both the side
frictional resistance and the end-bearing capacity are always taken into account.

Straight-shafted piers can also be extended into an underlying rock layer
(Figure 9.1d). In the calculation of the load-bearing capacity of such piers, the
end bearing and the shear stress developed along the pier perimeter and
rock interface can be taken into account. This is explained in more detail in
Section 9.7.

## 9.3

### Construction Procedures

One of the oldest methods of construction of drilled piers is the *Chicago*
method (Figure 9.2a). In this method, circular holes with diameters of 1.1 m or
more are excavated by hand for depths varying from 0.6–1.8 m at a time. The
sides of the excavated hole are then lined with vertical boards. These boards are
referred to as *laggings*. They are kept tightly in place by two circular steel rings.
After the placement of the rings, the excavation is continued for another
0.6–1.8 m. When the desired depth of excavation is reached, the bell of the pier
is excavated. Following the completion of the excavation, the hole is filled with
concrete.

In the *Gow* method of pier construction (Figure 9.2b), excavation of the
hole is done by hand. Telescopic metal shells are used to maintain the shaft.
The shells can be removed one section at a time as the concreting progresses.
The minimum diameter of a Gow drilled pier is about 1.22 m. Any given section

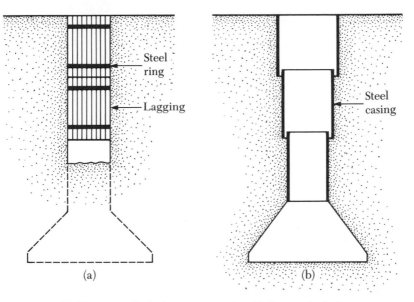

**Steel ring**

**Lagging**

**Steel casing**

(a)

(b)

**Figure 9.2** (a) Chicago method of pier construction; (b) Gow method of pier construction

of the shell is about 50 mm less in diameter than the section located immediately above it. Piers as deep as 30 m have been installed by this method.

Most of the pier shaft excavations nowadays are done mechanically rather than by hand. Open helix augers (flight augers) are common excavation tools. These augers have cutting edges or cutting teeth. The augers with cutting edges are used mostly for drilling in soft, homogeneous soil; those with cutting teeth are for drilling in hard soil and hard pan. The auger is attached to a square shaft referred to as the *Kelly* and pushed into the soil and rotated. When the flights are filled with soil, the auger is raised above the ground surface, and the soil is dumped into a pile by rotating the auger at high speed. These augers are available in various diameters; sometimes they may be as large as 3 m or more.

When the excavation is extended to the level of the load-bearing stratum, the auger is replaced by underreaming tools to shape the bell, if required. An underreamer essentially consists of a cylinder with two cutting blades that are hinged to the top of the cylinder (Figure 9.3). When the underreamer is lowered into the hole, the cutting blades stay folded inside the cylinder. When the bottom of the hole is reached, the blades are spread outward, and the underreamer is rotated. The loose soil falls inside the cylinder, which is raised periodically and emptied until the bell is completed. Most underreamers can cut bells with diameters as large as three times the diameter of the shaft.

Another common pier-drilling device is the *bucket type drill*. It is essentially a bucket with an opening and cutting edges at the bottom. The bucket is attached to the Kelly and rotated. The loose soil is collected in the bucket, which is periodically raised and emptied. Holes as large as 5–5.5 m in diameter can be drilled with this type of equipment.

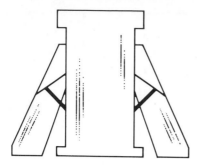

**Figure 9.3**   Underreamer

When rock is encountered during drilling, *core barrels* with *tungsten carbide teeth* attached to the bottom of the barrels are used. *Shot barrels* are also used for drilling into very hard rock. The principle of rock coring by a shot barrel is shown in Figure 9.4. The drill stem is attached to the shot barrel's plate. The barrel has some feeder slots through which chilled steel shots are supplied to the bottom of the bore hole. The steel shots cut the rock when the barrel is rotated. Water is supplied to the drill hole through the drill stem. Fine

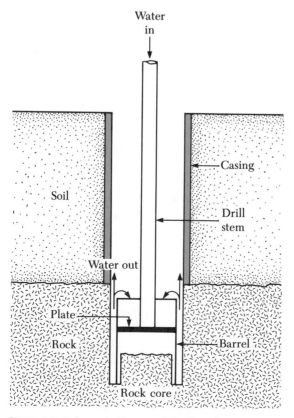

**Figure 9.4**   Schematic diagram of shot barrel

rock and steel particles (that are produced by the grinding of the steel shots) are washed upward, and they settle on the upper portion of the barrel.

The *Benoto machine* is another type of pier-drilling equipment that is generally used when the drilling conditions are difficult and many boulders are in the soil. It essentially consists of a steel tube that can be oscillated and pushed into the soil. A tool usually referred to as the *hammer grab*, which is fitted with cutting blades and jaws, is used to break up the soil and rock inside the tube and remove them.

### Use of Casings and Drilling Mud

When holes for piers are driven in soft clays, the soil tends to squeeze in and close the hole. In such situations, casings may be used to keep the hole open. The casings may have to be driven before excavation begins. Holes made in gravelly and sandy soils also tend to cave in. Excavation of pier holes in these soils can be continued either by casing as the hole progresses or by using *drilling mud*. As Chapter 2 pointed out, drilling mud is also used during field exploration.

### Inspection of the Bottom of the Hole

In many instances, the bottom of the hole must be inspected to ensure that the load-bearing stratum is what was anticipated and that the bell is properly done. For these reasons, an inspector must descend to the bottom of the hole. Several safety precautions must be observed during this procedure:

**1.** If a casing is not already in the hole, one should be lowered by crane. This will prevent the hole and the bell from collapsing.
**2.** The hole should be tested for the presence of poisonous or explosive gasses. This can be done by using a miner's safety lamp.
**3.** The inspector should wear a safety harness.
**4.** The inspector should also carry a safety lamp and an air tank.

## 9.4

### Other Design Considerations

For the design of ordinary drilled piers without casings, a minimum amount of vertical steel reinforcement is always desirable. Minimum reinforcement equals one percent of the gross cross-sectional area of the shaft. In the state of California, a reinforcing cage having a length of about 3.65 m is used in the top part of the caisson, and no reinforcement is provided at the bottom. This procedure helps in the construction process in that the cage is placed after most of the concreting is complete.

For piers with nominal reinforcement, most building codes suggest using a value of the design concrete strength $(f_c)$ of the order of $f'_c/4$. Thus, the minimum shaft diameter can be given as

$$f_c = 0.25f'_c = \frac{Q_w}{A_{gs}} = \frac{Q_w}{\frac{\pi}{4}D_s^2}$$

or

$$D_s = \sqrt{\frac{Q_w}{\left(\dfrac{\pi}{4}\right)(0.25)f'_c}} = 2.257 \sqrt{\frac{Q_w}{f'_c}} \tag{9.1}$$

where $D_s$ = diameter of the pier shaft
$\quad\quad f'_c$ = 28 days concrete strength
$\quad\quad Q_w$ = working load of the pier
$\quad\quad A_{gs}$ = gross cross-sectional area of the shaft

Depending on the loading conditions for a given pier, the reinforcement percentage may sometimes be too high. In that case, use of a *single rolled-steel section* at the center of the pier (Figure 9.5b) may be considered. In that case

$$Q_w = (A_{gs} - A_s)f_c + A_s\sigma_{all} \tag{9.2}$$

where $A_s$ = area of the steel section
$\quad\quad \sigma_{all}$ = allowable strength of steel $\approx 0.5\sigma_{yield}$

When a permanent steel casing is used for construction instead of a central rolled-steel section (Figure 9.5a), the preceding equation may also be used. However, $\sigma_{all}$ for steel should be of the order of $0.4\sigma_{yield}$.

If drilled piers are likely to be subjected to tensile loads, reinforcement should be continued for the entire length of the pier.

### Concrete Mix Design

The concrete mix design for piers is not very different from any other concrete structure. When a reinforcing cage is used, consideration should be

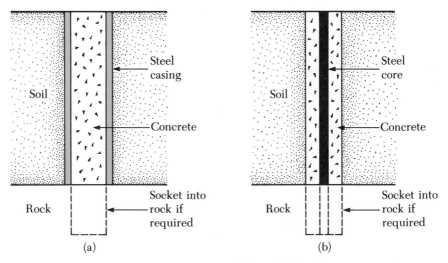

**Figure 9.5**  Drilled piers with (a) steel casing and (b) a central steel core

given to the ability of the concrete to flow through the reinforcement. In most cases, a concrete slump of about 150 mm is considered satisfactory. Also, the maximum size of the aggregate should be limited to about 20 mm.

## 9.5

## Estimation of Load-Bearing Capacity

The ultimate load-bearing capacity of a drilled pier (Figure 9.6) can be given as

$$Q_u = Q_p + Q_s \tag{9.3}$$

where $Q_u$ = ultimate load
$Q_p$ = ultimate load-carrying capacity at the base
$Q_s$ = frictional (skin) resistance

The ultimate base load can be given by the equation (similar to that for shallow foundations)

$$Q_p = A_p(cN_c^* + q'N_q^* + 0.3\gamma D_b N_\gamma^*) \tag{9.4a}$$

where $N_c^*, N_q^*, N_\gamma^*$ = the bearing capacity factors
$q'$ = vertical effective stress at the level of the bottom of the pier

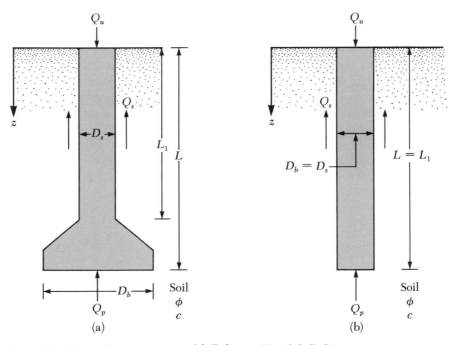

**Figure 9.6** Ultimate bearing capacity of drilled piers: (a) with bell; (b) straight shaft

$D_b$ = diameter of the base (see Figure 9.6a and b)
$A_p$ = area of the base = $\pi/4D_b^2$

In most cases, the last term (containing $N_\gamma^*$) is neglected except for relatively short piers, so

$$Q_p = A_p(cN_c^* + q'N_q^*) \tag{9.4b}$$

The net load-carrying capacity at the base (that is, the gross load minus the weight of the pier) can be approximated as

$$Q_{p(net)} = A_p(cN_c^* + q'N_q^* - q') = A_p[cN_c^* + q'(N_q^* - 1)] \tag{9.5}$$

The frictional, or skin, resistance, $Q_s$, can be expressed in a form similar to that for piles:

$$Q_s = \int_0^{L_1} pf\, dz \tag{9.6}$$

where $p$ = shaft perimeter = $\pi D_s$
$f$ = unit frictional (or skin) resistance

### Piers in Sand

For piers in sand, $c = 0$ and, hence, Eq. (9.5) can be simplified to

$$Q_{p(net)} = A_p q'(N_q^* - 1) \tag{9.7}$$

Determination of $N_q^*$ is always a problem for deep foundations, as was shown in the case of piles. Note, however, that all piers are *drilled*, unlike the majority of piles, which are *driven*. For similar initial soil conditions, the actual value of $N_q^*$ may be substantially lower for objects drilled and placed *in situ* compared to objects that are driven. Vesic (1967) has compared the theoretical results obtained by several investigators relating to the variation of $N_q^*$ with soil friction angle. These investigators include DeBeer, Meyerhof, Hansen, Vesic, and Terzaghi. The values of $N_q^*$ given by Vesic (1963) are approximately the lower bound, and, hence, they are recommended for use in this text (see Figure 9.7). It is recommended that Eq. (8.19) also be used for calculation of the ultimate point load, $Q_p$. Thus

$$Q_{p(net)} = A_p(\sigma'N_\sigma^* - q')$$

where $\sigma' = [(1 + 2K_o)/3]q'$

or

$$Q_{p(net)} = A_p\left[\frac{(1 + 2K_o)}{3}N_\sigma^* - 1\right]q' \tag{9.8}$$

Table 8.8 gave the values of $N_\sigma^*$ for various values of $I_{rr}$ and soil friction angles. For ease of calculation, the $N_\sigma^*$ values given in that table are plotted in Figure 9.8.

The frictional resistance at ultimate load, $Q_s$, developed in a drilled pier can be calculated from the relation given in Eq. (9.6). Referring to this equation

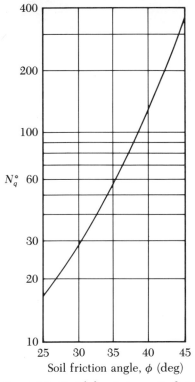

**Figure 9.7** Vesic's bearing capacity factor, $N_q^*$, for deep foundations

$$p = \text{perimeter of shaft} = \pi D_s$$

$$f = K\sigma_v' \tan \delta \tag{9.9}$$

where $K$ = earth pressure coefficient $\approx K_o = 1 - \sin \phi$

$\sigma_v'$ = effective vertical stress at any depth $z$

Thus

$$Q_s = \int_0^{L_1} Kf \, dz = \pi D_s (1 - \sin \phi) \int_0^{L_1} \sigma_v' \, dz \tag{9.10}$$

The value of $\sigma_v'$ will increase up to a depth of about $15D_s$ and will remain constant thereafter, as shown in Figure 8.14 on p. 359.

An appropriate factor of safety should be applied to the ultimate load to obtain the net allowable load, or

$$Q_{u(net)} = \frac{Q_{p(net)} + Q_s}{FS} \tag{9.11}$$

It is extremely important that a reliable estimate of the soil friction angle, $\phi$, is made in obtaining the net base resistance, $Q_{p(net)}$. Figure 9.9 shows a conservative correlation between the soil friction angle and the corresponding corrected standard penetration resistance numbers in granular soils. However,

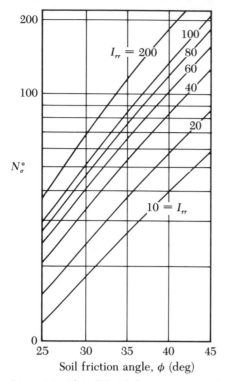

**Figure 9.8**   Plot of Vesic's bearing capacity factor, $N_\sigma^*$ (see Table 8.8)

these friction angles are valid for low confining pressures. At higher confining pressures, which occur in the case of deep foundations, the value of $\phi$ can decrease substantially for medium to dense sands. This affects the value of $N_q^*$ or $N_\sigma^*$ (that is, $I_{rr}$) to be used for estimation of $Q_p$. For example, Vesic (1977) has shown that for Chattahoochee River sand at a relative density of about 80%, the triaxial angle of friction is about 45° at a confining pressure of 70 kN/m². However, at a confining pressure of 10.35 MN/m², the friction angle is about 32.5°. This will ultimately result in a tenfold decrease of $N_q^*$ or $N_\sigma^*$. So, it is the opinion of the author that, for general working conditions of drilled piers, the estimated friction angle determined from Figure 9.9 should be reduced by about 10–15%. In general, the existing experimental values show the following range of $N_q^*$ for standard drilled piers (or cast-in-place piles).

| Sand type | Relative density of sand | Range of $N_q^*$ |
|---|---|---|
| Loose | 40 or less | 10–20 |
| Medium | 40–60 | 25–40 |
| Dense | 60–80 | 30–50 |
| Very dense | >80 | 75–90 |

Touma and Reese (1974) have suggested, based on the performance of bored piles in sand with an average diameter of 750 mm, the following pro-

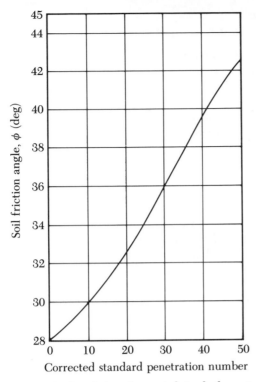

**Figure 9.9** Correlation of corrected standard penetration number with the soil friction angle

cedure for calculation of the allowable load-carrying capacity. This can also apply to drilled piers in sand.

For $L > 10D_b$ and a base movement of 25.4 mm

$$Q_{p(net)} = \frac{0.508A_p}{D_b} q_p \qquad (9.12)$$

where $Q_{p(net)}$ is in kN, $A_p$ is in m², $D_b$ is in m, and $q_p$ is the unit point resistance in kN/m².

The values of $q_p$ as recommended by Touma and Reese are as follows:

| Sand type | $q_p(kN/m^2)$ |
|---|---|
| Loose sand | 0 |
| Medium sand | 1530 |
| Very dense | 3830 |

For sands of intermediate densities, linear interpolation can be used. The shaft friction resistance can be calculated as

$$Q_s = \int_0^{L_1} (0.7)p\sigma'_v \tan \phi \, dz = 0.7(\pi D_s) \int_0^{L_1} \sigma'_v \tan \phi \, dz$$

$$= 2.2D_s \int_0^{L_1} \sigma'_v \tan \phi \, dz \qquad (9.13)$$

where $\phi$ = soil friction angle

$\sigma_v'$ = vertical effective stress at a depth $z$

With the base resistance and frictional resistance known, a suitable factor of safety (about 3) may be applied to determine the net allowable load.

### Piers in Clay

Referring to Eq. (9.5), in saturated clays with $\phi = 0$, the value of $N_q^*$ is equal to one; and, hence, the net base resistance becomes

$$Q_{p(net)} = A_p c_u N_c^* \tag{9.14}$$

where $c_u$ = undrained cohesion

The bearing capacity factor $N_c^*$ is usually taken to be equal to 9. It can be seen from Figure 7.9 on p. 319 that, when the $L/D_b$ ratio is 4 or more, $N_c^* = 9$. This is the condition in the case of most drilled piers. Experiments by Whitaker and Cooke (1966) have shown that, for belled piers, the full value of $N_c^* = 9$ is realized with a base movement of about 10–15% of $D_b$. Similarly, for piers with straight shafts (that is, $D_b = D_s$), the full value of $N_c^* = 9$ is obtained with a base movement of about 20% of $D_b$.

The skin resistance of piers in clay can be expressed by a relation similar to Eq. (8.37b), or

$$Q_s = \sum_{L=0}^{L=L_1} \alpha^* c_u p\, \Delta L \tag{9.15}$$

where $p$ = perimeter of the pier cross section

The value of $\alpha^*$ that can be used in Eq. (9.15) has not yet been fully established. However, the field test results available at this time indicate that $\alpha^*$ may vary from between 0.35 to 0.6. So, conservatively, one can assume that

$$\alpha^* = 0.4 \tag{9.16}$$

## 9.6

## Settlement of Piers at Working Load

The settlement of drilled piers at working load can be calculated in a manner similar to the one outlined in Section 8.9. In many cases, the load carried by shaft resistance is small compared to the load carried at the base. In such cases, the contribution of $s_3$ may be ignored. Note that, in Eqs. (8.51) and (8.52), the term $D$ should be replaced by $D_b$ for piers.

### Example
### 9.1

A soil profile is shown in Figure 9.10. A point bearing pier with a bell is proposed for placement in the dense sand and gravel layer. The working load, $Q_w$, is 2000 kN.

a. Assuming $f_c' = 21,000$ kN/m², determine the shaft diameter.

b. Using Eq. (9.7) and a factor of safety of 4, determine the bell diameter, $D_b$. Ignore the frictional resistance of the shaft.

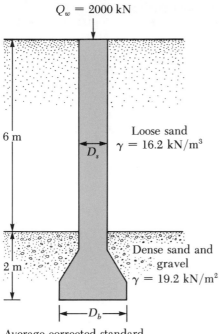

$Q_w = 2000$ kN

6 m

Loose sand
$\gamma = 16.2$ kN/m³

$D_s$

Dense sand and gravel
$\gamma = 19.2$ kN/m²

2 m

$D_b$

Average corrected standard
penetration number $= 40 = N_{cor}$

**Figure 9.10**

c. Using Eq. (9.12), obtain $D_b$ for a settlement of 25.4 mm. Ignore the frictional resistance of the shaft. Use $q_p = 3000$ kN/m².

d. Discuss the differences in the results obtained from parts b and c.

e. Choose a bell diameter for a settlement of 25.4 mm.

f. Estimate Young's modulus for the dense sand layer.

g. Calculate the possible settlement of the pier by using Eqs. (8.49), (8.50), and (8.51). [Assume $s_3$ in Eq. (8.49) to be zero.] Use $D_s$ determined in part a and $D_b$ in part e; $\mu_s = 0.3$, $E_c = 21 \times 10^6$ kN/m².

## Solution

**Part a: Determination of the Shaft Diameter, $D_s$**

From Eq. (9.1)

$$D_s = 2.257 \sqrt{\frac{Q_w}{f_c'}}$$

Given: $Q_w = 2000$ kN and $f_c' = 21,000$ kN/m². So

$$D_s = 2.257 \sqrt{\frac{2000}{21,000}} = 0.697 \text{ m}$$

Use $D_s = 1$ m.

Part b: Determination of the Bell Diameter Using Eq. (9.7)

$$Q_{p(net)} = A_p q'(N_q^* - 1)$$

Referring to Figure 9.9, for $N_{cor} = 40$, the value of $\phi$ is approximately equal to 39.5°. To be conservative, we will use a reduction of about 10%. So, assume $\phi = 35.6$. From Figure 9.6, $N_q^* \approx 60$.

$$q' = 6(16.2) + 2(19.2) = 135.6 \text{ kN/m}^2$$
$$Q_{p(net)} = (Q_w)(FS) = (2000)(4) = 8000 \text{ kN}$$
$$8000 = (A_p)(135.6)(60 - 1)$$
$$A_p = 1.0 \text{ m}^2$$

$$D_b = \sqrt{\frac{1.0}{\dfrac{\pi}{4}}} = \underline{1.13 \text{ m}}$$

Part c: Determination of Bell Diameter Using Eq. (9.12)

$$Q_{p(net)} = \frac{0.508 A_p}{D_b} q_p$$

Because the limit of settlement is 25.4 mm and Eq. (9.12) is for the limiting settlement, a factor of safety over $Q_w$ should not be used to obtain $Q_{p(net)}$. So

$$Q_w = Q_{p(net)}$$

Thus

$$Q_{p(net)} = Q_w = 2000 = \frac{0.508 A_p}{D_b} q_p = \frac{(0.508)\left(\dfrac{\pi}{4}\right)(D_b^2) q_p}{D_b}$$

$$= 0.399 D_b q_p$$

or

$$D_b = \frac{2000}{(0.399)(3000)} = \underline{1.67 \text{ m}}$$

Part d:

The value of $D_b$ determined in part b corresponds to an allowable bearing capacity that is based on the ultimate bearing capacity. The settlement has not been taken into consideration at all. The ultimate bearing capacity of drilled piers may occur at a settlement exceeding 10–15% of the bell diameter. The bell diameter in part c corresponds to a settlement of 25.4 mm.

Part e:

Use a bell diameter of 2 m to be conservative.

Part f:

The effective overburden pressure at the surface of the dense sand gravel layer is 135.6

$kN/m^2$. Substituting this value of effective overburden pressure and $N_{cor} = 40$ in Eq. (2.5), the uncorrected value of $N$ is about 41. Using Eq. (3.66), Young's modulus

$$E_s = 766N = 766(41) = 31,406 \text{ kN/m}^2$$

## Part g: Calculation of Settlement

From Eq. (8.50), with $\xi = 0$

$$s_1 = \frac{Q_w L}{A_s E_p} = \frac{(2000)(6)}{\frac{\pi}{4}(1)^2(21 \times 10^6)} = 0.000728 \text{ m} = 0.728 \text{ mm}$$

From Eq. (8.51)

$$s_2 = \frac{q_{wp} D_b}{E_s}(1 - \mu_s^2)I_{wp}$$

$$I_{wp} = 0.88 \text{ (Figure 3.13)}$$

$$q_{wp} = \frac{2000}{\frac{\pi}{4}D_b^2} = \frac{2000}{\frac{\pi}{4}(2)^2} = 636.6 \text{ kN/m}^2$$

$$s_2 = \frac{(636.6)(2)}{31,406}(1 - 0.3^2)(0.88) = 32.46 \text{ mm}$$

So, the total settlement is $32.46 + 0.728 = 33.19$ mm

This settlement is higher than 25.4 mm. However, the assumption of $E_s$ is approximate.

## Example 9.2

A drilled pier is shown in Figure 9.11. Given: $Q_w = 2800$ kN.

a. Assume $f_c' = 28,000 \text{ kN/m}^2$. Determine if the proposed diameter of the shaft is adequate or not.
b. Determine the net ultimate point load-carrying capacity.
c. Determine the ultimate skin resistance.
d. Calculate the factor of safety with respect to the working load, $Q_w$.
e. Estimate the total elastic settlement of the pier under the working load. Use Eqs. (8.50), (8.52), and (8.55). $E_p = 22 \times 10^6 \text{ kN/m}^2$.

## Solution
## Part a

From Eq. (9.1)

$$D_s = 2.257\sqrt{\frac{Q_w}{f_c'}} = 2.257\sqrt{\frac{2,800}{28,000}} = 0.714 \text{ m}$$

Proposed shaft diameter $D_s = 1 \text{ m} > 0.714 \text{ m—O.K.}$

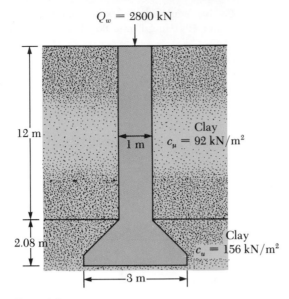

**Figure 9.11**

## Part b

$$Q_{p(net)} = c_u N_c^* A_p = (156)(9)\left[\left(\frac{\pi}{4}\right)(3)^2\right] = \underline{9924.3 \text{ kN}}$$

## Part c

According to Eq. (9.15)

$$Q_s = \pi D_s L_1 c_u \alpha^*$$

Assume $\alpha^* = 0.4$ [Eq. (9.16)]

$$Q_s = (\pi)(1)(12)(92)(0.4) = \underline{1387.3 \text{ kN}}$$

## Part d

Factor of safety =

$$\frac{Q_u}{Q_w} = \frac{9924.3 + 1387.3}{2800} = \underline{4.04}$$

## Part e

From Eq. (8.50)

$$s_1 = \frac{(Q_{wp} + \xi Q_{ws})L}{A_p E_p}$$

Assume full mobilization of the skin resistance. So, $Q_{ws} = 1387.3$ kN. Hence $Q_{wp} = 2800 - 1387.3 = 1412.7$ kN. Also, let $\xi = 0.6$.

$$A_p = \left(\frac{\pi}{4}\right)D_s^2 = \frac{\pi}{4}(1)^2 = 0.785 \text{ m}^2$$

So

$$s_1 = \frac{[1412.7 + (0.6)(1387.3)](14.08)}{(0.785)(22 \times 10^6)} = 0.00183 \text{ m} = \underline{1.83 \text{ mm}}$$

From Eq. (8.52)

$$s_2 = \frac{Q_{wp}C_p}{D_b q_p}$$

The value of $C_p$ for stiff clay is about 0.04 (Table 8.11).

$$q_p = cN_c^* = (156)(9) = 1404 \text{ kN/m}^2$$

So

$$s_2 = \frac{(1412.7)(0.04)}{(3)(1404)} = 0.01342 \text{ m} = \underline{13.42 \text{ mm}}$$

Again, from Eq. (8.5a)

$$s_3 = \frac{Q_{ws}C_s}{L_1 q_p}$$

$$C_s = \left(0.93 + 0.16\sqrt{\frac{L_1}{D_s}}\right)C_p = \left(0.93 + 0.16\sqrt{\frac{12}{1}}\right)0.04 = 0.0594$$

Hence

$$s_3 = \frac{(1387.3)(0.0594)}{(12)(1404)} = 0.00489 \text{ m} = \underline{4.89 \text{ mm}}$$

Total elastic settlement $s = s_1 + s_2 + s_3 = 1.83 + 13.42 + 4.89 = \underline{20.14 \text{ mm}}$

## 9.7
## Load-Bearing Capacity of Piers Extending into Rock

Section 9.1 noted that drilled piers can be extended into rock. This section discusses the principles of analysis of the load-bearing capacity of such drilled piers. Figure 9.12 shows a drilled pier whose depth of embedment in the rock is equal to $L$. Let the total load to be carried by the pier be equal to $Q_w$. Also, assume that (a) no load transfer takes place along the soil-pier interface and (b) under the pier loading condition, the *bond between the concrete and the rock is broken*. With these assumptions, the compressive stress along the cross section of the pier can be given as

$$\sigma_z = \sigma_{max} \exp\left[-\frac{2\mu_c \tan \phi_{rc}}{1 - \mu_c + (1 + \mu_r)\dfrac{E_c}{E_r}} \cdot \frac{2z}{D_b}\right] \tag{9.17}$$

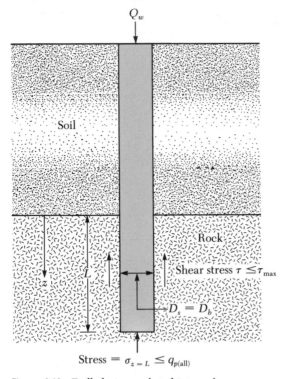

**Figure 9.12**   Drilled piers socketed into rock

where $\sigma_{max} = \dfrac{Q_{all}}{\dfrac{\pi}{4}D_b^2}$                                                                                 (9.18)

$\mu_c$, $\mu_r$ = Poisson's ratio in concrete and rock, respectively
$E_c$, $E_r$ = Young's modulus of concrete and rock, respectively
$\phi_{rc}$ = angle of friction along the rock-pier interface

The maximum shear stress developed along the rock and concrete pier perimeter can be given conservatively as (Goodman, 1980)

$$\tau_{max} = \frac{q_u}{20}$$                                                                                                          (9.19)

where $q_u$ = unconfined compression strength of concrete or rock, whichever is lower

The allowable shear stress can then be determined as

$$\tau_{all} = \frac{\tau_{max}}{FS}$$                                                                                                   (9.20)

where $FS$ = factor of safety

Once $\tau_{all}$ is known or assumed, the following step-by-step procedure should be adopted (Ladanyi, 1977) to determine the length, $L$, to which the pier should be embedded in rock.

**1.** Obtain the size of the shaft $D_s = D_b$ based on the load $Q_w$ and the strength of the concrete. If a rolled-steel section core is used, the diameter $D_s$ will depend on the strength of concrete and steel.

**2.** Assuming that the load carried at the pier tip is zero, determine the maximum required length of embedment ($L_1$) of the pier:

$$L_1 = \frac{Q_w}{\pi D_s \tau_{all}} \tag{9.21}$$

**3.** Now, assume another value of the length of embedment $L_2 < L_1$. Use Eq. (9.17) and calculate $\sigma_z$ at $z = L_2$. This value of $\sigma_z = L_2$ is the unit point resistance at the pier tip.

**4.** Compare the value of $\sigma_z = L_2$ obtained in step 3 with the allowable bearing capacity in rock, $q_{p(all)}$. This can be done by using Eq. (8.47). If $\sigma_{z=L_2} > q_{p(all)}$, go back to step 3 and assume another value of $L_2$.

**5.** If $\sigma_z = L_2 \leq q_{p(all)}$ in step 4, calculate $\tau$ developed along the perimeter of the pile as

$$\tau = \underbrace{\left[\left(1 - \frac{\sigma_{z=L_2}}{\sigma_{max}}\right)Q_w\right]}_{\substack{\text{load taken} \\ \text{by shaft}}} \frac{1}{\pi D_b L_2} \tag{9.22}$$

**6.** Compare the value of $\tau$ obtained from Eq. (9.22) to $\tau_{all}$ obtained from Eq. (9.20).

**7.** Repeat steps 3 through 6 to obtain the length $L$ that is most desirable with $\sigma_{z=L_2} \leq q_{p(all)}$ and $\tau \leq \tau_{all}$.

## Example 9.3

Refer to Figure 9.12. The soil located above the rock is soft clay. Neglect the skin resistance developed at the pier perimeter-clay interface. Given: $Q_w = 25,000$ kN; $E_r/E_c = 0.7$; $\mu_c = \mu_r = 0.3$; $\phi_{rc} = 38°$; $\tau_{all} = 490$ kN/m²; for rock, $q_{p(all)} = 2500$ kN/m²; for concrete $f_c' = 21,000$ kN/m².

Determine the required diameter and the length of embedment, $L$.

**Solution**

Calculation of $D_s = D_b$

Use $f_c = 0.25f_c' = (0.25)(21,000) = 5250$ kN/m². So, from Eq. (9.1)

$$D_s = 2.257\sqrt{\frac{Q_w}{f_c'}} = 2.257\sqrt{\frac{25,000}{21,000}} = 2.46 \text{ m}$$

Adopt $D_s = 2.5$ m

Calculation of $L_1$

From Eq. (9.21)

$$L_1 = \frac{Q_w}{\pi D_s \tau_{all}} = \frac{25,000}{(\pi)(2.5)(490)} = 6.496 \approx \underline{6.5 \text{ m}}$$

Calculation of $L$

Assume an embedment length $= L_2 = 5$ m. From Eq. (9.17), with $z = 5$ m

$$\frac{\sigma_{z=L_2}}{\sigma_{max}} = \exp\left[-\frac{2\mu_c \tan \phi_{rc}}{1 - \mu_c + (1 + \mu_r)\dfrac{E_c}{E_r}} \dfrac{2z}{D_b}\right]$$

$$= \exp\left\{-\left[\frac{(2)(0.3)(\tan 38°)}{1 - 0.3 + (1 + 0.3)\left(\dfrac{1}{0.7}\right)}\dfrac{(2)(5)}{2.5}\right]\right\} = 0.48$$

So

$$\sigma_{z=L_2} = (\sigma_{max})(0.48) = \left(\frac{Q_w}{\dfrac{\pi}{4}D_b^2}\right)(0.48) = \frac{(25,000)(0.48)}{(0.785)(2.5^2)}$$

$$\approx 2446 \text{ kN/m}^2 < 2500 \text{ kN/m}^2 = q_{all} \text{—O.K.}$$

The shear stress developed along the shaft-rock interface can now be checked by using Eq. (9.22) as

$$\tau = \left(1 - \frac{\sigma_{z=L_2}}{\sigma_{max}}\right)\frac{Q_w}{\pi D_b L_2} = (1 - 0.48)\frac{25,000}{(\pi)(2.5)(5)}$$

$$= 331 \text{ kN/m}^2 \leq \tau_{all} = 490 \text{ kN/m}^2$$

So, $L = L_2 = \underline{5 \text{ m}}$

Note that if the value of $\tau$ had been substantially lower than $\tau_{all}$, the value of $L_2$ could have been reduced. This would require a new set of calculations of $(\sigma_{z=L_2})/\sigma_{max}$, $\sigma_{z=L_2}$ and $\tau$. However, if $\tau$ or $\sigma_{z=L_2}$ becomes less than $\tau_{all}$ and $q_{p(all)}$, respectively, the length $L_2$ must be increased and checked again.

## 9.8
## Uplift Capacity of Piers

Sometimes drilled piers must resist uplifting loads. Field observations of these piers' uplift capacity are relatively scarce. The procedure for determination of the ultimate uplifting load for drilled piers without bells is similar to that for piles described in Chapter 8 (Section 8.10) and will not be described here. When a short drilled pier with a bell is subjected to an uplifting load, the nature of the failure surface in the soil will be like that shown in Figure 9.13. The net ultimate uplift capacity $T_{un}$ can be given as

$$T_{un} = T_{ug} - W \tag{9.23}$$

where $T_{ug}$ = gross ultimate uplift capacity
    $W$ = effective weight of the pier

The magnitude of $T_{un}$ for piers in sand can be estimated by the procedure outlined by Meyerhof and Adams (1968) and Das and Seeley (1975):

$$T_{un} = B_q A_p \gamma L \tag{9.24}$$

where $B_q$ = breakout factor
    $A_p = (\pi/4)D_b^2$
    $\gamma$ = unit weight of soil above the bell (*Note*: if the soil is submerged, effective unit weight should be used.)

The breakout factor can be given as

$$B_q = 2\frac{L}{D_b} K_u' \tan \phi \left(m\frac{L}{D_b} + 1\right) + 1 \tag{9.25}$$

where $K_u'$ = nominal uplift coefficient
    $\phi$ = soil friction angle
    $m$ = shape factor coefficient

The value of $K_u'$ may be taken as 0.9 for all values of $\phi$ varying from 30–45°. The variation of $m$ has been given by Meyerhof and Adams (1968) as follows:

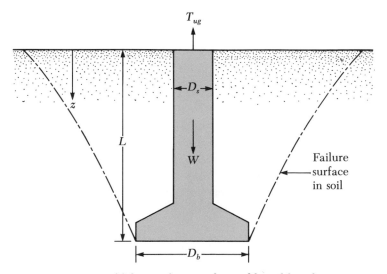

**Figure 9.13** Nature of failure surface in soil caused by uplifting force on belled piers

| Soil friction angle, $\phi$ (deg) | $m$ |
|---|---|
| 30 | 0.15 |
| 35 | 0.25 |
| 40 | 0.35 |
| 45 | 0.50 |

It has been shown experimentally that the value of $B_q$ increases with the $L/D_b$ ratio up to a critical value $(L/D_b)_{cr}$ and remains constant thereafter. The critical embedment ratio, $(L/D_b)_{cr}$, increases with the soil friction angle. The approximate ranges are given below:

| Soil friction angle, $\phi$ (deg) | $(L/D_b)_{cr}$ |
|---|---|
| 30 | 4 |
| 35 | 5 |
| 40 | 7 |
| 45 | 9 |

Hence, piers with $L/D_b \leq (L/D_b)_{cr}$ may be defined as *shallow foundations* with regard to the uplift. Piers with $L/D_b > (L/D_b)_{cr}$ are *deep foundations*. The nature of the failure surface in soil at failure as shown in Figure 9.13 is for the case of a shallow foundation. For deep foundations, local shear failure takes place, and the failure surface in soil *does not extend up to the ground surface*. Based on the preceding considerations, the variation of $B_q$ with $L/D_b$ is shown in Figure 9.14.

Following is a step-by-step procedure for the calculation of the net ultimate uplift capacity of drilled piers with bells in sand.

1. Determine $L$, $D_b$, and $L/D_b$.
2. Estimate $(L/D_b)_{cr}$ and, hence, $L_{cr}$.
3. If $(L/D_b) \leq (L/D_b)_{cr}$, obtain $B_q$ from Figure 9.14. Now,

$$T_{ug} = B_q A_p \gamma L + W$$

4. If $(L/D_b) > (L/D_b)_{cr}$

$$T_{ug} = B_q A_p \gamma L + W + \int_0^{L-L_{cr}} (\pi D_s) \sigma_v' K_u \tan \delta \, dz \tag{9.26}$$

The last term of the preceding equation is for the frictional resistance developed along the soil-shaft interface from $z = 0$ to $z = L - L_{cr}$. This is similar to Eq. (8.61). The term $\sigma_v'$ is the effective stress at any depth $z$, and $K_u$ and $\delta$ are to be obtained from Figure 8.25b and c, respectively (p. 381).

The net ultimate uplift capacity of belled piers in clay can be estimated according to the procedure outlined by Das (1980). Based on this method

$$T_{un} = \{c_u B_c + \gamma L\} A_p \tag{9.27}$$

where  $c_u$ = undrained cohesion
$B_c$ = breakout factor
$\gamma$ = unit weight of clay soil above the bell

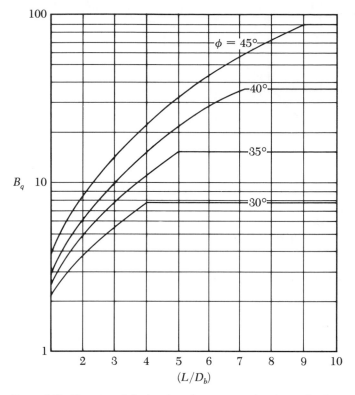

$\phi = 45°$

$40°$

$35°$

$B_q$ 10

$30°$

$(L/D_b)$

**Figure 9.14** Variation of the breakout factor, $B_q$, with $L/D_b$ and soil friction angle

As in the case of $B_q$, the value of $B_c$ increases with the embedment ratio up to a critical value of $L/D_b = (L/D_b)_{cr}$ and remains constant thereafter. Beyond the critical depth, the value of $B_c$ is approximately equal to 9. The critical embedment ratio can be related to the undrained cohesion by the relation

$$\left(\frac{L}{D_b}\right)_{cr} = 0.107c_u + 2.5 \leq 7 \tag{9.28}$$

where $c_u$ is in kN/m².

Following is a step-by-step procedure for determining the net ultimate uplift capacity of belled piers in clay:

**1.** Determine $c_u$, $L$, $D_b$, and $L/D_b$.
**2.** Obtain $(L/D_b)_{cr}$ from Eq. (9.28), and obtain $L_{cr}$.
**3.** If $L/D < (L/D_b)_{cr}$, refer to Figure 9.15 and obtain the value of $B_c$.
**4.** Use Eq. (9.27) to obtain $T_{un}$.
**5.** If $L/D \geq (L/D_b)_{cr}$, $B_c = 9$. The value of $T_{un}$ can be given by the following relation:

$$T_{un} = (9c_u + \gamma L)A_p + \Sigma(\pi D_s)(L - L_{cr})\alpha' c_u \tag{9.29}$$

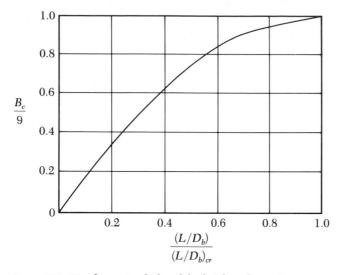

**Figure 9.15**   Nondimensional plot of the breakout factor, $B_c$

The last term of the preceding equation is the skin resistance obtained from the adhesion along the soil-shaft interface. This is similar to Eq. (8.57). The value of $\alpha'$ can be obtained from Eqs. (8.58a) and (8.58b).

## Example
## 9.4

Refer to Figure 9.13. A drilled pier with a bell has a shaft diameter of 0.76 m and a bell diameter of 1.85 m. Given: $L = 9.5$ m. The bell is supported by a dense sand ($z > 9.5$ m) layer. However, a fine, loose sand layer exists above the bell ($z = 0$–9.5 m). For this sand, $\gamma = 16.4$ kN/m$^3$, $\phi = 32°$, and the approximate relative density is 30%. The entire pier is located above the ground water table. Determine the net allowable uplift capacity of the pier with a factor of safety of 3.

### Solution

Given: $L = 9.5$ m and $D_b = 1.85$ m. So, $L/D_b = 9.5/1.85 = 5.14$. For $\phi = 30°$, $(L/D_b)_{cr} = 4$; for $\phi = 35°$, $(L/D_b)_{cr} = 5$. By interpolation, $(L/D_b)_{cr} \approx 4.2$ for $\phi = 32°$. So, $L_{cr} = (4.2)(D_b) = 7.77$. Because $L/D_b = 5.13 > (L/D_b)_{cr} = 4.2$, it is a deep foundation.

According to Eq. (9.25)

$$B_q = 2\left(\frac{L}{D_b}\right)_{cr} K_u' \tan\phi \left[m\left(\frac{L}{D_b}\right)_{cr} + 1\right] + 1$$

Note that $(L/D_b)_{cr}$ has been used in the preceding equation in place of $L/D_b$ because it is a deep foundation. The value of $m$ is approximately equal to 0.17 for $\phi = 32°$. Hence

$$B_q = (2)(4.2)(0.9)(\tan 32°)[(0.17)(4.2) + 1] + 1 = 9.09$$

From Eq. (9.26)

$$T_{un} = T_{ug} - W = B_q A_p \gamma L + \int_0^{L-L_{cr}} (\pi D_s)(\sigma_v' K_u \tan \delta) dz$$

$$= B_q A_p \gamma L + \frac{\pi}{2} \gamma D_s K_u \tan \delta (L - L_{cr})^2$$

$$A_p = \left(\frac{\pi}{4}\right)(D_b)^2 = \left(\frac{\pi}{4}\right)(1.85)^2 = 2.687 \ m^2$$

$$L - L_{cr} = 9.5 - 7.77 = 1.73 \ m$$

Also, from Figure 8.25b and c, for $\phi = 32°$ and relative density $= 30\%$, $K_u' = 1.5$ and $\delta/\phi \approx 0.73$. Hence

$$T_{un} = (9.09)(2.687)(16.4)(9.5) + \left(\frac{\pi}{2}\right)(16.4)(0.76)(1.5)$$

$$[\tan(0.73 \times 32)](1.73)^2$$

$$= 3805.4 + 37.96 = 3843.36 \approx 3843$$

So, net allowable capacity $= 3843/FS = 3843/3 = \underline{1281 \ kN}$

## Example 9.5

Consider the caisson described in Example Problem 9.4. If the soil above the bell is clay with an average value of the undrained shear strength equal to 95 kN/m², calculate the net ultimate uplift capacity. Given: $\gamma_{clay} = 17.9 \ kN/m^3$.

### Solution

From Eq. (9.28)

$$\left(\frac{L}{D_b}\right)_{cr} = 0.107c_u + 2.5 = (0.107)(95) + 2.5 = 12.67$$

This value is more than 7, so use $(L/D_b)_{cr} = 7$. Hence, $L_{cr} = (7)(1.85) = 12.95 \ m$. $L_{cr} = 12.95$ is greater than $L = 9.5 \ m$, so this is a shallow foundation for uplift consideration.

For shallow foundations [Eq. (9.27)]

$$T_{un} = (c_u B_c + \gamma L) A_p$$

The value of the breakout factor $B_c$ can be determined from Figure 9.15.

$$\frac{\left(\frac{L}{D_b}\right)}{\left(\frac{L}{D_b}\right)_{cr}} = \frac{\left(\frac{9.5}{1.85}\right)}{7} = 0.734$$

So, $B_c/9 = 0.92$, or $B_c = 8.28$. $A_p = (\pi/4)D_b^2 = (\pi/4)(1.85)^2 = 2.687 \ m^2$. Thus

$$T_{un} = [(95)(8.28) + (17.9)(9.5)]2.687 = \underline{2570.5 \ kN}$$

## 9.9

### Lateral Load-Carrying Capacity

The lateral load-carrying capacity of piers can be analyzed in a manner similar to that presented in Section 8.11 for piles. Therefore, it will not be repeated here.

## Caissons

## 9.10

### Types of Caisson

Caissons can be divided into three major types: (1) *open caissons* (Figure 9.16), (2) *box caissons* (or closed caissons; Figure 9.17), and (3) *pneumatic caissons* (Figure 9.18).

*Open caissons* are concrete shafts that remain open at the top and bottom during construction. The bottom of the caisson has a cutting edge. The caisson is sunk into place, and soil from the inside of the shaft is removed by grab buckets until the bearing stratum is reached. The shafts can be of any shape— circular, square, rectangular, or oval. Once the bearing stratum is reached, concrete is poured into the shaft (under water) to form a seal at its bottom. When the concrete seal hardens, the water inside the caisson shaft is pumped out. Concrete is then poured into the shaft to fill it. Open caissons can be

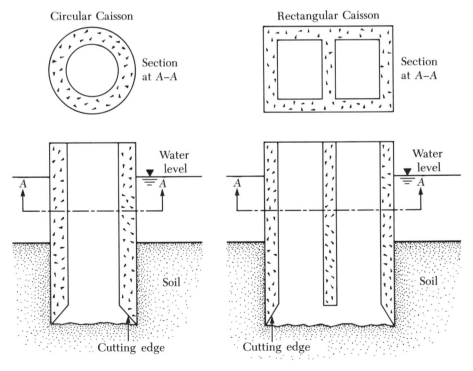

**Figure 9.16**  Open caisson

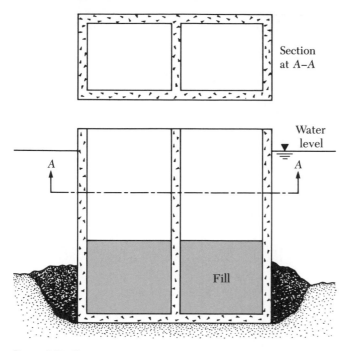

**Figure 9.17**  Box caisson

extended to great depths, and the cost of construction is relatively low. However, one of their major disadvantages is the lack of quality control of the
concrete poured into the shaft for the seal. Also, the bottom of the caisson
cannot be thoroughly cleaned out. An alternate method of open-caisson construction is to drive some sheet piles to form an enclosed area. The enclosed
area is filled with sand, and is generally referred to as a *sand island*. The caisson
is then sunk through the sand to the desired bearing stratum. This procedure
is somewhat analogous to sinking a caisson when the ground surface is above the
water table.

   *Box caissons* are caissons with their bottoms closed. They are constructed
on land and then transported to the construction site. They are gradually sunk
at the site by filling the inside with sand, ballast, water, or concrete. The cost
for this type of construction is low. However, the bearing surface should be
level. If the bearing surface is not level, it must be leveled by excavation.

   *Pneumatic caissons* are generally used for depths of about 15–40 m. This
type of caisson is necessary where an excavation cannot be kept open because
the soil flows into the excavated area faster than it can be removed. A pneumatic
caisson has a work chamber at the bottom that is at least 3 m high. In this
chamber, the workers excavate the soil and place the concrete. The air in the
chamber is kept at a specific pressure to prevent water and soil from entering.
The workers do not usually encounter any severe discomfort when the chamber
pressure is raised to about 100 kN/m² above the atmospheric pressure. Beyond

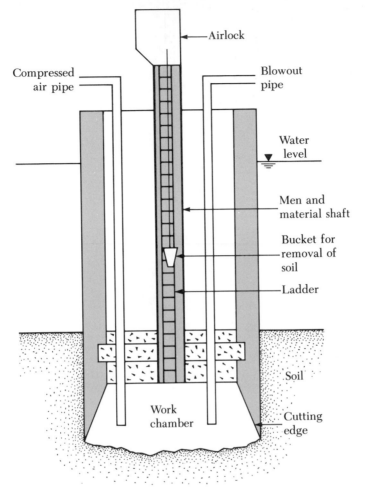

**Figure 9.18**   Pneumatic caisson

this pressure, decompression periods are required when the workers leave the chamber. When chamber pressures of about 300 kN/m² above the atmospheric pressure are required, the workers should not be kept inside for more than 1½–2 hours at a time. The workers enter and exit the work chamber through a steel shaft by means of a ladder. This shaft is also used for the removal of excavated soil and the placement of concrete. For large caisson construction, more than one shaft may be necessary; an airlock is provided for each one. Pneumatic caissons gradually sink with the progress of excavation work. When the bearing stratum is reached, the work chamber is filled with concrete.

## 9.11

### Load-Bearing Capacity of Caissons

Calculation of the load-bearing capacity of caissons is similar to that for drilled piers. Therefore, it will not be further discussed in this section.

## 9.12

## Thickness of Concrete Seal in Open Caissons

Section 9.10 mentioned that, before dewatering the caisson, a concrete seal is placed at the bottom of the shaft and allowed to cure for some time. The concrete seal should be thick enough to withstand an upward hydrostatic force from its bottom after dewatering is complete and before concrete fills the shaft. Based on the theory of elasticity, the thickness, $t$, (see Figure 9.19) can be given as (Teng, 1962)

$$t = 1.18R_i \sqrt{\frac{q}{f_c}} \quad \text{(circular caisson)} \tag{9.30}$$

and

$$t = 0.866B_i \sqrt{\frac{q}{f_c\left[1 + 1.61\left(\dfrac{L_i}{B_i}\right)\right]}} \quad \text{(rectangular caisson)} \tag{9.31}$$

where $R_i$ = inside radius of a circular caisson
$q$ = unit bearing pressure at the base of the caisson
$f_c$ = allowable concrete flexural stress ($\approx 0.1$–$0.2$ of $f'_c$; where $f'_c$ is the 28-day compressive strength of concrete)
$B_i$, $L_i$ = inside width and length, respectively, of rectangular caisson (Figure 9.19b)

According to Figure 9.19, the value of $q$ in Eqs. (9.30) and (9.31) can be approximated as

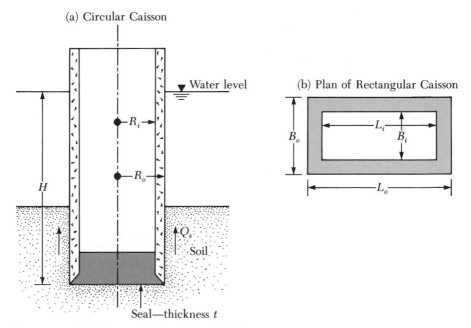

**Figure 9.19**  Calculation of the thickness of seal for an open caisson

$$q \approx H\gamma_w - t\gamma_c \tag{9.32}$$

where $\gamma_c$ = unit weight of concrete

The thickness of the seal calculated by Eqs. (9.30) and (9.31) will be sufficient to protect it from cracking immediately after dewatering. However, two other conditions should also be checked for safety. They are as follows:

1. *Check for Perimeter Shear at Contact Face of Seal and Shaft:* According to Figure 9.19, the net upward hydrostatic force from the bottom of the seal is equal to $A_i H\gamma_w - A_i t\gamma_c$ (where $A_i = \pi R_i^2$ for circular caissons, and $A_i = L_i B_i$ for rectangular caissons). So, the perimeter shear developed can be given by

$$v \approx \frac{A_i H\gamma_w - A_i t\gamma_c}{p_i t} \tag{9.33}$$

where $p_i$ = inside perimeter of the caisson

Note that

$$p_i = 2\pi R_i \quad \text{(for circular caissons)} \tag{9.34}$$

and

$$p_i = 2(L_i + B_i) \quad \text{(for rectangular caissons)} \tag{9.35}$$

The perimeter shear given by Eq. (9.33) should be less than the permissible shear stress, $v_u$, as given by Eq. (A.15a) (Appendix A), or

$$v(\text{MN/m}^2) \leq v_u(\text{MN/m}^2) = 0.17\phi\sqrt{f_c'}(\text{MN/m}^2) \tag{9.36}$$

where $\phi = 0.85$

2. *Check for Buoyancy:* If the shaft is completely dewatered, the buoyant upward force, $F_u$, can be given as

$$F_u = (\pi R_o^2)H\gamma_w \text{ (for circular caissons, as shown in Figure 9.19a)} \tag{9.37}$$

and

$$F_u = (B_o L_o)H\gamma_w \text{ (for rectangular caissons, as shown in Figure 9.19b)} \tag{9.38}$$

The downward force $(F_d)$ is caused by the weight of the caisson and the seal and by the skin friction at the caisson-soil interface, or

$$F_d = W_c + W_s + Q_s \tag{9.39}$$

where $W_c$ = weight of caisson
$W_s$ = weight of seal
$Q_s$ = skin friction

If $F_d > F_u$, the caisson is safe from buoyancy. However, if $F_d < F_u$, it will be unsafe to completely dewater the shaft. For that reason, the thickness of the seal should be increased by $\Delta t$ [over the thickness calculated by using Eqs. (9.30) or (9.31)], or

$$\Delta t = \frac{F_u - F_d}{A_i \gamma_c} \tag{9.40}$$

## Example 9.6

An open caisson (circular) is shown in Figure 9.20. Determine the thickness of the seal that will enable complete dewatering.

### Solution

From Eq. (9.30)

$$t = 1.18 R_i \sqrt{\frac{q}{f_c}}$$

For this problem, $R_i = 5/2 = 2.5$ m.

$$q \approx (15)(9.81) - t\gamma_c$$

With $\gamma_c = 23.58$ kN/m³, $q = 147.15 - 23.58t$

$$f_c = 0.1 f_c' = 0.1 \times 21 \times 10^3 \text{ kN/m}^2 = 2.1 \times 10^3 \text{ kN/m}^2$$

So

$$t = (1.18)(2.5) \sqrt{\frac{147.15 - 23.58t}{2.1 \times 10^3}}$$

or

$$t^2 + 0.0978t - 0.61 = 0$$

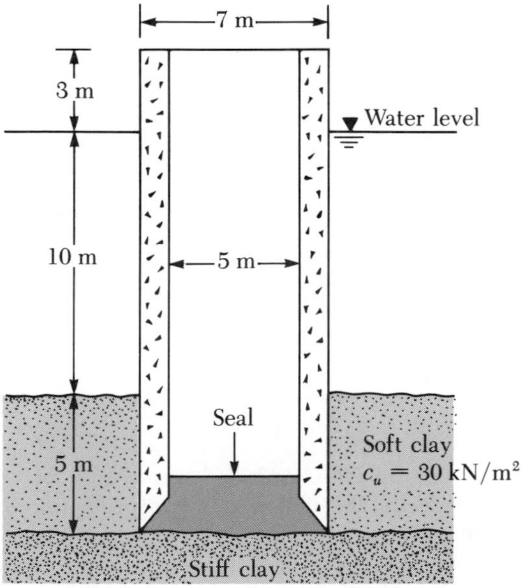

Figure 9.20

Solution to the preceding equation gives $t = 0.73$ m. So, let $t \approx 0.75$ m.

## Check for Perimeter Shear

According to Eq. (9.33)

$$v = \frac{(\pi)R_i^2 H\gamma_w - (\pi)R_i^2 t\gamma_c}{2\pi R_i t}$$

$$= \frac{(\pi)(2.5)^2[(15)(9.81) - (0.75)(23.58)]}{(2)(\pi)(2.5)(0.75)} = 215.77 \text{ kN/m}^2$$

However, the allowable shear stress

$$v_u = 0.17\phi\sqrt{f_c'} = (0.17)(0.85)\sqrt{21} = 0.662 \text{ MN/m}^2$$
$$v = 215.77 \text{ kN/m}^2 < v_u = 662 \text{ kN/m}^2 \text{---O.K.}$$

## Check Against Buoyancy

Buoyant upward force $= F_u = \pi R_o^2 H\gamma_w \cdot R_o = 3.5$ m. So

$$F_u = (\pi)(3.5)^2(15)(9.81) = 5663 \text{ kN}$$

Downward force $= F_d = W_c + W_s + Q_s$

$$W_c = \pi(R_o^2 - R_i^2)(\gamma_c)(18) = \pi(3.5^2 - 2.5^2)(23.58)(18) \approx 8000 \text{ kN}$$
$$W_s = (\pi R_i^2)t\gamma_c = (\pi)(2.5)^2(0.75)(23.58) = 347.24 \text{ kN}$$

Assume $Q_s \approx 0$. So

$$F_d = 8000 + 347.24 = 8347.24 \text{ kN}$$

Because $F_u < F_d$, it is safe. For design, assume $t = 1$ m.

---

## Problems

**9.1** A drilled pier is shown in Figure P9.1. Determine the net allowable point bearing capacity (factor of safety $= 4$). Do not reduce the soil friction angle $\phi$ of sand (use Figure 9.7).

**9.2** For the drilled pier described in Problem 9.1, what skin resistance would develop in the top twelve feet, which are in clay?

**9.3** Determine the net allowable point bearing capacity (factor of safety $= 3$) of the pier shown in Figure P9.1 by Vesic's method [Eq. (9.8)]. Given: $I_{rr} = 130$.

**9.4** Figure P9.4 shows a drilled pier without a bell. Determine:
a. the net ultimate point bearing capacity;
b. the ultimate skin resistance; and
c. the working load, $Q_w$ (factor of safety $= 3$).

**9.5** Refer to the soil profile shown in Figure P9.5. A drilled pier with a bell is to be constructed. Given: $f_c' = 21,000 \text{ kN/m}^2$. The pier has to support a working load, $Q_w = 900$ kN (factor of safety $= 3$).
a. Determine the minimum diameter of the pier shaft required [use Eq. (9.1)]. Make a reasonable assumption for the diameter to be used.
b. With the pier diameter determined in part (a), determine the diameter of the bell needed. Skin friction and point bearing capacity are to be considered.

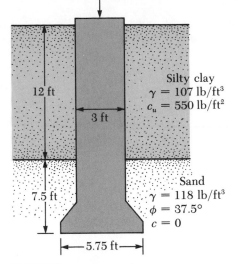

Silty clay
$\gamma = 107 \text{ lb/ft}^3$
$c_u = 550 \text{ lb/ft}^2$

12 ft

3 ft

7.5 ft

Sand
$\gamma = 118 \text{ lb/ft}^3$
$\phi = 37.5°$
$c = 0$

5.75 ft

**Figure P9.1**

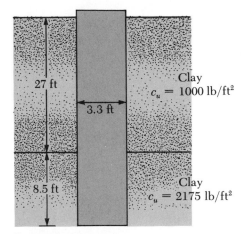

27 ft

3.3 ft

Clay
$c_u = 1000 \text{ lb/ft}^2$

8.5 ft

Clay
$c_u = 2175 \text{ lb/ft}^2$

**Figure P9.4**

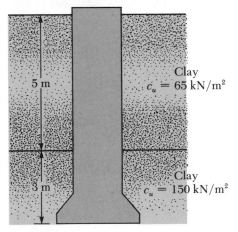

5 m

Clay
$c_u = 65 \text{ kN/m}^2$

3 m

Clay
$c_u = 150 \text{ kN/m}^2$

**Figure P9.5**

449

**9.6** For the pier described in Problem 9.4, estimate the total elastic settlement at working load. Use Eqs. (8.50), (8.52), and (8.54). Assume $E_p = 3 \times 10^6$ lb/in.$^2$, $\xi = 0.65$. Make other assumptions as necessary.

**9.7** For the pier described in Problem 9.5, estimate the total elastic settlement at working load. Use Eqs. (8.50), (8.52), and (8.54). $E_p = 19.5 \times 10^6$ kN/m$^2$, $\xi = 0.65$. Make other assumptions as necessary.

**9.8** A drilled pier in a medium sand is shown in Figure P9.8. Using the method proposed by Touma and Reese, determine:
a. the net point resistance for a base movement of 25.4 mm;
b. the shaft frictional resistance; and
c. the total load that can be carried by the pier for a total base movement of 25.4 mm.

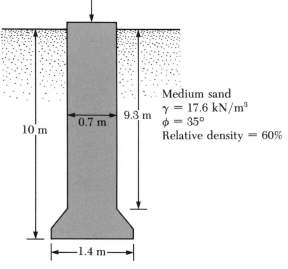

Medium sand
$\gamma = 17.6$ kN/m$^3$
$\phi = 35°$
Relative density $= 60\%$

9.3 m
0.7 m
10 m

—1.4 m—

**Figure P9.8**

**9.9** Assume the pier shown in Figure P9.1 to be a point bearing pier with a working load of 560 kip. Calculate the pier settlement by using Eqs. (8.49), (8.50), and (8.51). Given: $E_p = 3 \times 10^6$ lb/in$^2$., $\mu_s = 0.35$, and $E_s = 5070$ lb/in$^2$.

**9.10** Repeat Example Problem 9.1 for a working load $Q_w = 3000$ kN. Make any assumptions necessary.

**9.11** Consider the case of the drilled pier socketed into rock, as shown in Figure 9.12. Given: $Q_{all} = 18,000$ kN, $E_r/E_c = 0.6$, $\mu_c = \mu_r = 0.35$, $\phi_{rc} = 34°$, $\tau_{all} = 360$ kN/m$^2$, allowable point bearing capacity of rock $= 2200$ kN/m$^2$, and $f'_c$ for concrete $= 28,000$ kN/m$^2$. Determine the required length of embedment ($L$) of the pier in the rock.

**9.12** Refer to the drilled pier in Problem 9.11. If the allowable bearing capacity of rock is changed to 3000 kN/m$^2$ and other quantities remain the same, what will be the required length of embedment of the pier in the rock?

**9.13** Refer to Figure 9.13. Given: $D_s = 3$ ft, $D_b = 6$ ft, and $L = 21$ ft. The drilled pier is in a homogeneous sand with $\phi = 35°$, $\gamma = 111$ lb/ft$^3$. Determine the net ultimate uplift capacity ($T_{un}$) of the pier.

**9.14** For the drilled pier described in Problem 9.13, if $L = 30$ ft and other quantities remain the same, calculate the net ultimate uplift capacity.

**9.15** Refer to Figure 9.13. Given: $D_s = 0.7$ m, $D_b = 1.4$ m, and $L = 5.8$ m. If the drilled pier is located in a clay with $c_u = 72.4$ kN/m$^2$, estimate the net ultimate uplift capacity. The unit weight of the clay is 16.2 kN/m$^3$.

**9.16** In Problem 9.15, with other quantities remaining the same, if the undrained cohesion of clay is changed to 35 kN/m$^2$, what will be the net ultimate uplift capacity?

**9.17** Estimate the net ultimate uplift capacity of the drilled pier shown in Figure P9.4.

# References

Das, B. M. (1980). "A Procedure for Estimation of Ultimate Uplift Capacity of Foundations in Clay," *Soils and Foundations,* The Japanese Society of Soil Mechanics and Foundation Engineering, Vol. 20, No. 1, pp. 77–82.

Das, B. M., and Seeley, G. R. (1975). "Breakout Resistance of Shallow Vertical Anchors," *Journal of the Geotechnical Engineering Division,* American Society of Civil Engineers, Vol. 101, No. GT9, pp. 999–1003.

Goodman, R. E. (1980). *Introduction to Rock Mechanics,* Wiley, New York.

Ladanyi, B. (1977). "Discussion on Friction and Endbearing Tests on Bedrock for High Capacity Socket Design," *Canadian Geotechnical Journal,* Vol. 14, No. 1, pp. 153–156.

Meyerhof, G. G., and Adams, J. I. (1968). "The Ultimate Uplift Capacity of Foundations," *Canadian Geotechnical Journal,* Vol. 5, No. 4, pp. 225–244.

Teng, W. C. (1962). *Foundation Design,* Prentice-Hall, Englewood Cliffs, N.J.

Touma, F. T., and Reese, L. C. (1974). "Behavior of Bored Piles in Sand," *Journal of the Geotechnical Engineering Division,* American Society of Civil Engineers, Vol. 100, No. GT7, pp. 749–761.

Vesic, A. S. (1967). "Ultimate Load and Settlement of Deep Foundations in Sand," *Proceedings,* Symposium on Bearing Capacity and Settlement of Foundations, Duke University, Durham, N.C., p. 53.

Vesic, A. S. (1963). "Bearing Capacity of Deep Foundations in Sand," *Highway Research Record,* No. 39, Highway Research Board, National Academy of Science, Washington, D.C., pp. 112–153.

Vesic, A. S. (1977). "Design of Pile Foundations," *NCHRP No. 42,* Transportation Research Board, National Research Council, Washington, D.C.

Whitaker, T., and Cooke, R. W. (1966). "An Investigation of the Shaft and Base Resistance of Large Bored Piles in London Clay," *Proceedings,* Conference on Large Bored Piles, Institute of Civil Engineers, London, pp. 7–49.

# CHAPTER 10

# Foundations on Difficult Soils

## 10.1

### Introduction

In many areas of the United States and other parts of the world, there are soils that make construction of foundations extremely difficult. For example, expansive or collapsible soils may cause high differential movements in structures as the result of excessive heave or settlement. Similar problems can also arise when foundations are constructed over sanitary landfills. Foundation engineers must be able to identify difficult soils when they are encountered in the field. Although it is not possible to solve all the problems caused by all soils, preventive measures can be taken to reduce the possibility of damage to structures built on them.

This chapter comprises three major parts: *collapsible soils*, *expansive soils*, and *sanitary landfills*. It outlines the fundamental properties of these soil conditions, along with methods of careful foundation construction for each case.

## Foundations on Collapsing Soil

## 10.2

### Definition and Types of Collapsing Soil

*Collapsing soils*, which are sometimes referred to as *metastable soils*, are unsaturated soils that undergo a large volume change upon saturation. This volume change may or may not be the result of the application of additional load. In order to explain the behavior of collapsing soils under load, one needs to consider the typical nature of void ratio vs. pressure plot ($e$ vs. log $p$) for a collapsing soil, as shown in Figure 10.1. The branch $ab$ is the result of a consolidation test on a specimen at its natural moisture content. At a pressure level of $p_w$, the equilibrium void ratio is $e_1$. However, if water is introduced into the specimen for saturation, the soil structure will collapse. After saturation, the equilibrium void ratio at the same pressure level $p_w$ is $e_2$; $cd$ is the branch of $e$ vs. log $p$ curve under additional load after saturation. Foundations that are constructed on such soils may undergo large and sudden settlement if and when the soil under them becomes saturated with an unanticipated supply of moisture. This moisture may come from several sources, such as (a) broken water pipelines, (b) leaky sewers, (c) drainage from reservoirs and swimming pools, (d) slow increase of ground water table, and so on. This type of settlement generally causes considerable structural damage. Hence, identification of collapsing soils during field exploration is critical.

The majority of naturally occurring collapsing soils are *aeolian*—that is, wind-deposited sand and/or silts, such as loess, aeolic beaches, and volcanic dust deposits. These deposits have high void ratios and low unit weights and are cohesionless or only slightly cohesive. *Loess* is a type of deposit with silt-size particles. The cohesion in loess may be the result of the presence of clay coatings around the silt-size particles, which holds them in a rather stable condition in an unsaturated state. The cohesion may also be caused by the presence of chemical precipitates leached by rain water. When the soil becomes saturated, the clay binders lose their strength and, hence, undergo a structural collapse. Large parts of the midwestern states and arid western United States have such types of deposit. *Loess* deposits are also found over 15–20% of Europe and over large parts of China.

Many collapsing soils may be residual soils that are products of weathering of parent rocks. The weathering process produces soils with a large range of particle-size distribution. Soluble and colloidal materials are leached out by weathering, resulting in large void ratios and thus unstable structures. Many

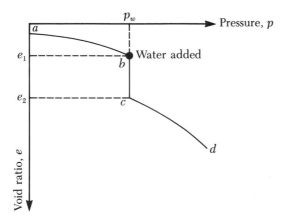

**Figure 10.1** Nature of void ratio vs. pressure variation for a collapsing soil

parts of South Africa and Rhodesia have residual soils that are decomposed granites. Sometimes collapsing soil deposits may be left by flash floods and mud flows. These deposits dry out and are very poorly consolidated. An excellent review of collapsing soils is given by Clemence and Finbarr (1981).

## 10.3

### Physical Parameters for Identification

Several investigators have proposed various methods to evaluate the physical parameters of collapsing soils for identification. Some of these will be briefly discussed in this section.

Jennings and Knight (1975) have suggested a procedure to describe the *collapse potential* of a soil, which is mostly a qualitative evaluation. The collapse potential can be determined by taking an undisturbed soil specimen at natural moisture content in a consolidation ring. Step loads are applied to the specimen up to a pressure level of 200 kN/m². (In Figure 10.1, this is $p_w$.) At this pressure ($p_w = 200$ kN/m²), the specimen is flooded for saturation and left for 24 hours. This test will provide the void ratios ($e_1$ and $e_2$) before and after flooding. The collapse potential, $C_p$, can now be calculated as

$$C_p = \frac{e_1 - e_2}{1 + e_o} \qquad (10.1)$$

where $e_o$ = natural void ratio of the soil

The foundation problems associated with a collapsible soil have been correlated with the collapse potential, $C_p$, by Jennings and Knight (1975). These have been summarized by Clemence and Finbarr (1981) and are given in Table 10.1.

Holtz and Hilf (1961) have suggested that a loessial soil that has a void ratio large enough to allow its moisture content to exceed its liquid limit upon saturation is susceptible to collapse. So, for collapse

$$w_{(saturated)} \geq LL \qquad (10.2)$$

However, for saturated soils

$$e_o = wG_s \qquad (10.3)$$

**Table 10.1** Relation of Collapse Potential to the Severity of Foundation Problems[a]

| $C_p(\%)$ | Severity of problem |
|-----------|---------------------|
| 0–1 | No problem |
| 1–5 | Moderate trouble |
| 5–10 | Trouble |
| 10–20 | Severe trouble |
| 20 | Very severe trouble |

[a]After Clemence and Finbarr (1981)

where $LL$ = liquid limit

$\quad\quad G_s$ = specific gravity of soil solids

Combining Eqs. (10.2) and (10.3), for collapsing soils

$$e_o \geq (LL)(G_s) \tag{10.4}$$

The natural dry unit weight, $\gamma_d$, of the soil for collapse may be given as

$$\gamma_d \leq \frac{G_s\gamma_w}{1 + e_o} = \frac{G_s\gamma_w}{1 + (LL)(G_s)} \tag{10.5}$$

Assuming an average value of $G_s$ = 2.65, the limiting values of $\gamma_d$ for various liquid limits may now be calculated from Eq. (10.5) as follows:

| Liquid limit (%) | Limiting values of $\gamma_d$ (kN/m³) |
|---|---|
| 10 | 20.56 |
| 15 | 18.60 |
| 20 | 16.99 |
| 25 | 15.64 |
| 30 | 14.48 |
| 35 | 13.49 |
| 40 | 12.62 |
| 45 | 11.86 |

Figure 10.2 shows a plot of the preceding limiting dry unit weights against the corresponding liquid limits. For any given soil, if the natural dry unit weight falls below the limiting line, the soil is likely to collapse.

Care should be taken to obtain undisturbed samples for determination of the collapse potentials and the dry unit weights just described. It is preferable to obtain block samples cut by hand. Samples obtained by thin wall tubes may undergo some compression during the sampling process. Note that if this procedure is used, the boreholes should be made *without water*.

## 10.4
## Procedure for Calculation of Collapse Settlement

Jennings and Knight (1975) have proposed the following laboratory procedure to determine the collapse settlement of structures upon saturation of soil.

**1.** Obtain *two* undisturbed soil specimens for tests in a standard consolidation test apparatus (oedometer).

**2.** Place the two specimens under 1 kN/m² pressure for 24 hours.

**3.** After 24 hours, saturate one specimen by flooding. The other specimen is kept at natural moisture content.

**4.** After 24 hours of flooding, resume the consolidation test for both specimens by doubling the load (same procedure as the standard consolidation test) to the desired pressure level.

**5.** Plot the $e$ vs. log $p$ graphs for both of the specimens (Figure 10.3a and b).

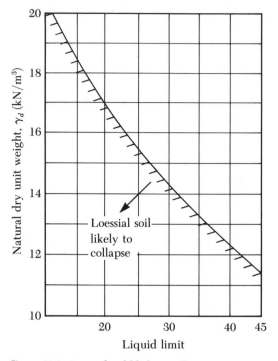

**Figure 10.2**   Loessial soil likely to collapse

**6.** Calculate the *in-situ* effective pressure, $p_o$. Draw a vertical line corresponding to the pressure $p_o$.

**7.** From the $e$ vs. log $p$ curve of the soaked sample, determine the preconsolidation pressure, $p_c$. If $p_c/p_o = 0.8–1.5$, the soil is normally consolidated; however, if $p_c/p_o > 1.5$, it is preconsolidated.

**8.** Determine $e'_o$, corresponding to $p_o$ from the $e$ vs. log $p$ curve of the soaked sample. (This procedure for normally consolidated and over-consolidated soils is shown in Figures 10.3a and b, respectively.)

**9.** Through the point $p_o$, $e'_o$, draw a curve that is similar to the $e$ vs. log $p$ curve obtained from the specimen tested at natural moisture content.

**10.** Determine the incremental pressure, $\Delta p$, on the soil caused by the construction of the foundation. Draw a vertical line corresponding to the pressure of $p_o + \Delta p$ in the $e$ vs. log $p$ curve.

**11.** Now, determine $\Delta e_1$ and $\Delta e_2$. The settlement of soil without change in the natural moisture content is

$$S_1 = \frac{\Delta e_1}{1 + e'_o}(H) \tag{10.6}$$

Also, the settlement caused by collapse in the soil structure

$$S_2 = \frac{\Delta e_2}{1 + e'_o}(H) \tag{10.7}$$

where $H$ = thickness of soil susceptible to collapse

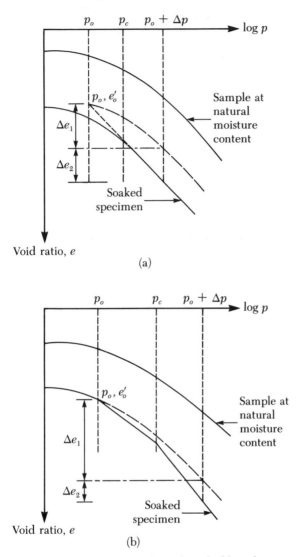

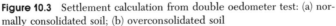

**Figure 10.3** Settlement calculation from double oedometer test: (a) normally consolidated soil; (b) overconsolidated soil

## 10.5

## Foundation Design in Soils Not Susceptible to Wetting

For actual foundation design purposes, some standard field load tests may also be conducted. Figure 10.4 shows the results of some field load tests in loess deposits in Nebraska and Iowa. Note that the load-settlement relationships are essentially linear up to a certain critical pressure, $p_{cr}$, at which there is a breakdown of the soil structure and, hence, a large settlement. Sudden breakdown of soil structure is more common with soils having a high natural moisture content.

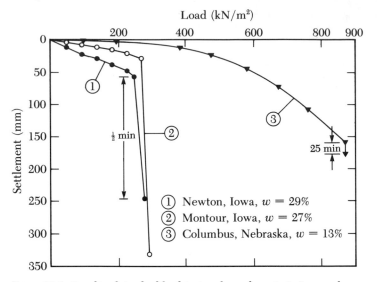

**Figure 10.4**  Results of standard load test on loess deposits in Iowa and Nebraska (adapted from *Foundation Engineering*, 2nd ed., by R. B. Peck, W. E. Hanson, and T. H. Thornburn. Copyright 1974 by John Wiley & Sons. Reprinted by permission)

If enough precautions are taken in the field to prevent moisture from increasing under structures, spread foundations and raft foundations may be built on potentially collapsible soils. However, the foundations must be proportioned in such a manner that the critical stresses (Figure 10.4) in the field are never exceeded. A factor of safety of about 2.5 to 3 should be used to calculate the allowable soil pressure; or

$$p_{all} = \frac{p_{cr}}{FS} \tag{10.8}$$

where $p_{all}$ = allowable soil pressure
$FS$ = factor of safety (about 2.5 to 3)

The differential and total settlements of these foundations should be similar to those of foundations designed for sandy soils.

In several instances, continuous foundations may be safer over collapsible soils than isolated foundations, in that they can effectively minimize differential settlement. A typical procedure for construction of continuous foundations is shown in Figure 10.5. This procedure uses footing beams and longitudinal load-bearing beams.

In the construction of heavy structures, such as grain elevators, over collapsible soils, settlements up to about 0.3 m are sometimes allowed (Peck, Hanson, and Thornburn, 1974). In this case, it is assumed that tilting of the foundation is not likely to occur; that is, there is no eccentric loading. The total expected settlement for such structures can be estimated from standard consolidation tests on samples at field moisture contents. If eccentric loading is

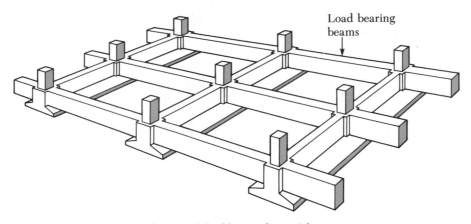

Load bearing
beams

**Figure 10.5** Continuous foundation with load-bearing beams (after
Clemence and Finbarr, 1981)

avoided, the foundations will exhibit uniform settlement over loessial deposits;
however, if the soil is of residual or colluvial nature, settlement may not be
uniform. This is due to the nonuniformity that is generally encountered in
residual soil profiles.

Extreme caution must be used in building heavy structures over collapsible
soils. If large settlements are expected, drilled-pier and pile foundations may be
considered. These foundations can transfer the load to a stronger load-bearing
stratum.

## 10.6

### Foundation Design in Soils Susceptible to Wetting

If it is suspected that the upper layer of soil may get wet and collapse at
some time after construction of the foundation, several design techniques may
be considered:

**1.** If the expected depth of wetting is about 1.5–2 m from the ground
surface, the soil may be moistened and recompacted by heavy rollers. Spread
footings and rafts may be constructed over the compacted soil.

**2.** If conditions are favorable, foundation trenches can be flooded with
solutions of sodium silicate and calcium chloride. This will chemically stabilize the soil. The soil will behave like a soft sandstone and resist collapse
upon saturation. This method is successful only if the solutions can penetrate
to the desired depth; thus it is most applicable to fine sand deposits. Silicates
are rather costly and are not generally used. However, in some parts of
Denver, silicates have been used very successfully.

**3.** When the soil layer is susceptible to wetting up to a depth of about
10 m, several techniques may be used to cause collapse of the soil *before*
foundation construction. Two of these are *vibroflotation* and *ponding* (also
called *flooding*). Vibroflotation is used successfully in free-draining soil (see

Chapter 12). The procedure of ponding—by construction of low dikes—is employed at sites that have no impervious layers. Note that, even after saturation and collapse of the soil by ponding, some additional settlement of the soil may occur after the placement of the foundation. Additional settlement may also be caused by incomplete saturation of the soil at the time of foundation construction. Ponding may be used successfully in the construction of earth dams.

**4.** If precollapsing of soil is not practical, foundations may be constructed beyond the zone of possible wetting. This may require drilled piers and piles. In the design of drilled piers and piles, one must consider the effect of negative skin friction, which is the result of the collapse of the soil structure and the associated settlement of the zone of subsequent wetting.

In some cases, a *rock column type of foundation (vibroreplacement)* may also be considered. Rock columns are built with large boulders that penetrate the potentially collapsible soil layer. They act like piles in transferring the load to a more stable soil layer.

## Foundations on Expansive Soil

### 10.7

Expansive Soils—General

There are many plastic clays that swell considerably when water is added to them and then shrink with the loss of water. Foundations constructed on these clays are subjected to large uplifting forces caused by the swelling. These forces will induce heaving, cracking, and breakup both of building foundations and slab-on-grade members. Expansive clays cover large parts of the United States, South America, Africa, Australia, and India. In the United States, these clays are predominant in Texas, Oklahoma, and in the upper Missouri Valley. In general, potentially expansive clays have liquid limits and plasticity indices greater than about 40 and 15, respectively.

As noted, an increase in moisture content causes clay to swell. The depth in a soil profile up to which periodic changes of moisture occur is usually referred to as the *active zone*. The depth of the active zone will vary depending on the location of the soil profile. Some typical active-zone depths in American cities are given in Table 10.2. In some clays and clay shales in the western United States, the depth of the active zone can be as much as 15 m. The

**Table 10.2** Typical Active-Zone Depths in Some U.S. Cities[a]

| City | Depth of active zone (m) |
|------|--------------------------|
| Houston | 1.5 to 3 |
| Dallas | 2.1 to 4.6 |
| San Antonio | 3 to 9 |
| Denver | 3 to 4.6 |

[a]After O'Neill and Poormoayed (1980)

active-zone depth can be easily determined by plotting the liquidity index vs. the depth of the soil profile over several seasons. Figure 10.6 shows an example of this from the Beaumont formation in the Houston area.

## 10.8

## Laboratory Measurement of Swell

To study the magnitude of possible swell in a clay, simple laboratory oedometer tests can be conducted on undisturbed specimens. Two common tests are the *unrestrained swell test* and *swelling pressure test*.

In the *unrestrained swell test*, the specimen is placed in an oedometer under a small surcharge of about 6.9 kN/m$^2$ (1 lb/in.$^2$). Water is then added to the sample, and the expansion of the volume of the specimen (that is, height; the area of cross section is constant) is measured until equilibrium is reached. The percent of free swell may be expressed as a ratio

$$S_{w(\text{free})}(\%) = \frac{\Delta H}{H}(100) \qquad\qquad (10.9)$$

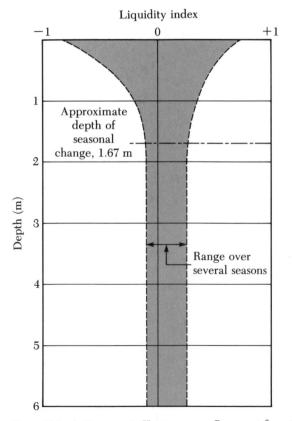

**Figure 10.6**  Active zone in Houston area—Beaumont formation (after O'Neill and Poormoayed, 1980)

where $s_{w(\text{free})}$ = free swell
$\Delta H$ = height of swell due to saturation
$H$ = original height of the specimen

Vijayvergiya and Ghazzaly (1973) have analyzed a number of soil test results conducted in this manner and have prepared a correlation chart of the free swell, liquid limit, and natural moisture content (see Figure 10.7). The free surface swell can be calculated by use of this chart (O'Neill and Poormoayed, 1980):

$$\Delta S_F = 0.0033 Z s_{w(\text{free})} \tag{10.10}$$

where $\Delta S_F$ = free surface swell
$Z$ = depth of active zone
$s_{w(\text{free})}$ = free swell, as a percent (Figure 10.7)

The *swelling pressure test* can be conducted by taking a specimen in a consolidation ring and applying a pressure equal to the effective overburden pressure $(p_o)$ plus the approximate anticipated surcharge caused by the foundation $(p_s)$. Water is then added to the specimen. As the specimen starts to swell,

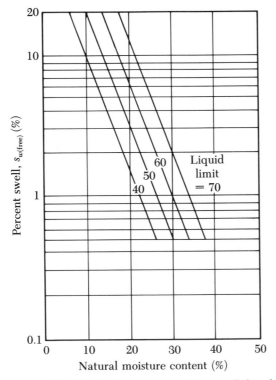

**Figure 10.7**   Relation between percent free swell, liquid limit, and natural moisture content (after Vijayvergiya and Ghazzaly, 1973)

pressure is applied in small increments to prevent swelling. This is continued until full swelling pressure is developed. At that time, the total pressure on the specimen is

$$p_T = p_o + p_s + p_1 \tag{10.11}$$

where $p_T$ = total pressure to prevent swelling, or zero swell pressure
$\quad\quad p_1$ = additional pressure added to prevent swelling after addition of water

A value of $p_T \approx$ 20–30 kN/m² is considered to be low, and a value of 1500–2000 kN/m² is considered to be very high. After zero swell pressure is attained, the soil specimen can be unloaded in steps up to the level of the overburden pressure, $p_o$. This unloading process will cause the specimen to swell. The equilibrium swell for each pressure level is also recorded. The nature of variation of the swell, in percent ($s_w$ %), and the applied pressure on the specimen will be like that shown in Figure 10.8.

The swelling pressure test can be used to determine the surface heave, $\Delta S$, for a given foundation as (O'Neill and Poormoayed, 1980)

$$\Delta S = \sum_{i=1}^{n} [s_{w(1)}\%](H_i)(0.01) \tag{10.12}$$

where $s_{w(1)}\%$ = swell, in percent, for layer $i$ under a pressure of $p_o + p_s$ (see Figure 10.8)
$\quad\quad \Delta H_i$ = thickness of layer $i$

The procedure of surface heave calculation using the preceding equation is shown in Example Problem 10.2.

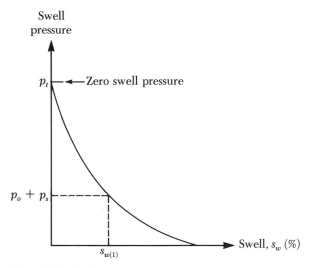

**Figure 10.8** Swell pressure test

## Example 10.1

A soil profile has an active zone of expansive soil of 2.0 m. The liquid limit and the average natural moisture content during the construction season are 50% and 20%, respectively. Determine the free surface swell.

### Solution

From Figure 10.7, for $LL = 50\%$ and $w = 20\%$, $s_{w(\text{free})} = 3\%$. From Eq. (10.10)

$$\Delta S_F = 0.0033 Z s_{w(\text{free})}$$

Hence

$$\Delta S_F = 0.0033(2)(3)(1000) = \underline{19.8 \text{ mm}}$$

## Example 10.2

A soil profile's active-zone depth is 3.5 m. If a foundation is to be placed at a depth of 0.5 m below the ground surface, what would be the estimated total swell? Given the following from laboratory tests:

| Depth, m | Swell under overburden and estimated foundation surcharge pressure, $s_{w(1)}(\%)$ |
|---|---|
| 0.5 | 2 |
| 1 | 1.5 |
| 2 | 0.75 |
| 3 | 0.25 |

### Solution

The values of $s_{w(1)}(\%)$ have been plotted with depth in Figure 10.9a. The area of this diagram will be the total swell. Using the trapezoidal rule

$$\Delta S = \frac{1}{100} \left[ \frac{1}{2}(1)(0 + 0.5) + \frac{1}{2}(1)(0.5 + 1.1) + \frac{1}{2}(1)(1.1 + 2) \right]$$

$$= 0.026 \text{ m} = \underline{26 \text{ mm}}$$

## Example 10.3

In Example Problem 10.2, if the allowable total swell is 10 mm, what would be the undercut necessary to reduce the total swell?

### Solution

Using the procedure outlined in Example Problem 10.2, the total swell at various depths below the foundation can be calculated as follows (from Figure 10.9a):

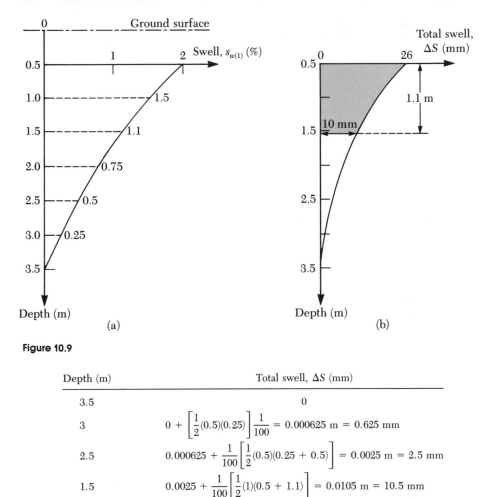

**Figure 10.9**

| Depth (m) | Total swell, $\Delta S$ (mm) |
|-----------|------------------------------|
| 3.5 | 0 |
| 3 | $0 + \left[\frac{1}{2}(0.5)(0.25)\right]\frac{1}{100} = 0.000625$ m $= 0.625$ mm |
| 2.5 | $0.000625 + \frac{1}{100}\left[\frac{1}{2}(0.5)(0.25 + 0.5)\right] = 0.0025$ m $= 2.5$ mm |
| 1.5 | $0.0025 + \frac{1}{100}\left[\frac{1}{2}(1)(0.5 + 1.1)\right] = 0.0105$ m $= 10.5$ mm |
| 0.5 | 26 mm |

These total settlements have been plotted in Figure 10.9b, and they show that a total swell of 10 mm corresponds to a depth of 1.6 m below the ground surface.

Hence, the undercut below the foundation is $1.6 - 0.5 = 1.1$ m.

This soil should be excavated, replaced by nonswelling soil, and recompacted.

## 10.9
## Classification of Expansive Soil

There are several classification systems for expansive soils based on the problems they can create in the construction of foundations. The classification system proposed by the U.S. Army Waterways Experiment Station (Snethen, Johnson, and Patrick, 1977) is the most recent one. It has also been summarized by O'Neill and Poormoayed (1980); see Table 10.3.

**Table 10.3**   Expansive Soil Classification System[a]

| Liquid limit | Plasticity index | Potential swell (%) | Potential swell classification |
|---|---|---|---|
| <50 | <25 | <0.5 | Low |
| 50–60 | 25–35 | 0.5–1.5 | Marginal |
| >60 | >35 | >1.5 | High |

Potential swell = vertical swell under a pressure equal to overburden pressure

[a]Compiled from O'Neill and Poormoayed (1980)

## 10.10

## Foundation Considerations on Expansive Soils

If a soil is classified as having a low swell potential, standard construction practices may be followed. However, if the soil possesses marginal or high swell potential, precautions need to be taken. This may entail the following:

1. replacing the expansive soil under the foundation;
2. changing the nature of the expansive soil by such measures as compaction control, prewetting, installation of moisture barriers, and chemical stabilization;
3. strengthening the structures to withstand heave, constructing structures that are flexible enough to withstand the differential soil heave without failure, or constructing isolated deep foundations below the depth of the active zone.

One particular method may not be the cure-all in every situation. It may become necessary to combine several techniques, and local construction experience should always be taken into consideration. Following are details about some of the commonly used techniques of dealing with expansive soils.

### Replacement of Expansive Soil

When moderately expansive soils of low thickness are present at the surface, they can be removed and replaced by less expansive soils and then compacted properly.

### Changing the Nature of Expansive Soil

1. *Compaction:* Heave of expansive soils is known to decrease substantially when the soil is compacted to a lower unit weight on the high side of the optimum moisture content (possibly 3–4% above the optimum moisture content). Even under such conditions, a slab-on-ground type of construction should not be encouraged where the total probable heave is expected to be about 35 mm or more. Figure 10.10a shows the recommended limits of soil compaction in the field for reduction of heave. Note that the recommended dry unit weights in Figure 10.10a are based on climatic ratings. According to the U.S. Weather Bureau data, a climatic rating of 15 represents an extremely unfavorable climatic

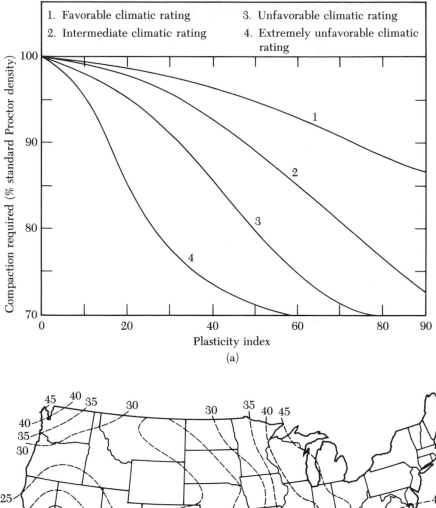

1. Favorable climatic rating
2. Intermediate climatic rating
3. Unfavorable climatic rating
4. Extremely unfavorable climatic rating

Compaction required (% standard Proctor density)

Plasticity index

(a)

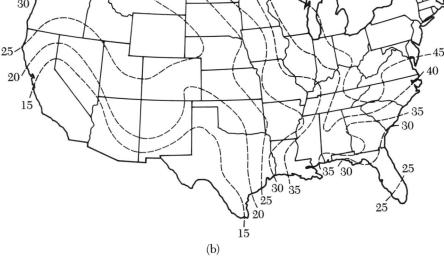

(b)

**Figure 10.10** (a) Soil compaction requirement based on climatic rating; (b) equivalent climatic rating of the United States (after Gromko, 1974)

condition; a value of 45 represents a favorable climatic rating. The isobars of climatic rating for the continental United States are shown in Figure 10.10b.

2. *Prewetting:* This is a technique for increasing the moisture content of the soil by ponding and, hence, achieving most of the heave before construction. However, this technique may be time consuming because the seepage of water through highly plastic clays is slow. After ponding, 4–5% of hydrated lime may be added to the top layer of the soil to make it less plastic and more workable (Gromko, 1974).

3. *Installation of Moisture Barriers:* The long-term effect of the differential heave can be reduced by controlling the moisture variation in the soil. This can be achieved by providing vertical moisture barriers about 1.5 m deep around the perimeter of slabs for the slab-on-grade type of construction. These moisture barriers may be constructed in trenches filled with gravel, lean concrete, or impervious membranes.

4. *Stabilization of Soil:* Chemical stabilization with the aid of lime and cement has often proved useful. A mix of about 5% lime is sufficient in most cases. Lime or cement and water are mixed with the top layer of soil and compacted. The addition of lime or cement will decrease the liquid limit, the plasticity index, and the swell characteristics of the soil. This type of stabilization work can be done to a depth of about 1–1.5 m. Hydrated high-calcium lime and dolomite lime are generally used for lime stabilization.

## 10.11
### Construction on Expansive Soils

One must be careful in choosing the type of foundation to be used on expansive soils. Table 10.4 shows some recommended construction procedures based on the total predicted heave ($\Delta S$) and the length-to-height ratio of the wall panels.

For example, Table 10.4 proposes the use of waffle slabs as an alternative in designing rigid buildings capable of tolerating movement. A schematic diagram of a waffle slab is shown in Figure 10.11. In this type of construction, the ribs hold the structural load. The waffle voids allow the expansion of soil.

Table 10.4 also suggests the use of foundation piers with a suspended floor slab for the construction of structures independent of movement. Figure 10.12a shows a schematic diagram of such an arrangement. The bottom of the piers should be placed below the active zone of the expansive soil. For the design of the piers, the uplifting force, $U$, may be estimated (Figure 10.12b) from the equation

$$U = \pi D_s Z p_t \tan \phi_{ps} \qquad (10.13)$$

where $D_s$ = diameter of the shaft of the pier
$\quad\quad Z$ = depth of the active zone
$\quad\quad \phi_{ps}$ = effective angle of plinth-soil friction
$\quad\quad p_t$ = pressure for zero horizontal swell (see Figure 10.8; $p_t = p_o + p_s + p_1$)

**Table 10.4** Construction Procedures on Expansive Clay Soils[a]

| Total predicted heave (mm) | | | Recommended construction | Method | Remarks |
|---|---|---|---|---|---|
| $L/H = 1.25$ | $L/H = 2.5$ | | | | |
| 0 to 6.35 | 12.7 | | No precaution | | |
| 6.35 to 12.7 | 12.7 to 50.8 | | Rigid building tolerating movement (steel reinforcement as necessary) | *Foundations:* Pads Strip footings Raft (waffle) | Footings should be small and deep, consistent with the soil-bearing capacity. Rafts should resist bending. |
| | | | | *Floor Slabs:* Waffle Tile | Slabs should be designed to resist bending and should be independent of grade beams. |
| | | | | *Walls:* | Walls on a raft should be as flexible as the raft. No rigid connections vertically. Brick works should be strengthened with tie bars or bands. |
| 12.7 to 50.8 | 50.8 to 101.6 | | Building damping movement | *Joints:* Clear Flexible | Contacts between structural units should be avoided; or flexible, waterproof material may be inserted in the joints. |
| | | | | *Walls:* Flexible Unit construction Steel frame | Walls or rectangular building units should heave as a unit. |
| | | | | *Foundations:* Three point Cellular Jacks | Cellular foundations allow slight soil expansion to reduce swelling pressure. Adjustable jacks can be inconvenient to owners. Three-point loading allows motion without duress. |
| >50.8 mm | >101.6 mm | | Building independent of movement | *Foundation Piers:* Straight shaft Bell bottom | Smallest-diameter and widely spaced piers compactable with load should be placed. Clearance should be allowed under grade beams. |
| | | | | *Suspended Floor:* | Floor should be suspended on grade beams 0.3 to 0.45 m above the soil. |

[a]After Gromko, 1974

469

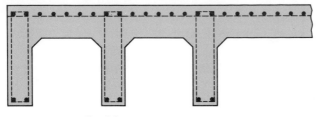

**Figure 10.11**   Waffle slab

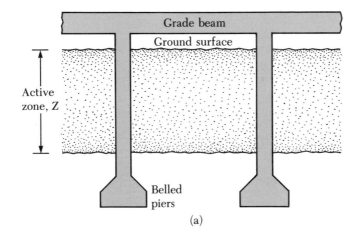

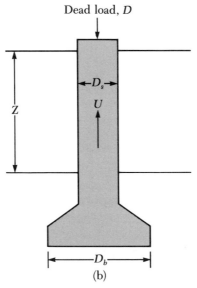

**Figure 10.12**   Construction of structures independent of movement: (a) belled piers and grade beam; (b) definition of parameter in Eq. (12.13)

In most cases, the value of $\phi_{ps}$ varies between $10°$ and $20°$. An average value of the zero horizontal swell pressure must be determined in the laboratory. In the absence of laboratory results, the value of $p_t \tan \phi_{ps}$ may be taken to be equal to the undrained shear strength of clay $(c_u)$ in the active zone.

The belled portion of the drilled pier will act like an anchor to resist the uplifting force. Ignoring the weight of the pier

$$Q_{net} = U - D \tag{10.14}$$

where $Q_{net}$ = net uplift load
$\quad\quad\quad D$ = dead load

Now

$$Q_{net} \approx \frac{c_u N_c}{FS} \cdot \frac{\pi}{4}(D_b^2 - D_s^2) \tag{10.15}$$

where $c_u$ = undrained cohesion of the clay in which the bell of the pier is located.

Combining Eqs. (10.14) and (10.15)

$$U - D = \frac{c_u N_c}{FS} \cdot \frac{\pi}{4}(D_b^2 - D_s^2) \tag{10.16}$$

where $N_c$ = bearing capacity factor
$\quad\quad FS$ = factor of safety
$\quad\quad D_b$ = diameter of the bell of the pier

Conservatively, the value of $N_c$ can be taken as [Table 3.2 and Eq. (3.20)]

$$N_c \approx N_{c(strip)}F_{cs} = N_{c(strip)}\left(1 + \frac{N_q B}{N_c L}\right) \approx 5.14\left(1 + \frac{1}{5.14}\right) = 6.14$$

An example of a drilled-pier design is given in Example Problem 10.4.

## Example 10.4

Figure 10.13 shows a drilled pier. The depth of the active zone is 5 m. The zero swell pressure of the swelling clay $(p_t)$ is $500 \text{ kN/m}^2$. For the pier:

Dead load = 600 kN
Live load = 300 kN

a. Determine the diameter of the bell, $D_b$.
b. Determine the reinforcement required for the pier shaft.
c. Check the bearing capacity of the pier, assuming that the uplift force is zero.

**Solution**
Part a: Determination of the bell diameter, $D_b$
The uplifting force is [Eq. (10.13)]

$$U = \pi D_s Z p_t \tan \phi_{ps}$$

Dead load + live load = 900 kN

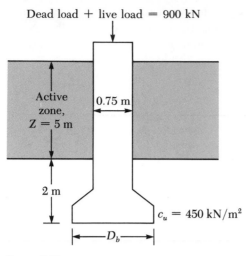

**Figure 10.13**

Assume $\phi_{ps} \approx 12°$, $Z = 5$ m, $p_t = 500$ kN/m². So

$$U = \pi(0.75)(5)(500) \tan 12° \approx 1252 \text{ kN}$$

Assume dead load and live load to be zero, and $FS$ in Eq. (10.16) to be 1.25. So, from Eq. (10.16)

$$U = \frac{c_u N_c}{FS} \cdot \frac{\pi}{4}(D_b^2 - D_s^2)$$

$$1252 = \frac{(450)(6.14)}{1.25} \cdot \frac{\pi}{4}(D_b^2 - 0.75^2)$$

$$D_b^2 \approx 1.28$$

or

$$D_b \approx 1.13. \text{ So use } \underline{1.25 \text{ m}}.$$

The factor of safety against uplift with the dead load should also be checked. A factor of safety of at least 3 is desirable. So, from Eq. (10.16)

$$FS = \frac{c_u N_c \left(\dfrac{\pi}{4}\right)(D_b^2 - D_s^2)}{U - D}$$

$$U - D = 1252 - 600 = 652 \text{ kN}$$

So

$$FS = \frac{c_u N_c \left(\dfrac{\pi}{4}\right)(D_b^2 - D_s^2)}{U - D} = \frac{(450)(6.14)\left(\dfrac{\pi}{4}\right)(1.25^2 - 0.75^2)}{652}$$

$$= \underline{3.33 > 3\text{—O.K.}}$$

## Part b: Reinforcement

The reinforcement should be provided for the total uplift load—that is, 1252 kN (assuming that dead load and live load are zero). So, the area of steel

$$A_s = \frac{U}{\left(\dfrac{\text{yield point of steel}}{\text{factor of safety}}\right)}$$

The preceding factor of safety may be taken as 1.25. The yield point of steel $\approx 275$ MN/m$^2$ ($\approx 40,000$ lb/in.$^2$). Hence, the area of steel

$$A_s = \frac{1252}{\left(\dfrac{275 \times 10^3}{1.25}\right)} = 5.69 \times 10^{-3} \text{ m}^2 = 5.69 \times 10^3 \text{ mm}^2 \text{ (8.82 in.}^2)$$

## Part c: Check for Bearing Capacity

Assume $U = 0$.

$$\text{Dead load} + \text{live load} = 600 + 300 = 900 \text{ kN}$$

$$\text{Downward load per unit area} = \frac{900}{\left(\dfrac{\pi}{4}\right)(D_b^2)} = \frac{900}{\left(\dfrac{\pi}{4}\right)(1.25)^2} = 733.4 \text{ kN/m}^2$$

$$\text{Net bearing capacity of the soil under the bell} = q_u =$$

$$c_u N_c = 450(6.14) = 2763 \text{ kN/m}^2$$

Hence, factor of safety against bearing capacity failure

$$= \frac{2763}{733.4} = 3.77 > 3\text{—O.K.}$$

# Sanitary Landfill

## 10.12

### Sanitary Landfills—General

Sanitary landfills provide a way to dispose of refuse on land without causing danger to public health. Almost all countries use sanitary landfills with varying degrees of success. The refuse disposed in sanitary landfills can contain materials like wood, paper, and fibrous wastes or demolition wastes like bricks and stones. The refuse is dumped and compacted at frequent intervals and then covered with a layer of soil (Figure 10.14). In the compacted state, the average unit weight of the refuse may vary between 5–10 kN/m$^3$. A typical city in the United States, with a population of one million, can generate about $3.8 \times 10^6$ m$^3$ of compacted landfill material per year.

As property value continues to increase in densely populated areas, it becomes more and more tempting to construct structures over sanitary landfills. In some instances, a visual site inspection may not be enough to detect an old sanitary landfill. However, construction of foundations over sanitary landfills is

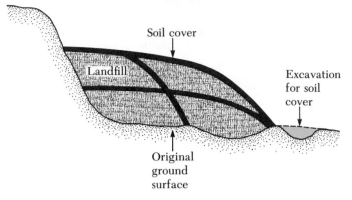

**Figure 10.14**  Schematic diagram of sanitary landfill in progress

generally problematic because of poisonous gases such as methane, excessive settlement, and low inherent bearing capacity. Few studies are available on this topic. Based on the information available, the next two sections will discuss settlement and bearing capacity problems associated with sanitary landfills.

## 10.13

### Settlement of Sanitary Landfills

Sanitary landfills undergo large continuous settlements over a long period of time. Yen and Scanlon (1975) have documented the settlement of several landfill sites in California. Based on their analysis, the settlement rate after the completion of the landfill (Figure 10.15) can be expressed as

$$m = \frac{\Delta H_f}{\Delta t} \tag{10.17}$$

where $m$ = settlement rate
  $H_f$ = maximum height of the sanitary landfill

Based on several field observations, the following empirical correlations for the settlement rate have been determined (Yen and Scanlon, 1975).

$$m = 0.0268 - 0.0116 \log t_1 \text{ (for fill heights ranging}$$
$$\text{from } 12\text{--}24 \text{ m)} \tag{10.18}$$

$$m = 0.038 - 0.0155 \log t_1 \quad \text{(for fill heights ranging}$$
$$\text{from } 24\text{--}30\text{m)} \tag{10.19}$$

$$m = 0.0433 - 0.0183 \log t_1 \text{ (for fill heights greater}$$
$$\text{than } 30 \text{ m)} \tag{10.20}$$

where $m$ is in m/month and $t_1$ is the median fill age, in months. The medium fill age can be defined as (Figure 10.15)

$$t_1 = t - \frac{t_c}{2} \tag{10.21}$$

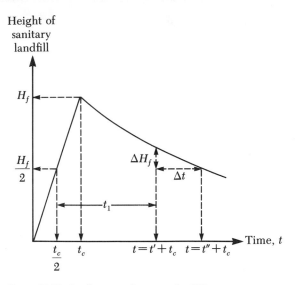

**Figure 10.15**   Settlement of sanitary landfills

where $t$ = time from the beginning of landfill
$t_c$ = time for completion of the landfill

Equations (10.18), (10.19), and (10.20) were based on field data from landfills for which the value of $t_c$ varied from 70 to 82 months. To get an idea of the approximate length of time required for a sanitary landfill to undergo complete settlement, consider Eq. (10.18). For a fill of height of 12 m and $t_c = 72$ months

$$m = 0.0268 - 0.0116 \log t_1$$

or

$$\log t_1 = \frac{0.0268 - m}{0.0116}$$

If $m = 0$ (that is, for zero settlement rate), the $\log t_1 = 2.31$, or $t_1 \approx 200$ months. Thus, the settlement of the fill will continue for a period of $t_1 - t_c/2 = 200 - 36 = 164$ months ($\approx 14$ years) after its completion. This is a fairly long time. This calculation shows that one must pay close attention to the settlement of foundations constructed on sanitary landfills.

A comparison of Eqs. (10.18) to (10.20) for rates of settlement shows that the value of $m$ increases with the increase of the height of the fill. However, for fill heights greater than about 30 m, the rate of settlement should not be much different than that from Eq. (10.20). This is because decomposition of the organic matter close to the surface is mainly the result of anaerobic environment. For deeper fills, the decomposition is slower. Hence, for fill heights greater than about 30 m, the rate of settlement does not accelerate any faster than those that are about 30 m in height.

Sowers (1973) has also proposed a relation for calculation of the settlement of a sanitary landfill:

$$\Delta H = \frac{\alpha H_f}{1 + e} \log\left(\frac{t''}{t'}\right)$$

where $H_f$ = height of the fill
$e$ = void ratio
$\alpha$ = a coefficient for settlement
$t''$, $t'$ = times (see Figure 10.15)
$\Delta H$ = settlement between times $t'$ and $t''$

The coefficients $\alpha$ are between the following limits

$$\alpha = 0.09e \text{ (for conditions favorable to decomposition)} \qquad (10.23)$$

and

$$\alpha = 0.03e \text{ (for conditions unfavorable to decomposition)} \qquad (10.24)$$

Equation (10.22) is similar to the equation for secondary consolidation settlement.

## 10.14
### Bearing Capacity of Foundations on Sanitary Landfills

Shallow foundations constructed on sanitary landfills with a compacted soil cover may fail in two ways (Sowers, 1968): by punching shear, or by rotational shear, as shown in Figure 10.16. Punching shear failure (Figure 10.16a) occurs when the width of the foundation, $B$, is relatively small compared to the thickness of the soil cover, $D_C$. However, when the thickness of the soil cover is relatively small compared to the foundation width and where the strength of the soil cover is low, rotational shear failure may occur (Figure 10.16b). The allowable bearing capacity of shallow foundations for light residential or office buildings over sanitary landfills should not be greater than about 20–40 kN/m².

Sometimes the allowable bearing capacity of shallow foundations can be increased by increasing the thickness of the compacted soil cover such that $D_C \geq 1.5$–$2B$. Note, however, that this excess weight of the fill and the compaction process may eventually increase the ultimate settlement of the structure.

When the fill is relatively homogeneous and light structures are constructed over it, the settlement pattern will be somewhat like that in Figure 10.16c. However, when the fill is nonhomogeneous in nature with irregular hard zones, such as boulders, the settlement is nonuniform, as shown in Figure 10.16d. Serious damage to structures is usually caused by this type of settlement.

In several instances, the use of continuous foundations, such as that shown in Figure 10.5, may help reduce the differential settlement problem. If the estimated settlement of a structure is not tolerable, or heavier structures are to be built, pile or drilled-pier foundations are alternatives. In this case, it will be

Bearing Capacity Failure

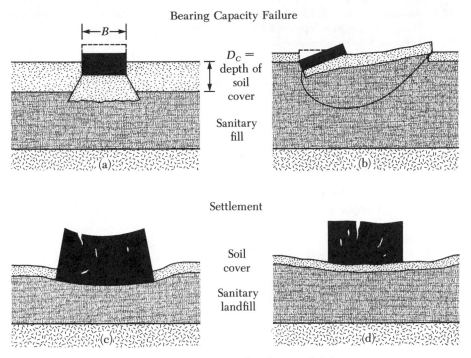

**Figure 10.16** Nature of bearing capacity failure and settlement of shallow foundations on sanitary landfills (redrawn after Sowers, 1968)

necessary to use noncorroding materials, because sanitary fills with moisture may corrode metal piles and may also damage concrete. If pile or pier foundations are used, the floor slab should be poured independently of the grade beams to avoid cracking due to differential settlement; otherwise, a structural slab will be necessary.

## Problems

**10.1** Refer to Figure 10.2, which was based on Eq. (10.5) and $G_s = 2.65$. Draw a similar curve of $\gamma_d$ (in $kN/m^3$) vs. liquid limit with $G_s = 2.7$, and show the zone in which the loessial soils are likely to collapse on saturation.

**10.2** A collapsible soil layer in the field has a thickness of 11 ft. The average effective overburden pressure on the soil layer is 1775 lb/ft². An undisturbed specimen of this soil was subjected to a double oedometer test (refer to Figure 10.3). The preconsolidation pressure of the specimen as determined from the soaked specimen was 2190 lb/ft². Is the soil in the field normally consolidated or preconsolidated?

**10.3** An expansive soil has an active-zone thickness of 21 ft. The natural moisture content of the soil is 14.5%, and its liquid limit is 40. Calculate the free surface swell of the expansive soil upon saturation.

**10.4** Repeat Problem 10.3 assuming that the liquid limit of the soil is 50. All other quantities are the same.

**10.5**  An expansive soil profile has an active zone thickness of 5.2 m. A shallow foundation is to be constructed at a depth of 1.2 m below the ground surface. Based on a swell pressure test, the following are given:

| Depth from ground surface (m) | Swell under overburden and estimated foundation surcharge pressure, $s_{w(1)}$ (%) |
|---|---|
| 1.2 | 3 |
| 2.2 | 2 |
| 3.2 | 1.2 |
| 4.2 | 0.55 |
| 5.2 | 0 |

Estimate the total possible swell of soil under the foundation.

**10.6**  Repeat Problem 10.5 with the following (active zone thickness = 17 ft; depth of shallow foundation = 4 ft):

| Depth from ground surface (ft) | Swell under overburden pressure and estimated foundation surcharge pressure, $s_{w(1)}$ (%) |
|---|---|
| 4 | 1.75 |
| 6.5 | 1.2 |
| 10 | 0.7 |
| 13.5 | 0.3 |
| 15 | 0.15 |
| 17 | 0 |

**10.7**  Refer to Problem 10.5. If the allowable total swell is 25 mm, what would be the necessary undercut?

**10.8**  Refer to Problem 10.6. If the allowable total swell is 0.6 in., what would be the necessary undercut?

**10.9**  Refer to Figure 10.12b. For the drilled pier, given:

Thickness of active zone, $Z$ = 7.6 m
Dead load = 1000 kN
Live load = 320 kN
Diameter of pier shaft, $D_s$ = 1 m
Zero swell pressure for the clay in active zone = 650 kN/m²
Average angle of plinth-soil friction, $\phi_{ps}$ = 10°
Average undrained cohesion of the clay around the bell = 150 kN/m²

Determine the diameter of the pier bell, $D_b$. A factor of safety of 1.5 against uplift is required with the assumption that dead load plus live load is equal to zero.

**10.10**  Refer to Problem 10.9. If an additional requirement is placed such that the factor of safety against uplift is at least 3 with the dead load on (live load = 0), what should be the diameter of the bell?

# References

Clemence, S. P., and Finbarr, A. O. (1981). "Design Considerations for Collapsible Soils," *Journal of the Geotechnical Engineering Division*, American Society of Civil Engineers, Vol. 107, No. GT3, pp. 305–317.

Gromko, G. J. (1974). "Review of Expansive Soils," *Journal of the Geotechnical Engineering Division*, American Society of Civil Engineers, Vol. 100, No. GT6, pp. 667–687.

Holtz, W. G., and Hilf, J. W. (1961). "Settlement of Soil Foundations Due to Saturation," *Proceedings*, Fifth International Conference on Soil Mechanics and Foundation Engineering, Paris, Vol. 1, 1961, pp. 673–679.

Jennings, J. E., and Knight, K. (1975). "A Guide to Construction On or With Materials Exhibiting Additional Settlements Due to 'Collapse' of Grain Structure," *Proceedings*, Sixth Regional Conference for Africa on Soil Mechanics and Foundation Engineering, Johannesburg, pp. 99–105.

O'Neill, M. W., and Poormoayed, N. (1980). "Methodology for Foundations on Expansive Clays," *Journal of the Geotechnical Engineering Division*, American Society of Civil Engineers, Vol. 106, No. GT12, pp. 1345–1367.

Peck, R. B., Hanson, W. E., and Thornburn, T. B. (1974). *Foundation Engineering*, Wiley, New York.

Snethen, D. R., Johnson, L. D., and Patrick, D. M. (1977). "An Evaluation of Expedient Methodology for Identification of Potentially Expansive Soils," *Report No. FHWA-RD-77-94*, U.S. Army Engineers Waterways Experiment Station, Vicksburg, Miss.

Sowers, G. F. (1968). "Foundation Problems in Sanitary Landfills," *Journal of the Sanitary Engineering Division*, American Society of Civil Engineers, Vol. 94, No. SA1, pp. 103–116.

Sowers, G. F. (1973). "Settlement of Waste Disposal Fills," *Proceedings*, Eighth International Conference on Soil Mechanics and Foundation Engineering, Moscow, pp. 207–210.

Vijayvergiya, V. N., and Ghazzaly, O. I. (1973). "Prediction of Swelling Potential of Natural Clays," *Proceedings*, Third International Research and Engineering Conference on Expansive Clays, pp. 227–234.

Yen, B. C., and Scanlon, B. (1975). "Sanitary Landfill Settlement Rates," *Journal of the Geotechnical Engineering Division*, American Society of Civil Engineers, Vol. 101, No. GT5, pp. 475–487.

# CHAPTER 11

# Reinforced Earth Structures

## 11.1

### Introduction

The use of reinforced earth is a recently developed method of designing foundations and earth-retaining structures. *Reinforced earth* is a construction material consisting of soil that has been strengthened by metal rods and/or strips, nonbiodegradable fabrics, and the like. The fundamental idea of reinforcing soil is not new; in fact, it goes back to biblical times. However, the present concept of systematic analysis and design was initiated by a French engineer named Vidal (1966). The French Road Research Laboratory has done an extensive amount of research on the applicability and the beneficial effects of the use of reinforced earth as a construction material. This research has been documented in detail by Darbin (1970), Schlosser and Long (1974), and Schlosser and Vidal (1969).

Several retaining walls with reinforced earth have been constructed in France since Vidal initiated his work. The first reinforced earth retaining wall in the United States was constructed in 1972 in Southern California.

This chapter will discuss in detail the fundamentals of the theory and design of shallow continuous foundations and retaining walls with reinforced earth. However, it will limit the discussion to the use of metal reinforcing strips as a reinforcement material.

The beneficial effects of soil reinforcement derive from (a) the soil's increased tensile strength and (b) the shear resistance developed from the friction at the soil-reinforcement interfaces. This is comparable to the reinforcement of concrete structures. At this time, the design of reinforced earth is done with *free-draining granular soil only*. Thus one avoids the effect of pore water pressure development in cohesive soils, which will, in turn, control the cohesive bond at the soil-reinforcement interfaces.

In most instances, galvanized steel strips are used as reinforcement in soil. However, galvanized steel is subject to corrosion. The rate of corrosion will depend on several environmental factors. Binquet and Lee (1975b) have suggested that the average rate of corrosion of galvanized steel strips varies between 0.025 mm/year to 0.050 mm/year. So, in the actual design of reinforcements, allowances must be made to take into account the rate of corrosion. Thus

$$t_c = t_{\text{design}} + r \text{ (life span of structure)}$$

where $t_c$ = actual thickness of reinforcing strips to be used in construction
$t_{\text{design}}$ = thickness of strips determined from design calculations
$r$ = rate of corrosion

Further research needs to be done on corrosion-resistant materials such as fiberglass before they can be used as reinforcing strips.

## Bearing Capacity of Continuous Shallow Foundations

### 11.2

#### Modes of Failure

The bearing capacity of shallow foundations resting on reinforced earth has been studied in detail by Binquet and Lee (1975a, b), who have proposed a rational design method that will be presented in the following sections.

The nature of bearing capacity failure of a shallow strip foundation resting on a compact and homogeneous soil mass was presented in Figure 3.1a. In contrast, if layers of reinforcing strips (sometimes referred to as *ties*) are placed in the soil under a shallow strip foundation, the nature of failure in the soil mass will be like that shown in Figure 11.1a, b, and c.

The nature of failure in the soil mass, shown in Figure 11.1a, generally occurs when the first layer of reinforcement is located at a depth, $d$, greater than about 2/3 $B$ ($B$ = width of the foundation). If the reinforcements in the first layer are strong and their concentration is sufficiently large, they may act like a rigid base located at a limited depth. The bearing capacity of foundations in such cases can be evaluated by the theory presented by Mandel and Salencon (1972). Experimental laboratory results for the bearing capacity of shallow foundations resting on a sand layer with a rigid rough base at a limited depth have also been provided by Meyerhof (1974), Pfeifle and Das (1979), and Das (1981).

The type of failure shown in Figure 11.1b could occur if $d/B$ is less than about 2/3 and the number of layers of reinforcement ($N$) is less than about 2–3. In this type of failure, reinforcement tie pullout occurs.

The most beneficial effect of reinforced earth is obtained when $d/B$ is less than about 2/3 and the number of reinforcement layers is greater than 4 but no more than 6–7. (The ties must be sufficiently long; see section 11.4 for length calculation.) In this case, the soil mass fails when the upper ties break (see

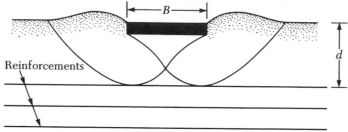

(a) $d/B > 2/3$ shear above reinforcement

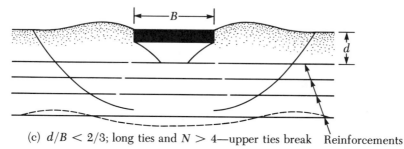

(b) $d/B < 2/3$; $N < 2$ or 3 or short ties—tie pullout   Reinforcements

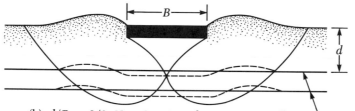

(c) $d/B < 2/3$; long ties and $N > 4$—upper ties break   Reinforcements

**Figure 11.1**   Three modes of bearing capacity failure in reinforced earth (redrawn after Binquet and Lee, 1975b)

Figure 11.1c). The following section discusses this type of bearing capacity failure in detail.

## 11.3

### Determination of the Force Induced in Reinforcement Ties

#### Location of Failure Surface

When designing shallow strip foundations, one must estimate the force that develops in the reinforcing ties as a result of the foundation load. This section presents the analytical procedure for estimation proposed by Binquet and Lee (1975b).

Figure 11.2 shows an idealized condition for the development of the failure surface in soil for the condition shown in Figure 11.1c. It consists of a central

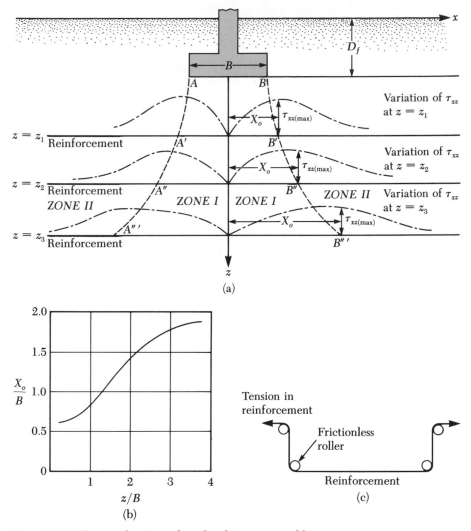

**Figure 11.2** Failure mechanism under a foundation supported by reinforced earth [part (b) after Binquet and Lee, 1975b]

zone—Zone I—immediately below the foundation that settles along with the foundation with the application of load. On each side of Zone I, the soil is pushed outward and upward—this is Zone II. The points $A'$, $A''$, $A'''$, . . . and $B'$, $B''$, $B'''$, . . . , which define the limiting lines between Zones I and II, can be obtained by considering the shear stress distribution, $\tau_{xz}$, in the soil caused by the foundation load. The term $\tau_{xz}$ refers to the shear stress developed at a depth $z$ below the foundation at a distance $x$ measured from the center line of the foundation. If integration of Boussinesq's equation is used, $\tau_{xz}$ can be given by the relation

$$\tau_{xz} = \frac{4bq_R x z^2}{\pi[(x^2 + z^2 - b^2)^2 + 4b^2 z^2]} \tag{11.1}$$

where $b$ = half-width of the foundation = $B/2$
$\quad\quad B$ = width of foundation
$\quad\quad q_R$ = load per unit area on the foundation

The nature of variation of $\tau_{xz}$ at any given depth $z$ is shown by the broken lines in Figure 11.2a. The points $A'$, $B'$ refer to the points at which the value of $\tau_{xz}$ is maximum at $z = z_1$. Similarly, $A''$, $B''$ refer to the points at which $\tau_{xz}$ is maximum at $z = z_2$. The distances $x = X_o$ at which the maximum value of $\tau_{xz}$ occurs can be given in a nondimensional form; this is shown in Figure 11.2b.

### Other Assumptions

Other assumptions needed to obtain the tie force at any given depth are as follows:

**1.** Under the application of bearing pressure by the foundation, the reinforcing ties at points $A'$, $A''$, $A'''$, . . . and $B'$, $B''$, $B'''$, . . . take the shape shown in Figure 11.2c. This means that the ties take two right angle turns on each side of Zone I around two frictionless rollers.

**2.** For a given number of reinforcing layers $(N)$, the ratio of the load per unit area on the foundation supported by reinforced earth $(q_R)$ to the load per unit area on the foundation supported by unreinforced earth $(q_o)$ is constant irrespective of the settlement level, $s$ (see Figure 11.3). Binquet and Lee (1975a) have proved this in laboratory experiments.

### Derivation of Equation

Figure 11.4a shows a continuous foundation supported by unreinforced soil and subjected to a load of $q_o$ per unit area. Similarly, Figure 11.4b shows a continuous foundation supported by a reinforced soil layer (one layer of reinforcement—that is, $N = 1$) and subjected to a load of $q_R$ per unit area. (Due to symmetry only one-half of the foundation is shown in Figure 11.4.) In both cases—that is, in Figures 11.4a and 11.4b—let the settlement be equal to $s$. For

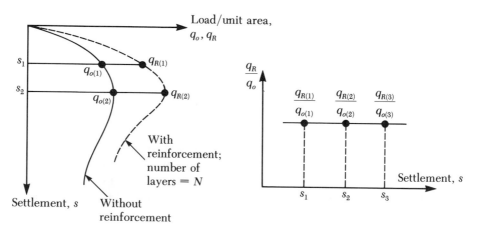

**Figure 11.3**  Relationship between load per unit area vs. settlement for foundations resting on reinforced and unreinforced soil

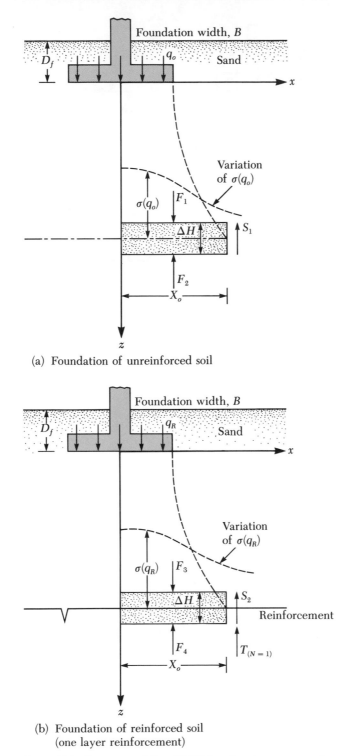

(a) Foundation of unreinforced soil

(b) Foundation of reinforced soil
(one layer reinforcement)

**Figure 11.4** Derivation of Eq. (11.20) (continued on next page)

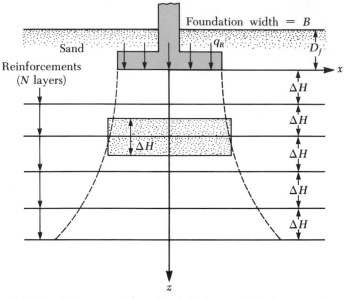

(c) Foundation on reinforced soil ($N$ layers of reinforcement)

**Figure 11.4**   (Continued)

one-half of each foundation under consideration, the forces per unit length on a soil element of thickness $\Delta H$ located at a depth $z$ are as follows:

*Unreinforced case:* $F_1$ and $F_2$ are the vertical forces, and $S_1$ is the shear force. Hence, for equilibrium

$$F_1 - F_2 - S_1 = 0 \tag{11.2}$$

*Reinforced case:* $F_3$ and $F_4$ are the vertical forces, $S_2$ is the shear force, and $T_{(N = 1)}$ is the tensile force developed in the reinforcement. The force $T_{(N = 1)}$ is vertical due to the assumption made for the deformation of reinforcement as shown in Figure 11.2c. So

$$F_3 - F_4 - S_2 - T_{(N = 1)} = 0 \tag{11.3}$$

If the settlement of foundation ($s$) is the same in both cases

$$F_2 = F_4 \tag{11.4}$$

Subtracting Eq. (11.2) from Eq. (11.3) and using the relationship given in Eq. (11.4), one obtains

$$T_{(N = 1)} = F_3 - F_1 - S_2 + S_1 \tag{11.5}$$

Note that the force $F_1$ is caused by the vertical stress, $\sigma$, on the soil element under consideration as a result of the load $q_o$ on the foundation. Similarly, $F_3$ is caused by the vertical stress imposed on the soil element as a result of the load $q_R$. Hence

$$F_1 = \int_0^{X_o} \sigma(q_o) \cdot dx \tag{11.6}$$

$$F_3 = \int_0^{X_o} \sigma(q_R) \cdot dx \tag{11.7}$$

$$S_1 = \tau_{xz}(q_o) \cdot \Delta H \tag{11.8}$$

$$S_2 = \tau_{xz}(q_R) \cdot \Delta H \tag{11.9}$$

where $\sigma(q_o)$ and $\sigma(q_R)$ are the vertical stresses at a depth $z$ caused by the loads $q_o$ and $q_R$ on the foundation; and $\tau_{xz}(q_o)$ and $\tau_{xz}(q_R)$ are the shear stresses at a depth $z$ and at a distance $X_o$ from the center line caused by the loads $q_o$ and $q_R$. Using integration of Boussinesq's solution

$$\sigma(q_o) = \frac{q_o}{\pi}\left[ \tan^{-1}\frac{z}{x-b} - \tan^{-1}\frac{z}{x+b} - \frac{2bz(x^2 - z^2 - b^2)}{(x^2 + z^2 - b^2)^2 + 4b^2z^2} \right]$$
$$\tag{11.10}$$

$$\sigma(q_R) = \frac{q_R}{\pi}\left[ \tan^{-1}\frac{z}{x-b} - \tan^{-1}\frac{z}{x+b} - \frac{2bz(x^2 - z^2 - b^2)}{(x^2 + z^2 - b^2)^2 + 4b^2z^2} \right]$$
$$\tag{11.11}$$

$$\tau_{xz}(q_o) = \frac{4bq_oX_oz^2}{\pi[X_o{}^2 + z^2 - b^2)^2 + 4b^2z^2]} \tag{11.12}$$

$$\tau_{xz}(q_R) = \frac{4bq_RX_oz^2}{\pi[(X_o{}^2 + z^2 - b^2)^2 + 4b^2z^2]} \tag{11.13}$$

where $b = B/2$

The procedure for derivation of Eqs. (11.10) to (11.13) will not be presented here; for this information readers are referred to a soil mechanics text (for example, Das, 1983). Proper substitution of Eqs. (11.10) to (11.13) into Eqs. (11.6) to (11.9) and simplification will yield the following:

$$F_1 = A_1q_oB \tag{11.14}$$

$$F_3 = A_1q_RB \tag{11.15}$$

$$S_1 = A_2q_o\ \Delta H \tag{11.16}$$

$$S_2 = A_2q_R\ \Delta H \tag{11.17}$$

where $A_1$ and $A_2 = f(z/B)$

The variations of $A_1$ and $A_2$ with nondimensional depth $z$ are given in Figure 11.5.

Substitution of Eqs. (11.14) through (11.17) into Eq. (11.5) gives

$$\begin{aligned}
T_{(N\ =\ 1)} &= A_1q_RB - A_1q_oB - A_2q_R\ \Delta H + A_2q_o\ \Delta H \\
&= A_1B(q_R - q_o) - A_2\ \Delta H(q_R - q_o) \\
&= q_o\left(\frac{q_R}{q_o} - 1\right)(A_1B - A_2\ \Delta H) \tag{11.18}
\end{aligned}$$

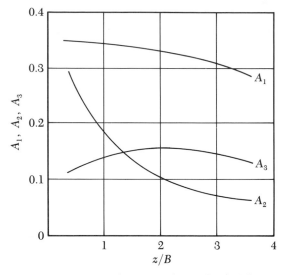

**Figure 11.5** Variation of $A_1$, $A_2$, and $A_3$ with $z/B$ (after Binquet and Lee, 1975b)

Note that Eq. (11.18) has been derived with the assumption that there is only one layer of reinforcement under the foundation shown in Figure 11.4b. However, if there are $N$ layers of reinforcement under the foundation with center-to-center spacings of $\Delta H$, as shown in Figure 11.4c, it can be assumed that

$$T_{(N)} = \frac{T_{(N\,=\,1)}}{N} \tag{11.19}$$

Combining Eqs. (11.18) and (11.19)

$$T_{(N)} = \frac{1}{N}\left[ q_o\left( \frac{q_R}{q_o} - 1 \right)(A_1 B - A_2\,\Delta H) \right] \tag{11.20}$$

The unit of $T_{(N)}$ in Eq. (11.20) is kN/m length of foundation.

## 11.4
## Factor of Safety of Ties Against Breaking and Pullout

Once the tie forces that develop in each layer as the result of the foundation load are determined by Eq. (11.20), one must determine whether or not the ties at any given depth $z$ will fail either by *breaking* or by *pullout*. The factor of safety against tie breaking at any depth $z$ below the foundation can be calculated as

$$FS_{(B)} = \frac{wtnf_y}{T_{(N)}} \tag{11.21}$$

where $FS_{(B)}$ = factor of safety against tie breaking
$\quad\quad\quad w$ = width of a single tie
$\quad\quad\quad\; t$ = thickness of each tie

$n$ = number of ties per unit length of the foundation
$f_y$ = yield or breaking strength of the tie material

The term $w \cdot n$ may be defined as the *linear density ratio (LDR)*, so

$$FS_{(B)} = \left[\frac{tf_y}{T_{(N)}}\right](LDR) \tag{11.22}$$

The resistance against the tie being pulled out derives from the frictional resistance between the soil and the ties at any given depth. From the fundamental principles of statics, we know that the frictional force per unit length of the foundation resisting tie pullout at any depth $z$ is equal to (Figure 11.6)

$$F_B = 2 \tan \phi_\mu [\text{normal force}]$$

$$= 2 \tan \phi_\mu [\underbrace{(LDR)\int_{X_o}^{L_o} \sigma(q_R)dx}_{} + \underbrace{(LDR)(\gamma)(L_o - X_o)(z + D_f)]}_{} \tag{11.23}$$

Two sides     Due to foundation         Due to effective
of tie,          load = $F_5$                 overburden
i.e., top                                  pressure = $F_6$
and bottom

where $\gamma$ = unit weight of soil
$D_f$ = depth of foundation
$\phi_\mu$ = tie-soil friction angle

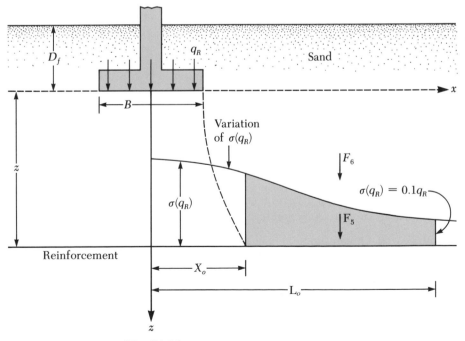

**Figure 11.6** Derivation of Eq. (11.24)

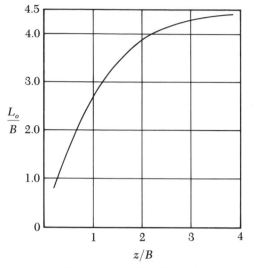

**Figure 11.7**    Variation of $L_o/B$ with $z/B$ (after Binquet and Lee, 1975b)

The relation for $\sigma(q_R)$ has been defined in Eq. (11.11). The value of $x = L_o$ is generally assumed to be the distance at which the value of $\sigma(q_R)$ is equal to $0.1q_R$. The value of $L_o$ as a function of depth $z$ is given in Figure 11.7. Equation (11.23) may be simplified as

$$F_B = 2 \tan \phi_\mu (LDR)[A_3 Bq_o\left(\frac{q_R}{q_o}\right) + \gamma(L_o - X_o)(z + D_f)] \qquad (11.24)$$

where $A_3$ is a nondimensional quantity and can be expressed as a function of depth $(z/B)$ (see Figure 11.5).

The factor of safety against tie pullout, $FS_{(P)}$, can be given as

$$FS_{(P)} = \frac{F_B}{T_{(N)}} \qquad (11.25)$$

## 11.5

### Design Procedure

Following is a step-by-step procedure for the design of a continuous foundation supported by reinforced earth:

**1.** Obtain the total load to be supported per unit length of the foundation. Also obtain the following quantities:

   a. Soil friction angle, $\phi$
   b. Soil-tie friction angle, $\phi_\mu$
   c. Factor of safety against bearing capacity failure
   d. Factor of safety against tie break, $FS_{(B)}$
   e. Factor of safety against tie pullout, $FS_{(P)}$
   f. Breaking strength of reinforcement ties, $f_y$

g. Unit weight of soil, $\gamma$
h. Young's modulus of soil, $E_s$
i. Poisson's ratio of soil, $\mu_s$
j. Allowable settlement of foundation, $s$
k. Depth of foundation, $D_f$

**2.** Assume a width of foundation, $B$, and also $d$ and $N$. The value of $d$ should be less than $2/3\ B$. Also, the distance from the bottom of the foundation to the lowest layer of the reinforcement should be about $2B$ or less. Calculate $\Delta H$.

**3.** Assume a value of $LDR$.

**4.** For the assumed value of $B$ (Step 2), determine the ultimate bearing capacity $(q_u)$ for the condition of unreinforced soil [Eq. (3.3); *note: c = 0*]. Determine $q_{all(1)}$ as

$$q_{all(1)} = \frac{q_u}{FS \text{ against bearing capacity failure}} \tag{11.26}$$

**5.** Calculate the allowable load $[q_{all(2)}]$ based on tolerable settlement $(s)$, assuming that the soil is not reinforced [Eq. (3.63)].

$$s = \frac{Bq_{all(2)}}{E_s}(1 - \mu_s^2)\alpha_r$$

For $L/B = \infty$, the value of $\alpha_r$ may be taken to be equal to 2, or

$$q_{all(2)} = \frac{E_s s}{B(1 - \mu_s^2)\alpha_r} \tag{11.27}$$

(The allowable load for a given settlement, $s$, could have been determined from equations in Chapter 3 that relate to standard penetration resistances.)

**6.** Determine the lower of the two values of $q_{all}$ as obtained from Steps 4 and 5. The lower value of $q_{all}$ is equal to $q_o$.

**7.** Calculate the value of $q_R$ for the foundation supported by reinforced earth as

$$q_R = \frac{\text{load on foundation per unit length}}{B} \tag{11.28}$$

**8.** Calculate the tie force, $T_{(N)}$, in each layer of reinforcement by using Eq. (11.20) (*note: unit of $T_{(N)}$ as kN/m of foundation*).

**9.** Calculate the frictional resistance of ties for each layer per unit length of foundation $(F_B)$ by using Eq. (11.24). For each layer, determine if $F_B/T_{(N)}$ is equal to or more than the required factor of safety against tie pullout $[FS_{(P)}]$. If $F_B/T_{(N)}$ is less than the required value of $FS_{(P)}$, the length of the reinforcing strips for any given layer may be increased. This will increase the value of $F_B$ and thus $FS_{(P)}$, and so Eq. (11.24) will need to be rewritten as

$$F_B = 2 \tan \phi_\mu (LDR) \left[ A_3 Bq_o \left( \frac{q_R}{q_o} \right) + \gamma(L - X_o)(z + D_f) \right] \tag{11.29}$$

where $L$ = the required length to obtain the desired value of $F_B$

**10.** Use Eq. (11.22) to obtain the tie thickness for each layer. Some allow-ance should be made to take into account the corrosion effect of the reinforce-ments during the life span of the structure.

**11.** If the design is unsatisfactory, the entire process may be repeated—that is, Steps 2 through 10.

The following example demonstrates the steps outlined above.

## Example 11.1

Design a continuous foundation that will carry a load of 1.8 MN/m. Given the following:

*Soil:* $\gamma = 17.3$ kN/m³; $\phi = 35°$; $E_s = 3 \times 10^4$ kN/m²; $\mu_s = 0.35$

*Reinforcement ties:* $f_y = 2.5 \times 10^5$ kN/m²; $\phi_\mu \doteqdot 28°$; $FS_{(B)} = 3$; $FS_{(P)} = 2.5$

*Foundation:* $D_f = 1$ m; factor of safety against bearing capacity failure $= 3$; tolerable settlement $= s = 25$ mm; desired life span $= 50$ years

**Solution**

Let

$$B = 1 \text{ m}$$

$d$ = depth from the bottom of the foundation to the first reinforcing layer = 0.5 m

$$\Delta H = 0.5 \text{ m}$$

$$N = 5$$

$$LDR = 65\%$$

If the reinforcing strips used are 75 mm wide, then

$$w \cdot n = LDR$$

or

$$n = \frac{LDR}{w} = \frac{0.65}{0.075 \text{ m}} = \underline{8.67/\text{m}}$$

Hence, there will be 8.66 strips in each layer per meter length of the foundation.

**Determination of $q_o$**

For an unreinforced foundation

$$q_u = \gamma D_f N_q + \tfrac{1}{2}\gamma B N_\gamma$$

From Table 3.2, for $\phi = 35°$, $N_q = 33.30$ and $N_\gamma = 48.03$. Thus

$$q_u = (17.3)(1)(33.3) + \tfrac{1}{2}(17.3)(1)(48.03)$$

$$= 576.09 + 415.46 = 991.55 \approx 992 \text{ kN/m}^2$$

$$q_{all(1)} = \frac{q_u}{FS} = \frac{992}{3} = 330.7 \text{ kN/m}^2$$

From Eq. (11.27)

$$q_{all(2)} = \frac{(E_s)(s)}{B(1 - \mu_s^2)\alpha_r} = \frac{(30{,}000 \text{ kN/m}^2)(0.025 \text{ m})}{(1 \text{ m})(1 - 0.35^2)(2)} = 427.35 \text{ kN/m}^2$$

Since $q_{all(1)} < q_{all(2)}$, $q_o = q_{all(1)} = \underline{330.7 \text{ kN/m}^2}$

Determination of $q_R$

From Eq. (11.28)

$$q_R = \frac{1.8 \text{ MN/m}}{B} = \frac{1.8 \times 10^3}{1} = \underline{1.8 \times 10^3 \text{ kN/m}^2}$$

Calculation of Tie Force

From Eq. (11.20)

$$T_{(N)} = \left(\frac{q_o}{N}\right)\left(\frac{q_R}{q_o} - 1\right)(A_1 \cdot B - A_2 \cdot \Delta H)$$

The tie forces for each layer are given in the following table.

| Layer No. | $\left(\frac{q_o}{N}\right)\left(\frac{q_R}{q_o} - 1\right)$ | $z$ (m) | $\frac{z}{B}$ | $A_1 B$ | $A_2\,\Delta H$ | $A_1 B - A_2\,\Delta H$ | $T_{(N)}$ (kN/m) |
|---|---|---|---|---|---|---|---|
| 1 | 293.7 | 0.5 | 0.5 | 0.35 | 0.125 | 0.225 | 66.08 |
| 2 | 293.7 | 1.0 | 1.0 | 0.34 | 0.09 | 0.25 | 73.43 |
| 3 | 293.7 | 1.5 | 1.5 | 0.34 | 0.065 | 0.275 | 80.77 |
| 4 | 293.7 | 2.0 | 2.0 | 0.33 | 0.05 | 0.28 | 82.24 |
| 5 | 293.7 | 2.5 | 2.5 | 0.32 | 0.04 | 0.28 | 82.24 |

$A_1$ from Figure 11.5; $B$ = 1 m; $\Delta H$ = 0.5 m; $A_2$ from Figure 11.5; $q_R/q_o$ = 1.8 × $10^3$/330.7 ≈ 5.45

Calculation of Tie Resistance Due to Friction, $F_B$

This can be done by using Eq. (11.24):

$$F_B = 2 \tan \phi_\mu(LDR)\left[A_3 B q_o\left(\frac{q_R}{q_o}\right) + \gamma(L_o - X_o)(z + D_f)\right]$$

The following table shows the value of $F_B$ for each layer (continued on next page).

| Quantity | Layer No. | | | | |
|---|---|---|---|---|---|
| | 1 | 2 | 3 | 4 | 5 |
| $2 \tan \phi_\mu(LDR)$ | 0.691 | 0.691 | 0.691 | 0.691 | 0.691 |
| $A_3$ | 0.125 | 0.14 | 0.15 | 0.15 | 0.15 |
| $A_3 B q_o\,(q_R/q_o)$ | 225.0 | 252.0 | 270.0 | 270.0 | 270.0 |
| $z$ (m) | 0.5 | 1.0 | 1.5 | 2.0 | 2.5 |

| Quantity | Layer No. | | | | |
|---|---|---|---|---|---|
| | 1 | 2 | 3 | 4 | 5 |
| $z/B$ | 0.5 | 1.0 | 1.5 | 2.0 | 2.5 |
| $L_o(m)$ | 1.55 | 2.6 | 3.4 | 3.85 | 4.2 |
| $X_o(m)$ | 0.55 | 0.8 | 1.1 | 1.4 | 1.65 |
| $L_o - X_o(m)$ | 1.0 | 1.8 | 2.3 | 2.45 | 2.55 |
| $z + D_f(m)$ | 1.5 | 2.0 | 2.5 | 3.0 | 3.5 |
| $\gamma(L_o - X_o)(z + D_f)$ | 25.95 | 62.28 | 99.48 | 127.16 | 154.4 |
| $F_B(kN/m)$ | 173.4 | 217.2 | 255.1 | 274.4 | 293.3 |
| $FS_{(P)} = F_B/T_{(N)}$ | 2.62 | 2.96 | 3.16 | 3.34 | 3.57 |

$A_3$ from Figure 11.5; $X_o$ from Figure 11.2; $L_o$ from Figure 11.7; $T_{(N)}$ from the preceding table.

The minimum factor of safety is greater than the required value of $FS_{(P)}$, which is 2.5.

## Calculation of Tie Thickness to Resist Tie Break

From Eq. (11.22)

$$FS_{(B)} = \frac{tf_y}{T_{(N)}}(LDR)$$

or

$$t = \frac{FS_{(B)}T_{(N)}}{(LDR)(f_y)}$$

Given: $f_y = 2.5 \times 10^5\ kN/m^2$; $LDR = 0.65$; $FS_{(B)} = 3$. So

$$t = \left[\frac{3}{(2.5 \times 10^5)(0.65)}\right]T_{(N)} = (1.846 \times 10^{-5})T_{(N)}$$

So, for Layer 1

$$t = (1.846 \times 10^{-5})(66.08) = 0.00122\ m = \underline{1.22\ mm}$$

For layer 2

$$t = (1.846 \times 10^{-5})(73.43) = 0.00136\ m = \underline{1.36\ mm}$$

Similarly, for layer 3

$$t = 0.00149 = \underline{1.49\ mm}$$

For layer 4

$$t = \underline{1.52\ mm}$$

For layer 5

$$t = \underline{1.52\ mm}$$

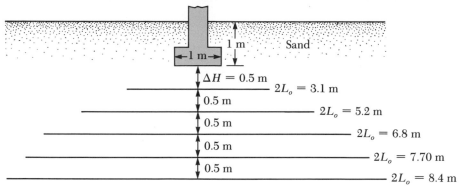

**Figure 11.8**

So, in each layer, ties with a thickness of 1.6 mm will be sufficient. However, if galvanized steel is used, the rate of corrosion is about 0.025 mm/year. So, $t$ should be equal to $1.6 + (0.025)(50) = 2.85$ mm.

## Calculation of Minimum Length of Ties

The minimum length of ties in each layer should be equal to $2L_o$. Following is the length of ties in each layer.

| Layer No. | Minimum length of tie, $2L_o$ (m) |
|-----------|-----------------------------------|
| 1 | 3.1 |
| 2 | 5.2 |
| 3 | 6.8 |
| 4 | 7.7 |
| 5 | 8.4 |

Figure 11.8 is a diagram of the foundation with the ties. The design could have been changed by varying $B$, $d$, $N$, and $\Delta H$ to determine the most economical combination.

## Example 11.2

Refer to Example Problem 11.1. For the loading given, determine the width of the foundation that is needed for an unreinforced earth condition. Note that the factor of safety against bearing capacity failure is 3, and the tolerable settlement is 25 mm.

### Solution

#### Bearing Capacity Consideration

For continuous foundation

$$q_u = \gamma D_f N_q + \tfrac{1}{2} \gamma B N_\gamma$$

For $\phi = 35°$, $N_q = 33.3$ and $N_\gamma = 48.03$. So

$$q_{all} = \frac{q_u}{FS} = \frac{1}{FS}\left[\gamma D_f N_q + \frac{1}{2} \gamma B N_\gamma\right]$$

or

$$q_{all} = \frac{1}{3}\left[(17.3)(1)(33.3) + \frac{1}{2}(17.3)(B)(48.03)\right]$$                     (a)

$$= 192.03 + 138.5B$$

However

$$q_{all} = \frac{1.8 \times 10^3 \text{ kN}}{(B)(1)}$$                     (b)

Equating the right sides of Eqs. (a) and (b)

$$\frac{1800}{(B)(1)} = 192.0\overline{3} + 138.5B$$

From the solution of the preceding equation, $B \approx 3$ m. So, with $B = 3$ m, $q_{all}$ is equal to 600 kN/m².

### Settlement Consideration

For a friction angle of $\phi = 35°$ the average standard penetration number is about 10–15 (Table 2.4). Let us assume the higher value, $N = 15$. From Eq. (3.99b)

$$q_{all} = 11.98N\left(\frac{3.28B + 1}{3.28}\right)^2\left(1 + \frac{0.33D_f}{B}\right) \text{ for a settlement of about 25 mm}$$

Now, we can make a few trials as shown below:

| Assumed B | $q_{all} = 11.98N\left(\dfrac{3.28B + 1}{3.28B}\right)^2\left(1 + \dfrac{0.33D_f}{B}\right)$ | $Q = (B)(q_{all})$ |
|---|---|---|
| (m) (1) | (kN/m²) (2) | = Col. 1 × Col. 2 (kN/m) |
| 6 | 209 | 1254 |
| 9 | 199 | 1791* |

$D_f = 1$ m; *required 1800 kN/m

If $N = 15$, the width of the foundation should be 9 m or more. Based on the consideration of bearing capacity failure and tolerable settlement, the latter settlement criteria will control. So, $B$ is about 9 m.

---

*Note:* At first, the results of this calculation may show the use of reinforced earth for foundation construction to be desirable. However, several factors must be considered before a final decision is made. For example, reinforced earth needs overexcavation and backfilling. It is possible, under many circumstances, that proper material selection and compaction may make the construction of foundations on *unreinforced* soils more economical.

Binquet and Lee (1975b) analyzed the costs of constructing a foundation to carry a load of 2408 kN/m for a maximum allowable settlement of 25.4 mm. The analysis was done using the 1974 unit prices of Southern California. It showed that, with no corrosion allowance, a reinforced earth foundation costs about half as much as a conventional foundation. However, with a corrosion allowance for a 50-year life, a reinforced earth foundation costs about 5% more than a conventional foundation.

## Retaining Walls

## 11.6

### Reinforced Earth Retaining Walls—General

Reinforced earth walls are flexible walls. Their main components are

1. *backfill*, which is granular soil;
2. *reinforcing strips*, which are thin, wide strips placed at regular intervals; and
3. a *cover* on the front face, which is referred to as the *skin*.

Figure 11.9 is a diagram of a reinforced earth wall. Note that, at any given depth, the reinforcing strips are placed with a horizontal spacing of $S_H$ center-to-center; the vertical spacing of the strips (center-to-center) is $S_V$. The skin can be constructed with sections of relatively flexible thin material. Lee, Adams, and Vagneron (1973) have shown that, with a conservative design, a 5-mm thick galvanized steel skin would be enough to hold a wall about 14–15 m high. In some cases, precast concrete slabs can also be used as skin. The slabs are grooved to fit into each other so that soil does not flow out between the joints. When metal skins are used, they are bolted together, and reinforcing strips are placed between the skins.

Reinforced earth walls are now used for construction of retaining walls,

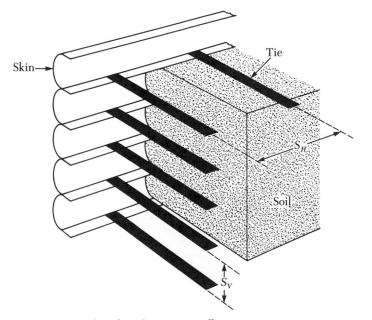

**Figure 11.9** Reinforced earth retaining wall

bridge abutments, waterfront walls, and so forth. There are three basic ways to design ties that will resist the lateral earth pressure:

1. The Rankine method
2. The Coulomb force method
3. The Coulomb moment method

The Rankine method is the simplest and most widely used of the three design methods. The following section will discuss this procedure in detail.

## 11.7

### Design of Reinforcement Ties

#### Maximum Tie Force

Figure 11.10a shows a retaining wall section with reinforcement ties located at depths $z = 0$, $S_V$, $2S_V$, ... , $NS_V$. The height of the wall is equal to $NS_V = H$.

According to the Rankine active pressure theory (Section 5.3)

$$\sigma_a = \sigma_v K_a - 2_c \sqrt{K_a} \tag{5.11}$$

where $\sigma_a$ = Rankine active pressure at any depth $z$

For dry granular soils with no surcharge at the top, $c = 0$, $\sigma_v = \gamma z$, and $K_a = \tan^2(45 - \phi/2)$. Thus

$$\sigma_a = \gamma z K_a \tag{11.30}$$

Figure 11.10b shows the active pressure distribution against the wall. According to this design procedure, the tie force developed at any depth $z$ can be given as

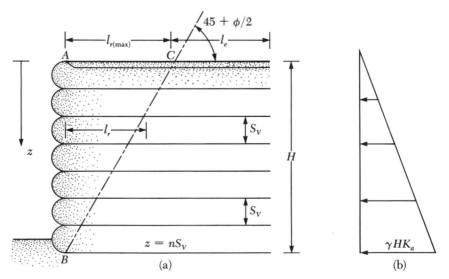

(a)

(b)

**Figure 11.10** Analysis of a reinforced earth retaining wall

$T$ = (active earth pressure at depth $z$) (area of the wall to be supported by the tie)

So

$$T = (\gamma z K_a)(S_V S_H) \tag{11.31}$$

The maximum tie force will be developed at the bottommost ties—that is, at $z = H$. So

$$T_{max} = (\gamma H K_a)(S_V S_H) \tag{11.32}$$

where $T_{max}$ = maximum force to which the tie will be subjected

## Factor of Safety Against Tie Failure

The reinforcement ties and, thus, the walls could fail by either (a) *tie breaking* or (b) *tie pullout*.

The factor of safety against tie break can be determined as

$$FS_{(B)} = \frac{\text{yield or breaking strength of each tie}}{\text{maximum tie force in any tie}}$$

$$= \frac{wtf_y}{\gamma H K_a S_V S_H} \tag{11.33}$$

A factor of safety of about 2.5–3 is generally recommended.

The reinforcing ties at any given depth, $z$, will fail by pullout if the frictional resistance developed along their surfaces is less than the force to which the ties are being subjected. The *effective length* of the ties along which the frictional resistance is developed may be conservatively taken as the length that extends *beyond the limits of the Rankine active failure zone*. According to Figure 5.5, this is zone $ABC$ in Figure 11.10a. The line $BC$ in Figure 11.10a makes an angle of $45 + \phi/2$ with the horizontal. Now, the maximum friction force $F_R$ that can be realized for a given tie at a depth $z$ is equal to

$$F_R = 2l_e w \sigma_v \tan \phi_\mu \tag{11.34}$$

where $l_e$ = effective length
$\sigma_v$ = effective vertical pressure at a depth $z$

So

$$F_R = 2l_e w \gamma z \tan \phi_\mu \tag{11.35}$$

So, the factor of safety against tie pullout at a given depth $z$ can be given as

$$FS_{(P)} = \frac{F_R}{T} \tag{11.36}$$

where $FS_{(P)}$ = factor of safety against tie pullout

Substituting Eqs. (11.31) and (11.35) into Eq. (11.36)

$$FS_{(P)} = \frac{2l_e w \gamma z \tan \phi_\mu}{\gamma z K_a S_V S_H} = \frac{2l_e w \tan \phi_\mu}{K_a S_V S_H} \tag{11.37}$$

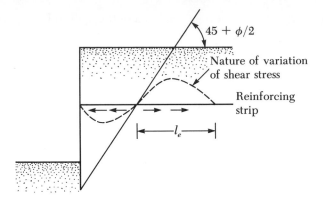

**Figure 11.11** Nature of shear stress variation assumed in Eq. (11.38)

Al-Hussaini and Perry (1978) have shown that the factor of safety against tie pullout can be given as

$$FS_{(P)} = \frac{4l_e w \tan \phi_\sigma}{3K_a S_V S_H} \tag{11.38}$$

Note that Eq. (11.38) is a modification of Eq. (11.37). This is due to the nature of shear stress distribution along the reinforcing ties (see Figure 11.11).

In construction, the total length of ties at all depths is usually kept constant. The total length of ties, $L$, is the sum of

1. the length within the Rankine failure zone behind the wall, $l_r$, and
2. the effective length, $l_e$, or

$$L = l_r + l_e \tag{11.39}$$

For a given $FS_{(P)}$, from Eq. (11.37)

$$l_e = \frac{FS_{(P)} K_a S_V S_H}{2w \tan \phi_\mu} \tag{11.40}$$

The maximum value of $l_r$ will occur at the top of the wall—that is, $AC$ in Figure 11.10a. So

$$l_{r(max)} = \overline{AC} = \frac{H}{\tan\left(45 + \dfrac{\phi}{2}\right)} \tag{11.41}$$

Combining Eqs. (11.30), (11.40), and (11.41)

$$L = H \tan\left(45 - \frac{\phi}{2}\right) + \frac{FS_{(P)} K_a S_V S_H}{2w \tan \phi_\mu} \tag{11.42}$$

## 11.8

### Design Procedure

Following is a step-by-step procedure for the design of reinforced earth retaining walls:

1. Determine the height of the wall, $H$, and also the properties of the granular backfill material, such as unit weight ($\gamma$) and angle of friction ($\phi$).
2. Obtain the soil-tie friction angle, $\phi_\mu$, and also the required values of $FS_{(B)}$ and $FS_{(P)}$.
3. Assume values for the horizontal and vertical spacings of the ties. Also assume the width of reinforcing strip ($w$) to be used.
4. Calculate $K_a = \tan^2(45 - \phi/2)$.
5. Calculate the maximum tie force, $T_{(max)}$ [Eq. (11.32)].
6. With the known value of $FS_{(B)}$, calculate the thickness of ties ($t$) required to resist the tie breakout, or

$$T_{(max)} = (\gamma H K_a)(S_V S_H) = \frac{wtf_y}{FS_{(B)}}$$

or

$$t = \frac{(\gamma H K_a S_V S_H)[FS_{(B)}]}{wf_y} \tag{11.43}$$

7. With the known values of $\phi_\mu$ and $FS_{(P)}$, determine the length, $L$, of the ties from Eq. (11.42).
8. The values of $S_V$, $S_H$, $t$, and $w$ can be changed to obtain the most economical design.

## Example
## 11.3

A reinforced earth retaining wall is proposed to be 8 m high. The properties of the backfill material are as follows: $\gamma = 16.6 \text{ kN/m}^3$ and $\phi = 30°$. Galvanized steel ties are to be used for the construction of the wall. Design the reinforcements with $FS_{(B)} = 3$, $FS_{(P)} = 3$, $f_y = 2.4 \times 10^5 \text{ kN/m}^2$, and $\phi_\mu = 20°$.

### Solution

Let $S_V = 0.5 \text{ m}$, $S_H = 1 \text{ m}$, and $w = 75 \text{ mm}$. Given: soil friction angle, $\phi$, is 30°. So, $K_a = \tan^2(45 - \phi/2) = \tan^2(45 - 30/2) = 1/3$. So, from Eq. (11.32)

$$T_{max} = \gamma H K_a S_V S_H = (16.6)(8)\left(\frac{1}{3}\right)(0.5)(1) = 22.14 \text{ kN}$$

From Eq. (11.43)

$$t = \frac{(\gamma H K_a S_V S_H)[FS_{(B)}]}{wf_y} = \frac{(22.14)(3)}{\left(\dfrac{75}{1000}\right)2.4 \times 10^5} = 0.00369 \text{ m} = 3.69 \text{ mm}$$

If the rate of corrosion is 0.025 mm/year and the life span of the structure is 50 years, then the actual thickness, $t$, of the ties will be

$$t = 3.69 + (0.025)50 = 4.94 \text{ mm}$$

So, a tie thickness of 5 mm would be enough.

## Determination of the Length of the Ties

From Eq. (11.42)

$$L = H \tan \left( 45 - \frac{\phi}{2} \right) + \frac{FS_{(P)}K_aS_VS_H}{2w \tan \phi_\mu}$$

$$= (8)(0.577) + \frac{(3)\left(\dfrac{1}{3}\right)(0.5)(1)}{(2)(0.075) \tan 20°} = 4.62 + 9.16 = 13.78 \text{ m}$$

So, there will be 16 layers of strips, each having a length of 13.78 m, at vertical spacings of 0.5 m.

## 11.9
## Comments on Reinforced Earth Walls

Reinforced earth retaining walls are economical when they are higher than 10–12 m. Lee, Adams, and Vagneron (1973) have shown this by preparing cost analyses of reinforced and unreinforced walls of similar height. Table 11.1 shows a breakdown of cost on a percentage basis for various items of reinforced earth wall design.

Reinforced earth walls are flexible walls, and they must be simultaneously built along with the backfill. Because they are flexible, these walls can take a large amount of differential settlement—for example, when the underlying soil is highly compressible.

The possibility of wall failure as the result of loss of bearing capacity of the foundation soil should be checked. This is especially important when walls are built over soft to medium clays.

**Table 11.1**  Distribution of Estimated Costs of Reinforced Earth Walls[a]

| Item (1) | No corrosion (% of total cost) (2) | Corrosion at the rate of 0.050 mm/year (% of total cost) (3) |
|---|---|---|
| Material and special fabrication | 20–30 | 40–45 |
| Erection and labor | 10–20 | 5–15 |
| Backfill | 65 | 45–50 |

[a] After Lee, Adams, and Vagneron (1973)

## 11.10

Other Studies on Reinforced Earth

  Construction of foundations, retaining walls, and other structures on or with reinforced earth is a relatively new concept. The design procedures will definitely be improved and/or modified with further research in the future.

  In addition to the studies mentioned in this chapter, important research on retaining wall design with reinforced earth has been done by Schlosser and Long (1974), Holtz (1975), Al-Hussaini (1977), Bell and Steward (1977), Broms (1977, 1978), Juran and Schlosser (1978), and Nicholls (1981). Schlosser and Long (1974) have also presented, based on the results of reduced-scale model tests, an empirical equation for distribution of stress under foundation rafts supported by reinforced earth. Akinmusuru and Akinbolande (1981) have presented some laboratory model test results for the bearing capacity of square foundations supported by reinforced earth. This study generally complements the works of Binquet and Lee (1975a, b).

## Problems

**11.1** Figure 11.4c shows a continuous foundation on reinforced soil. Given: $B = 0.9$ m; $D_f = 1$ m; number of layers of reinforcement, $N = 5$; $\Delta H = 0.4$ m. Make the necessary calculations and plot the lines on both sides of the foundation that define the point of maximum shear stress $[\tau_{xz(max)}]$ on the reinforcements.

**11.2** The tie forces under a continuous foundation are given by Eq. (11.20). For the foundation described in Problem 11.1, given: $q_o = 200$ kN/m$^2$ and $q_R/q_o = 4.5$. Determine the tie forces, $T_{(N)}$, in kN/m for each layer of reinforcement.

**11.3** Repeat Problem 11.2 with $q_o = 300$ kN/m$^2$ and $q_R/q_o = 6$.

**11.4** A continuous foundation (see Figure 11.4c) is to be built on reinforced earth to carry a load of 82.3 kip/ft. Given:

  *Foundation:* $B = 4$ ft, $D_f = 2.6$ ft, factor of safety against bearing capacity failure = 3, tolerable settlement = 0.8 in.
    *Soil:* $\gamma = 116$ lb/ft$^3$, $\phi = 37°$, $E_s = 5200$ lb/in.$^2$, $\mu_s = 0.30$
  *Reinforcement:* $\Delta H = 1.3$ ft, $N = 5$, $LDR = 70\%$, width of reinforcement strips = 0.23 ft

Calculate:
a. The number of reinforcement strips per foot length of the foundation.
b. The allowable load per unit area of the foundation ($q_o$) without reinforcement.
c. The ratio of $q_R/q_o$.
d. The tie forces for each layer of reinforcement under the foundation (kip/ft).

**11.5** Refer to Problem 11.4. For the reinforcements, given: $f_y = 38,000$ lb/in.$^2$, $\phi_\mu = 25°$, factor of safety against tie break = 2.5, factor of safety against tie pullout = 2.5. Calculate:
a. The minimum thickness of ties needed to resist tie break.
b. The minimum length of ties necessary for each layer of reinforcement.

**11.6** A reinforced earth retaining wall (Figure 11.10) is to be 30 ft high. Given the following:

*Backfill:* unit weight = 119 lb/ft³; soil friction angle = 34°
*Reinforcement:* vertical spacing, $S_V$ = 3 ft; horizontal spacing, $S_H$ = 4 ft; width of rein-forcement = 4.75 in.; $f_y$ = 38,000 lb/in.²; $\phi_\mu$ = 25°; factor of safety against tie pull = 3; factor of safety against tie break = 3

Determine:
a. The required thickness of ties.
b. The required maximum length of ties.

**11.7** Redo Problem 11.6 for a retaining wall with a height of 21 ft.

**11.8** Redo Problem 11.6 and change $S_V$ to 1.5 ft.

# References

Akinmusuru, J. O., and Akinbolande, J. A., (1981). "Stability of Loaded Footings on Reinforced Soil," *Journal of the Geotechnical Engineering Division*, American Society of Civil Engineers, Vol. 107, No. GT6, pp. 819–827.

Al-Hussaini, M. (1977). "Field Experiment of Fabric Reinforced Earth Wall," *Proceedings*, International Conference on the Use of Fabric in Geotechnics, Paris, Vol. 1, pp. 113–118.

Al-Hussaini, M., and Perry, E. (1978). "Field Experiment of Reinforced Earth Walls," *Symposium on Earth Reinforcement*, American Society of Civil Engineers, pp. 127–156.

Bell, J. R., and Steward, J. E. (1977). "Construction and Observation of Fabric Retaining Soil Walls," *Proceedings*, International Conference on the Use of Fabric in Geotechnics, Paris, Vol. 1, pp. 123–128.

Binquet, J., and Lee, K. L. (1975a). "Bearing Capacity Tests on Reinforced Earth Mass," *Journal of the Geotechnical Engineering Division*, American Society of Civil Engineers, Vol. 101, No. GT12, pp. 1241–1255.

Binquet, J., and Lee, K. L. (1975b). "Bearing Capacity Analysis of Reinforced Earth Slabs," *Journal of the Geotechnical Engineering Division*, American Society of Civil Engineers, Vol. 101, No. GT12, pp. 1257–1276.

Broms, B. B. (1977). "Polyester Fabric as Reinforcement in Soil," *Proceedings*, International Conference on the Use of Fabric in Geotechnics, Paris, Vol. 1, pp. 129–135.

Broms, B. B. (1978). "Design of Fabric Reinforced Retaining Structures," *Symposium on Earth Reinforcement*, American Society of Civil Engineers, pp. 282–304.

Darbin, M. (1970). "Reinforced Earth for Construction of Freeways" (in French), *Revue Générale des Routes et Aerodromes*, No. 457, Sept.

Das, B. M. (1981). "Bearing Capacity of Eccentrically Loaded Surface Footings on Sand," *Soils and Foundations*, Vol. 21, No. 1, pp. 115–119.

Das, B. M. (1983). *Advanced Soil Mechanics*, McGraw-Hill, New York.

Holtz, R. D. (1975). "Recent Developments in Reinforced Earth," *Proceedings*, 7th Scandinavian Geotechnical Meeting, Copenhagen, pp. 281–291.

Juran, I., and Schlosser, F. (1978). "Theoretical Analysis of Failure in Reinforced Earth Structures," *Symposium on Earth Reinforcement*, American Society of Civil Engineers, pp. 528–555.

Lee, K. L., Adams, B. D., and Vagneron, J. J. (1973). "Reinforced Earth Retaining Walls," *Journal of the Soil Mechanics and Foundations Division*, American Society of Civil Engineers, Vol. 99, No. SM10, pp. 745–763.

Mandel, J., and Salencon, J. (1972). "Force portante d'un sol sur une assise rigide (étude theorizué)," *Geotechnique*, Vol. 22, No. 1, pp. 79–93.

Meyerhof, G. G. (1974). "Ultimate Bearing Capacity of Footings on Sand Layer Overlying Clay," *Canadian Geotechnical Journal*, Vol. 11, No. 2, pp. 223–229.

Nicholls, R. L. (1981). "Comparison of Fabric Sheet-Reinforced Earth Design Method," *Canadian Geotechnical Journal*, Vol. 18, No. 4, pp. 585–591.

Pfeifle, T. W., and Das, B. M. (1979). "Bearing Capacity of Surface Footings on Sand Layer Resting on a Rigid Rough Base," *Soils and Foundations*, Vol. 19, No. 1, pp. 1–11.

Schlosser, F., and Long, N. (1974). "Recent Results in French Research on Reinforced Earth," *Journal of the Construction Division*, American Society of Civil Engineers, Vol. 100, No. CO3, pp. 113–237.

Schlosser, F., and Vidal, H. (1969). "Reinforced Earth" (in French), *Bulletin de Liaison des Laboratoires Routier*, Ponts et Chaussées, Paris, France, Nov., pp. 101–144.

Vidal, H. (1966). "La terre Armee," *Anales de l'Institut Technique du Bâtiment et des Travaux Publiques*, France, July-August, pp. 888–938.

CHAPTER 12

# Soil Improvement

## 12.1

### Introduction

The existing soil at a construction site may not always be totally suitable for supporting structures such as buildings, bridges, highways, and dams. For example, in granular soil deposits the *in-situ* soil may be very loose and could present a large elastic settlement. In such a case, the soil needs to be densified to increase its unit weight and thus the shear strength.

Sometimes the top soil is undesirable and must be removed and replaced with a better quality soil, on which the structural foundation can be built. The soil that is used as a fill needs to be well compacted to sustain the desired structural load. Compacted fills may also be required in low-lying areas to raise the ground elevation for foundation construction.

Soft saturated clay layers are often encountered at shallow depths below the foundation(s). Depending on the structural load and the depth of the clay layer(s), unusually large consolidation settlement may occur. Special soil-improvement techniques are required to overcome such settlement problems.

Chapter 10 mentioned that the properties of expansive soils could be substantially altered by the addition of stabilizing agents such as lime. Improving *in-situ* soils by using additives is usually referred to as *stabilization*.

This chapter briefly discusses various techniques for improving the soil to:

1. reduce the settlement of structures;
2. improve the shear strength of soil and thus increase the bearing capacity of shallow foundations;

3. increase the factor of safety for possible slope failure of embankments and earth dams;
4. reduce the shrinkage and swelling characteristics of soils.

The chapter is divided into six major parts: (1) compaction, (2) vibroflotation, (3) precompression, (4) sand drains, (5) soil stabilization by admixtures, and (6) use of geotextiles.

## 12.2

### Compaction—General Principles

If a small amount of water is added to a soil that is then subjected to compaction by a given amount of energy, the soil will be compacted to a certain unit weight. If the moisture content of the same soil is gradually increased and the compaction is done in the same way, the dry unit weight of the compaction will gradually increase. This is because water behaves as a lubricant between the soil particles; under compaction it helps rearrange the solid particles into a denser state. The increase in dry unit weight with increase of moisture content for a given soil will reach a limiting value beyond which further addition of water to the soil will result in a *reduction* of dry unit weight. The moisture content at which the *maximum dry unit weight* is obtained is referred to as the *optimum moisture content*.

The standard laboratory tests that are used for evaluation of maximum dry unit weights and optimum moisture contents for various soils are (a) the *standard Proctor test* [ASTM designation D-698; Title: Moisture Density Relations of Soils and Soil-Aggregate Mixtures Using 5.5 lb (2.49 kg) Rammer and 12 in. (305 mm) Drop] and (b) the *modified Proctor test* [ASTM designation D-1557; Title: Moisture Density Relations of Soils and Soil-Aggregate Mixtures Using 10 lb (4.54 kg) Rammer and 18 in. (457 mm) Drop]. Both of these tests are conducted in a mold having a volume of $0.944 \times 10^{-3} \text{ m}^3$ ($1/30 \text{ ft}^3$). The soil is compacted in several layers by a hammer. The moisture content of the soil ($w$) is changed, and the dry unit weight ($\gamma_d$) of compaction for each test is determined. The maximum dry unit weight of compaction and the corresponding optimum moisture content is determined by plotting a graph of $\gamma_d$ vs. $w$ (%). The standard specifications for the two types of Proctor test are given in Table 12.1.

Figure 12.1 shows the plot of $\gamma_d$ vs. $w$ (%) for a clayey silt as obtained for standard and modified Proctor tests. From this figure, the following observations can be made:

1. The maximum dry unit weight and the optimum moisture content are dependent on the level of energy of compaction.
2. The higher the energy of compaction, the higher is the maximum dry unit weight.
3. The higher the energy of compaction, the lower is the optimum moisture content.
4. No portion of the compaction curve can lie to the right side of the zero-air-void line. The zero-air-void dry unit weight ($\gamma_{zav}$) at a given

**Table 12.1**   Specifications for Standard and Modified Proctor Tests

| No. | Item | Specifications | |
| --- | --- | --- | --- |
| | | Standard Proctor | Modified Proctor |
| 1 | Volume of mold | $0.944 \times 10^{-3} \, m^3$ (1/30 ft³) | $0.944 \times 10^{-3} \, m^3$ (1/30 ft³) |
| 2 | Mass of hammer | 2.495 kg (5.5 lb) | 4.536 kg (10 lb) |
| 3 | Height of drop of the hammer | 304.8 mm (12 in.) | 457 mm (18 in.) |
| 4 | Number of hammer blows per layer of soil | 25 | 25 |
| 5 | Number of layers of compaction | 3 | 5 |
| 6 | Energy of compaction | 593 kiloJoules/m³ (12,375 ft-lb/ft³) | 2695 kiloJoules/m³ (56,250 ft-lb/ft³) |

From "Factors That Influence Field Compaction of Soils," by A. W. Johnson and J. R. Sallberg, *Bulletin No. 262*, Highway Research Board, 1960. Reprinted by permission.

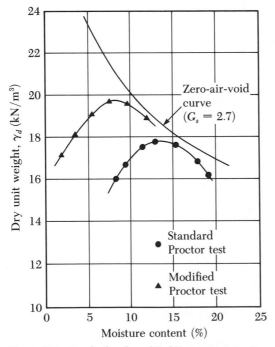

**Figure 12.1**   Standard and modified Proctor compaction curves for a clayey silt

moisture content is the theoretical maximum value of $\gamma_d$, which means that all of the void spaces of the compacted soil are filled with water. This can be given by the equation

$$\gamma_{zav} = \frac{\gamma_w}{\dfrac{1}{G_s} + w}$$                                  (12.1)

where $\gamma_w$ = unit weight of water
             $G_s$ = specific gravity of the soil solids
             $w$ = moisture content

**5.** The values of the maximum dry unit weight of compaction and the corresponding optimum moisture content will vary from soil to soil.

Using the results of the laboratory compaction ($\gamma_d$ vs. $w$), specifications can be written for the compaction of a given soil in the field. In most cases, the contractor is required to achieve a relative compaction of 90% or more on the basis of a specific laboratory test (that is, standard or modified Proctor compaction test). Relative compaction ($RC$) is defined as

$$RC = \frac{\gamma_{d(\text{field})}}{\gamma_{d(\text{max})}}$$                                  (12.2)

Chapter 1 introduced the concept of relative density (for compaction of granular soils). Relative density was defined as

$$D_r = \left[ \frac{\gamma_d - \gamma_{d(\text{min})}}{\gamma_{d(\text{max})} - \gamma_{d(\text{min})}} \right] \frac{\gamma_{d(\text{max})}}{\gamma_d}$$

where $D_r$ = relative density
             $\gamma_d$ = dry unit weight of compaction in the field
        $\gamma_{d(\text{max})}$ = maximum dry unit weight of compaction as determined in the laboratory
        $\gamma_{d(\text{min})}$ = minimum dry unit weight of compaction as determined in the laboratory

For granular soils in the field, the degree of compaction obtained is often measured in terms of relative density. Comparing the expressions for relative density and relative compaction, one can see that

$$RC = \frac{A}{1 - D_r(1 - A)}$$                                  (12.3)

where $A = \dfrac{\gamma_{d(\text{min})}}{\gamma_{d(\text{max})}}$

Lee and Singh (1971) have reviewed 47 different soils, and, based on that, they have given the following correlation:

$$D_r(\%) = \frac{(RC - 80)}{0.2}$$                                  (12.4)

## One-Point Method of Obtaining $\gamma_{d(max)}$

The state highway department of Ohio has developed a family of standard curves of their various soil types; these are shown in Figure 12.2. Note that they are plots of *wet unit weight* ($\gamma$) vs. moisture content ($w\%$). These curves can be used to obtain $\gamma_{d(max)}$ in the field. This technique is referred to as the *one-point method*; it serves as a rapid means of field compaction control. In this method one first conducts a standard Proctor test in the field with the soil in use and determines the moist unit weight of compaction and the corresponding moisture content. Then one plots the values of $\gamma$ and $w$ (%) according to Figure 12.2 and identifies the compaction curve number that corresponds to the test results. Using this curve number, one can refer to Table 12.2 to determine the maximum dry unit weight and the corresponding optimum moisture content.

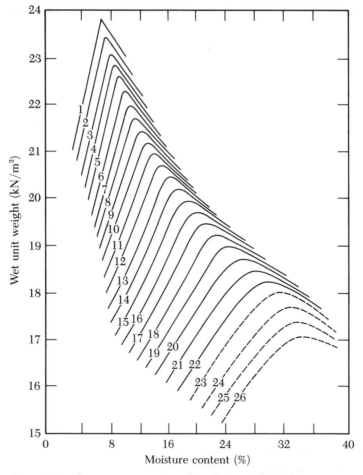

**Figure 12.2**   Ohio compaction curves (from "Factors That Influence Field Compaction of Soils," by A. W. Johnson and J. R. Sallberg, *Bulletin No. 262*, Highway Research Board, 1960. Reprinted by permission.)

**Table 12.2**  Maximum Dry Unit Weight and Optimum Moisture Content for the Compaction Curves in Figure 12.2

| Curve No. | Maximum dry unit weight (kN/m³) | Optimum moisture content (%) | Curve No. | Maximum dry unit weight (kN/m³) | Optimum moisture content (%) |
|---|---|---|---|---|---|
| 1 | 22.29 | 6.6 | 14 | 17.23 | 16.9 |
| 2 | 21.87 | 7.2 | 15 | 16.84 | 18.1 |
| 3 | 21.43 | 7.9 | 16 | 16.46 | 19.2 |
| 4 | 21.08 | 8.5 | 17 | 16.10 | 20.3 |
| 5 | 20.75 | 9.0 | 18 | 15.71 | 21.5 |
| 6 | 20.33 | 9.7 | 19 | 15.31 | 22.7 |
| 7 | 19.90 | 10.5 | 20 | 14.87 | 24.4 |
| 8 | 19.53 | 11.2 | 21 | 14.48 | 25.8 |
| 9 | 19.13 | 11.9 | 22 | 14.13 | 27.4 |
| 10 | 18.76 | 12.7 | 23 | 13.76 | 29.5 |
| 11 | 18.39 | 13.5 | 24 | 13.36 | 30.5 |
| 12 | 18.02 | 14.6 | 25 | 13.05 | 31.5 |
| 13 | 17.61 | 15.8 | 26 | 12.75 | 32.5 |

From "Factors That Influence Field Compaction of Soils," by A. W. Johnson and J. R. Sallberg, *Bulletin No. 262*, Highway Research Board, 1960. Reprinted by permission.

The one-point method just described appears to be simple and easy to use. However, this may not always be the case. During the past several years researchers have determined that not all soils yield the bell-shaped compaction curves shown in Figure 12.2. Lee and Suedkamp (1972) performed 700 compaction tests on 35 soil samples. Their results show that, depending on the property of the soil, the plot of $\gamma_d$ vs. $w$ (%) may exhibit one of four different shapes. These shapes are shown in Figure 12.3 and are marked Types I, II, III, and IV. Type I is a standard bell-shaped curve. Type II is a curve showing one and one-half peaks. Type III is a double-peak curve. Type IV is an oddly shaped curve that shows no distinct optimum moisture content. Lee and Suedkamp (1972) developed the following guidelines to help predict the nature of compaction curves that may be obtained from various soils.

| Liquid limit of soil | Nature of compaction curve to be expected |
|---|---|
| 30 to 70 | Type I |
| Less than 30 | Types II and III |
| Greater than 70 | Types III and IV |

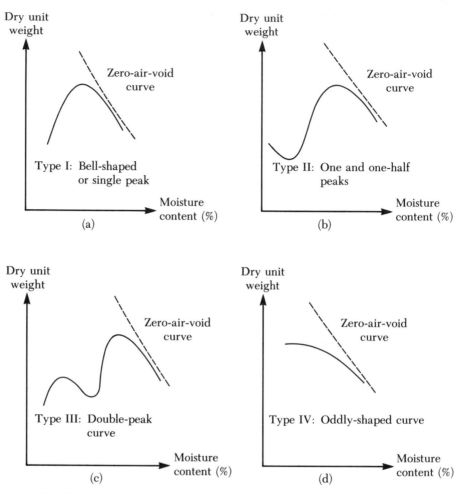

**Figure 12.3**   Various types of compaction curve

## Example
## 12.1

A soil was compacted in the field by the standard Proctor test procedure. Following are the results:

Weight of compacted wet soil in the mold  = 17.2 N
Moisture content = 14%

Using Figure 12.2, determine $\gamma_{d(\max)}$ and the optimum moisture content.

### Solution

Moist unit weight of compaction =

$$\gamma = \frac{17.2 \text{ N}}{\text{volume of Proctor mold}} = \frac{17.2}{0.944 \times 10^{-3}} = 18.22 \text{ kN/m}^3$$

According to Figure 12.2 (with $\gamma = 18.22$ kN/m$^3$ and $w = 14\%$), it appears that the soil falls between the curve numbers 15 and 16. Now, according to Table 12.2, $\gamma_{d(max)}$ is between 16.84 kN/m$^3$ and 16.48 kN/m$^3$, and the optimum moisture content is between 18.1–19.2%. So, for this soil

$$\gamma_{d(max)} \approx \underline{16.6 \text{ kN/m}^3}$$

and

$$\text{optimum moisture content} \approx \underline{18.5\%}$$

## Field Compaction

Ordinary compaction in the field is done by means of rollers. Several types of roller are used, and the most common types include:

1. Smooth wheel rollers (or smooth drum rollers)
2. Pneumatic rubber-tired rollers
3. Sheepsfoot rollers
4. Vibratory rollers

Figure 12.4 shows a *smooth wheel roller* that can also create vertical vibration during compaction. Smooth wheel rollers are suitable for proofrolling

**Figure 12.4**  Vibratory smooth wheel rollers (courtesy of Tampo Manufacturing Co., Inc., San Antonio, Texas)

subgrades and for finishing the construction of fills with sandy or clayey soils. They provide 100% coverage under the wheels, and the contact pressure can be as high as 300–400 kN/m². However they do not produce uniform unit weight of compaction when used on deep layers.

*Pneumatic rubber-tired rollers* (Figure 12.5) are better in many respects than smooth wheel rollers. These rollers, which may weigh up to 2000 kN, consist of a heavily loaded wagon with several rows of tires. These tires are closely spaced—about four to six in a given row. The contact pressure under the tires may range up to 600–700 kN/m², and they produce about 70–80% coverage. Pneumatic rollers can be used for sandy and clayey soil compaction. They produce a combination of pressure and kneading action.

*Sheepsfoot rollers* (Figure 12.6) consist basically of drums with large numbers of projections. The area of each of the projections may be 25–85 cm². These rollers are most effective in compacting cohesive soils. The contact pressure under the projections may range from 1500 to 7500 kN/m². During compaction in the field, the initial passes compact the lower portion of a lift. Later, compaction is produced in the middle and top of the lift.

*Vibratory rollers* are efficient in compacting granular soils. Vibrators can be attached to smooth wheel, pneumatic rubber-tired, or sheepsfoot rollers to

**Figure 12.5**  Pneumatic rubber-tired roller (courtesy of Tampo Manufacturing Co., Inc., San Antonio, Texas)

**Figure 12.6**  Vibratory sheepsfoot roller (courtesy of Tampo Manufacturing Co., Inc., San Antonio, Texas)

send vibrations into the soil under compaction. Figures 12.4 and 12.6 show vibratory smooth wheel rollers and a vibratory sheepsfoot roller.

In general, compaction in the field depends on several factors, such as the type of compactor, soil type, moisture content, lift thickness, towing speed of the compacter, and the number of roller passes. Figure 12.7 shows the variation of the dry unit weight of a heavy clay with the number of passes of pneumatic-tired rollers. Table 12.3 gives the details of the variables for the three curves shown in Figure 12.7.

Figure 12.8 shows the variation of the unit weight of compaction with depth for a poorly graded dune sand that was compacted by a vibratory drum roller. Vibration was produced by mounting an eccentric weight on a single rotating shaft within the drum cylinder. The weight of the roller used for this compaction was 55.7 kN and the drum diameter was 1.19 m. The lifts were kept at 2.44 m. Note that, at any given depth, the dry unit weight of compaction increases with the number of roller passes. However, the rate of increase of unit weight gradually decreases after about 15 passes. Note also (Figure 12.8) the nature of variation of dry unit weight with depth for any given number of roller passes. The dry unit weight and, hence, the relative density $(D_r)$ reaches a maximum value at a depth of about 0.5 m and then gradually decreases as the depth decreases. This is due to the lack of confining pressure toward the surface. Once the depth vs. relative density (or dry unit weight) relation for a given soil with a given number of roller passes is determined, it is easy to estimate the approximate thickness of each lift. This procedure is shown in Figure 12.9 on page 518 (D'Appolonia, Whitman, and D'Appolonia, 1969).

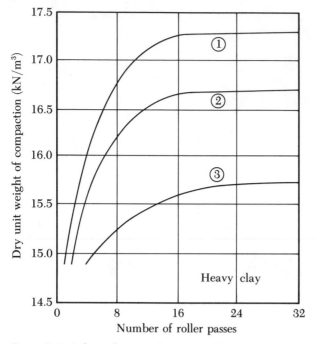

**Figure 12.7**   Relation between dry unit weight of compaction for the upper 150 mm of soil and the number of passes of pneumatic-tired roller (from "Factors That Influence Field Compaction of Soils," by A. W. Johnson and J. R. Sallberg, *Bulletin No. 262,* Highway Research Board, 1960. Reprinted by permission.)

**Table 12.3**   Details of the Variables for the Three Curves Shown in Figure 12.7[a]

| Curve No. | 1 | 2 | 3 |
|---|---|---|---|
| Moisture content as rolled (%) | 19.0 | 20.0 | 24.0 |
| Optimum moisture content--standard Proctor test (%) | 22.8 | 22.8 | 22.8 |
| Roller rating (kN) | 416.0 | 416.0 | 120.0 |
| Wheel load (kN) | 99.6 | 49.8 | 13.3 |
| Tire pressure (kN/m²) | 966.0 | 621.0 | 248.4 |
| Loose lift thickness (mm) | 305.0 | 305.0 | 229.0 |

[a]After Johnson and Sallberg (1960)

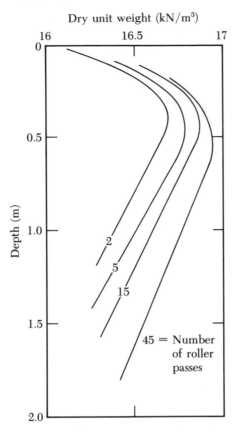

**Figure 12.8**  Vibratory compaction of a sand—variation of dry unit
weight with depth and number of roller passes. Lift thickness = 2.44 m
(after D'Appolonia, Whitman, and D'Appolonia, 1969)

## 12.3

### Vibroflotation

*Vibroflotation* is a technique developed in Germany in the 1930s for *in-situ* densification of thick layers of loose granular soil deposits. Vibroflotation was first used in the United States about ten years later. The process involves the use of a *vibroflot* (called the *vibrating unit*), as shown in Figure 12.10, which is about 2 m in length. This vibrating unit has an eccentric weight inside it and can develop a centrifugal force. The weight enables the vibrating unit to vibrate horizontally. There are openings at the bottom and top of the vibrating unit for water jets. The vibrating unit is attached to a follow-up pipe. Figure 12.10 shows all the vibroflotation equipment necessary for conducting field compaction.

The entire compaction process in the field can be divided into four stages (Figure 12.11, p. 520):

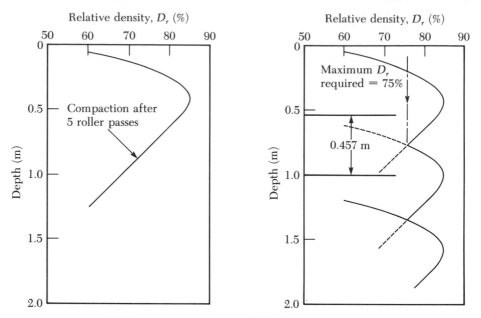

**Figure 12.9**  A method for estimating compaction lift thickness. Minimum relative density required is 75% after five roller passes (after D'Appolonia, Whitman, and D'Appolonia, 1969)

**Stage 1:** The jet at the bottom of the vibroflot is turned on, and the vibroflot is lowered into the ground.

**Stage 2:** The water jet creates a quick condition in the soil, which allows the vibrating unit to sink.

**Stage 3:** Granular material is poured into the top of the hole. The water from the lower jet is transferred to the jet located at the top of the vibrating unit. This water carries the granular material down the hole.

**Stage 4:** The vibrating unit is gradually raised in about 0.3-m lifts and held vibrating for about 30 seconds at a time. This process compacts the soil to the desired unit weight.

Table 12.4 gives the details of various types of vibroflot unit used in the United States. Note that the 30-horsepower electric units have been used since the latter part of the 1940s. The 100-horsepower units were introduced in the early 1970s.

The zone of compaction around a single probe will vary according to the type of vibroflot used. The cylindrical zone of compaction will have a radius of about 2 m for a 30-hp unit. This radius may extend to about 3 m for a 100-hp unit.

Compaction by vibroflotation is done in various probe spacings, depending on the zone of compaction mentioned above (see Figure 12.12). Mitchell (1970) and Brown (1977) have listed several successful cases of foundation design using vibroflotation.

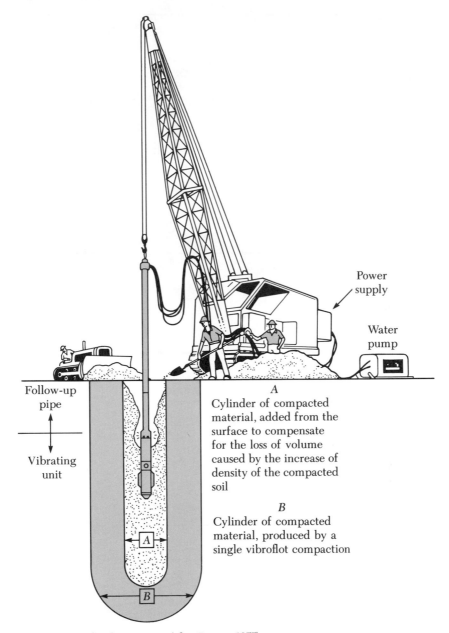

**Power supply**

**Water pump**

Follow-up pipe

Vibrating unit

A
Cylinder of compacted material, added from the surface to compensate for the loss of volume caused by the increase of density of the compacted soil

B
Cylinder of compacted material, produced by a single vibroflot compaction

A

B

**Figure 12.10**  Vibroflotation unit (after Brown, 1977)

The capacity of successful densification of *in-situ* soil depends on several factors, the most important of which is the grain-size distribution of the soil and also the nature of backfill used to fill the holes during the withdrawal period of the vibroflot. The range of the grain-size distribution of *in-situ* soil marked Zone 1 in Figure 12.13 is most suitable for compaction by vibroflotation. Soils that

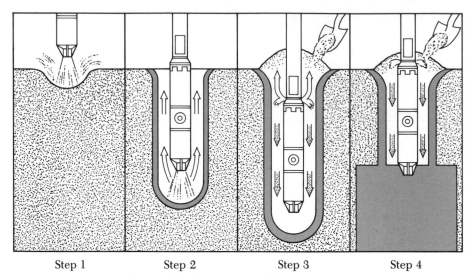

|  | Step 1 | Step 2 | Step 3 | Step 4 |

**Figure 12.11**  Compaction by the vibroflotation process (after Brown, 1977)

**Table 12.4**  Types of Vibroflot Units[a]

| | 100-hp electric and hydraulic motors | 30-hp electric motors |
|---|---|---|
| **(a) Vibrating Tip** | | |
| Length (m) | 2.1 | 1.86 |
| Diameter (mm) | 406.4 | 381.0 |
| Weight (kN) | 17.8 | 17.8 |
| Maximum movement when full (mm) | 12.45 | 7.62 |
| Centrifugal force (kN) | 160.0 | 89.0 |
| **(b) Eccentric** | | |
| Weight (kN) | 1.16 | 0.76 |
| Offset (mm) | 38.1 | 31.75 |
| Length (mm) | 610.0 | 387.0 |
| Speed (rpm) | 1800.0 | 1800.0 |
| **(c) Pump** | | |
| Operating flow rate (m³/min) | 0–1.6 | 0.6 |
| Pressure (kN/m²) | 690–1035 | 690–1035 |
| **(d) Lower Follow-up Pipe and Extensions** | | |
| Diameter (mm) | 305.0 | 305.0 |
| Weight (kN/m) | 3.65 | 3.65 |

[a]After Brown (1977)

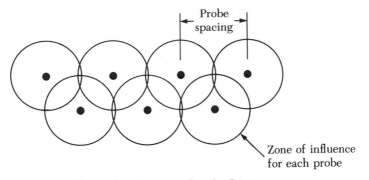

Figure 12.12   Nature of probe spacing for vibroflotation

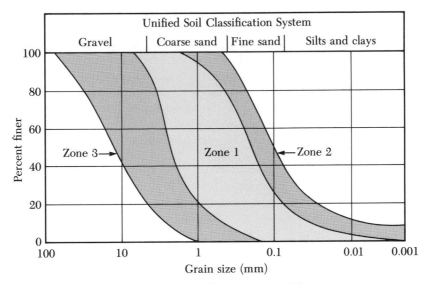

**Figure 12.13**   Effective range of grain-size distribution of soil for vibroflotation

contain excessive amounts of fine sand and silt-size particles are difficult to compact; in these cases, considerable effort is needed to reach proper relative density of compaction. Zone 2 in Figure 12.13 is the approximate lower limit of grain-size distribution for compaction by vibroflotation. Soil deposits whose grain-size distribution falls in Zone 3 contain appreciable amounts of gravel. For these soils, the rate of probe penetration may be rather slow, and so compaction by vibroflotation might prove to be uneconomical in the long run.

The grain-size distribution of the backfill material is one of the factors that control the rate of densification. Brown (1977) has defined a quantity called *suitability number* ($S_N$) for rating a backfill material, or

$$S_N = 1.7 \sqrt{\frac{3}{(D_{50})^2} + \frac{1}{(D_{20})^2} + \frac{1}{(D_{10})^2}} \tag{12.5}$$

where $D_{50}$, $D_{20}$, and $D_{10}$ are the diameters (in mm) through which, respectively, 50%, 20%, and 10% percent of the material is passing.

The smaller the value of $S_N$, the more desirable is the backfill material. Following is a backfill rating system as proposed by Brown (1977):

| Range of $S_N$ | Rating as backfill |
|---|---|
| 0–10 | Excellent |
| 10–20 | Good |
| 20–30 | Fair |
| 30–50 | Poor |
| > 50 | Unsuitable |

## 12.4

### Precompression—General Considerations

When highly compressible, normally consolidated clayey soil layers are located at a limited depth, and large consolidation settlements are expected as the result of the construction of large buildings, highway embankments, or earth dams, precompression of soil may be used to eliminate postconstruction settlement problems. The principles of precompression can be explained by means of Figure 12.14. As shown in Figure 12.14a, let the proposed structural load per unit area be $\Delta p_{(p)}$ and the thickness of the clay layer undergoing consolidation be $H_c$. Hence, the maximum primary consolidation settlement caused by the structural load, $S_{(p)}$, can be given as

$$S_{(p)} = \frac{C_c H_c}{1 + e_o} \log \frac{p_o + \Delta p_{(p)}}{p_o} \tag{12.6}$$

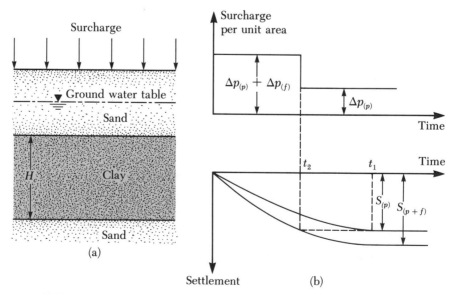

**Figure 12.14**  Principles of precompression

The nature of the settlement-time relationship under the structural load will be like that shown in Figure 12.14b. However, if a surcharge of $\Delta p_{(p)} + \Delta p_{(f)}$ is placed on the ground, the primary consolidation settlement $S_{(p + f)}$ would be

$$S_{(p + f)} = \frac{C_c H_c}{1 + e_o} \log \frac{p_o + [\Delta p_{(p)} + \Delta p_{(f)}]}{p_o} \tag{12.7}$$

The settlement-time relationship under a surcharge of $\Delta p_{(p)} + \Delta p_{(f)}$ is also shown in Figure 12.14b. Note that a total settlement of $S_{(p)}$ would occur at a time $t_2$. Time $t_2$ is much smaller than $t_1$. So, if a temporary total surcharge of $\Delta p_{(f)} + \Delta p_{(p)}$ is applied on the ground surface for a time $t_2$, the settlement would be equal to $S_{(p)}$. At that time, if the surcharge is removed and a structure with a permanent load per unit area of $\Delta p_{(p)}$ is built, no appreciable settlement would occur. The procedure just described is *precompression*. The total surcharge $\Delta p_{(p)} + \Delta p_{(f)}$ can be applied by means of temporary fills.

## Derivation of Equations to Obtain $\Delta p_{(f)}$ and $t_2$

Figure 12.14b shows that, under a surcharge of $\Delta p_{(p)} + \Delta p_{(f)}$, the degree of consolidation at time $t_2$ after load application can be expressed as

$$U = \frac{S_{(p)}}{S_{(p + f)}} \tag{12.8}$$

Substitution of Eqs. (12.6) and (12.7) into Eq. (12.8) yields

$$U = \frac{\log \left[ \dfrac{p_o + \Delta p_{(p)}}{p_o} \right]}{\log \left[ \dfrac{p_o + \Delta p_{(p)} + \Delta p_{(f)}}{p_o} \right]} = \frac{\log \left[ 1 + \dfrac{\Delta p_{(p)}}{p_o} \right]}{\log \left\{ 1 + \dfrac{\Delta p_{(p)}}{p_o} \left[ 1 + \dfrac{\Delta p_{(f)}}{\Delta p_{(p)}} \right] \right\}} \tag{12.9}$$

Figure 12.15 gives the values of $U$ for various combinations of $\Delta p_{(p)}/p_o$ and $\Delta p_{(f)}/\Delta p_{(p)}$. The degree of consolidation referred to in Eq. (12.9) is actually the average degree of consolidation at time $t_2$, as shown in Figure 12.14b. However, if the average degree of consolidation is used to determine time $t_2$, then there might be some construction problems. This is because, after the removal of the surcharge, the portion of clay close to the drainage surface will continue to swell, and the soil close to the midplane will continue to settle (Figure 12.16). In some cases, net continuous settlement might result. A conservative approach may solve this problem—that is, assume that $U$ in Eq. (12.9) is the midplane degree of consolidation (Johnson, 1970a). Now, from Eq. (1.71)

$$U = f(T_v)$$

where $T_v$ = time factor = $C_v t_2 / H^2$ (1.67)
$\quad C_v$ = coefficient of consolidation
$\quad t_2$ = time
$\quad H$ = maximum drainage path ( = $H_c/2$ for two-way drainage and equal to $H_c$ for one-way drainage)

The variation of $U$ (midplane degree of consolidation) with $T_v$ is given in Figure 12.17.

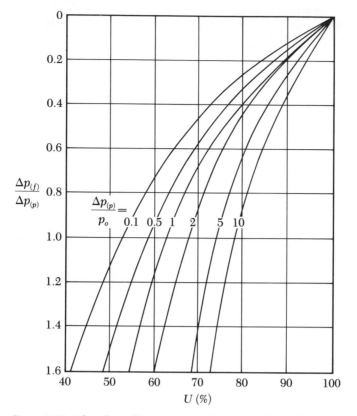

**Figure 12.15**  Plot of $\Delta p_{(f)}/\Delta p_{(p)}$ against $U$ for various values of $\Delta p_{(p)}/p_o$—Eq. (12.9)

## Procedure for Obtaining Precompression Parameters

Two problems may be encountered by engineers during precompression work in the field:

**1.** The value of $\Delta p_{(f)}$ is given; but one must obtain $t_2$. In such a case, obtain $p_o$, $\Delta p_{(p)}$, and solve for $U$ using Eq. (12.9) or Figure 12.15. For this value of $U$, obtain $T_v$ from Figure 12.17. Now

$$t_2 = \frac{T_v H^2}{C_v} \tag{12.10}$$

**2.** One may need to obtain $\Delta p_{(f)}$ for a specified value of $t_2$. In such a case, calculate $T_v$. Then refer to Figure 12.17 to obtain the midplane degree of consolidation, $U$. With the estimated value of $U$, go to Figure 12.15 to get the required $\Delta p_{(f)}/\Delta p_{(p)}$ from which the required value of $\Delta p_{(f)}$ can be calculated.

## Examples of Precompression and General Comments

Johnson (1970a) has presented an excellent review of the use of precompression for improving foundation soils for several projects—for example,

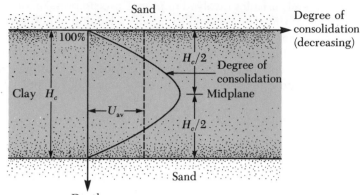

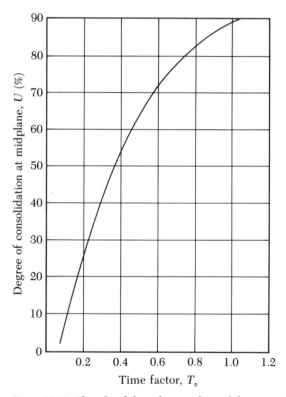

**Figure 12.16**

Figure 12.17   Plot of midplane degree of consolidation against $T_v$

the Morganza Floodway Control Structure near Baton Rouge, Louisiana; the Old River Low-Sill Control Structure near Natchez, Mississippi; and the Old River Overbank Control Structure, Port Elizabeth Marine Terminal, New York.

There are several variables involved in proper precompression design and performance. The information obtained from only a handful of borings is used in the calculation of both the surcharge load and the time necessary for removal of the surcharge. For that reason, precise numbers for precompression design may be difficult to obtain. Settlement observations should be continued during the period of surcharge application. These observations may dictate changes in the design.

## Example
## 12.2

Refer to Figure 12.14. During the construction of a highway bridge, it is expected that the average permanent load on the clay layer will increase by about 115 kN/m². The average effective overburden pressure at the middle of the clay layer is 210 kN/m². Given: $H_c = 6$ m, $C_c = 0.28$, $e_o = 0.9$, and $C_v = 0.36$ m²/month. The clay is normally consolidated.

    a. Determine the total primary consolidation settlement of the bridge without precompression.

    b. Determine the surcharge $\Delta p_{(f)}$ that will be required to eliminate by precompression the entire primary consolidation settlement in 9 months.

**Solution**
**Part a**

The total primary consolidation settlement can be calculated from Eq. (12.6).

$$S_{(p)} = \frac{C_c H_c}{1 + e_o} \log \left[ \frac{p_o + \Delta p_{(p)}}{p_o} \right] = \frac{(0.28)(6)}{1 + 0.9} \log \left[ \frac{210 + 115}{210} \right]$$

$$= 0.1677 \text{ m} = \underline{167.7 \text{ mm}}$$

**Part b**

$$T_v = \frac{C_v t_2}{H^2}$$

$$C_v = 0.36 \text{ m}^2/\text{month}$$

$$H = 3 \text{ m (two-way drainage)}$$

$$t_2 = 9 \text{ months}$$

Hence

$$T_v = \frac{(0.36)(9)}{3^2} = 0.36$$

According to Figure 12.17, for $T_v = 0.36$, the value of $U$ is 47%. Now

$$\Delta p_{(p)} = 115 \text{ kN/m}^2$$

$$p_o = 210 \text{ kN/m}^2$$

So

$$\frac{\Delta p_{(p)}}{p_o} = \frac{115}{210} = 0.548$$

According to Figure 12.15, for $U = 47\%$ and $\Delta p_{(p)}/p_o = 0.548$, $\Delta p_{(f)}/\Delta p_{(p)} \approx 1.8$. So

$$\Delta p_{(f)} = (1.8)(115) = \underline{\underline{207 \text{ kN/m}^2}}$$

## 12.5

## Sand Drains

The use of *sand drains* is another way to accelerate the consolidation settlement of soft, normally consolidated clay layers and achieve precompression before foundation construction. Sand drains are constructed by drilling holes through the clay layer(s) in the field at regular intervals. The holes are then backfilled with highly permeable sand (see Figure 12.18a). After

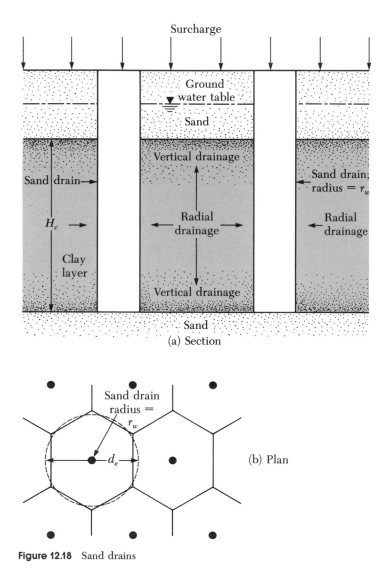

**Figure 12.18** Sand drains

backfilling the drill holes with sand, a surcharge is applied at the ground surface. This surcharge will increase the pore water pressure in the clay. The excess pore water pressure in the clay will be dissipated by drainage—both in the vertical direction and in the radial direction to the sand drains, which accelerates the rate of settlement of the clay layer.

Note that the radius of the sand drains is equal to $r_w$ (Figure 12.18a). Figure 12.18b shows the plan of the layout of the sand drains. As shown in this figure, the effective zone from which the radial drainage will be directed toward a given sand drain can be approximated to be cylindrical with a diameter equal to $d_e$.

To determine the surcharge that needs to be applied at the ground surface and the length of time that it has to be maintained, refer to Figure 12.14 and use the corresponding equation (12.9):

$$U_{v,r} = \frac{\log\left[1 + \frac{\Delta p_{(p)}}{p_o}\right]}{\log\left\{1 + \frac{\Delta p_{(p)}}{p_o}\left[1 + \frac{\Delta p_{(f)}}{\Delta p_{(p)}}\right]\right\}} \tag{12.11}$$

The notations $\Delta p_{(p)}$, $p_o$, and $\Delta p_{(f)}$ are the same as in the case of Eq. (12.9). However, unlike Eq. (12.9), the left side of the preceding equation is the *average degree* of consolidation instead of the degree of consolidation at mid-plane. This average degree of consolidation is due to the contribution of *radial* and *vertical* drainage. If $U_{v,r}$ can be determined for any given time $t_2$ (see Figure 12.14b), then the total surcharge $\Delta p_{(f)} + \Delta p_{(p)}$ can be easily obtained from Figure 12.15. The procedure for determination of the average degree of consolidation $(U_{v,r})$ is given in the following sections.

The successful use of sand drains has been described in detail by Johnson (1970b). As in the case of precompression technique described earlier, constant field settlement observations may be necessary during the period of surcharge application.

## Average Degree of Consolidation due to Radial Drainage Only

The theory for equal-strain consolidation due to radial drainage only (with no smear) was developed by Barron (1948). It assumes that there is *no drainage in the vertical direction*. According to this theory

$$U_r = 1 - \exp\left(\frac{-8T_r}{m}\right) \tag{12.12}$$

where $U_r$ = average degree of consolidation due to radial drainage only

$$m = \left(\frac{n^2}{n^2 - 1}\right) \ln(n) - \frac{3n^2 - 1}{4n^2} \tag{12.13}$$

$$n = \frac{d_e}{2r_w} \tag{12.14}$$

$T_r$ = nondimensional time factor for radial drainage only

$$= \frac{C_{vr}t_2}{d_e^2} \tag{12.15}$$

$$C_{vr} = \text{coefficient of consolidation for radial drainage}$$

$$= \frac{k_h}{\left[\dfrac{\Delta e}{\Delta p(1 + e_{av})}\right]\gamma_w} \tag{12.16}$$

Note that the preceding equation is similar to Eq. (1.65). In Eq. (1.65), $k$ was the coefficient of permeability in the vertical direction of the clay layer. In Eq. (12.16), $k$ has to be replaced by $k_h$, which is the coefficient of permeability for flow in the horizontal direction. In some cases, $k_h$ can be assumed to be equal to $k$; however, for soils like varved clay, $k_h > k$.

Figure 12.19 shows the plot of $U_r$ against $T_r$ for various values of $n$.

### Average Degree of Consolidation due to Vertical Drainage Only

The average degree of consolidation due to vertical drainage only can be obtained from Eqs. (1.72) and (1.73) (or Figure 1.20), or

$$T_v = \frac{\pi}{4}\left[\frac{U_v(\%)}{100}\right] \text{ (for } U_v = 0\text{–}60\%) \tag{1.72}$$

and

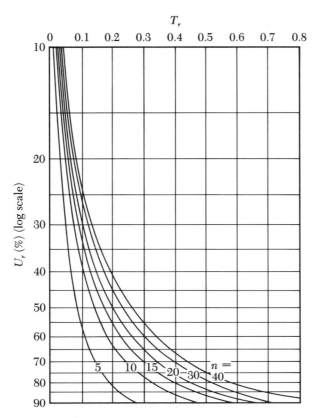

**Figure 12.19** Average degree of consolidation for radial drainage only—Eq. (12.12)

$$T_v = 1.781 - 0.933 \log(100 - U_v\%) \quad \text{(for } U_v > 60\%\text{)} \tag{1.73}$$

where $U_v$ = average degree of consolidation due to vertical drainage only

$$T_v = \frac{C_v t_2}{H^2} \tag{1.67}$$

$C_v$ = coefficient of consolidation for vertical drainage

## Average Degree of Consolidation due to Vertical and Radial Drainage

For a given surcharge and duration $t_2$, the average degree of consolidation due to drainage in the vertical and radial directions can be given by the expression

$$U_{v,r} = 1 - (1 - U_r)(1 - U_v) \tag{12.17}$$

## Example
## 12.3

Redo Example Problem 12.2 with the addition of some sand drains. Assume $r_w = 0.1$ m, $d_e = 3$ m, and $C_v = C_{vr}$.

### Solution
### Part a

The total primary consolidation settlement will be 167.7 mm as before.

### Part b

From Example Problem 12.2, $T_v = 0.36$. The value of $U_v$, from Figure 1.20, is about 67%. From Eq. (12.14)

$$n = \frac{d_e}{2r_w} = \frac{3}{2 \times 0.1} = 15$$

Again,

$$T_r = \frac{C_{vr} t_2}{d_e^2} = \frac{(0.36)(9)}{(3)^2} = 0.36$$

From Figure 12.19, with $n = 15$ and $T_r = 0.36$, the value of $U_r$ is about 77%. Hence

$$U_{v,r} = 1 - (1 - U_v)(1 - U_r) = 1 - (1 - 0.67)(1 - 0.77)$$
$$= 0.924 = 92.4\%$$

Now, referring to Figure 12.15, with $\Delta p_{(p)}/p_o = 0.548$ and $U_{v,r} = 92.4\%$, the value of $\Delta p_{(f)}/\Delta p_{(p)} \approx 0.12$. Hence

$$\Delta p_{(f)} = (115)(0.12) = \underline{13.8 \text{ kN/m}^2}$$

## Wick Drains

The *wick drain* was recently developed as an alternative to the sand drain for inducing vertical drainage in saturated clay deposits. They appear to be better, faster, and cheaper. Wick drains essentially consist of paper or plastic strips that are held in a long tube. The tube is pushed into the soft clay deposit, then withdrawn, leaving behind the strips. These strips will act as vertical drains and induce rapid consolidation. Wick drains can be placed at desired spacings like sand drains. The main advantage of wick drains over sand drains is that they do not require drilling, and, thus, the process of installation is much faster.

## 12.6
## Soil Stabilization by Admixtures

As mentioned in Section 12.1, admixtures are occasionally used to stabilize soils in the field—particularly fine-grained soils. Most common of these admixtures are *lime, lime-fly ash, cement,* and *asphalt*. The main purposes of soil stabilization are to (a) modify the soil, (b) expedite construction, and (c) improve the strength and durability of the soil. The general principles of lime, cement, and lime-fly ash stabilization are discussed in the following sections. (Asphalt stabilization is no longer frequently used.)

### Lime Stabilization

The limes commonly used for stabilization of fine-grained soils are hydrated high-calcium lime [$Ca(OH)_2$], calcitic quick lime (CaO), monohydrated dolomitic lime [$Ca(OH)_2 \cdot MgO$], and dolomitic quick lime. The quantity of lime used for stabilization of most soils usually varies in the range of 5–10%. When lime is added to clayey soils, several chemical reactions are initiated. These reactions are *cation exchange* and *flocculation-agglomeration*, and they are also *pozzolanic*. In the cation exchange and flocculation-agglomeration reactions, the *monovalent* cations generally associated with clays are replaced by the *divalent* calcium ions. The cations, based on their affinity for exchange, can be arranged in a series as shown below:

$$Al^{+++} > Ca^{++} > Mg^{++} > NH_4^+ > K^+ > Na^+ > Li^+$$

Based on this series, any given cation can replace the ions to the right of it. For example, calcium ions can replace potassium and sodium ions from a clay. Flocculation and agglomeration produce a change in the texture of clay soils. The clay particles tend to clump together to form larger-sized particles. These reactions tend to (a) decrease the liquid limit, (b) increase the plastic limit, (c) decrease the plasticity index, (d) increase the shrinkage limit, (e) increase the workability, and (f) improve the strength and deformation properties of a soil.

Pozzolanic reaction between soil and lime involves reaction between lime and the silica and alumina of soil to form cementing material. For example

$$Ca(OH)_2 + \underset{\underset{\text{Clay silica}}{\uparrow}}{SiO_2} \rightarrow CSH$$

where $C = CaO$
$\quad\quad\ S = SiO_2$
$\quad\quad\ H = H_2O$

The pozzolanic reaction may continue for a long period of time.

Figure 12.20 shows the variation of the liquid limit, the plasticity index, and the shrinkage limit of a clay with percentage of lime admixture. The first 2–3% lime (on the dry weight basis) has a substantial influence in improving the workability and the property of the soil. Addition of lime to clayey soils has an effect on their compaction characteristics.

Figure 12.21a shows the results of standard Proctor tests for Vicksburg clay without any additives and also with 4% high-calcium hydrated lime additive (uncured). Note that the addition of lime has an influence in reducing the maximum compacted dry unit weight and increasing the optimum moisture content. Figure 12.21b also shows the change of the unconfined compressive strength $(q_u)$ of uncured Vicksburg clay with the percentage of high-calcium hydrated lime. The value of $q_u$ with 6% lime is about six times the value obtained with no additive. Note that the specimens prepared for the determination of $q_u$ were all at a moisture content of 29–29.5%. This is shown as the molding moisture content in Figure 12.21a.

Arman and Munfakh (1972) also evaluated the lime stabilization of organic clays found in Louisiana. Figure 12.22a shows the change of plasticity with $Ca(OH)_2$ content for an organic soil with 22% organic material. The curing time for these soil specimens was 48 hours. The effects of lime are generally similar to those seen in Figure 12.20. The change of the unconfined compression strength of the same soil with lime additives is shown in Figure 12.22b. Based on this study, Arman and Munfakh concluded that (a) the presence of organic matter does not block the pozzolanic reaction that helps in changing the fundamental soil properties and making the soil more workable, and (b) about 2% of lime is sufficient to satisfy the base exchange capacity of organic matters.

Lime stabilization in the field can be done in three ways:

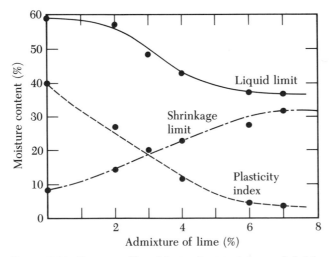

**Figure 12.20**  Variation of liquid limit, plasticity index, and shrinkage limit of a clay with lime additive

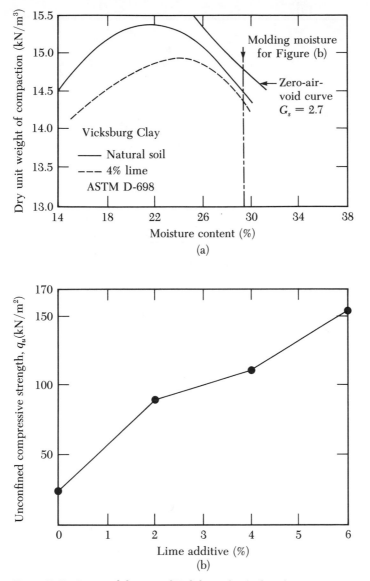

**Figure 12.21** Lime stabilization of Vicksburg clay (% less than $2\mu$ size = 46; liquid limit = 59, plasticity index = 30, A-7-6 (20), pH = 6). (a) laboratory dry unit weight vs. moisture content curve; (b) change of unconfined compression strength with percent of lime. *Note:* The specimens for (b) were molded at a moisture content shown in (a) (from "Stability Properties of Uncured Lime-Treated Fine Grained Soil," by C. H. Neubauer and M. R. Thompson. In *Highway Research Record No. 381*, Highway Research Board, 1972, pp. 20–26. Reprinted by permission).

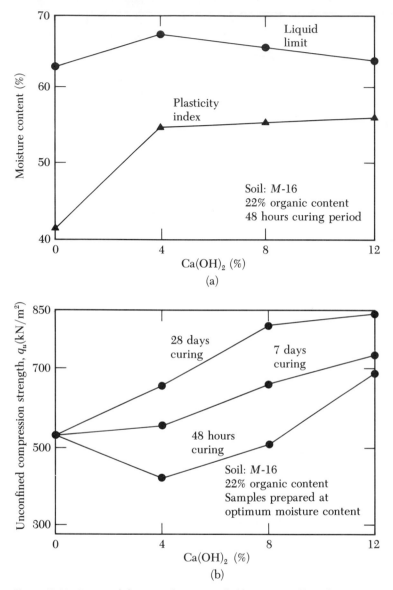

**Figure 12.22**  Lime stabilization of organic soil: (a) variation of liquid limit and plasticity index with 48 hours curing; (b) variation of unconfined compression strength (from "Lime Stabilization of Organic Soil," by A. Arman and G. A. Munfakh. In *Highway Research Record No. 381*, Highway Research Board, 1972, pp. 37–45. Reprinted by permission.)

1. The *in-situ* material and/or the borrowed material can be mixed with the proper amount of lime at the site and then compacted after the addition of moisture.
2. The soil can be mixed with the proper amount of lime and water at a plant and then hauled back to the site for compaction.

**3.** Lime slurry can be pressure injected into the soil to a depth of 2–3.5 m at spacings of 2.5–2 m. This is a useful technique to control swelling of expansive soils and also to improve unstable soils under building sites. If this procedure is adopted, the top soil (about 0.25 m) layer should be compacted by conventional procedures.

Because the addition of hydrated lime to soft clayey soils results in an immediate increase of plastic limit, thus changing the soil from plastic to solid and making it appear to "dry up," limited amounts of it can be thrown on muddy and troublesome construction sites. This improves the trafficability at the construction sites and may save money and time.

Quick limes have also been successfully used in drill holes having diameters of 100 mm to 150 mm for stabilization of subgrades and slopes. For this type of work, holes are drilled in a grid pattern and then filled with quick lime.

## Cement Stabilization

Cement is increasingly used as a stabilizing material for soil, particularly for the construction of highways and earth dams. The first controlled soil-cement construction in the United States was carried out near Johnsonville, South Carolina, in 1935. Cement can be used to stabilize sandy and clayey soils. As in the case of lime, cement has an effect in decreasing the liquid limit and increasing the plasticity index and workability of clayey soils. For clayey soils, cement stabilization is effective when fine fractions (passing No. 200 sieve) are less than about 40%, liquid limit is less than 45–50, and plasticity index is less than about 25. The optimum requirements of cement by volume for effective stabilization of various types of soil are given in Table 12.5.

Like lime, cement helps increase the strength of soils; and, strength increases with curing time. Table 12.6 presents some typical values of the unconfined compressive strength of various types of untreated soil and soil-cement mixture made with approximately 10% cement by weight.

On the basis of the preceding discussion and the data presented in Tables 12.5 and 12.6, it is obvious that granular soils and clayey soils with low plasticity are more suitable for cement stabilization. Experience has shown that calcium clays are more easily stabilized by addition of cement, whereas sodium and hydrogen clays, which are expansive in nature, respond better to lime stabilization. For these reasons, proper care should be given in the selection of the stabilizing material.

**Table 12.5** Cement Requirement by Volume for Effective Stabilization of Various Soils[a]

| Soil type | | Percent cement by volume |
|---|---|---|
| AASHTO classification | Unified classification | |
| A-2 and A-3 | GP, SP, and SW | 6–10 |
| A-4 and A-5 | CL, ML, and MH | 8–12 |
| A-6 and A-7 | CL, CH | 10–14 |

[a]After Mitchell and Frietag (1959)

**Table 12.6** Typical Compressive Strengths of Soils and Soil-Cement Mixtures[a]

| Material | Unconfined compressive strength range $(kN/m^2)$[b] |
|---|---|
| *Untreated soil:* | |
| Clay, peat | Less than 350 |
| Well-compacted sandy clay | 70–280 |
| Well-compacted gravel, sand, and clay mixtures | 280–700 |
| *Soil-cement (10% cement by weight):* | |
| Clay, organic soils | Less than 350 |
| Silts, silty clays, very poorly graded sands, slightly organic soils | 350–1050 |
| Silty clays, sandy clays, very poorly graded sands, and gravels | 700–1730 |
| Silty sands, sandy clays, sands, and gravels | 1730–3460 |
| Well-graded sand-clay or gravel-sand-clay mixtures and sands and gravels | 3460–10,350 |

[a]After Mitchell and Freitag (1959)

[b]Rounded off

For field compaction, the proper amount of cement can be mixed with soil, either at the site or in a mixing plant (and then carried to the site). The soil is compacted to the required unit weight with a predetermined amount of water.

Similar to lime injection, cement slurry made of Portland Cement and water (water-cement ratio = 0.5 to 5) can be used for pressure grouting of poor soils under foundations of buildings and other structures. Grouting decreases the permeability of soils and increases the strength and the load-bearing capacity. For design of low-frequency machine foundations subjected to vibrating forces, it is sometimes required to stiffen the foundation soil by grouting and thereby increase the resonant frequency.

## 12.7

### Fly Ash Stabilization

*Fly ash* is a by-product of the pulverized coal combustion process usually associated with electric power generating plants. Fly ash is a fine-grained dust and is primarily composed of silica, alumina, and various oxides and alkalies. It is pozzolanic in nature and can react with hydrated lime to produce cementitious products. For that reason, lime-fly ash mixtures can be used for stabilization of highway bases and subbases. Effective mixes can be prepared with 10–35% of fly ash and 2–10% of lime. Soil-lime-fly ash mixes are compacted under controlled conditions with proper amounts of moisture to obtain stabilized soil layers.

A certain type of fly ash is obtained from the burning of coal primarily from the western United States, and it is referred to as *"Type C"* fly ash. It contains a fairly large proportion (up to about 25%) of free lime that, with the addition of water, will react with other fly-ash compounds to form cemetitious products. This may eliminate the need to add manufactured lime.

## 12.8

### Geotextiles

*Geotextile* is a generic name used for permeable fabrics that are used in soil reinforcement. Chapter 11 discussed earth reinforcement with thin metal strips for the construction of shallow foundations and earth-retaining structures. Geotextiles are similarly used in foundation engineering work. The fabrics are usually petroleum products—that is, polyester, polyethylene, and polypropylene. They may be made from fiberglass. Geotextiles are not prepared from natural fabrics because these decay too quickly. Geotextiles can be *woven, nonwoven,* and *knitted*.

Until now, geotextiles have been used primarily for the construction of highways and embankments. For highway construction, the fabric is unrolled like a carpet over soft subgrade, and the overburden is laid directly on it and then compacted.

The use of geotextiles is a rather new approach in foundation engineering, and so far only a relatively small amount of research has been done in this area. However, it is gaining popularity rapidly. There are four major uses of geotextiles in foundation engineering:

1. *Drainage:* The fabrics can rapidly channel water from soil to various outlets and thereby provide a higher shear strength of soil and hence stability.
2. *Filtration:* When placed between two soil layers, one coarse grained and the other fine grained, the fabric allows free seepage of water from one layer to the other. However, it protects the fine-grained soil from being washed into the coarse-grained soil.
3. *Separation:* Geotextiles help keep various soil layers separate after construction and during the projected service period of the structure. For example, in the construction of highways, a clayey subgrade can be kept separate from a granular base course.
4. *Reinforcement:* The tensile strength of the geofabrics increases the load-bearing capacity of the soil.

Table 12.7 lists various types of geotextile presently available in North America, along with their manufacturers.

The tension to which a given geotextile can be subjected depends to a high degree on the modulus, $E_g$; or

$$t_g = \epsilon_g E_g \tag{12.18}$$

where $t_g$ = tension in geotextiles (kN/m width)
$\epsilon_g$ = strain

Efforts are now being made by several investigators to develop proper test

**Table 12.7** Geotextiles Manufactured in North America[a]

| Manufacturer or distributor | Fabric trade name | Structure | Fiber | Width (m) | Mass (g/m²) |
|---|---|---|---|---|---|
| Du Pont | Typar 3201, 3341, 3401, 3601 | Nonwoven | Polypropylene | 3.84 to 5 | 67.8, 115.3 135.6, 203.4 |
| Du Pont | Reepav | Nonwoven | Polypropylene | 3.76 | 101.7 |
| Texel | Texel 7607, 7609 7612, 7618, 7643 | Nonwoven | Polyester | Up to 12.19 | 240 to 2000 |
| Nicolon | 40/30 | Woven | Polypropylene | 3.66 and multiples of 3.66 | |
| Nicolon | 70/06, 70/20, 100/08 | Woven | Polypropylene | 1.83 multiples | |
| Nicolon | 100/05 | Woven | Polyvinylidene chloride | 1.83 multiples | |
| Nicolon | Geolon 200 | Woven | Polypropylene | 3.81 and 5.03 | |
| Nicolon | Geolon 400 | Woven | Polypropylene | 3.66 multiples | |
| Nicolon | Geolon 1250 | Woven | Polypropylene | 5.09 | |
| Enka | Enkamat | | Nylon | 0.99 | 265 and 405 |
| Enka | Enkadrain | | Nylon and polyester | 0.99 | 735 |
| Enka | Stabilenka | Nonwoven | Polyester | 1.07 and 2.14 | 67.8, 101,7, 135.6 |
| Synflex | Permealiner | Woven and nonwoven | Polypropylene | | |
| Exxon | Industrial Fabrics Style 49862 | Woven | Polypropylene | 3.81 and 4.72 | 115.3 |
| Celanese | Mirafi 500X | Woven | Polypropylene | 3.81 and 5.33 | |
| Celanese | Mirafi 140N | Nonwoven | Polypropylene | 4.57 | |
| Celanese | Mirafi 600X | Woven | Polypropylene | 3.81 | |
| Celanese | Mirafi 700X | Woven | Polypropylene | 1.83 and 3.66 | |
| Hoechst | Trevira | Nonwoven | Polyester | up to 4.88 | |
| Crown Zellerbach | Fibretex 150, 200, 300, 400 | Nonwoven | Polypropylene | 3.96 and 5.33 | |
| Crown Zellerbach | Fibretex Ten-1 | Woven | Polypropylene | 3.81 and 4.72 | 115.3 |

| | | | | | |
|---|---|---|---|---|---|
| Dominion Textile | Penroad 50, 150, 250, 350, 450, 550 | Nonwoven | Polyester | 1.11 | 200, 270, 340, 400, 540, 680 |
| Dominion Textile | Penroad 100,500 | Woven | Polypropylene | 1.16 | 125 and 150 |
| Amoco | ProPex 4545, 4551, 4553 | Nonwoven | Polypropylene | 4.42 | 152.6, 203.4, 271.2 |
| Amoco | ProPex 1325, 1369 | Nonwoven | Polypropylene | 4.42 | 220.4 |
| Amoco | ProPex 2002 | Woven | Polypropylene | 4.42 | 152.6 |
| Collins & Aikman | Diamond 8 | Woven | Nylon or polyester | 1.52 | 389.9 |
| Collins & Aikman | Contain 4 | Woven | Polyester | 1.52 | 466.1 |
| Collins & Aikman | Retain | Woven | Nylon | 1.83 | 305.1 |
| Collins & Aikman | SB-375 | Stitchbound | Polyester | 1.42 to 3.66 | 127.1 and 271.2 |
| Phillips | Petromat | Nonwoven | Polypropylene | 1.91 and 3.81 | 145.8 |
| Phillips | Supac | Nonwoven | Polypropylene | Up to 4.57 | 135.6, 169.5, 542.4 |
| Carthage Mills | Poly-Filter X | Woven | Polypropylene | 1.83 to 2.13 (custom fab.) | 244.1 |
| Carthage Mflls | Poly-Filter GB | Woven | Polypropylene | 1.83 to 21.95 (custom fab.) | 223.7 |
| Carthage Mills | Filter-X | Woven | Polyvinylidene chloride | | 393.2 |
| Rhone-Poulene | Bidim | Nonwoven | Polyester | 2.7, 4.2, 5.3 | 50 to 100 |
| Foss | F-55 | Nonwoven | Polyester | 0.91, 1.91, 3.81, 4.72 (other widths up to 7.42 available on request) | 186.5 |
| Foss | F-65 | Nonwoven | Polyester | 0.91, 1.91, 3.81, 4.72 (other widths up to 7.42 available on request) | 220.4 |
| Foss | F-90 | Nonwoven | Polyester | 0.91, 1.91, 3.81, 4.72 (other widths up to 7.42 available on request) | 305.1 |
| Foss | F-180 | Nonwoven | Polyester | 0.91, 1.91, 3.81, 4.72 (other widths up to 7.42 available on request) | 610.2 |

aAfter Industrial Fabric Products Review, July, 1981, pp. 25–26

procedures to define the properties of geotextiles. Kinney (1979) has recommended the following values of $E_g$ from his experimental work:

$$E_g \approx 524 \text{ N/m per \% strain (for Mirafi 140)}$$

and

$$E_g \approx 1926 \text{ N/m per \% strain (for Mirafi 500X)}$$

For the other limited amount of information presently available, readers are referred to Bender and Barenberg (1978), Kinney (1979), Barenberg (1980), and Giroud (1980).

### Examples of Application of Geotextiles to Foundation Engineering Problems

Several examples of the use of geotextiles to solve foundation engineering problems were discussed during the Second International Conference on Geotextiles held in Las Vegas in August, 1982. (The papers presented during the conference are now available in three volumes, and they may serve as the basis for future research work.) Following are brief descriptions of some of these uses of geotextiles.

Geotextiles can be used for the construction of railroads over weak subgrades. Under ordinary conditions, the ballast under railroad tracks tends to penetrate the weak subgrades (Figure 12.23a), and the fines present in the

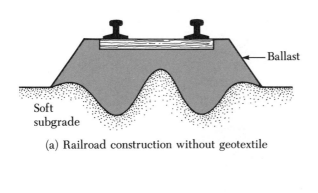

(a) Railroad construction without geotextile

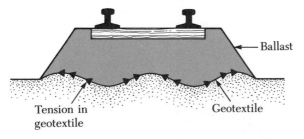

(b) Railroad construction with the use of geotextile

**Figure 12.23** Use of geotextile in the construction of railroads

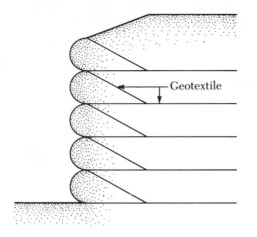

**Figure 12.24**  Use of geotextile in retaining wall construction

subgrade gradually intrude into the ballast. This process will cause rapid deterioration of the railroad tracks. The serviceability of tracks can be improved by stabilizing the subgrade materials. However, instead of using conventional stabilization techniques, geotextile can be used over the soft subgrade to economically improve the performance of railroad tracks. This is shown in Figure 12.23b. More discussion on this topic can be found in Friedli and Anderson (1982), Newby (1982), Raymond (1982), and Saxena and Chiu (1982).

Low-cost unpaved roads over soft subgrades have been economically and successfully built in several places by applying layers of geotextiles over the subgrades. Steward and Mohney (1982) have given detailed accounts of such construction by the United States Forest Service in the Pacific Northwest region.

Construction of retaining walls with thin metallic reinforcement strips has been discussed in Chapter 11. Geotextiles can be used in a similar manner in the construction of economical retaining walls (Figure 12.24). Steward and Mohney (1982) have also presented some examples of this type of construction by the United States Forest Service.

Geotextiles are now being increasingly used in drainage works, such as the stabilization of landslides or unstable slopes. They are also used in partial substitution of traditional graded filters (Chapter 1) made of selected natural soils.

## Problems

**12.1** Make the necessary calculations and prepare the zero-air-void unit weight (in lb/ft³) curves as related to a Proctor compaction test for $G_s$ = 2.6, 2.65, 2.7, and 2.75.

**12.2** The maximum dry unit weight of a soil was determined by a Proctor compaction test to be 113.3 lb/ft³. If the same soil is used in the compaction of an embankment to a unit weight of 95.3 lb/ft³, what would be the relative compaction?

**12.3** A soil was compacted in the field by the standard Proctor test procedure. For the test, given:

$$\text{Moist unit weight of compaction} = 18 \text{ kN/m}^3$$

$$\text{Moisture content} = 12\%$$

Estimate $\gamma_{d(\max)}$ and the optimum moisture content for the soil according to the Ohio one-point-method.

**12.4** According to the Ohio one-point-method, a soil will have a maximum dry unit weight of 99.9 lb/ft³ at an optimum moisture content of 21.5%. Estimate the dry unit weight of the soil when compacted to a moisture content of 18%.

**12.5** For a vibroflotation work, the backfill to be used has the following characteristics:

$$D_{50} = 1 \text{ mm}$$
$$D_{20} = 0.5 \text{ mm}$$
$$D_{10} = 0.08 \text{ mm}$$

Determine the suitability number of the backfill. How would you rate the material?

**12.6** Refer to Figure 12.14. For the construction of an airport, a large fill operation is required. For this work, the average permanent load $[\Delta p_{(p)}]$ on the clay layer will increase by about 1460 lb/ft². The average effective overburden pressure on the clay layer before the fill operation is 1985 lb/ft². For the clay layer, which is normally consolidated and drained at top and bottom, given:

$$H_c = 16.75 \text{ ft} \qquad e_o = 0.81$$
$$C_c = 0.24 \qquad C_v = 4.73 \text{ ft}^2/\text{month}$$

Determine:
a. the primary consolidation settlement of the clay layer caused by the additional permanent load, $\Delta p_{(p)}$;
b. the time required for 90% of primary consolidation settlement under the additional permanent load only;
c. the temporary surcharge, $\Delta p_{(f)}$, that will be required to eliminate the entire primary consolidation settlement in 5 months by precompression technique.

**12.7** Redo Part c of Problem 12.6. Given: the time for elimination of primary consolidation settlement is 7 months.

**12.8** For Example Problem 12.2, if the temporary surcharge $\Delta p_{(f)}$ is 100 kN/m², how long will it be before the entire primary consolidation settlement is eliminated by precompression technique?

**12.9** The diagram of a sand drain is shown in Figure 12.18. If $r_w = 0.25$ m, $d_e = 4$ m, $C_v = C_{vr} = 0.28$ m²/month, and $H_c = 8.4$ m, determine the degree of consolidation caused by a sand drain only after 6 months of surcharge application.

**12.10** Estimate the degree of consolidation for the clay layer described in Problem 12.9 that is caused by the combination of vertical drainage (drained on top and bottom) and radial drainage after 6 months of the application of surcharge.

# References

Arman, A., and Munfakh, G. A. (1972). "Lime Stabilization of Organic Soils," *Highway Research Record, No. 381,* National Academy of Sciences, pp. 37–45.

Barenberg, E. (1980). "Design Procedures for Soil-Fabric-Aggregate Systems with Mirafi 500X," Civil Engineering Studies, *Report No. UILU-Eng-80-2019,* Department of Civil Engineering, University of Illinois, Urbana, Oct.

Barron, R. A. (1948). "Consolidation of Fine-Grained Soils by Drain Wells," *Transactions,* American Society of Civil Engineers, Vol. 113, pp. 718–754.

Bender, D., and Barenberg, E. (1978). "Design and Behavior of Soil-Fabric-Aggregate Systems," *Transportation Research Record No.671,* National Academy of Sciences, pp. 40–44.

Brown, R. E. (1977). "Vibroflotation Compaction of Cohesionless Soils," *Journal of the Geotechnical Engineering Division,* American Society of Civil Engineers, Vol. 103, No. GT12, pp. 1437–1451.

D'Appolonia, D. J., Whitman, R. V., and D'Appolonia, E. (1969). "Sand Compaction with Vibratory Rollers," *Journal of the Soil Mechanics and Foundations Division,* American Society of Civil Engineers, Vol. 95, No. SM1, pp. 263–284.

Friedli, P., and Anderson, D. G. (1982). "Behavior of Woven Fabrics under Simulated Railroad Loading," *Proceedings,* Second International Conference on Geotextiles, Las Vegas, Vol. II, pp. 473–478.

Giroud, J. P. (1980). "Introduction to Geotextiles and Their Application," *Proceedings,* First Canadian Symposium on Geotextiles, Sept., pp. 1–32.

Industrial Fabric Association International (1981). "Geotextile Directory," *Industrial Fabric Products,* Vol. 58, No. 3, St. Paul, pp. 25–26.

Johnson, A. W., and Sallberg, J. R. (1960). "Factors That Influence Field Compaction of Soils," *Bulletin No. 272,* Highway Research Board, National Academy of Sciences, Washington, D.C.

Johnson, S. J. (1970a). "Precompression for Improving Foundation Soils," *Journal of the Soil Mechanics and Foundations Division,* American Society of Civil Engineers, Vol. 96, No. SM1, pp. 114–144.

Johnson, S. J. (1970b). "Foundation Precompression with Vertical Sand Drains," *Journal of the Soil Mechanics and Foundations Division,* American Society of Civil Engineers, Vol. 96, No. SM1, pp. 145–175.

Kinney, T. (1979). *Fabric Induced Changes in High Deformation Soil-Fabric-Aggregate Systems,* Ph.D. Thesis, University of Illinois, Urbana, January.

Lee, K. L., and Singh, A. (1971). "Relative Density and Relative Compaction," *Journal of the Soil Mechanics and Foundations Division,* American Society of Civil Engineers, Vol. 97, No. SM7, pp. 1049–1052.

Lee, P. Y., and Suedkamp, R. J. (1972). "Characteristics of Irregularly Shaped Compaction Curves of Soils," *Highway Research Record No. 381,* National Academy of Sciences, Washington, D.C., pp. 1–9.

Mitchell, J. K. (1970). "In-Place Treatment of Foundation Soils," *Journal of the Soil Mechanics and Foundations Division,* American Society of Civil Engineers, Vol. 96, No. SM1, pp. 73–110.

Mitchell, J. K., and Freitag, D. R. (1959). "A Review and Evaluation of Soil-Cement Pavements," *Journal of the Soil Mechanics and Foundations Division,* American Society of Civil Engineers, Vol. 85, No. SM6, pp. 49–73.

Neubauer, C. H., Jr., and Thompson, M. R. (1972). "Stability Properties of Uncured

Lime-Treated Fine-Grained Soils," *Highway Research Record No. 381*, National Academy of Sciences, pp. 20–26.

Newby, J. E. (1982). "Southern Pacific Transportation Company Utilization of Geotextiles in Railroad Subgrade," *Proceedings*, Second International Conference on Geotextiles, Las Vegas, Vol. II, pp. 467–472.

Raymond, G. (1982). "Geotextiles in Railroad Bed Rehabilitation," *Proceedings*, Second International Conference on Geotextiles, Las Vegas, Vol. II, pp. 479–484.

Saxena, S. K., and Chiu, D. (1982). "Evaluation of Fabric Performance in a Railroad System," *Proceedings*, Second International Conference on Geotextiles, Las Vegas, Vol. II, pp. 485–490.

Steward, J., and Mohney, J. (1982). "Trial Use Results and Experience Using Geotextiles for Low-Volume Forest Roads," *Proceedings*, Second International Conference on Geotextiles, Vol. II, pp. 491–495.

# Reinforced Concrete Design

## Fundamentals of Reinforced Concrete Design

At the present time, most reinforced concrete designs are based on the recommendations of the building code prepared by the American Concrete Institute—that is, ACI 318-77. The basis for this code is the *ultimate strength design* or *strength design*. Some of the fundamental recommendations of the code are briefly summarized in the following sections.

### Load Factors

According to ACI Code Section 9.2, the ultimate load-carrying capacity of a structural member should be as follows:

$$U = 1.4D + 1.7L \tag{A.1}$$

where $U$ = ultimate load-carrying capacity of a member
$D$ = dead load
$L$ = live load

If the wind load is to be considered for design, then

$$U = 0.75(1.4D + 1.7L + 1.7W) \tag{A.2}$$

where $W$ = wind load

If there is no live load, then

$$U = 0.9D + 1.3W \tag{A.3}$$

However, in no case should the value of $U$ be less than that given in Eq. (A.1).

## Strength Reduction Factor

The design strength provided by a structural member is equal to the theoretical strength times a strength reduction factor, $\phi$, or

$$\text{design strength} = \phi(\text{theoretical strength})$$

The reduction factor, $\phi$, takes into account the inaccuracies in the design assumptions, changes in property or strength of the construction materials, and so on. Following are some of the recommended values of $\phi$ (ACI Code Section 9.3):

| Condition | Value of $\phi$ |
|---|---|
| a. Axial tension; flexure with or without axial tension | 0.9 |
| b. Shear or torsion | 0.85 |
| c. Axial compression with or without flexure, spiral reinforcement | 0.75 |
| d. Axial compression with or without flexure, tied reinforcement | 0.7 |
| e. Bearing on concrete | 0.7 |
| f. Flexure in plain concrete | 0.65 |

## Design Concepts for a Rectangular Section in Bending

Figure A.1a shows a section of a concrete beam having a width $b$ and a depth $h$. The assumed stress distribution across the section at ultimate load is shown in Figure A.1b. The following notations have been used in this figure:

$f_c'$ = compressive strength of concrete at 28 days
$A_s$ = area of steel tension reinforcement
$f_y$ = yield stress of reinforcement in tension
$d$ = effective depth
$l$ = location of the neutral axis measured from the top of the compression face
$a = \beta l$

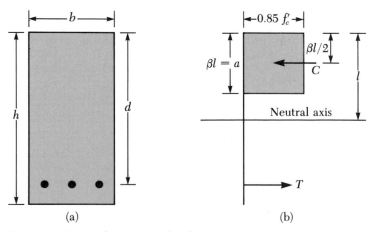

(a)                              (b)

**Figure A.1**   Rectangular section in bending

$\beta = 0.85$ for $f'_c$ of 28 MN/m² (4000 lb/in.²) or less and decreases at the rate of 0.05 for every 7 MN/m² (1000 lb/in.²) increase of $f'_c$. However, it cannot be less than 0.65 in any case (ACI Code Section 10.2.7).

From the principles of statics, for the section

$$\Sigma \text{ compressive force, } C = \Sigma \text{ tensile force, } T$$

Thus

$$0.85 f'_c ab = A_s f_y$$

or

$$a = \frac{A_s f_y}{0.85 f'_c b} \tag{A.4}$$

Also, for the beam section, the nominal ultimate moment can be given as

$$M_n = A_s f_y \left( d - \frac{a}{2} \right) \tag{A.5}$$

where $M_n$ = theoretical ultimate moment

The design ultimate moment, $M_u$, can be given as

$$M_u = \phi A_s f_y \left( d - \frac{a}{2} \right) \tag{A.6}$$

Combining Eqs. (A.4) and (A.6)

$$M_u = \phi A_s f_y \left[ d - \left( \frac{1}{2} \right) \frac{A_s f_y}{0.85 f'_c b} \right] = \phi A_s f_y \left( d - \frac{0.59 \, A_s f_y}{f'_c b} \right) \tag{A.7}$$

The steel percentage is defined by the equation

$$s = \frac{A_s}{bd} \tag{A.8}$$

In a balanced beam, failure would occur by sudden simultaneous yielding of tensile steel and crushing of concrete. The balanced percentage of steel (for Young's modulus of steel, $E_s = 200$ MN/m²) can be given as

$$s_b = \frac{0.85 f'_c}{f_y} (\beta) \left( \frac{600}{600 + f_y} \right) \tag{A.8a}$$

where $f'_c$ and $f_y$ are in MN/m².

In conventional English units (with $E_s = 29 \times 10^6$ lb/in.²)

$$s_b = \frac{0.85 f'_c}{f_y} (\beta) \left( \frac{87,000}{87,000 + f_y} \right) \tag{A.8b}$$

where $f'_c$ and $f_y$ are in lb/in.²

To avoid sudden failure without warning, ACI Code Section 10.3.3 recommends that the maximum steel percentage ($s_{\text{max}}$) should be limited to $0.75 s_b$, or

$$s_{max} = 0.75s_b \tag{A.9}$$

Table A.1 gives the values of $s_{max}$ for various values of $f_c'$ and $f_y$.

The nominal or theoretical shear strength of a section, $V_n$, can be given as

$$V_n = V_c + V_s \tag{A.10}$$

where $V_c$ = nominal shear strength of concrete
$V_s$ = nominal shear strength of reinforcement

The permissible shear strength, $V_u$, can be given by

$$V_u = \phi V_n = \phi(V_c + V_s) \tag{A.11}$$

The values of $V_c$ can be given by the following equations (ACI Code Sections 11.3 and 11.11)

$$V_c = 0.17\sqrt{f_c'}\ bd \quad \text{(for member subjected to shear and flexure)} \tag{A.12a}$$

and

$$V_c = 0.34\sqrt{f_c'}\ bd \quad \text{(for member subjected to diagonal tension)} \tag{A.12b}$$

where $f_c'$ is in $MN/m^2$, $V_c$ is in MN, and $b$ and $d$ are in m.

In conventional English units, Eqs. (A.12a) and (A.12b) take the following form:

$$V_c = 2\sqrt{f_c'}\ bd \tag{A.13a}$$
$$V_c = 4\sqrt{f_c'}\ bd \tag{A.13b}$$

where $V_c$ is in lb, $f_c'$ is in lb/in.$^2$, and $b$ and $d$ are in inches.

Note that

$$v_c = \frac{V_c}{bd} \tag{A.14}$$

where $v_c$ is the shear stress

Now, combining Eqs. (A.11), (A.12a), and (A.14), one obtains

$$\text{permissible shear stress} = v_u = \frac{V_u}{bd} = 0.17\phi\sqrt{f_c'} \tag{A.15a}$$

Similarly, from Eqs. (A.11), (A.12b), and (A.14)

**Table A.1**   Values of $s_{max}$ [Eq. (A.9)]

| $f_y$ | | $f_c'$ | | | |
|---|---|---|---|---|---|
| MN/m² | lb/in.² | 21 MN/m² ($\approx$3000 lb/in.²) | 28 MN/m² ($\approx$4000 lb/in.²) | 35 MN/m² ($\approx$5000 lb/in.²) | 42 MN/m² ($\approx$6000 lb/in.²) |
| 276 | 40,000 | 0.0284 | 0.0378 | 0.0445 | 0.0501 |
| 345 | 50,000 | 0.0209 | 0.0279 | 0.0329 | 0.0370 |
| 414 | 60,000 | 0.0163 | 0.0217 | 0.0255 | 0.0287 |
| | | $\beta = 0.85$ | $\beta = 0.85$ | $\beta = 0.80$ | $\beta = 0.75$ |

$$v_u = 0.34\phi \sqrt{f_c'}$$ (A.15b)

Table A.2 on page 550 gives the values of $v_u/\phi$ for various values of $f_c'$.

## A.2
## Reinforcing Bars

The nominal sizes of reinforcing bars commonly used in the United States are given in Table A.3 on page 550.

The details regarding standard metric bars used in Canada are as follows:

| Bar number | Diameter, mm | Area, mm$^2$ |
|:----------:|:------------:|:------------:|
| 10 | 11.3 | 100 |
| 15 | 16.0 | 200 |
| 20 | 19.5 | 300 |
| 25 | 25.2 | 500 |
| 30 | 29.9 | 700 |
| 35 | 35.7 | 1000 |
| 45 | 43.7 | 1500 |
| 55 | 56.4 | 2500 |

Reinforcing-bar sizes in the metric system have been recommended by UNESCO (1971):

| Bar diameter, mm | Area, mm$^2$ |
|:----------------:|:------------:|
| 6 | 28 |
| 8 | 50 |
| 10 | 79 |
| 12 | 113 |
| 14 | 154 |
| 16 | 201 |
| 18 | 254 |
| 20 | 314 |
| 22 | 380 |
| 25 | 491 |
| 30 | 707 |
| 32 | 804 |
| 40 | 1256 |
| 50 | 1963 |
| 60 | 2827 |

This appendix uses the standard bar diameters recommended by UNESCO.

## A.3
## Development Length

The development length, $L_d$, is the length of embedment required to develop the yield stress in the tension reinforcement for a section in flexure. ACI Code Section 12.2.2 lists the basic development lengths for tension reinforcement as

a. 35-mm bar and smaller    $0.019A_b f_y/\sqrt{f_c'}$
    but not less than          $0.058 d_b f_y$
b. 43-mm bar               $26 f_y/\sqrt{f_c'}$
c. 57-mm bar               $34 f_y/\sqrt{f_c'}$

where $A_b$ = area of the individual bar (mm$^2$)
      $d_b$ = nominal diameter of the bar (mm)

**Table A.2**   Values of $v_u/\phi$ [Eqs. (A.15a) and (A.15b)]

|                    | $f_c'$ (MN/m²) |       |      |      |
| ------------------ | -------------- | ----- | ---- | ---- |
| Equation<br>number | 21             | 28    | 35   | 42   |
| Eq. (A.15a)        | 0.78           | 0.90  | 1.0  | 1.1  |
| Eq. (A.15b)        | 1.56           | 1.80  | 2.0  | 2.2  |

**Table A.3**   Nominal Sizes of Reinforcing Bars Used in the United States

| Bar<br>No. | Diameter |        | Area of cross section |          |
| ---------- | -------- | ------ | --------------------- | -------- |
|            | (mm)     | (in.)  | (mm²)                 | (in.²)   |
| 3          | 9.52     | 0.375  | 71                    | 0.11     |
| 4          | 12.70    | 0.500  | 129                   | 0.20     |
| 5          | 15.88    | 0.625  | 200                   | 0.31     |
| 6          | 19.05    | 0.750  | 284                   | 0.44     |
| 7          | 22.22    | 0.875  | 387                   | 0.60     |
| 8          | 25.40    | 1.000  | 510                   | 0.79     |
| 9          | 28.65    | 1.128  | 645                   | 1.00     |
| 10         | 32.26    | 1.270  | 819                   | 1.27     |
| 11         | 35.81    | 1.410  | 1006                  | 1.56     |
| 14         | 43.00    | 1.693  | 1452                  | 2.25     |
| 18         | 57.33    | 2.257  | 2580                  | 4.00     |

The units $f_y$ and $f_c'$ in the preceding expressions are in MN/m², and $L_d$ is in millimeters.

   In conventional English units, $L_d$ is expressed in inches. The expressions for $L_d$ are as follows:

a.  No. 11 bar and smaller     $0.04A_b f_y / \sqrt{f_c'}$
    but not less than          $0.0004 d_b f_y$

b.  No. 14 bar                 $0.085 f_y / \sqrt{f_c'}$

c.  No. 18 bar                 $0.11 f_y / \sqrt{f_c'}$

The units of $A_b$ and $d_b$ are in in.² and in., respectively. The values of $f_y$ and $f_c'$ are expressed in lb/in².

   The basic development length must be multiplied by appropriate factors given by ACI Code Sections 12.2.3 and 12.2.4:

a.  Top reinforcement                              1.4

b.  Reinforcement with $f_y > 414$ MPa             $2 - \dfrac{414}{f_y (\text{MN/m}^2)}$

    In English units, for $f_y > 60,000$ lb/in.²   $2 - \dfrac{60,000}{f_y (\text{lb/in.}^2)}$

c.  For lightweight concrete                       1.33

d.  Reinforcement spaced at least 152 mm
    on center and at least 76 mm in from
    all sides                                      0.8

e. Reinforcement in excess of that required $\left(\dfrac{\text{as required}}{\text{as provided}}\right)$

In any case, the basic development length should not be less than 305 mm (12 in.).

## A.4

## Summary of ACI 318-77 Code Requirements

For convenience, Table A.4 gives a summary of the strength design principles that apply to foundation design. The table has been divided into two parts: (1) general design principles that apply to strength design and (2) principles that are specifically applicable to foundation design.

**Table A.4** Summary of ACI 318-77 Code Requirements

| Principles | Design items | Code requirements | Code section |
|---|---|---|---|
| General | Load | $U = 1.4D + 1.7L$<br>$U = 0.75(1.4D + 1.7L + 1.7W)$<br>$U = 0.9D + 1.3W$<br>(See Article A.1.1 for explanations) | 9.2 |
| | Load factor, $\phi$ | Flexure: 0.9<br>Shear and torsion: 0.85<br>Bearing: 0.7<br>Flexure in plain concrete: 0.65<br>(See Article A.1.2) | 9.3 |
| | Flexure | $M_u = \phi A_s f_y \left[ d - \dfrac{0.59 A_s f_y}{f_c' b} \right]$<br>[See Eq. (A.7)] | 10.2 |
| | Maximum flexure reinforcement | $s_{\max} = 0.75 s_b$<br>[See Eqs. (A.8a) and (A.8b)] | 10.3.3 |
| | Minimum flexure reinforcement— steel percentage | $s_{\min} = \dfrac{1.4}{f_y (\text{MN/m}^2)}$<br><br>$s_{\min} = \dfrac{200}{f_y (\text{lb/in.}^2)}$<br><br>Uniform thickness: use steel percentage equal to that required for shrinkage and temperature | 10.5 |
| | Shrinkage and temperature reinforcement— steel percentage | For $f_y = 275$ MN/m² or 345 MN/m²:<br>$s_s = (0.002)$(gross concrete area)<br>For $f_y = 414$ MN/m²: $s_s = (0.0018)$<br>(gross concrete area) | 7.12 |
| | $\beta$ factor<br>(See Fig. A.1) | $\beta = 0.85$ for $f_c' \leq 28$ MN/m² (4000 lb/in.²) and reduces by 0.05 for every 7 MN/m² (1000 lb/in.² in excess of 28 MN/m²). *Minimum value = 0.65* | 10.2.7 |
| | Shear reinforcement | Refer to ACI Code | 11.11 |
| | Development length, $L_d$ | See Section A.3 | 12.2 |

**Table A.4**  (Continued)

| Principles | Design items | Code requirements | Code section |
|---|---|---|---|
| | Reinforcement spacing | Clear distance not less than diameter of bar or 25.4 mm (1 in.) | 7.6.1 |
| | | Walls and slabs: not to be spaced farther apart than 3 times the wall or slab thickness or 457 mm (18 in.) | 7.6.5 |
| | Minimum reinforcement cover | 76 mm (3 in.) for concrete cast against and permanently exposed to earth | 7.7.1 |
| | Modulus of elasticity of concrete, $E_c$ | *SI system:* $$E_c(\text{MN/m}^2) = (W_c^{1.5})(0.043)\sqrt{f_c'\,(\text{MN/m}^2)}$$ $W_c$ = density of concrete (for 1440 kg/m³ to 2480 kg/m³ $$E_c(\text{MN/m}^2) = 4730\sqrt{f_c'\,(\text{MN/m}^2)}\text{ (for normal weight concrete)}$$ *English system:* $$E_c(\text{lb/in.}^2) = (W_c^{1.5})(33)\sqrt{f_c'\,(\text{lb/in.}^2)}$$ for $W_c$ = 90 to 155 lb/ft³ $$E_c(\text{lb/in.}^2) = 57000\sqrt{f_c'\,(\text{lb/in.}^2)}\text{ (for normal weight concrete)}$$ | 8.5 |
| | Shear strength | $$v_n = \frac{V_n}{bd}$$ | |
| | | *Wide beam:* $$v_c(\text{MN/m}^2) = 0.17\sqrt{f_c'\,(\text{MN/m}^2)}$$ $$v_c(\text{lb/in.}^2) = 2\sqrt{f_c'\,(\text{lb/in.}^2)}$$ | 11.3 |
| | | *Diagonal tension:* $$v_c(\text{MN/m}^2) = 0.34\sqrt{f_c'\,(\text{MN/m}^2)}$$ $$v_c(\text{lb/in.}^2) = 4\sqrt{f_c'\,(\text{lb/in.}^2)}$$ [See Eqs. (A.12) and (A.13)] | 11.11 |
| | Bearing strength | Bearing strength = $0.85\,\phi f_c'\,A_1$ $\phi = 0.7$ *Exceptions:* when supporting surface is wider on all sides than the loaded area, the bearing strength on the loaded area is equal to $0.85\,\phi f_c'A_1\sqrt{A_2/A_1}$. Limit of $\sqrt{A_2/A_1} \le 2$. $A_1$ = loaded area; $A_2$ = area of the portion of the supporting surface that is concentric and geometrically similar to the loaded area | 10.16 |

*FOUNDATIONS*

| Footings | General considerations | See ACI Code | 15 |
|---|---|---|---|
| | Maximum moment | See ACI Code | 15.4.2 |
| | Shear | 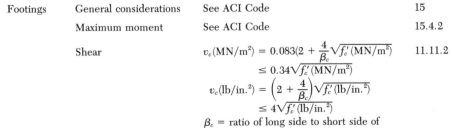 | 11.11.2 |

$$v_c(\text{MN/m}^2) = 0.083\left(2 + \frac{4}{\beta_c}\right)\sqrt{f_c'\,(\text{MN/m}^2)}$$
$$\le 0.34\sqrt{f_c'\,(\text{MN/m}^2)}$$
$$v_c(\text{lb/in.}^2) = \left(2 + \frac{4}{\beta_c}\right)\sqrt{f_c'\,(\text{lb/in.}^2)}$$
$$\le 4\sqrt{f_c'\,(\text{lb/in.}^2)}$$
$\beta_c$ = ratio of long side to short side of concentrated load or reaction area

**Table A.4** (Continued)

| Principles | Design items | Code requirements | Code section |
|---|---|---|---|
| | Minimum footing depth | Not less than 152 mm (6 in.) above the bottom of reinforcement for footing on soil. Not less than 305 mm (12 in.) for footing on piles<br>Not less than 203 mm (8 in.) for unreinforced concrete footing | 15.7<br><br>15.11.4 |
| | Plain concrete pedestals and footings | *Maximum stress:*<br>Flexure $(\text{MN/m}^2) = 0.42\sqrt{f_c'}\,(\text{MN/m}^2)$<br>Flexure $(\text{lb/in.}^2) = 5\sqrt{f_c'}\,(\text{lb/in.}^2)$<br>Shear stress for beam action $(\text{MN/m}^2) = 0.17\sqrt{f_c'}\,(\text{MN/m}^2)$<br>Shear stress for beam action $(\text{lb/in.}^2) = 2\sqrt{f_c'}\,(\text{lb/in.}^2)$<br>Shear stress for two-way action $(\text{MN/m}^2) = 0.34\sqrt{f_c'}\,(\text{MN/m}^2)$<br>Shear stress for two-way action $(\text{lb/in.}^2) = 4\sqrt{f_c'}\,(\text{lb/in.}^2)$ | 15.11.1 |
| | Transfer of force at base of column or reinforced pedestal | Area of reinforcement $\geq 0.005$ gross area of supported member with at least four dowels | 15.8.4.1 |
| | Round columns | Treat as square columns with same area for location of critical sections for moment, shear, and development of reinforcement in footings | 15.3 |
| WALLS | General considerations | Refer to ACI Code | 14 |
| | Minimum thickness | Not less than 152 mm (6 in.) for uppermost 4.6 m (15 ft) of wall height. Increase by 25.4 mm (1 in.) for each successive 7.5 m (25 ft) downward for bearing walls<br>Exterior basement walls, foundation walls not less than 203 mm (8 in.) | 14.2.6<br><br><br><br><br>14.2.7 |
| | Grade beam | See ACI Code | 14.3 |
| | Reinforcement | Horizontal: $A_s \geq 0.0025A_g$ of wall<br>Vertical: $A_s \geq 0.0015A_g$ of wall<br>$A_s$ = area of reinforcement<br>$A_g$ = gross area of the wall | 14.2.10<br>14.2.11 |

## A.5

## Design Example of a Continuous Wall Foundation

Let it be required to design a load-bearing wall with the following data:

Dead load = $D$ = 43.8 kN/m

Live load = $L$ = 17.5 kN/m

Gross allowable bearing capacity of soil = 94.9 kN/m$^2$

Depth of the top of foundation from the ground surface = 1.2 m

$f_y$ = 413.7 MN/m$^2$

$f_c'$ = 20.68 MN/m$^2$

$$\text{Unit weight of soil} = \gamma = 17.27 \text{ kN/m}^3$$
$$\text{Unit weight of concrete} = \gamma_c = 22.97 \text{ kN/m}^3$$

## General Considerations

For this design, assume the foundation thickness to be 0.3 m. Refer to ACI Code Section 7.7.1, which recommends a minimum cover of 76 mm over steel reinforcement, and assume that the steel bars to be used are 12 mm in diameter (Figure A.2a)

$$d = 300 - 76 - \frac{12}{2} = 218 \text{ mm}$$

Also, weight of the foundation $= (0.3)\gamma_c = (0.3)(22.97) = 6.89 \text{ kN/m}^2$

$$\text{weight of soil above the foundation} = (1.2)\gamma = (1.2)(17.27)$$
$$= 20.72 \text{ kN/m}^2$$

So, net allowable soil bearing capacity

$$q_{\text{net(all)}} = 94.9 - 6.89 - 20.72 = 67.29 \text{ kN/m}^2$$

Hence, required width of foundation =

$$B = \frac{D + L}{q_{\text{net(all)}}} = \frac{43.8 + 17.5}{67.29} = 0.91 \text{ m}$$

So, assume $B = 1$ m.
According to ACI Code Section 9.2

$$U = 1.4D + 1.7L = (1.4)(43.8) + (1.7)(17.5) = 91.07 \text{ kN/m}$$

Converting the net allowable soil pressure to an ultimate (factored) value

$$q_s = \frac{U}{(B)(1)} = \frac{91.07}{(1)(1)} = 91.07 \text{ kN/m}^2$$

## Investigation of Shear Strength of the Foundation

The critical section for shear occurs at a distance $d$ from the face of the wall (ACI Code Sections 15.2.2 and 11.11.1.1), as shown in Figure A.2b. So, shear at critical section

$$V_u = (0.35 - d)q_s = (0.35 - 0.218)(91.07) = 12.02 \text{ kN/m}$$

From Eq. (A.12a)

$$V_c = 0.17\sqrt{f_c'}\, bd = 0.17\sqrt{20.68}\,(1)(0.218) = 0.1685 \text{ MN/m} \approx 168 \text{ kN/m}$$

Also, from Eq. (A.11)

$$\phi V_c = (0.85)(168) = 142.8 \text{ kN/m} > V_u = 12.02 \text{ kN/m—O.K.}$$

Because $V_u < \phi V_c$, the total thickness of the foundations could be reduced to 250 mm. So, the modified

$$d = 250 - 76 - \frac{12}{2} = 168 \text{ mm} > 152 \text{ mm} = d_{\min}\,(\text{ACI Code Section 15.7})$$

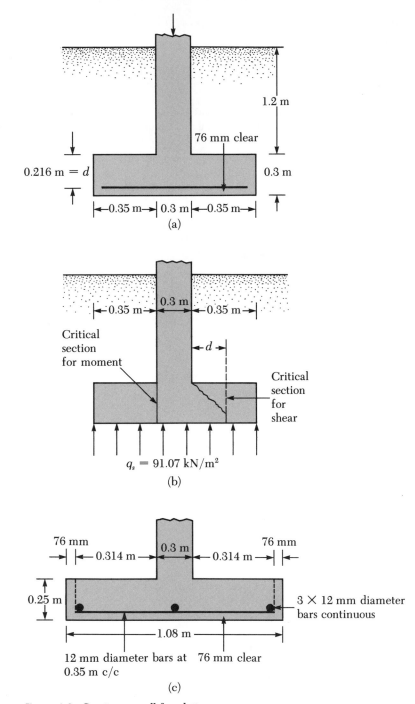

**Figure A.2** Continuous wall foundation

If $d = 168$ mm

$$\phi V_c = (0.85)(0.17)\sqrt{20.68}\,(1)(0.168) = 0.1104 \text{ MN}$$
$$= 110.4 \text{ kN} > V_u\text{—O.K.}$$

## Flexural Reinforcement

For steel reinforcement, factored moment at the face of the wall has to be determined (ACI Code Section 15.4.2). The bending of the foundation will be in one direction only. So, according to Figure A.2b, the design ultimate moment

$$M_u = \frac{q_s l^2}{2}$$

$$l = 0.35 \text{ m}$$

So

$$M_u = \frac{(91.07)(0.35)^2}{2} = 5.58 \text{ kN-m/m}$$

From Eqs. (A.4) and (A.5)

$$M_n = A_s f_y \left( d - \frac{a}{2} \right)$$

$$a = \frac{A_s f_y}{0.85 f_c' b} = \frac{(A_s)(413.7)}{(0.85)(20.68)(1)} = 23.5351 A_s$$

Thus

$$M_n = (A_s)(413.7)\left( 0.168 - \frac{23.5351}{2} A_s \right)$$

or

$$M_n (\text{MN-m/m}) = 69.5 A_s - 4868.24 A_s^2$$

Again, from Eq. (A.6)

$$M_u = \phi M_n$$

where $\phi = 0.9$

Thus

$$5.58 \times 10^{-3} (\text{MN-m/m}) = 0.9(69.5 A_s - 4868.24 A_s^2)$$

Solving for $A_s$, one gets

$$A_{s(1)} = 0.0143 \text{ m}^2; \ A_{s(2)} = 0.00009 \text{ m}^2$$

Hence, steel percentage with $A_{s(1)}$ is

$$s_1 = \frac{A_{s(1)}}{bd} = \frac{0.0143}{(1)(0.168)} = 0.0851$$

Similarly, steel percentage with $A_{s(2)}$ is

$$s_2 = \frac{A_{s(2)}}{bd} = \frac{0.00009}{(1)(0.168)} = 0.00054 < s_{min} = 0.0018 \text{ (ACI Code Section 7.12.2)}$$

The maximum steel percentage that can be provided is given in Eqs. (A.8a) and (A.9). Thus

$$s_{max} = (0.75)(0.85)\frac{f'_c}{f_y}\beta\left(\frac{600}{600 + f_y}\right)$$

Note that $\beta = 0.85$. Substituting the proper values of $\beta$, $f'_c$, and $f_y$ in the preceding equation, one obtains

$$s_{max} = 0.016$$

Note that $s_1 = 0.0851 > s_{max} = 0.016$. So use $s = s_{min} = 0.0018$. So,

$$A_s = (s_{min})(b)(d) = (0.0018)(1)(0.168) = 0.000302 \text{ m}^2 = 302 \text{ mm}^2$$

Use 12-mm diameter bars @ 350 mm c/c. Hence

$$A_s \text{ (provided)} = \frac{1000}{350}\left(\frac{\pi}{4}\right)(12)^2 = 323 \text{ mm}^2$$

### Development Length of Reinforcement Bars

According to section A.3 of the appendix, for the 12-mm bars, $L_d = 0.019A_b(f_y/\sqrt{f'_c})$ but not less than $0.058 d_b f_y$ or 305 mm. For this case, $A_b = 113 \text{ mm}^2$ and $d_b = 12$ mm. So

$$L_d = 0.019A_b\frac{f_y}{\sqrt{f'_c}} = (0.019)(113)\left(\frac{413.7}{\sqrt{20.68}}\right) = 195 \text{ mm}$$

$$L_d = 0.058d_b f_y = (0.058)(12)(413.7) = 288 \text{ mm}$$

Hence, $L_d = 305$-mm controls.

Now, assuming a 76-mm cover to be on the sides of the foundation, the development length would be 350 mm − 76 mm = 274 mm < 305 mm. To achieve the proper development length, increase the width of the foundation ($B$) to 1080 mm. This will give 390-mm cantilevers, as shown in Figure A.2c. This increase of $B$ will not greatly affect the structural steel provided. It will reduce the value of soil-bearing pressure ($q_s$) while the shear and moment capacities of the cross section remain constant.

Minimum reinforcement should be furnished in the long direction to offset shrinkage and temperature effects (ACI Code Section 7.12). So

$$A_s = (0.0018)(b)(d) = (0.0018)[(0.390)(2) + 0.3](0.168)$$
$$= 0.000327 \text{ m}^2 = 327 \text{ mm}^2$$

Provide 3 × 12-mm diameter bars ($A_s = 339 \text{ mm}^2$).
The final design sketch is shown in Figure A.2c.

## A.6

### Design Example of a Square Foundation for a Column

Figure A.3a shows a square column foundation with the following conditions:

Live load $= L = 675$ kN

Dead load $= D = 1125$ kN

Allowable gross soil-bearing capacity $= q_{all} = 145$ kN/m²

Column size $= 0.5$ m $\times$ 0.5 m

$f_c' = 20.68$ MN/m²

$f_y = 413.7$ MN/m²

The design of the foundation can be done in the following manner.

### General Considerations

Let the average unit weight of concrete and soil above the base of the foundation be 21.97 kN/m³. So, the net allowable soil-bearing capacity

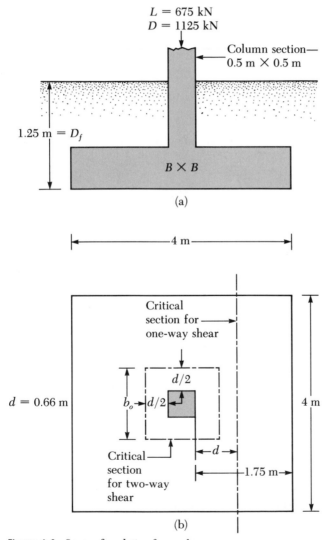

(a)

(b)

**Figure A.3**   Square foundation for a column

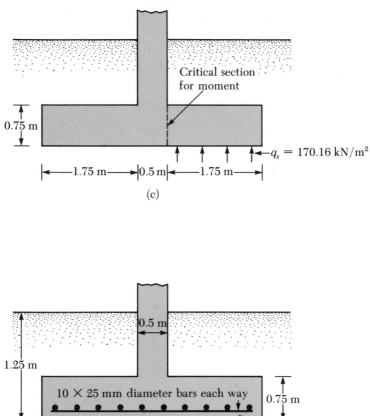

$$q_s = 170.16 \text{ kN/m}^2$$

(c)

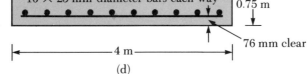

(d)

**Figure A.3** (Continued)

$$q_{\text{all(net)}} = 145 - (D_f)(21.97) = 145 - (1.25)(21.97) = 117.54 \text{ kN/m}^2$$

Hence, the required foundation area =

$$A = B^2 = \frac{D + L}{q_{\text{all(net)}}} = \frac{675 + 1125}{117.54} = 15.31 \text{ m}^2$$

Use a foundation with dimensions $(B)$ of 4 m $\times$ 4 m.
    The factored load for the foundation

$$U = 1.4D + 1.7L = (1.4)(1125) + (1.7)(675) = 2722.5 \text{ kN}$$

Hence, the factored soil pressure

$$q_s = \frac{U}{B^2} = \frac{2722.5}{16} = 170.16 \text{ kN/m}^2$$

Assume the thickness of the foundation to be equal to 0.75 m. With a clear cover of 76 m over the steel bars and an assumed bar diameter of 25 mm

$$d = 0.75 - 0.076 - \frac{0.025}{2} = 0.6615 \text{ m}$$

## Check for Shear

As we have seen in Section A.5, $V_u$ should be equal to or less than $\phi V_c$. For one-way shear

$$V_u \leq \phi(0.17)\sqrt{f_c'}\, bd$$

The critical section for one-way shear is located at a distance $d$ from the edge of the column (ACI Code Section 11.11.1.1) as shown in Figure A.3b. So

$$V_u = q_s \times \text{critical area} = (170.16)(4)(1.75 - 0.6615) = 740.9 \text{ kN}$$

Also

$$\phi V_c = (0.85)(0.17)(\sqrt{20.68})(4)(0.6615)(1000) = 1738.7 \text{ kN}$$

So

$$V_u = 749.9 \text{ kN} \leq \phi V_c = 1738.7 \text{ kN—O.K.}$$

For two-way shear, the critical section is located at a distance of $d/2$ from the edge of the column (ACI Code Section 11.11.1.2). This is shown in Figure A.3b. For this case

$$\phi V_c = \phi(0.34)\sqrt{f_c'}\, b_o d$$

The term $b_o$ is the perimeter of the critical section for two-way shear. Or, for this design

$$b_o = 4[0.5 + 2(d/2)] = 4[0.5 + 2(0.3308)] = 4.65 \text{ m}$$

Hence

$$\phi V_c = (0.85)(0.34)(\sqrt{20.68})(4.65)(0.6615) = 4.042 \text{ MN} = 4042 \text{ kN}$$

Also

$$V_u = (q_s)(\text{critical area})$$
$$\text{Critical area} = (4 \times 4) - (0.5 + 0.6615)^2 = 14.65 \text{ m}^2$$

So

$$V_u = (170.16)(14.65) = 2492.84 \text{ kN}$$
$$V_u = 2492.84 \text{ kN} < \phi V_c = 4042 \text{ kN—O.K.}$$

So, the assumed depth of foundation is more than adequate.

## Flexural Reinforcement

According to Figure A.3c, the moment at critical section (ACI Code Section 15.4.2) is

$$M_u = (q_s B)\left(\frac{1.75}{2}\right)^2 = \frac{[(170.16)(4)](1.75)^2}{2} = 1042.23 \text{ kN-m}$$

From Eq. (A.4)

$$a = \frac{A_s f_y}{0.85 f'_c b} \quad (note: b = B)$$

or

$$A_s = \frac{0.85 f'_c B a}{f_y} = \frac{(0.85)(20.68)(4)a}{413.7} = 0.17a$$

From Eq. (A.6)

$$M_u \le \phi A_s f_y \left( d - \frac{a}{2} \right)$$

With $\phi = 0.9$ and $A_s = 0.17a$

$$M_u = 1042.23 = (0.9)(0.17a)(413700) \left( 0.6615 - \frac{a}{2} \right)$$

Solution of the preceding equation gives $a = 0.0254$ m. Hence

$$A_s = 0.17a = (0.17)(0.0254) = 0.0043 \text{ m}^2$$

Percentage of steel

$$s = \frac{A_s}{bd} = \frac{A_s}{Bd} = \frac{0.0043}{(4)(0.6615)} = 0.00163 < s_{min}$$
$$= 0.0018 \text{ (ACI Code Section 7.12.2)}$$

So

$$A_{s(min)} = (0.0018)(B)(d) = (0.0018)(4)(0.6615)$$
$$= 0.004762 \text{ m}^2 = 47.62 \text{ cm}^2$$

Provide $10 \times 25$-mm diameter bars each way $[A_s = (4.91)(10) = 49.1 \text{ cm}^2]$.

## Check for Development Length

From Section A.3, the development length

$$L_d = 0.019 A_b \frac{f_y}{\sqrt{f'_c}} = (0.019)(491 \text{ mm}^2) \left( \frac{413.7}{\sqrt{20.68}} \right) = 848.68 \text{ mm}$$

Also

$$L_d \ge 0.058 \, d_b f_y = (0.058)(25)(413.7) = 599.87 \text{ mm}$$

So, $L_d = 848.68$-mm controls. Actual $L_d$ provided is $(4 - 0.5/2) - 0.076$ (cover) $=$ 1.674 m $> 599.87$ mm—O.K.

## Check for Bearing Strength

Table A.4 indicates that the bearing strength should be at least $0.85 \phi f'_c A_1 \sqrt{A_2/A_1}$ with a limit of $\sqrt{A_2/A_1} \le 2$. For this problem, $\sqrt{A_2/A_1} = \sqrt{(4 \times 4)/(0.5 \times 0.5)} = 8$. So, use $\sqrt{A_2/A_1} = 2$. Also, $\phi = 0.7$. Hence, the design bearing strength $= (0.85)(0.7)$

$(20.68)(0.5 \times 0.5)(2) = 6.15$ MN $= 6150$ kN. However, the factored column load $U = 2722.5$ kN $< 6150$ kN—O.K.

The final design section is shown in Figure A.3d.

## A.7

## Design Example of a Rectangular Foundation for a Column

This section describes the design of a rectangular foundation to support a column having dimensions of 0.4 m $\times$ 0.4 m in cross section. Other details are as follows:

Dead load  $= D = 290$ kN

Live load   $= L = 110$ kN

Depth from the ground surface to the top of the foundation $= 1.2$ m

Allowable gross soil-bearing capacity $= 120$ kN/m$^2$

Maximum width of foundation $= B = 1.5$ m

$f_y = 413.7$ MN/m$^2$

$f_c' = 20.68$ MN/m$^2$

Unit weight of soil $= \gamma = 17.27$ kN/m$^3$

Unit weight of concrete $= \gamma_c = 22.97$ kN/m$^3$

### General Considerations

For this design, let us assume a foundation depth of 0.45 m (Figure A.4a). The weight of foundation/m$^2 = 0.45 \, \gamma_c = (0.45)(22.97) = 10.34$ kN/m$^2$, and the weight of soil above the foundation/m$^2 = (1.2)\gamma = (1.2)(17.27) = 20.72$ kN/m$^2$. Hence, the net allowable soil-bearing capacity $[q_{net(all)}] = 120 - 10.34 - 20.72 = 88.94$ kN/m$^2$.

The required area of the foundation $= (D + L)/q_{net(all)} = (290 + 110)/88.94 = 4.5$ m$^2$. Hence, the length of the foundation is 4.5 m$^2$/$B = 4.5/1.5 = 3$ m.

The factored column load $= 1.4D + 1.7L = 1.4(290) + 1.7(110) = 593$ kN.

The factored soil-bearing capacity, $q_s =$ factored load/foundation area $= 593/4.5 = 131.78$ kN/m$^2$.

### Shear Strength of Foundation

Assume that the steel bars to be used have a diameter of 16 mm. So, the effective depth $d = 450 - 76 - 16 = 358$ mm. (Note that the assumed clear cover is 76 mm.)

Figure A.4a shows the critical section for one-way shear (ACI Code Section 11.11.1.1). According to this figure

$$V_u = \left(1.5 - \frac{0.4}{2} - 0.358\right) Bq_s = (0.942)(1.5)(131.78) = 186.21 \text{ kN}$$

The nominal shear capacity of concrete for one-way beam action

$$V_c = 0.17 \sqrt{f_c'} \, Bd = 0.17 \, (\sqrt{20.68})(1.5)(0.358) = 0.4152 \text{ MN} = 415.2 \text{ kN}$$

Now

$$V_u = 184.22 \leq \phi V_c = (0.85)(415.2) = 352.92 \text{ kN—O.K.}$$

The critical section for two-way shear is also shown in Figure A.4a. This is based on the recommendations given by ACI Code Section 11.11.1.2. For this section

$$V_u = q_s[(1.5)(3) - 0.758^2] = 517.3 \text{ kN}$$

The nominal shear capacity of the foundation can be given as (ACI Code Section 11.11.2; also see Table A.4 of the text)

$$V_c = v_c b_o d = 0.083 \left( 2 + \frac{4}{\beta_c} \right) \sqrt{f_c'} b_o d$$

where $b_o$ = perimeter of the critical section
$\beta_c$ = length of foundation/width of foundation

or

$$V_c = 0.083 \left[ 2 + \frac{4}{(3/1.5)} \right] \sqrt{20.68} \, (4 \times 0.758)(0.358) = 1.638 \text{ MN}$$

So, for two-way shear condition

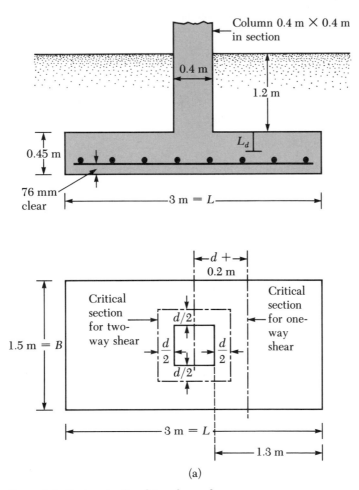

**Figure A.4** Rectangular foundation for a column

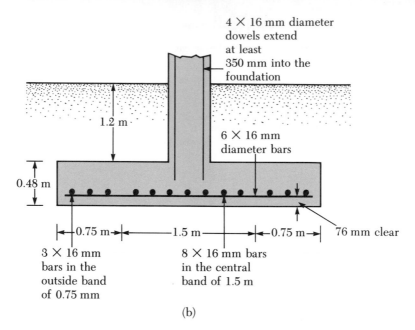

4 × 16 mm diameter
dowels extend
at least
350 mm into the
foundation

1.2 m

6 × 16 mm
diameter bars

0.48 m

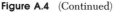

|←0.75 m→|←————1.5 m————|←0.75 m→|  76 mm clear

3 × 16 mm
bars in the
outside band
of 0.75 mm

8 × 16 mm bars
in the central
band of 1.5 m

(b)

**Figure A.4**   (Continued)

$$V_u = 517.3 \text{ kN} < \phi V_c = (0.85)(1639) = 1393 \text{ kN}$$

Therefore the section is adequate.

## Check for Bearing Capacity of Concrete Column at the Interface with Foundation

According to ACI Code Section 10.16.1 (also see Table A.4), bearing strength is equal to $0.85\phi f'_c A_1 (\phi = 0.7)$. For this problem, $U = 593$ kN < bearing strength = $(0.85)(0.7)(20.68)(0.4)^2 = 1.969$ MN.

So, a minimum area of dowels should be provided across the interface of the column and the foundation (ACI Code Section 15.8.4). Based on ACI Code Section 15.8.4b

$$\text{minimum area of steel} = (0.005)(\text{area of column})$$
$$= (0.005)(400^2) = 800 \text{ mm}^2$$

So use 4 × 16-mm diameter bars as dowels.

The minimum required length of development $(L_d)$ of dowels into the foundation is $(0.24 f_y d_b)/\sqrt{f'_c}$, but not less than $0.044 f_y d_b$ (ACI Code Section 12.3.2). So

$$L_d = \frac{0.24 f_y d_b}{\sqrt{f'_c}} = \frac{(0.24)(413.7)(16)}{\sqrt{20.68}} = 349.33 \text{ mm}$$

Also

$$L_d = 0.044 f_y d_b = (0.044)(413.7)(16) = 291.25 \text{ mm}$$

Hence, $L_d = 349.33$-mm controls.

Available depth for the dowels (Figure A.4a) is $450 - 76 - 16 - 16 = 342$ mm. Since hooks cannot be used, the foundation depth must be increased. Let the new depth be equal to 480 mm to accommodate the required $L_d = 349.33$ mm. Hence, the new value of $d$ is equal to $480 - 76 - 16 = 388$ mm.

## Flexural Reinforcement in the Long Direction

According to Figure A.4a, the design moment about the column face =

$$M_u = \frac{(q_s B)1.3^2}{2} = \frac{(131.78)(1.5)(1.3)^2}{2} = 167.07 \text{ kN-m}$$

From Eq. (A.4)

$$a = \frac{A_s f_y}{0.85 f_c' b} = \frac{(A_s)(413.7)}{(0.85)(20.68)(1.5)} = 15.69 A_s$$

Again, from Eq. (A.6)

$$M_u = \phi M_n = \phi A_s f_y \left( d - \frac{a}{2} \right)$$

or

$$167.07 = (0.9)(A_s)(413.7 \times 10^3) \left[ 0.388 - \frac{15.69}{2}(A_s) \right]$$

$$167.07 = 144{,}464 \, A_s - 2{,}920{,}928 A_s^2$$

The solution of the preceding equation gives

$$A_{s(1)} = 0.0483 \text{ m}^2 \left[ \text{that is, steel percentage} = \frac{A_{s(1)}}{Bd} = \frac{0.0483}{(1.5)(0.388)} \right.$$
$$\left. = 0.0829 = s_1 \right]$$

and

$$A_{s(2)} = 0.0012 \text{ m}^2 \left[ \text{that is, steel percentage} = \frac{A_{s(2)}}{Bd} = \frac{0.00120}{(1.5)(0.388)} \right.$$
$$\left. = 0.00206 = s_2 \right]$$

Section A.5 (for similar values of $f_c'$ and $f_y$) indicated that $s_{max} = 0.016$. Also, from ACI Code Section 7.12.2, $s_{min} = 0.0018$. Note that $s_1 > s_{max}$ and $s_2 > s_{min}$. So, $A_s = A_{s(2)}$ may be used. Hence, provide 6 × 16-mm diameter bars ($A_s$ provided is 0.001206 m²).

## Flexural Reinforcement in the Short Direction

According to Figure A.4a, the moment at the face of the column =

$$M_u = \frac{(q_s L)(0.55)^2}{2} = \frac{(131.78)(3)(0.55)^2}{2} = 59.8 \text{ kN-m}$$

From Eq. (A.4)

$$a = \frac{A_s f_y}{0.85 f_c' b} = \frac{(A_s)(413.7)}{(0.85)(20.68)(3)} = 7.845 A_s$$

From Eq. (A.6)

$$M_u = \phi A_s f_y \left( d - \frac{a}{2} \right)$$

or

$$59.8 = (0.9)(A_s)(413.7 \times 10^3) \left[ 0.388 - \frac{7.845}{2}(A_s) \right]$$

The solution of the preceding equation gives

$A_{s(1)} = 0.0985 \text{ m}^2$      (thus $s_1 > s_{max}$)

$A_{s(2)} = 0.0004 \text{ m}^2$      (thus $s_2 < s_{min}$)

So, use $s = s_{min}$, or

$$A_s = s_{min} bd = (0.0018)(3)(0.388) \approx 0.0021 \text{ m}^2$$

Use $12 \times 16$-mm diameter bars.

### Final Design Sketch

According to ACI Code Section 15.4.4, a portion of the reinforcement in the short direction shall be distributed uniformly over a bandwidth equal to the smallest dimension of the foundation. The remainder of the reinforcement should be distributed uniformly outside the central band of the foundation. The reinforcement in the central band can be given to be equal to $2/(\beta_c + 1)$ (where $\beta_c = L/B$). For this problem, $\beta_c = 2$. Hence, 2/3 of the reinforcing bars (that is, 8 bars) should be placed in the center band of the foundation. The remaining 4 bars should be placed outside the central band. However, one needs to check the steel percentage in the outside band, or

$$s = \frac{A_s}{bd} = \frac{(2)(201 \text{ mm}^2)}{\left( \dfrac{3000 - 1500}{2} \right)(388)} = 0.0014 < s_{min} = 0.0018$$

So, use $A_s = (s_{min})(b)(d) = (0.0018)(750)(388) = 523.8 \text{ mm}^2$. Hence, $3 \times 16$-mm diameter bars on each side of the central band will be sufficient.

The final design sketch is shown in Figure A.4b.

## A.8

### Design Example of a Retaining Wall

In Example Problem 5.7 (refer to Figure 5.24), the overall stability of a cantilever retaining wall was checked. This section uses the same retaining wall for the design of stem, toe, and heel. It will be assumed that $f_y = 413.7 \text{ MN/m}^2$ and $f_c' = 20.68 \text{ MN/m}^2$.

### Stem Design

The lateral earth pressure distribution behind the stem of the wall given in Figure 5.24 is shown in Figure A.5a. Note that, at any given depth $z$ from the top of the wall

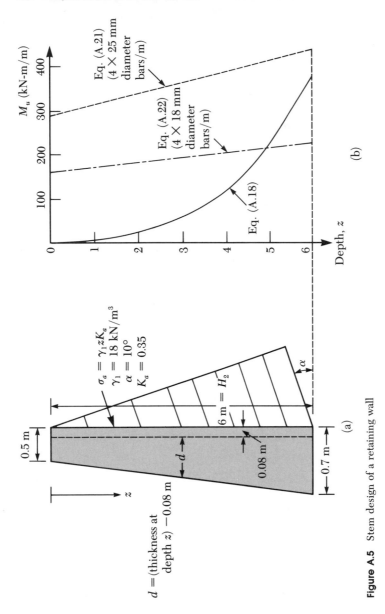

**Figure A.5**   Stem design of a retaining wall

$$\sigma_a = \gamma_1 z K_a \tag{5.27}$$

The horizontal component of the lateral pressure can be given as

$$\sigma_{a(H)} = \gamma_1 z K_a \cos \alpha \tag{A.16}$$

Hence the moment at any depth $z$ is

$$M = \frac{1}{6}\gamma_1 z^3 K_a \cos \alpha \,. \tag{A.17}$$

The ultimate design moment (ACI Code Section 9.2.4) can be given as

$$M_u = 1.7M = \frac{1.7}{6}\gamma_1 z^3 K_a \cos \alpha \tag{A.18}$$

The variations of $M_u$ at $z = 0$, 2, 4, and 6 m have been calculated by using Eq. (A.18) and are tabulated in Table A.5. From Eq. (A.4)

$$A_s = \frac{0.85 a f_c' b}{f_y} = \frac{(0.85)(a)(20.68)(1)}{413.7} = 0.0425a \tag{A.19}$$

Also, from Eq. (A.6)

$$M_u = \phi A_s f_y \left( d - \frac{a}{2} \right)$$

With $\phi = 0.9$ and $A_s = 0.0425a$

$$M_u = (0.9)(0.0425a)(413.7 \times 10^3 \text{ kN/m}^2)\left( d - \frac{a}{2} \right)$$

or

$$M_u = 15{,}824.03ad - 7912.01a^2 \tag{A.20}$$

Assume that the value of $d$ at any given depth $z$ is equal to the thickness of the stem at that depth minus a cover of 80 mm. These values of $d$ at $z = 0$, 2, 4, and 6 m are shown in Table A.5. Now, with known values of $d$ and $M_u$ at any given depth, the magnitude of $a$ can be obtained from Eq. (A.20) and thus $A_s$ from Eq. (A.19). This is also shown in Table A.5. Note that at $z = 0$, 2, and 4 m, $A_s$ is less than $A_{s(min)}$. Therefore use (ACI Code Section 14.2.11)

$$A_s = A_{s(min)} = (0.0015)(\text{gross wall area}) = (0.0015)(1)(0.7)$$
$$= 0.00105 \text{ m}^2 = 10.5 \text{ cm}^2$$

For selection of reinforcement bars, try $4 \times 25$-mm diameter bars per meter for the entire height of the stem. Thus, $A_s = (4)(\pi/4)(2.5)^2 = 19.63 \text{ cm}^2$. For this reinforcement, from Eq. (A.19)

$$a = \frac{A_s}{0.0425} = \frac{19.64 \times 10^{-4} \text{ m}^2}{0.0425} = 0.0462 \text{ m}$$

From Eq. (A.20)

$$M_u = 15824.03ad - 7912.01a^2 = (15824.03)(0.0462)d - (7912.01)(0.0462)^2$$
$$= 731.07d - 16.888 \tag{A.21}$$

**Table A.5** Stem Design for a Retaining Wall

| $z$ (m) | Thickness of stem (m) | $d$ (m) | $M_u$ (kN-m/m) | $a$ (m) | $A_s$ (cm$^2$) |
|---------|----------------------|---------|----------------|---------|----------------|
| 0 | 0.5 | 0.42 | 0 | 0 | 0* |
| 2 | 0.57 | 0.49 | 14.06 | 0.0018 | 0.77* |
| 4 | 0.64 | 0.56 | 112.5 | 0.0128 | 5.44* |
| 6 | 0.7 | 0.62 | 379.7 | 0.0400 | 17.00 |

$d$ = thickness of stem − 0.08 m; $M_u$ from Eq. (A.18); $a$ from Eq. (A.20); $A_s$ from Eq. (A.19);
*$A_s < A_{s(min)}$; so use $A_{s(min)}$ = 10.5 cm$^2$

In a similar manner, 4 × 18-mm diameter bars $[A_s = 10.15 \text{ cm}^2 \approx A_{s(min)}]$ can be tried for the entire height of the wall. Next one obtains

$$M_u = 378.19d - 4.52 \tag{A.22}$$

Figure A.5b shows the variation of actual $M_u$ from Table A.5 and also the variation of $M_u$ from the relations given by Eqs. (A.21) and (A.22). From this figure, it can be seen that, for economy, the 25-mm bars should be cut off at a close distance from the base of the stem. In order to lap the 18-mm diameter bars to the 25-mm diameter bars, refer to ACI Code Section 12.16.1. Because 100% of the bars are to be lapped, the lap splice class is $C$. Thus

$$\text{lap distance} = 1.7L_d \tag{A.23}$$

where $L_d$ = tensile development length (see Section A.3)

$$= 0.019 \, A_b f_y / \sqrt{f_c'}$$

So, lap distance $= (1.7)(0.019)\left(\dfrac{\pi}{4} \cdot 25^2\right)\dfrac{413.7}{\sqrt{20.68}} = 1440 \text{ mm} = 1.44 \text{ m}.$

Therefore, the 25-mm diameter bars must be extended beyond the cutoff point by 1.44 m. The cutoff point can be determined by combining Eqs. (A.18) and (A.22), or

$$\frac{1.7}{6} \gamma_1 z^3 K_a \cos \alpha = 378.19d - 4.52 \tag{A.24}$$

where $d = 0.42 + 0.0333z$ (A.25)

Substituting $\gamma_1 = 18 \text{ kN/m}^3$, $K_a = 0.35$, and $\alpha = 10°$ in Eq. (A.24) and combining Eqs. (A.24) and (A.25), one obtains $z = 4.98$ m. So, extend the 25-mm bars to a height of $6 - 4.98 + 1.44 = 2.46$ m above the base of the stem.

The cutting off of bars in the tension zone (ACI Code Section 12.11.5) also needs to be checked. According to the ACI Code, in that section $V_u$ should be less than or equal to 2/3 of the shear capacity of the section. The shear capacity at a distance of $z = 3.54$ m above the base of the stem is $0.17 \, \phi \, \sqrt{f_c'} \, bd = 0.17(0.85)(\sqrt{20.68})(1)(0.538) = 0.3535 \text{ MN/m} = 353.5 \text{ kN/m}$. Also, at that depth

$$V_u = (1.7)\left(\frac{1}{2}\right)\gamma_1 z^2 K_a \cos \alpha = (1.7)\left(\frac{1}{2}\right)(18)(3.54)^2(0.35)(\cos 10°)$$

$$= 66.08 \text{ kN} < \left(\frac{2}{3}\right)353.5 \text{ kN—O.K.}$$

*Determination of the Length of Development of the Main Reinforcement Bars into the Foundation:* According to ACI Code Section 12.2.2, the development length

$$L_d = 0.019A_b \frac{f_y}{\sqrt{f_c'}}$$

For the 25-mm diameter bars

$$L_d = (0.019)\left(\frac{\pi}{4}25^2\right)\frac{413.7}{\sqrt{20.68}} = 848.7 \text{ mm}$$

The $L_d$ = 848.7 mm calculated above is more than the thickness of the base slab, which is 700 mm (Figure 5.22). So, in order to have the proper development length in tension, standard hooks may be used. According to ACI Code Section 12.5.1, a standard hook develops a tensile stress that may be given as

$$f_h = \xi\sqrt{f_c'} \tag{A.26}$$

For a 25-mm diameter bar, $\xi$ = 30 (refer to ACI Code Table 12.5.1). So

$$f_h = 30\sqrt{20.68} = 136.43 \text{ MN/m}^2$$

Hence, the remaining stress to be developed is equal to $f_y$ − 136.43 = 413.7 − 136.43 = 277.27 MN/m². The extra embedment length required to develop the stress of 277.27 MN/m² is $(277.27/f_y)L_d = (277.27/413.7)(848.7) = 569$ mm ≈ 0.57 m. Figure A.6a shows a sketch of the hook and the embedded length of a bar. Note that the distance from the top of the base slab to the bottom of the hook is 0.645 m. If a cover of 76 mm is to be provided, the minimum thickness of the base slab should be 0.645 + 0.076 = 0.721 m. So, for final design purposes, the thickness of the base slab should be increased to 0.75 m (from 0.7 m, as shown in Figure 5.24).

*Check for Shear Strength at the Base of the Wall:* From Eq. (A.16), the shear at the base of the wall can be given as

$$V_u = (1.7)(V) = (1.7)(\tfrac{1}{2}\gamma_1 H_2^2 K_a \cos \alpha)$$
$$= (1.7)(\tfrac{1}{2})(18)(6)^2(0.35)(\cos 10°) = 189.9 \text{ kN/m}$$

A shear key, usually 50 mm × 100 mm, is provided at the base of the stem. The shear stress in the key should not exceed a nominal stress of $0.2f_c'$ or 5.52 MN/m². So

$$\frac{V_u}{A_{\text{key}}} = \frac{189.9 \text{ kN}}{(10 \times 10^{-2} \text{ m})(1)} = 1899 \text{ kN/m}^2 < 0.2\phi f_c'$$
$$= (0.2)(0.85)20.68 = 3.52 \text{ MN/m}^2\text{—O.K.}$$

So, provide a 50 mm × 100-mm shear key at the base of the stem.

*Temperature and Shrinkage Steel:* Horizontal temperature and shrinkage steel must be provided in the wall; according to ACI Code Section 14.2.10,

$$A_s = (0.0025)(\text{gross area of the wall})$$
$$= (0.0025)(1)(0.7) = 0.00175 \text{ m}^2/\text{m} = 1750 \text{ mm}^2/\text{m of wall}$$

Hence, provide 18-mm diameter bars at 250 mm center-to-center at each face of the wall ($A_s$ = 2032 cm²). Also provide 14-mm diameter bars at 300 mm center-to-center at the outside face of the wall.

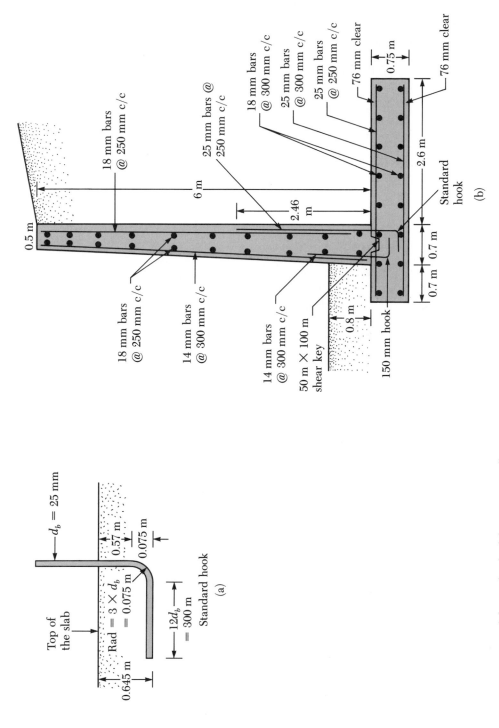

**Figure A.6**  Final design sketch of the retaining wall

Heel Design

For the design of the heel, refer to Figure A.7. Note that the thickness of the base slab has been increased to 0.75 m, per the findings from the stem design. (In Figure 5.22, it is 0.7 m). The equations for load, shear, and moment for the heel of the retaining wall are as follows:

$$q = q_1 + q_2 - q_3 \qquad \qquad (A.27)$$

where $q_1$ = load caused by the soil above the heel = $\gamma H_{av}$
    = $18(6 + 6.458)/2 = 112.12$ kN/m$^2$
  $q_2$ = load caused by the concrete slab = $\gamma_c(0.75) = (23.58)(0.75)$
    = $17.69$ kN/m$^2$
  $q_3$ = soil reaction (refer to Problem 5.7) = $q_{heel} + mx$
    = $45.99 + 35.8x$

Substitution of proper values of $q_1$, $q_2$, and $q_3$ into Eq. (A.27) gives

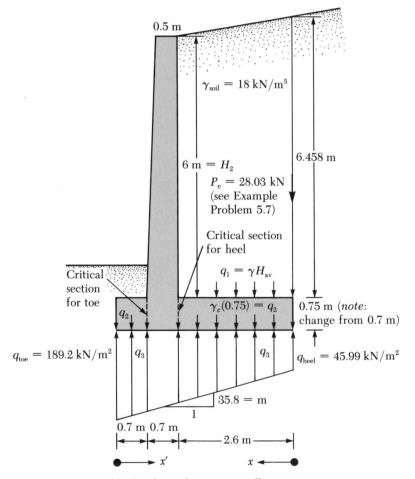

**Figure A.7**   Design of heel and toe of a retaining wall

$$q = 112.12 + 17.69 - 45.99 - 35.8x = 83.82 - 35.8x \tag{A.28}$$

$$\text{Shear, } V = \int q\, dx = 83.82x - \frac{35.8x^2}{2} + C_1$$

where $C_1$ = a constant. At $x = 0$, $V = P_v$ (see Figure A.7) = 28.03 kN/m.

So

$$V = 83.82x - \frac{35.8x^2}{2} + 28.03 \tag{A.29}$$

Also,

$$\text{Moment, } M = \int V\, dx = \frac{83.82x^2}{2} - \frac{35.8x^3}{6} + 28.03x \tag{A.30}$$

The critical section for shear and moment is at $x = 2.6$ m. At the critical section, from Eqs. (A.29) and (A.30)

$$V = 124.96 \text{ kN/m}$$

and

$$M = 251.32 \text{ kN-m/m}$$

Hence

$$V_u = (1.7)V = (1.7)(124.96) = 212.43 \text{ kN/m}$$

and

$$M_u = (1.7)M = (1.7)(251.32) = 427.24 \text{ kN-m/m}$$

*Check for Shear:* From Eq. (A.12a)

$$V_c = 0.17\sqrt{f_c'}\, bd$$

With $b = 1$ m, $d = 0.75 - 0.076 - 0.025/2 \approx 0.662$ m

$$V_c = (0.17)(\sqrt{20.68})(1)(0.662) = 0.5118 \text{ MN/m} = 511.8 \text{ kN/m}$$

Also

$$\phi V_c = (0.85)(511.8 \text{ kN/m}) = 435.03 \text{ kN/m} > V_u = 212.43 \text{ kN/m}$$

Hence, the section is adequate for shear.

*Flexural Reinforcement:* From Eq. (A.6)

$$M_u = \phi A_s f_y \left(d - \frac{a}{2}\right)$$

For this design, $M_u = 427.24$ kN-m/m and $a = [(A_s)(413.7)]/[(0.85)(20.68)(1)]$. Or

$$A_s = 0.0425a$$

Thus

$$427.24 = (0.9)(0.0425a)(413.7 \times 10^3)\left(0.662 - \frac{a}{2}\right)$$

Solution of the preceding equation gives $a = 0.0461$ m. Hence, $A_s = (0.0425)a = (0.0425)(0.0421) = 0.00178$ m$^2$. This gives a steel percentage that is higher than $s_{min}$. So, provide 25-mm diameter bars at 250 mm center-to-center, which gives $A_s = 1000/250 \times (\pi/4)(2.5)^2 = 19.63$ cm$^2$.

## Toe Design

As in the case of the heel design, refer to Figure A.7. The equation for load, shear, and moment at any point of toe can be written

$$q = -q_2 + q_3 \tag{A.31}$$

where $q_2 =$ load caused by the concrete slab $= \gamma_c(0.75) = (23.58)(0.75)$
$\quad = 17.69$ kN/m$^2$
$q_3 =$ soil reaction $= 189.2 - mx' = 189.2 - 35.8x'$

Note that, in Eq. (A.31), the weight of the soil above the toe has been ignored; thus,

$$q = 189.2 - 35.8x' - 17.69 = 171.51 - 35.8x' \tag{A.32}$$

$$\text{Shear force, } V = \int q \, dx' = 171.51x' - \frac{35.8x'^2}{2} \tag{A.33}$$

$$\text{Moment, } M = \int V \, dx' = \frac{171.51x'^2}{2} - \frac{35.8x'^3}{6} \tag{A.34}$$

For critical section, $x' = 0.7$ m. So, at the critical section,

$$V = 111.29 \text{ kN/m}$$

and

$$V_u = 1.7V = 189.2 \text{ kN/m}$$
$$M = 39.98 \text{ kN-m/m}$$

and

$$M_u = 1.7M = 67.97 \text{ kN-m/m}$$

*Check for Shear:* Because $V_u$ at the critical section of the toe is less than $V_u$ at the critical section of the heel, the section is adequate for shear.

*Flexural Reinforcement:* It has been shown in the design of the heel that

$$A_s = 0.0425a$$

and

$$M_u = (0.9)(0.0425a)(413.7 \times 10^3)\left(0.6 - \frac{a}{2}\right)$$

The solution of the preceding equation gives

$$a = 0.0072 \text{ m}$$

So

$$A_s = 0.0425a = (0.0425)(0.0072) = 0.000306 \text{ m}^2$$

Hence, the steel percentage

$$s = \frac{A_s}{bd} = \frac{0.000306}{(1)(0.662)} = 0.000462 < s_{min}$$

Use $A_{s(min)} = s_{min}bd = (0.0018)(1)(0.75) \approx 0.00135$ m$^2$ = 13.5 cm$^2$

Hence, provide 25-mm diameter bars at 300 mm center-to-center, which gives $A_s = 16.35$ cm$^2$.

*Shrinkage and Temperature Reinforcement for Heel and Toe:* For shrinkage and temperature, minimum steel should be used. So

$$A_s = s_{min}bd = (0.0018)(1)(0.75) = 0.00135 \text{ m}^2$$

So, provide 18-mm diameter bars at 300 mm center-to-center.

The final design sketch of a retaining wall is shown in Fig. A.6.

## A.9

## Computer Applications to Design Problems

In the late 1950s, civil engineers recognized the numerical computing power of electronic digital computers and their usefulness in solving analysis and design problems. This early adoption of computers has resulted in their current use in all major branches of civil engineering. During the early years, only those civil engineers employed by government, large companies, or universities had easy access to computing facilities. However, during the past decade, the development of the microprocessor has made in-house computer systems affordable to even the smallest firms.

Problems are solved on digital computers by numerical methods, with addition, subtraction, multiplication, and division being accomplished by the use of discrete numbers in sequential operations. The digital computer consists of four basic elements: (1) a storage unit for storing data and instructions, (2) an input unit for entering information into the storage unit, (3) a central processing unit for processing data according to the stored instructions, and (4) an output unit for displaying the computed results.

By 1950 the basic design for electronic digital computers was established. Programming the early computers was not simple, because the problem application had to be stated in the language of the computer and not the language of the problem. Consequently, in 1954, IBM began work on a problem-oriented language and, in 1957, introduced the now famous FORTRAN language. Numerous programming languages have been developed since; their simplicity allows today's engineer to develop many of his or her own programs.

In 1957, the Portland Cement Association (PCA) began computer development of engineering design aids. PCA released a series of reinforced concrete design programs developed for the IBM 1130 in the late 1960s. Those programs are made available to users through licensing agreements.

During the 1960s, the computing options available to engineering firms consisted of an in-house computer system, a time-sharing terminal in the office, or a walk-in computer service bureau. The IBM 1130 was the most popular computer for those civil engineering firms that were large enough to afford an in-house system. A computer-users group called *CEPA* (Civil Engineering Program Applications) was founded for the purpose of sharing computer programs. The CEPA library grew rapidly and acquired programs in all major areas of civil engineering.

Since the mid-1970s, new electronic technology has made it possible for even the smallest civil engineering firm to afford sophisticated, microprocessor-based in-house computers. Now that hardware is readily available, the problem for small firms is obtaining effective software, or programs. Microcomputer software may be purchased from vendors, or it may be developed in-house. Most of the in-house programs are written in BASIC, which is the most common microcomputer language.

Typical programs for foundation analysis and design include column, wall, and ring footings; pile foundations; sheet piles and braced excavations; retaining walls; and slope stability programs. Common, but less typical, programs include footings on elastic foundations and two- and three-dimensional finite element programs.

Typical descriptions for some of the programs found in almost all in-house libraries are as follows:

## Concrete Footing Design

Program will design a reinforced concrete footing in accordance with ACI-318 for a single column subjected to an axial load plus biaxial bending. User input includes both factored and unfactored column loads and bending moments, allowable soil pressure, and material properties. Computer output includes footing dimensions, reinforcing, and material quantities.

## Concrete Retaining Wall

Program will provide a complete design and analysis of a cantilever retaining wall. Input includes material properties, wall height, and surcharge loading. Output includes a completely dimensioned wall and footing that have been sized to provide required safety factors against sliding and overturning. Reinforcing steel sizes, spacing, and cutoffs are provided along with material quantities.

## Slope Stability

Basic program computes the critical factor of safety for a given slope. Input consists of slope geometry, soil properties, and assumed failure circles. Output includes the safety factor for each assumed failure circle.

An example of a more sophisticated program is a FORTRAN program developed by PCA. The title and a brief description of the program follow.

## Analysis and Design of Foundation Mats and Combined Footing

The program analyzes and designs foundation mats and combined footings as plates on elastic foundations. An elastic analysis is made using the theory of thin plate bending and the finite element method. Three service-load and four ultimate-load cases are generated by the program in accordance with ACI 318-77. Deflections and contact pressures are calculated for each service-loading case, and required reinforcement is calculated using strength design procedures for each ultimate-load case.

The ACI 318-77 Code requires that footings be sized on the basis of allowable soil pressure. This means that the designer must know the unfactored column loads. If the designer wants to design the concrete elements by the strength method, which is the most commonly used method, then he or she must also know the factored column loads. This does not create a problem, because most structural analysis computer programs in use today allow the user to specify multiple load combinations.

In addition to performing conventional designs, the computer is increasingly being used as a word processor to prepare specifications. Another growing use of the computer, which is still in its infancy, is in design/drafting (CADD) systems. These systems will be very important to the foundation engineer in the near future.

## References

American Concrete Institute (1977). *ACI Standard—Building Code Requirements for Reinforced Concrete, ACI 318-77*, Detroit.

UNESCO (1971). *Reinforced Concrete: An International Manual*, Butterworth, London.

# Conversion Factors, Equations, and Tables in English Units

**B.1**

Conversion Factors from English to SI Units

| Length: | |
|---|---|
| | 1 ft  = 0.3048 m |
| | 1 ft  = 30.48 cm |
| | 1 ft  = 304.8 mm |
| | 1 in. = 0.0254 m |
| | 1 in. = 2.54 cm |
| | 1 in. = 25.4 mm |

| Area: | |
|---|---|
| | $1 \text{ ft}^2 = 929.03 \times 10^{-4} \text{ m}^2$ |
| | $1 \text{ ft}^2 = 929.03 \text{ cm}^2$ |
| | $1 \text{ ft}^2 = 929.03 \times 10^2 \text{ mm}^2$ |
| | $1 \text{ in.}^2 = 6.452 \times 10^{-4} \text{ m}^2$ |
| | $1 \text{ in.}^2 = 6.452 \text{ cm}^2$ |
| | $1 \text{ in.}^2 = 645.16 \text{ mm}^2$ |

| Volume: | |
|---|---|
| | $1 \text{ ft}^3 = 28.317 \times 10^{-3} \text{ m}^3$ |
| | $1 \text{ ft}^3 = 28{,}317 \text{ cm}^3$ |
| | $1 \text{ in.}^3 = 16.387 \times 10^{-6} \text{ m}^3$ |
| | $1 \text{ in.}^3 = 16.387 \text{ cm}^3$ |

| Section Modulus: | |
|---|---|
| | $1 \text{ in.}^3 = 0.16387 \times 10^5 \text{ mm}^3$ |
| | $1 \text{ in.}^3 = 0.16387 \times 10^{-4} \text{ m}^3$ |

Coefficient of
Permeability:

| | |
|---|---|
| 1 ft/min | = 0.3048 m/min |
| 1 ft/min | = 30.48 cm/min |
| 1 ft/min | = 304.8 mm/min |
| 1 ft/sec | = 0.3048 m/sec |
| 1 ft/sec | = 304.8 mm/sec |
| 1 in./min | = 0.0254 m/min |
| 1 in./sec | = 2.54 cm/sec |
| 1 in./sec | = 25.4 mm/sec |

Coefficient of
Consolidation:

| | |
|---|---|
| 1 in.$^2$/sec | = 6.452 cm$^2$/sec |
| 1 in.$^2$/sec | = 20.346 × 10$^3$ m$^2$/year |
| 1 ft$^2$/sec | = 929.03 cm$^2$/sec |

Force:

| | |
|---|---|
| 1 lb | = 4.448 N |
| 1 lb | = 4.448 × 10$^{-3}$ kN |
| 1 lb | = 0.4536 kgf |
| 1 kip | = 4.448 kN |
| 1 U.S. ton | = 8.896 kN |
| 1 lb | = 0.4536 × 10$^{-3}$ metric ton |
| 1 lb/ft | = 14.593 N/m |

Stress:

| | |
|---|---|
| 1 lb/ft$^2$ | = 47.88 N/m$^2$ |
| 1 lb/ft$^2$ | = 0.04788 kN/m$^2$ |
| 1 U.S. ton/ft$^2$ | = 95.76 kN/m$^2$ |
| 1 kip/ft$^2$ | = 47.88 kN/m$^2$ |
| 1 lb/in.$^2$ | = 6.895 kN/m$^2$ |

Unit Weight:

| | |
|---|---|
| 1 lb/ft$^3$ | = 0.1572 kN/m$^3$ |
| 1 lb/in.$^3$ | = 271.43 kN/m$^3$ |

Moment:

| | |
|---|---|
| 1 lb ft | = 1.3558 Nm |
| 1 lb in. | = 0.11298 Nm |

Energy:

1 ft lb = 1.3558 Joules

Moment of
Inertia:

| | |
|---|---|
| 1 in.$^4$ | = 0.4162 × 10$^6$ mm$^4$ |
| 1 in.$^4$ | = 0.4162 × 10$^{-6}$ m$^4$ |

**B.2**

## Conversion Factors from SI to English Units

Length:

| | |
|---|---|
| 1 m | = 3.281 ft |
| 1 cm | = 3.281 × 10$^{-2}$ ft |
| 1 mm | = 3.281 × 10$^{-3}$ ft |
| 1 m | = 39.37 in. |
| 1 cm | = 0.3937 in. |
| 1 mm | = 0.03937 in. |

Area:

| | |
|---|---|
| 1 m$^2$ | = 10.764 ft$^2$ |
| 1 cm$^2$ | = 10.764 × 10$^{-4}$ ft$^2$ |

$$1 \text{ mm}^2 = 10.764 \times 10^{-6} \text{ ft}^2$$
$$1 \text{ m}^2 \ \ = 1550 \text{ in.}^2$$
$$1 \text{ cm}^2 \ = 0.155 \text{ in.}^2$$
$$1 \text{ mm}^2 = 0.155 \times 10^{-2} \text{ in.}^2$$

Volume:
$$1 \text{ m}^3 \ \ = 35.32 \text{ ft}^3$$
$$1 \text{ cm}^3 = 35.32 \times 10^{-4} \text{ ft}^3$$
$$1 \text{ m}^3 \ \ = 61023.4 \text{ in.}^3$$
$$1 \text{ cm}^3 = 0.061023 \text{ in.}^3$$

Force:

| | |
|---|---|
| 1 N | = 0.2248 lb |
| 1 kN | = 224.8 lb |
| 1 kgf | = 2.2046 lb |
| 1 kN | = 0.2248 kip |
| 1 kN | = 0.1124 U.S. ton |
| 1 metric ton | = 2204.6 lb |
| 1 N/m | = 0.0685 lb/ft |

Stress:
$$1 \text{ N/m}^2 \ = 20.885 \times 10^{-3} \text{ lb/ft}^2$$
$$1 \text{ kN/m}^2 = 20.885 \text{ lb/ft}^2$$
$$1 \text{ kN/m}^2 = 0.01044 \text{ U.S. ton/ft}^2$$
$$1 \text{ kN/m}^2 = 20.885 \times 10^{-3} \text{ kip/ft}^2$$
$$1 \text{ kN/m}^2 = 0.145 \text{ lb/in.}^2$$

Unit Weight:
$$1 \text{ kN/m}^3 = 6.361 \text{ lb/ft}^3$$
$$1 \text{ kN/m}^3 = 0.003682 \text{ lb/in.}^3$$

Moment:
$$1 \text{ Nm} = 0.7375 \text{ lb ft}$$
$$1 \text{ Nm} = 8.851 \text{ lb in.}$$

Energy:
$$1 \text{ Joule} = 0.7375 \text{ ft lb}$$

Moment of
Inertia:
$$1 \text{ mm}^4 = 2.402 \times 10^{-6} \text{ in.}^4$$
$$1 \text{ m}^4 \ \ = 2.402 \times 10^6 \text{ in.}^4$$

Section
Modulus:
$$1 \text{ mm}^3 = 6.102 \times 10^{-5} \text{ in.}^3$$
$$1 \text{ m}^3 \ \ = 6.102 \times 10^4 \text{ in.}^3$$

Coefficient of
Permeability:

| | |
|---|---|
| 1 m/min | = 3.281 ft/min |
| 1 cm/min | = 0.03281 ft/min |
| 1 mm/min | = 0.003281 ft/min |
| 1 m/sec | = 3.281 ft/sec |
| 1 mm/sec | = 0.03281 ft/sec |
| 1 m/min | = 39.37 in./min |
| 1 cm/sec | = 0.3937 in./sec |
| 1 mm/sec | = 0.03937 in./sec |

Coefficient of
Consolidation:
$$1 \text{ cm}^2/\text{sec} = 0.155 \text{ in.}^2/\text{sec}$$
$$1 \text{ m}^2/\text{year} = 4.915 \times 10^{-5} \text{ in.}^2/\text{sec}$$
$$1 \text{ cm}^2/\text{sec} = 1.0764 \times 10^{-3} \text{ ft}^2/\text{sec}$$

# B.3

## Equations in English Units

| Chapter 2 | Eq. (2.1): |
|---|---|

$$D_b(\text{ft}) = 10S^{0.7}$$

Eq. (2.2):

$$D_b(\text{ft}) = 20S^{0.7}$$

Eq. (2.4):

$$N_{\text{cor}} = \frac{4N_F}{1 + 2\sigma_v'} \text{ (for } \sigma_v' < 1.5 \text{ kip/ft}^2)$$

$$\sigma_v' \text{ in kip/ft}^2$$

Eq. (2.5):

$$N_{\text{cor}} = \frac{4N_F}{3.25 + 0.5\sigma_v'} \text{ (for } \sigma_v' > 1.5 \text{ kip/ft}^2)$$

Eq. (2.6):

$$N_{\text{cor}} = N_F \text{ (for } \sigma_v' = 1.5 \text{ kip/ft}^2)$$

Eq. (2.7):

$$N_{\text{cor}} = 0.77N_F \log\left(\frac{20}{\sigma_v'}\right) \text{(for } \sigma_v' > 0.25 \text{ ton/ft}^2)$$

$$\sigma_v' \text{ in ton/ft}^2$$

Eq. (2.8):

$$N_{\text{cor}} = N_F \text{ (for } \sigma_v' = 1 \text{ ton/ft}^2)$$

Chapter 3     Eq. (3.66):

$$E(\text{ton/ft}^2) = 8N$$

Eq. (3.98a):

$$q_{\text{net(all)}}(\text{kip/ft}^2) = \frac{N}{4}(\text{for } B \leq 4 \text{ ft})$$

Eq. (3.98b):

$$q_{\text{net(all)}}(\text{kip/ft}^2) = \frac{N}{6}\left(\frac{B + 1}{B}\right)^2 \text{ (for } B > 4 \text{ ft)}$$

Eq. (3.99a):

$$q_{\text{net(all)}}(\text{kip/ft}^2) = \frac{N}{2.5}F_dS \text{ (for } B \leq 4 \text{ ft)}$$

$$S = \text{tolerable settlement, in inches}$$

Eq. (3.99b):

$$q_{\text{net(all)}}(\text{kip/ft}^2) = \frac{N}{4}\left(\frac{B + 1}{B}\right)^2 F_dS \text{ (for } B > 4 \text{ ft)}$$

$$S = \text{tolerable settlement, in inches}$$

Eq. (3.101a):

$$q_{\text{(net)(all)}}(\text{lb/ft}^2) = \frac{q_c(\text{lb/ft}^2)}{15} \text{ (for } B \leq 4 \text{ ft and settlement of 1 in.)}$$

Eq. (3.101b):

$$q_{(net)(all)}(lb/ft^2) = \frac{q_c(lb/ft^2)}{25}\left(\frac{B + 1}{B}\right)^2 \text{ (for } B > 4 \text{ ft and settlement of } 1 \text{ in.)}$$

$B$ is in ft

Eq. (3.105):

$$S = S_P\left(\frac{B_F}{B_P}\right)^2\left(\frac{B_P + 1}{B_F + 1}\right)^2$$

$B_P$ and $B_F$ are in ft

**Chapter 4**

Eq. (4.12):

$$q_{all(net)}(kip/ft^2) = 0.25N\left(1 + 0.33\frac{D_f}{B}\right)(S, \text{ in.}) \leq (0.33N)(S, \text{ in.})$$

Eq. (4.13):

$$q_{all(net)}(kip/ft^2) \approx 0.5N$$

Eq. (4.30):

$$U(lb) = b_o d[4\phi\sqrt{f_c'(lb/in.^2)}]$$

$b_o$ and $d$ are in in.

Eq. (4.38):

$$k = k_1\left(\frac{B + 1}{2B}\right)^2$$

$k_1$ and $k$ = coefficients of subgrade reaction of footings measuring 1 ft $\times$ 1 ft and $B$ ft $\times$ $B$ ft (unit—lb/in.$^2$)

Eq. (4.39):

$$k(lb/in.^3) = k_1(lb/in.^3)\left[\frac{1 \text{ (ft)}}{B \text{ (ft)}}\right]$$

Eq. (4.41):

$$k_1 \text{ (ton/ft}^3\text{)} = 6N$$

$k_1$ = coefficient of subgrade reaction of footing measuring 1 ft $\times$ 1 ft

**Chapter 6**

Eq. (6.42):

$$f = \frac{H'^4}{EI}$$

$H$ is in ft; $E$ is in lb/in.$^2$; $I$ is in in.$^4$/ft of wall.

**Chapter 8**

Eq. (8.13):

$$q_l \text{ (lb/ft}^2\text{)} = 1000 N_q^* \tan \phi$$

Eq. (8.14):

$$q_p(lb/ft^2) = 800 N(L/D) \leq 8000N$$

Eq. (8.32):
$$f_{av} \ (\text{lb/ft}^2) = 40\overline{N}$$

Eq. (8.33):
$$f_{av} \ (\text{lb/ft}^2) = 20\overline{N}$$

Eq. (8.105):
$$S_{g(e)} \ (\text{in.}) = \frac{2q\sqrt{B_g}I}{N_{cor}}$$
$$q \text{ in ton/ft}^2; \ B_g \text{ in ft}$$

Chapter 9    Eq. (9.12):
$$Q_{p(net)} = \frac{A_p}{0.6D_b} \ q_p$$
$$Q_{p(net)} \text{ is in lb}; A_p \text{ is in ft}^2; D_b \text{ is in ft}; q_p \text{ is in lb/ft}^2.$$

Eq. (9.36):
$$v \ (\text{lb/in.}^2) \le v_u \ (\text{lb/in.}^2) = 2\phi\sqrt{f_c' \ (\text{lb/in.}^2)}$$

## B.4

## Tables in English Units

**Table B.1**    Common H-Pile Sections Used in the United States (see Table 8.1)

| Designation size (in.) × weight (lb/ft) | Depth, $d_1$ (in.) | Section area (in.$^2$) | Flange and web thickness $w$ (in.) | Flange width $d_2$ (in.) | Moment of inertia (in.$^4$) | |
|---|---|---|---|---|---|---|
| | | | | | $I_{xx}$ | $I_{yy}$ |
| HP  8 × 36 | 8.02 | 10.6 | 0.445 | 8.155 | 119 | 40.3 |
| HP 10 × 57 | 9.99 | 16.8 | 0.565 | 10.225 | 294 | 101 |
| × 42 | 9.70 | 12.4 | 0.420 | 10.075 | 210 | 71.7 |
| HP 12 × 84 | 12.28 | 24.6 | 0.685 | 12.295 | 650 | 213 |
| × 74 | 12.13 | 21.8 | 0.610 | 12.215 | 570 | 186 |
| × 63 | 11.94 | 18.4 | 0.515 | 12.125 | 472 | 153 |
| × 53 | 11.78 | 15.5 | 0.435 | 12.045 | 394 | 127 |
| HP 13 × 100 | 13.15 | 29.4 | 0.766 | 13.21 | 886 | 294 |
| × 87 | 12.95 | 25.5 | 0.665 | 13.11 | 755 | 250 |
| × 73 | 12.74 | 21.6 | 0.565 | 13.01 | 630 | 207 |
| × 60 | 12.54 | 17.5 | 0.460 | 12.9 | 503 | 165 |
| HP 14 × 117 | 14.21 | 34.4 | 0.805 | 14.89 | 1220 | 443 |
| × 102 | 14.01 | 30.0 | 0.705 | 14.78 | 1050 | 380 |
| × 89 | 13.84 | 26.1 | 0.615 | 14.70 | 904 | 326 |
| × 73 | 13.61 | 21.4 | 0.505 | 14.59 | 729 | 262 |

**Table B.2** Selected Pipe Pile Sections

| Outside diameter (in.) | Wall thickness (in.) | Area of steel (in.$^2$) |
|---|---|---|
| $8\frac{5}{8}$ | 0.125 | 3.34 |
| | 0.188 | 4.98 |
| | 0.219 | 5.78 |
| | 0.312 | 8.17 |
| 10 | 0.188 | 5.81 |
| | 0.219 | 6.75 |
| | 0.250 | 7.66 |
| 12 | 0.188 | 6.96 |
| | 0.219 | 8.11 |
| | 0.250 | 9.25 |
| 16 | 0.188 | 9.34 |
| | 0.219 | 10.86 |
| | 0.250 | 12.37 |
| 18 | 0.219 | 12.23 |
| | 0.250 | 13.94 |
| | 0.312 | 17.34 |
| 20 | 0.219 | 13.62 |
| | 0.250 | 15.51 |
| | 0.312 | 19.30 |
| 24 | 0.250 | 18.7 |
| | 0.312 | 23.2 |
| | 0.375 | 27.8 |
| | 0.500 | 36.9 |

# Answers to Selected Problems

**4.3** $255.4$ kN/m$^2$

**4.5** $19.63$ ft

**4.7** $\Delta p = 124.5$ kN/m$^2$
$S = 0.243$ m

**4.9** $0.2$ m

**4.11** $d = 15.5$ in.
$h = 19.5$ in.
No. 6 bars 12 in. center-to-center

**4.13** $21.36$ lb/in.$^3$

## CHAPTER 5

**5.1** $P_o = 12{,}485$ lb/ft
$\bar{z} = 6.13$ ft

**5.3** $P_o = 330.41$ kN/m
$\bar{z} = 1.92$ m

**5.5** a. $\sigma_p = 1.26$ kip/ft$^2$ at top, and $\sigma_p = 3.64$ kip/ft$^2$ at bottom
b. $P_p = 51.49$ kip/ft; $\bar{z} = 8.8$ ft

**5.7** $1079.68$ kN/m

**5.9** $81.76$ kip/ft

**5.11** $1965.6$ kN/m at a distance of 1.67 m from the bottom wall and inclined at an angle of 20° to the back face of the wall

**5.13** a.

| $z$ (m) | $\sigma_a$ (kN/m$^2$) |
|---------|------------------------|
| 2.0 | 11.32 |
| 4.0 | 22.64 |
| 6.0 | 33.96 |
| 7.5 | 42.45 |

b. $P_a = 159.19$ kN/m inclined at an angle of 5° to the back face of the wall at a distance of 2.5 m from the bottom

**5.15** $FS_{(overturning)} = 3.55$ (*note:* $\gamma_{concrete} = 23.58$ kN/m$^3$)
$FS_{(sliding)} = 1.54$ [neglecting $P_p$ in Eq. (5.60); $k_1$ and $k_2 = 2/3$]
$FS_{(bearing\ capacity)} = 5.65$

**5.17** $FS_{(overturning)} = 2.82$
($\gamma_{concrete} = 23.58$ kN/m$^3$)
$FS_{(sliding)} = 1.56$ [neglecting $P_p$ in Eq. (5.60); $k_1$ and $k_2 = 2/3$]
$FS_{(bearing\ capacity)} = 3.17$

**5.19** $FS_{(overturning)} = 2.66$
($\gamma_{concrete} = 150$ lb/ft$^3$)
$FS_{(sliding)} = 1.7$ [neglecting $P_p$ in Eq.

(5.60); $k_1$ and $k_2 = 2/3$]
$FS_{(bearing\ capacity)} = 3.48$

## CHAPTER 6

**6.1** a. $23.52$ ft
b. Similar to Figure 6.5(a) with
$p_1 = 0.217$ kip/ft,
$p_2 = 0.4103$ kip/ft,
$p_4 = 9.415$ kip/ft, $L_3 = 2.52$ ft,
$L_4 = 21$ ft, $L_5 = 4.6$ ft
c. $30.6$ ft

**6.3** a. $27.34$ ft
b. Similar to Figure 6.5(a) with
$p_1 = 0.322$ kip/ft, $p_2 = 0.606$ kip/ft,
$p_4 = 10.78$ kip/ft, $L_3 = 3.338$ ft,
$L_4 = 24$ ft, $L_5 = 5.64$ ft
c. $35.54$ ft

**6.5** Section modulus $= 0.009031$ m$^3$/m

**6.7** a. $D = 7$ m
b. Similar to Figure 6.6 with $p_1 = 12.56$ kN/m$^2$, $p_2 = 24.04$ kN/m$^2$,
$p_6 = 43.87$ kN/m$^2$, $p_7 = 188.13$ kN/m$^2$
c. $9.8$ m

**6.9** $M_{max} = 367.04$ kN-m
$S = 2.13 \times 10^{-3}$ m$^3$/m

**6.11** a. $4.1$ m
b. Similar to Figure 6.8 with $p_1 = 16.27$ kN/m$^2$, $p_2 = 35.97$ kN/m$^2$,
$L_3 = 1.39$ m, $L_4 = 2.71$ m
c. $114.87$ kN/m

**6.13** a. $11.23$ ft
b. Similar to Figure 6.8 with $p_1 = 0.264$ kip/ft$^2$, $p_2 = 0.730$ kip/ft$^2$,
$L_3 = 3.23$ ft, $L_4 = 8$ ft
c. $8.07$ kip/ft

**6.15** a. $2.2$ m
b. $41.56$ kN/m

**6.17** a. $15.7$ kN
b. $21.5$ kN
c. $28.3$ kN

## CHAPTER 7

**7.1** $p_a = 603.44$ lb/ft$^2$
Strut loads: $A = 90.52$ kip; $B = 47.07$ kip; $C = 57.93$ kip

**7.3** $p_a = 549.7$ lb/ft$^2$

Strut loads: $A = 68.71$ kip; $B = 35.73$ kip; $C = 43.98$ kip

**7.5** a. $c_{av} = 14.98$ kN/m$^2$
$\gamma_{av} = 17.07$ kN/m$^3$
b. Similar to Figure 7.4(b) with $p_a = 42.53$ kN/m$^2$

**7.7** Earth pressure envelope similar to Figure 7.4(b) with $p_a = 36.75$ kN/m$^2$
Strut loads: $A = 306.5$ kN;
$B = 405.6$ kN; $C = 413.5$ kN

**7.9** Earth pressure diagram similar to Figure 7.4(c) with $p_a = 0.3\gamma H = 36.75$ kN/m$^2$
Strut loads: $A = 306.5$ kN;
$B = 396.3$ kN; $C = 262$ kN

**7.11** 3.8

## CHAPTER 8

**8.1** a. 56.5 kip
b. 256 kip

**8.3** 235 kip

**8.5** a. $Q_p = 625$ kip
b. $Q_s = 575$ kip

**8.7** 60 kip

**8.9** a. 1423 kN
b. 1645 kN
c. 1536 kN

**8.11** 52 kip

**8.13** 116 kip

**8.15** 123 kip

**8.17** 70.7%

**8.19** 60.6%

**8.21** 3793 kN

**8.23** 281 mm

**8.25** 5584 lb

**8.27** 188 kN

## CHAPTER 9

**9.1** 1022 kip

**9.3** 855 kip

**9.5** $D_s \simeq 0.5$ m
$D_b = 1.5$ m (use $L_1 = 7$ m)

**9.7** Assume $L_1 = 7$ m; $Q_{ws} = 393$ kN (that is, full skin friction mobilized);

$D_s \simeq 0.5$ m; $D_b = 1.5$ m
$s = 13.71$ mm

**9.9** 1.62 in.

**9.11** 6.5 ft

**9.13** 626 kip

**9.15** 985 kN

**9.17** 81.5 kip

## CHAPTER 10

**10.1**

| Liquid limit | $\gamma_d$(kN/m$^3$) Below which collapse is likely to occur |
|---|---|
| 20 | 17.19 |
| 25 | 15.81 |
| 30 | 14.63 |
| 35 | 13.61 |
| 40 | 12.73 |

**10.3** 3.33 in.

**10.5** 52.5 mm

**10.7** 1.4 m

**10.9** 2.38 m

## CHAPTER 11

**11.1**

| Layer No. | $X_o$ (m) |
|---|---|
| 1 | 0.59 |
| 2 | 0.72 |
| 3 | 0.95 |
| 4 | 1.25 |
| 5 | 1.45 |

**11.3**

| Layer No. | $T_n$ (kN/m) |
|---|---|
| 1 | 62.3 |
| 2 | 70.7 |
| 3 | 76.4 |
| 4 | 78.6 |
| 5 | 79.3 |

**11.5** a. 0.02 in. without corrosion effect

b.

| Layer No. | Length (ft) |
|---|---|
| 1 | 14* |
| 2 | 16 |
| 3 | 21.2 |
| 4 | 25.6 |
| 5 | 28.8 |

*Length has to be increased [Eq. (11.29)]

**11.7** $t \simeq 0.05$ in. without corrosion
$L \simeq 39$ ft

## CHAPTER 12

**12.1**

| $G_s$<br>$w$ → | $\gamma_{\text{zav}}$ (kN/m³) | | | |
|---|---|---|---|---|
| | 2.6 | 2.65 | 2.7 | 2.75 |
| 5 | 22.57 | 22.95 | 23.34 | 23.72 |
| 10 | 20.24 | 20.55 | 20.86 | 21.16 |
| 15 | 18.35 | 18.60 | 18.85 | 19.1 |
| 20 | 16.78 | 16.99 | 17.19 | 17.4 |

**12.3** $\gamma_{d(\max)} = 16.48$ kN/m³
$w = 19.2\%$

**12.5** $S_n = 21.72$
Rating = fair

**12.7** $\Delta p_f = 1460$ lb/ft²

**12.9** 46%

# Index